Composite Materials

VOLUME 4

Metallic Matrix Composites

Composite Materials

Edited by

LAWRENCE J. BROUTMAN
Illinois Institute of Technology
Chicago, Illinois

AND

RICHARD H. KROCK
P. R. Mallory & Co., Inc.
Laboratory for Physical Science
Burlington, Massachusetts

1. Interfaces in Metal Matrix Composites
   Edited by ARTHUR G. METCALFE
2. Mechanics of Composite Materials
   Edited by G. P. SENDECKYJ
3. Engineering Applications of Composites
   Edited by BRYAN R. NOTON
4. Metallic Matrix Composites
   Edited by KENNETH G. KREIDER
5. Fracture and Fatigue
   Edited by LAWRENCE J. BROUTMAN
6. Interfaces in Polymer Matrix Composites
   Edited by EDWIN P. PLUEDDEMANN
7. Structural Design and Analysis, Part I
   Edited by C. C. CHAMIS
8. Structural Design and Analysis, Part II
   Edited by C. C. CHAMIS

VOLUME 4

Metallic Matrix Composites

Edited by

KENNETH G. KREIDER

Institute of Applied Technology
National Bureau of Standards
Gaithersburg, Maryland

ACADEMIC PRESS New York and London 1974

A Subsidiary of Harcourt Brace Jovanovich, Publishers

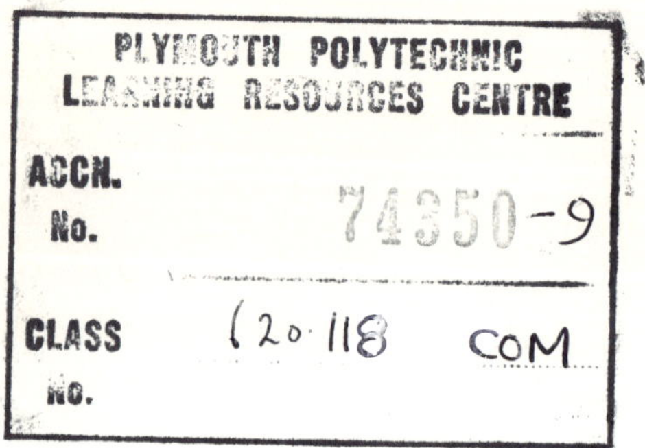

ACADEMIC PRESS, INC.
111 Fifth Avenue, New York, New York 10003

United Kingdom Edition published by
ACADEMIC PRESS, INC. (LONDON) LTD.
24/28 Oval Road, London NW1

Library of Congress Cataloging in Publication Data

Kreider, Kenneth G
Metallic matrix composites.

(Composite materials, v. 4)
Includes bibliographies.
1. Metallic composites. I. Title. II. Series.
TA481.K73 620.1'6 74-1302
ISBN 0-12-136504-2

PRINTED IN THE UNITED STATES OF AMERICA

Contents

3 Directionally Solidified Eutectic Superalloys

E. R. Thompson and F. D. Lemkey

4 Nickel Alloys Reinforced with α-Al_2O_3 Filaments

R. L. Mehan and M. N. Noone

5 Wire-Reinforced Superalloys

Robert A. Signorelli

6 Fiber-Reinforced Titanium Alloys

Arthur G. Metcalfe

List of Contributors

Numbers in parentheses indicate the pages on which the authors' contributions begin.

E. G. KENDALL (319), The Aerospace Corporation, El Segundo, California

KENNETH G. KREIDER* (1,399), United Aircraft Research Laboratories, East Hartford, Connecticut

F. D. LEMKEY (101), United Aircraft Research Laboratories, East Hartford, Connecticut

ALBERT P. LEVITT (37), Army Materials and Mechanics Research Center, Watertown, Massachusetts

R. L. MEHAN (159), Research Development Center, General Electric Company, Schenectady, New York

ARTHUR G. METCALFE (269), Solar Division of International Harvester Company, San Diego, California

M. N. NOONE (159), Space Sciences Laboratory, General Electric Company, King of Prussia, Pennsylvania

K. M. PREWO (399), United Aircraft Research Laboratories, East Hartford, Connecticut

ROBERT A. SIGNORELLI (229), Lewis Research Center, Cleveland, Ohio

E. R. THOMPSON (101), United Aircraft Research Laboratories, East Hartford, Connecticut

E. S. WRIGHT (37), Army Materials and Mechanics Research Center, Watertown, Massachusetts

*Present address: Institute of Applied Technology, National Bureau of Standards, Gaithersburg, Maryland.

Foreword

The development of composite materials has been a subject of intensive interest for at least 15 years, but the concept of using two or more elemental materials combined to form the constitutent phases of a composite solid has been employed ever since materials were first used. From the earliest uses, the goals for composite development have been to achieve a combination of properties not achievable by any of the elemental materials acting alone; thus a solid could be prepared from constituents which, by themselves, could not satisfy a particular design requirement. Because physical, chemical, electrical, and magnetic properties might be involved, input from investigators of various disciplines was required. In the various volumes of this treatise references to specific materials have generally only included the man-made or synthetic composites, but certainly the broad definition of composite materials must include naturally occurring materials such as wood. Chapters dealing with analytical studies of course can apply equally to synthetic or naturally occurring composites.

While composites have been used in engineering applications for many years, the severe operating conditions at which materials have to function (in the space age) led to the science of composite materials as we know it today. The efforts of scientists and engineers working on government research and development programs created entirely new materials, fabrication techniques, and analytical design tools within a short span of time to serve a limited market, but one with constantly demanding requirements. At end of the 1960s, a sharp reduction in the level of government expenditures in these areas and redistribution or re-emphasis of much of the personnel and institutions that had been involved in the development of composite materials raised the possibility that no complete reference work would be available to record many of these important developments and techniques. It was also apparent that this great bulk of technology, if properly digested and evaluated, could be employed in industrial and consumer applications for advantages of economy, performance, and design simplicity. For these reasons, the Editors and Academic Press have prepared this treatise detailing the major aspects of the science and technology of composite materials. We believe that the wide representation of contributors and the diversity of subject matters contained in the treatise assure the complete coverage of this field.

We intend that the volumes be used for reference purposes, or for text supplement purposes, but particularly to serve as a bridge in transferring the bulk of composite materials technology to industrial and consumer applications.

The Editors are indebted for the cooperation and enthusiasm they have received from the Editor of each volume and the individual contributors who worked diligently and as a unit to complete this task. We are also grateful to the staff at Academic Press who provided constant support and advice for the project.

Finally, the Editors wish to thank the management of P. R. Mallory and Co., Inc. and the administration of Illinois Institute of Technology for providing a key element in the successful completion of this work through their support and encouragement.

LAWRENCE J. BROUTMAN
RICHARD H. KROCK

Preface to Volume 4

The intent of this book is to provide an in-depth report and a reference on the technology of metal-matrix composites. The chapters represent separate materials systems and have been prepared by materials scientists from the materials technology viewpoint. These authors have considerable experience with composite materials and specific familiarity with the composite material system of the chapter.

The use of the composite system is discussed with respect to the need for improved materials in specific applications. From this need a technical statement of the materials problem is presented which relates to the materials selection. The objectives of materials research in the composite system are discussed and goals for the application of specific materials are defined.

In the materials research discussed in this volume, the method of approach to solving the materials problems has generally emphasized two major thrusts: a development of useful composite materials fabrication techniques, and an evaluation of the physical and mechanical properties of the newly developed composite material. Many of the fabrication techniques relate to those used in the aerospace industry. This is because the authors are more familiar with this industry, and because it has been generally assumed that the first major applications of this technology would be in the aerospace industry. Fortunately, for the purpose of this volume, the development of the fabrication techniques for metal-matrix composites has been described either in open literature or government reports, thus allowing rather complete disclosures of up-to-date technology. This is no doubt in part due to the significant support of the developing technology sponsored by the U.S. Department of Defense, specifically the U.S. Air Force Materials Laboratory at Wright-Patterson Air Force Base.

The evaluation of mechanical and physical properties of metal-matrix composites has also been supported in part by the U.S. Air Force. This technology has been multidisciplinary and has demonstrated some of the most exemplary teamwork of experts in materials science, fracture mechanics, applied mechanics, and structural or design mechanics. The latter fields are discussed in much more depth in Volumes 2, 5, 7, and 8 of this treatise.

The materials evaluation discussed in this volume relates primarily to an application for which the composite system is being designed. For example, in the fiber-reinforced composites the complex stiffness coefficients and ultimate tensile strengths are the primary design data required. In contrast, the primary property evaluation with wire-reinforced composites and eutectics has been in the field of high-temperature strength and corrosion.

Each of the chapters also presents an analysis of the status of the particular composite material system. This is not only to review the accomplishments and present technology but also to discuss the prime needs for future work toward improving the system or relating the system to future applications. An extended introductory chapter is presented to give the reader a background in the concepts and language of the technology of metal-matrix composites. In addition, some attention is given in this chapter to discussing parts of the field which are not covered in the ensuing chapters.

I hope that this volume is useful to both students of composite materials as well as to engineers and scientists active in this field.

KENNETH G. KREIDER

Contents of Previous Volumes

Volume 1

Volume 2

Volume 3

Composite Materials

VOLUME 4

Metallic Matrix Composites

1

Introduction to Metal-Matrix Composites

KENNETH G. KREIDER

Institute of Applied Technology
National Bureau of Standards
Gaithersburg, Maryland

I. Introduction

Modern technology has placed increasing demands on materials. This need for better materials is particularly acute in the area of dynamic structures, where not only high strength is required but also light weight. The efficiency of dynamic structures such as aircraft, high-speed manufacturing machinery, power-generating equipment, and aerospace equipment could be affected by an improvement in the structural efficiency of the materials. The structural efficiency of materials is becoming even more important as "hardware" and capital equipment become larger and larger. Examples of these large structures include the 747 airliner, Apollo manned-space-flight equipment, giant paper-making and textile-manufacturing mills, and the enormous Maine Yankee steam-turbine electric generating plant at Wiscasset, Maine. A key problem in designing these structures pertains to the square–cube relationship: that is, the strength and stiffness of a structure increase with the square of the linear dimension (cross-sectional area), whereas the weight increases with the cube of the linear dimension. In order to maintain the stiffness and strength, new design techniques must be used and high-strength, stiffer materials are required.

One of the key properties in structural design is stiffness or rigidity. The corresponding significant material property is specific modulus (the ratio of elastic modulus to density). Engineering materials such as steel, aluminum, titanium, and glass have similar specific moduli, and organic materials have lower modulus-to-density ratios. This has meant that the designer must generally use lower-density materials and increase the section size to gain flexural or bending stiffness without excessive weight. In many structures however, this choice is not practical, and a material having an increased stiffness-to-density ratio is required. Boron and carbon, which are covalently bonded, have a dramatically higher specific modulus of 15×10^8 cm (6×10^8 in.) compared to materials that have metallic or ionic bonding. Other materials that have high covalent-bonding fractions, such as boron carbide, silicon carbide, and aluminum oxide, also have high specific moduli. Not only is the high fraction of covalent bonding desired in the solid, but the material should have a low atomic number in order to have the highest specific modulus. Engineering metals such as aluminum, cobalt, copper, chromium, iron, magnesium, nickel, and titanium have specific moduli in the range $1.3–3.5 \times 10^7$ cm. Organic polymers have generally much lower specific moduli.

Unfortunately, high-specific-modulus materials such as boron and carbides cannot be fabricated into large engineering structures. Moreover, they are brittle and therefore very sensitive to cracks and flaws, which make them weak in large sections. In fact, high specific modulus cannot be generally used unless high strength is also attained in the engineering structure. Strength and toughness under tensile loading together with the reproducibility of properties have led to the general acceptance of metal alloys as the prime material for large dynamic structures.

In addition to the critical stiffness problem in design, many structures are limited in durability under service conditions. These limitations relate to the strength of the material under fatigue loading, high-temperature stress rupture, stress corrosion, and crack growth around notches and flaws.

Although metal alloys have been strengthened considerably by various hardening mechanisms for improvements in static properties, they often sacrifice toughness and durability under dynamic service conditions. One of the most important goals in the design of composite materials is to lower the sensitivity to cracks and flaws while increasing the static and dynamic strength. Improved response to dynamic loads is attained by absorbing the loads with the elastic member of the composite rather than the plastic (ductile metal) member, which would undergo cumulative damage. Lowered crack sensitivity is attained by redirecting cumulative damage in the material in directions that do not reduce the load-carrying ability.

In summary, the goals of the composite system are to take advantage of the high specific moduli of certain covalently bonded materials and to minimize the effect of cumulative damage in inelastic solids under dynamic loading conditions that lead to structural failure. The concept of a composite material is to combine certain assets of the components or members of the composite system and suppress the shortcomings of the members in order to give the newly synthesized composite material unique and useful properties.

For our purposes, the definition of a composite material includes the following: A high-strength or high-modulus material called the reinforcement is combined with a second material called the matrix, which permits fabrication into the desired engineering structure and transfers the environmental loads onto the carrying reinforcement. Typical examples of such composites include steel-reinforced concrete, fiber-glass reinforced plastic, nylon-reinforced rubber, and wood. The last of these is naturally occurring lignin reinforced with fibrous cellulose. Magnificent examples of these composite structures are the 300-ft-tall California redwoods. The use of the word composites in this volume is not intended to mean materials that are protected from the environment by a coating, such as porcelain-coated steel or gold-clad copper, nor is the word intended to indicate a filled material in which the second material is added primarily to lower the cost, such as a filled epoxy or phenolic.

To define further the use of the term composite material in this work, the following criteria should be met:

1. The components of the composite material have been selected and designed intentionally.
2. The composite material itself is man made, not naturally occurring.
3. The composite material contains at least two separate and distinct chemical phases.
4. The properties of the composite will depend on each of the component phases present, which must be present in significant amounts (at least 20% for the second phase).

5. The components of the composite must be present in repeating geometry, so that at a large enough scale, the material could be considered homogeneous. This criterion is used to exclude structures such as a honeycomb sandwich or a coaxial cable, since they can be thought of better as structures than as materials.

6. The composite material should be created in order that the new material has useful properties not possessed by the individual components.

II. Metal-Matrix Composites

Composite materials should include component materials that complement each other and are compatible. Resin-matrix composites have received the widest recognition and acceptance in the area of high-modulus composites. With these composites, the high-modulus reinforcement is combined with a matrix that has been selected for its ease of fabrication into structural hardware. In addition, there is little chemical or mechanical interaction between the two phases, which simplifies matrix-reinforcement compatibility problems. Generally, the chemical inertness of the reinforcement (usually a fiber) at the modest resin-fabrication temperatures and the large elastic compliance of the matrix are the cornerstones of the compatibility solution. With metal-matrix composites, the temperature of composite fabrication is generally much higher, and the matrix has elastic-modulus coefficients that are one or two orders of magnitude higher that those of the organic-resin matrices. Therefore, both chemical compatibility and mechanical (stress) compatibility are much more serious problems. Both of these compatibility problems will be treated in some depth in the chapters on the individual systems. At present, it is sufficient to recognize the significance of the problems and the fact that they arise from intrinsic properties of the matrix and reinforcement.

Chemical compatibility has been solved in metal-matrix composites in two ways: either by using low-temperature (solid-state) fabrication techniques or by selecting thermodynamically stable component phases that are at equilibrium with each other. A corresponding thermal–mechanical compatibility problem has been solved either by using a ductile matrix that yields and takes up all the differential strain necessary in thermal cycling or by selecting a matrix and a reinforcement having nearly matching thermal-expansion coefficients.

The metal-matrix composite, on the other hand, has a series of advantages that are very important in the utilization of structural materials. These advantages relate to the same metallic properties that have led to the general primacy of metal alloys for use in dynamic engineering structures. They include the combination of the following properties:

1. high strength;
2. high modulus;
3. high toughness and impact properties;
4. low sensitivity to changes in temperature or thermal shock;
5. high surface durability and low sensitivity to surface flaws;
6. high electrical and thermal conductivity;
7. excellent reproducibility of properties;
8. excellent technological background with respect to (a) design, (b)

manufacture, (c) shaping and forming, (d) joining, (e) finishing, and (f) service durability information.

The high strength of engineering alloys compared to structural ceramics or organic materials can be utilized in composite materials. The matrix strength is particularly important with respect to composite properties at some angle away from the reinforcement direction. Properties such as transverse strength, torsional strength, and interlaminar shear strength are generally matrix controlled. This is true in creep and fatigue loading as well as under static loading. The strength of joints or bonding to composite materials is matrix controlled also. The higher strength of metal can also be used to conserve the amount of high-cost reinforcement necessary for a given structure.

The high moduli of metal alloys compared to those of organic materials are particularly significant in high-modulus composites. Figure 1 shows a comparison of several fiber-reinforced composite materials on the basis of specific modulus. Notice that although the very highest specific moduli are recorded with unidirectional boron–epoxy in the direction of the reinforcement, the overall specific modulus (pseudoisotropic 0 ± 60 deg) is considerably lower than that of BORSIC†—aluminum. Also, the shear specific modulus is higher with the fiber-reinforced metal. These stiffness ratios are very important in dynamic structures such as turbine-engine fan blades and large airfoils.

The high toughness and impact properties of metal alloys are very important in composite materials, since the reinforcement is generally a linear elastic solid and does not have good impact properties by itself. Ductile-metal matrices such as aluminum, titanium, or nickel-chromium alloys undergo energy-absorbing plastic deformation under impact, which is very important in many dynamic structural applications. The ductile matrix also

† BORSIC is a registered trademark of United Aircraft Corp.

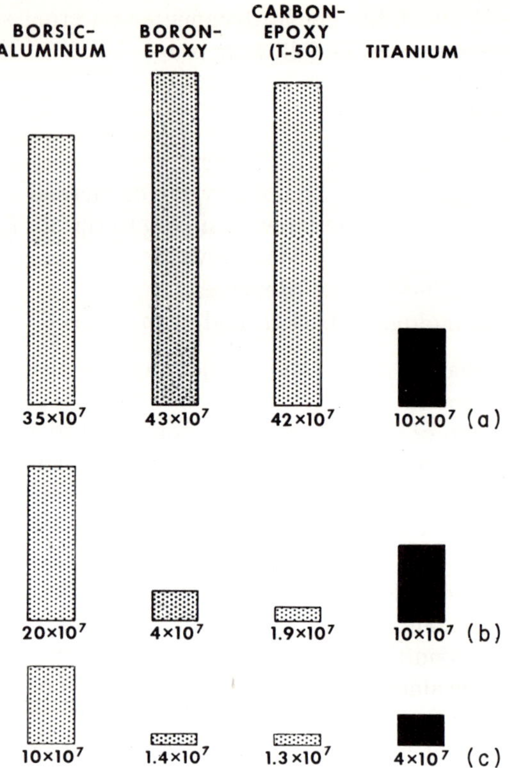

Fig. 1. Specific moduli (in inches) of several aerospace materials; composite materials are unidirectionally reinforced and have 50 v/o fiber. (a) Longitudinal modulus E_{11}; (b) transverse modulus E_{22}; (c) shear modulus G_{12}.

permits the blunting of cracks and stress concentrations by plastic deformation and gives the material improved fracture toughness.

The insensitivity of metal properties to changes in temperature is an important asset of metals as matrix materials for high-modulus structural composites. The poor thermal-shock resistance of ceramic materials compared to that of metals has often deterred their use. With organic-resin matrices, the properties are quite sensitive to temperature changes, especially when approaching the glass transition temperature of the polymer. Not only do the resins tend to soften at moderately elevated temperatures, but their resistance to oxidation, corrosion, and erosion drops off significantly at elevated temperatures. Engineering aluminum, titanium, and nickel-based alloys are much less sensitive to temperature changes or thermal cycling.

Structural engineering alloys are generally less sensitive to surface flaws than ceramics or organic resins, and the surfaces are more durable. Small cracks in ceramics started by erosion, abrasion, or corrosion can drastically lower the strength of a ceramic material because of the high elastic coefficients and the inability of the ceramic to blunt these flaws and notches by plastic flow. The small cracks are significant stress risers and weaken the material. With organic polymers, various mechanisms, including abrasion, erosion, and corrosion, can have marked surface effects due to the relatively low hardness and strength of the material. Many resins are also moisture sensitive and have a tendency toward porosity. In addition, resins are often more sensitive to moderate-temperature oxidation and ultraviolet radiation than alloys of aluminum or titanium.

Metal-matrix alloys have high electrical and thermal conductivity, which permits the diffusion and elimination of high thermal and electrical concentrations from point sources. Problems such as lightning strikes and hot-gas impingement are less severe if the impacting energy (electrical or thermal) can be conducted away rapidly.

One of the most important assets of metal matrices is the excellent reproducibility of wrought-metal properties. No other engineering material can be as precisely controlled to mechanical and physical specifications as the metallurgical alloys we depend on for dynamic structures. This characteristic of the matrix alloys is also very important in high-modulus composites, where not only are the matrix alloy properties important but also the bonding and interfacial properties should be consistent and reproducible.

The metal-matrix alloys that have been chosen for use in high-modulus composites have a significant asset in that an excellent technological background has been developed. This technology relates to their present use in the design of engineering structures, in the manufacture of complex hardware, including shaping, forming, and joining operations, and in the development of finishing and coating techniques, together with comprehensive information on service durability.

In summary, metal-matrix composites require a sophisticated technological development in order to achieve the advantages they can yield in engineering structures. In the development of these composite materials, one must consider carefully problems of chemical and mechanical compatibility of the two phases. Because of the high strength and modulus of the matrix, there is more interaction between the matrix and reinforcement than is the case in resin-matrix composites. However, many of the same properties of metal alloys that have made them useful in engineering structures make them attractive for use as the matrix of structural engineering composites.

III. Types of Metal-Matrix Composites

Metal-matrix composites can be reinforced by strong second phases of three-dimensional shapes (particulate), two-dimensional shapes (laminar), or one-dimensional shapes (fibrous). Each of these classes of reinforcement has its own advantages and characteristics. Generally, these classes of composites fall into considerably different areas of application, not only because of the differing mechanical properties of the classes but also because of the contrasts in fabrication techniques.

A. Particle-Reinforced Composites

Particle-reinforced composites are intended herein to include those composites having more than 20 v/o of the hard reinforcing dispersed phase and do not include the class of dispersion-hardened metals, which have a considerably lower volume fraction of dispersoid. In addition, the diameter of the particles and the interparticle spacing are much greater in the composites—normally greater that 1 μm compared to the 0.01–0.1-μm interparticle distances that are best suited for the dislocation friction utilized in dispersion-hardened metals. With particle-reinforced composites such as tungsten-carbide–cobalt, the reinforcing phase is the principal load-bearing phase, and the matrix is used for transferring the load and for ease of fabrication, as is the case for the other composites considered in this volume. High matrix-constraint factors produced by the hard reinforcement are used to prevent yielding in the matrix, and the composite strength generally increases linearly with decreasing volume fraction of the matrix. However, yielding and plastic flow can occur, and fracture elongations can range up to 30% which is considerably more than that possible with laminar or fibrous composites stressed in the direction of the brittle reinforcement. This difference is present because elongation of the reinforcement itself is needed in fibrous or laminar composites in order for composite elongation to take place.

The three-dimensional reinforcement can lead to isotropic material properties, since the material is symmetrical across the three orthogonal planes. The particulate composite material is not homogeneous, however, and material properties not only are sensitive to the constituent properties but also are very sensitive to the interfacial properties and geometric shapes of the array. Strength of the particulate composites normally depends on the diameter of the particles, interparticle spacing, and the volume fraction of the reinforcement. Matrix properties, including the work-hardening coefficient, which increases the effectiveness of the reinforcement constraint, are also important. A review of inorganic particulate composites has been given by

Broutwau and Krock (1967). They discuss the methods of preparation, with emphasis on sintering-type processes and properties of particulate composites, including a discussion of the theoretical aspects of the performance of these composites. Although these materials make up a large class of composite metal-matrix composites, they are not treated in this volume. Rather, the oriented reinforced composites are emphasized, which represent in many ways a new approach in materials technology.

B. Laminated Composites

For the purposes of this volume, laminated composite materials are considered to be reinforced by a repeating lamellar reinforcement of high modulus and strength, which is contained in the more ductile and formable metallic matrix material. Spacing of the lammelae is microscopic, so that in the structural component the material can be considered as an anisotropic, homogeneous material at the proper scale. These composites are structural composites and therefore do not include the many types of coated materials in which the lamella can be considered as a structural member with an environmental-protection coating as a second constituent of the structural material. Examples of structural laminated composite materials include the boron-carbide–titanium composites, in which the repeating reinforcing structural constituent consists of chemical-vapor-deposited boron-carbide films of 5–25 μm thickness. A second example would be the eutectic composites such as Ni–Mo and Al–Cu, in which the two phasis solidify in a lamellar array. Both of these eutectic composites consist of a ductile metallic matrix reinforced with a stronger and higher-modulus lamellar phase.

The elastic constants of a structural lamellar composite have been predicted by laminate theory (Tsai, 1966). In either of the principal directions of the reinforcing plates, the elastic modulus is given by the rule of mixture:

$$E_R V_R + E_M V_M = E_C \tag{1}$$

where E_R, E_M, and E_C are the elastic moduli of the reinforcement, matrix, and composite respectively, and V refers to the volume fraction.

Other elastic constants of the anisotropic material are somewhat more complex to derive but can be accurately predicted.

The strength of laminated composite materials relates more closely to the properties of the bulk reinforcement than to the small-volume (one-dimensional) properties that can be developed in whiskers or filaments. Since the reinforcing lamellae can have two dimensions that are comparable in size to the structural part, flaws in the reinforcement can nucleate cracks of lengths similar to that of the part. This action is in contrast to the action of a flaw that nucleates a crack in a filamentary material. In a filament loaded in axial tension, the crack propagates through the cross section of the filament, which is quite small in area in relation to the whole piece. Since the most important reinforcing materials are brittle in nature, their strength is related to the statistical population of their flaw density and intensity. Therefore, they obey classical fracture mechanics as stated by Griffith (1921), and their strength is related inversely to their size. This relationship of size to strength of brittle materials has been treated by Weibull and is discussed at some length in Volume 5 of this treatise. The mechanics of fracture of brittle materials is not treated in this volume; however, the effect of size on strength should be recognized by materials engineers.

The strength of laminated structural composites has been limited by the somewhat lower strengths of film reinforcements compared to filamentary reinforcements. In addition, the low strain to failure of the brittle reinforcement limits the elongation and ductility of the composite in all directions in the plane of reinforcement. The reinforcement strength and modulus, however, are present in all directions of the plane and offer a significant advantage compared to the unidirectional reinforcement of a filament array.

C. *Fiber-Reinforced Composites*

This section includes a discussion of the stress–strain behavior, strength, and fracture behavior of fiber-reinforced metal-matrix composites. Emphasis is placed on systems having relatively ductile low-yield-strength matrices and high-strength, high-modulus, quasi-brittle fibers. Early studies by McDaniels *et al.* (1959, 1963) and by Kelly and Tyson (1965) on copper–tungsten, by Cratchley and Baker (1964) on stainless-steel wires in aluminum, and by Dow and Rosen (1964) on glass-reinforced plastic have provided a significant empirical base for the understanding of this subject.

1. Stress–Strain Behavior

Four stages in the stress–strain behavior of a metal-matrix composite with continuous uniaxially aligned fibers when it is stressed parallel to the fibers were described by McDaniels *et al.* (1963a) and by Kelly and Davies (1965).

The first stage is defined as including elastic deformation of both the fibers and the matrix and is depicted in Fig. 2. The elastic modulus has been found experimentally to be

$$E_C = E_F V_F + E_M V_M \tag{2}$$

where V stands for the volume fraction of the matrix M and the fiber F. Rosen *et al.* (1964) demonstrated that Eq. (2) is the theoretical lower bound

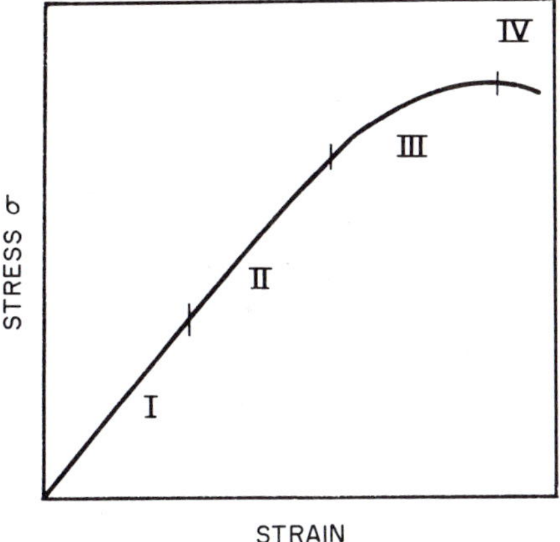

Fig. 2. Schematic curve showing the stress–strain behavior of a metal-matrix composite: (I) fiber elastic, matrix elastic; (II) fiber elastic, matrix plastic; (III) fiber plastic, matrix plastic; (IV) fiber fractured.

to the modulus where the Poisson ratios of the two phases are nearly equal. This same expression holds in elastic–elastic compression. Higher values can be measured with mismatches in the Poisson ratios. Thus, the simple additive rule of mixtures as applied to elastic impedance is very useful in predicting composite performance.

However, any other phase stressed in parallel (or in series) with the fiber and matrix must also be considered. Perhaps the most important of these phases is porosity, which can drastically lower the elastic modulus of the composite. Several empirical equations have been developed for determining the effect of porosity on modulus. McAdam (1951), studying iron powder metallurgy, used the expression

$$E_V = E_0(1 - C_V)^{3/4} \tag{3}$$

to express the modulus E_V in terms of void content C_V. Cohen and Ishai (1967) have used the expression

$$E_V = E_0(1 - C_V{}^{2/3}) \tag{4}$$

which has a low porosity sensitivity. For example, at 10 v/o porosity, E_V would be $0.70E_0$ for Eq. (3) and $0.78E_0$ for Eq. (4).

The second stage in Fig. 2 comprises the region in which the fiber is extending elastically and the matrix is extending plastically. Since the fiber is usually in a high volume fraction and has a considerably higher elastic modulus than the matrix, E_2 is nearly equal to E_1. The modulus of the composite would be the addition of the effective slopes on the stress–strain curve of the two phases times their volume fractions:

$$E_{11} = E_F V_F + (d\sigma_M/d\varepsilon_M)_G V_M \tag{5}$$

where $d\sigma_M/d\varepsilon_M$ is the effective strain-hardening coefficient of the matrix phase, which is normally much less than the modulus of the fiber and can be neglected. Therefore, the modulus of the elastic–plastic portion of the stress–strain curve frequently can be given as

$$E_{11} = E_F V_F \tag{6}$$

Notice that although this "modulus," E_{11}, is nearly equal to E_1 (85% in 50 v/o aluminum–boron), it is not a truly elastic modulus. Straining the composite into region II results in some permanent elongation, and relaxing the applied stress leaves the fibers elongated elastically and the matrix under residual compressive stresses. Strain cycling through this region includes a hysteresis effect and changes the effects of any previous thermomechanical treatment.

The third stage is observed when both fiber and matrix can undergo plastic deformation, and includes normal plastic extension of the two phases. This deformation mode may deviate from the performance of the constituents alone with respect to necking or other inhomogeneous plastic flow. In some ductile fibers, it has been found that the onset of necking has been delayed by the constraint of the matrix. This effect was pointed out by Piehler (1965) in the silver–steel system and led to strengths higher than the summation of the strengths of the constituents.

Stage IV in the stress–strain performance of the composite described in Fig. 2 includes the fracturing of the high-strength fibers. During this stage, the matrix transfers load from broken fiber ends to unbroken segments and flows around the opening pores or cracks. Fracture of the composite normally terminates stage IV. If the matrix has sufficient ductility and the dynamic loads associated with multiple fiber fractures do not lead to catastrophic failure, the composite may remain whole when the volume fraction fiber is less than the critical value. The fourth deformation stage is similar to that of a composite containing discontinuous fibers. With the fiber ends present, the matrix transfers load by shear stresses on the surface of the fiber near the broken end. This shearing stress is generated by the gradient of load on the fiber, which has a higher effective stress–strain slope than the matrix from a balance of forces in the x direction as given in Fig. 3,

$$dP/dx = 2\pi r \tau_{rx} \tag{7}$$

where P is the tensile load on the fiber in the x direction and τ_{rx} is the shear stress at the fiber–matrix interface.

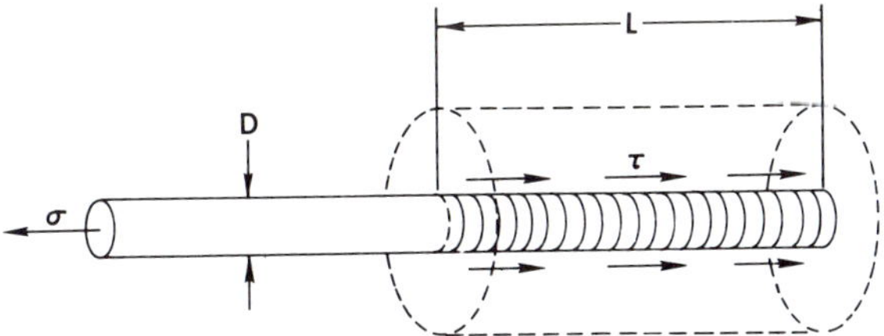

Fig. 3. Forces acting on a reinforcing fiber: σ is the tensile stress acting over area $\pi D^2/4$; τ is the shear stress acting over area πDL.

Solutions for this shear-stress load transfer have been developed for the elastic case by Rosen *et al.* (1964) where the matrix material is deforming elastically. They found

$$P = E_F A_F \left[1 - \frac{(\cosh \beta)\,(l/2 - 1)}{(\cosh \beta)\, l/2} \right] \qquad 0 < x < l/2 \tag{8}$$

where P, l, and E are the load, length, and tensile strain on the fiber, respectively. A_F is area of fiber, and β is the aspect ratio. However, studies by Schuster and Scala (1968) have indicated the shear-load transfer rate is significantly affected by the geometry of the ends of the fiber.

Perhaps a much more significant case for metal-matrix composites includes that in which the fiber behaves elastically but the matrix deforms plastically. In this case, the elastic strains in the matrix are limited by the tensile or shear flow stress of the matrix. If the work-hardening effects are small, one can assume that τ_{rx} is independent of x. Therefore, Eq. (7) integrates to

$$P = 2\pi r_0 x \tau \tag{9}$$

or

$$\sigma_x = 2\tau x / r_0 \tag{10}$$

The load builds up linearly from the ends with a constant shear stress as given in Fig. 4. In the case of discontinuous-fiber composites, a minimum critical reinforcing length can be defined at which stresses can be built up high enough to fracture the fibers. If $\sigma_F{}^*$ is the ultimate tensile strength of the fibers, then the critical reinforcing length l_C can be determined using the matrix shear yield stress τ. Using Eq. (10) to solve for l_C, which includes a shear-stress region at each end of length x,

$$l_C = r_0 \sigma / \tau \tag{11}$$

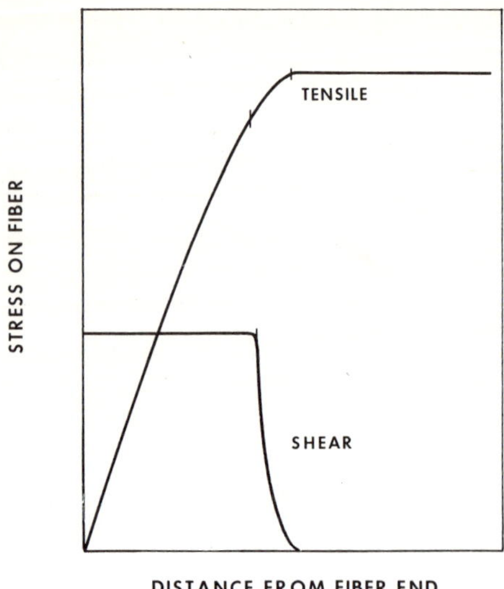

Fig. 4. Buildup of tensile stress at fiber ends.

Notice that, as shown in Fig. 4, once the tensile stress σ_ε on the broken fiber has reached that of adjacent unbroken fibers, the shear stresses go to zero as in a composite without fiber ends present.

The transverse and shear properties of fiber-reinforced composite materials are considerably more sensitive to the matrix behavior than the longitudinal properties are. With ductile metal matrices reinforced with high-strength brittle fibers, yielding and plastic flow of the matrix are primary characteristics of the behavior of the composite. However, considerable reinforcement of the elastic modulus of the composite is generated by the fiber—typically, a factor of 2 in 50 v/o boron–aluminum at 90° to the reinforcement—and comparable increases in the shear modulus are also effected. The performance of metal-matrix composites loaded off-axis from the principal reinforcing direction is treated in Section IV.B.

IV. Strength of Metal-Matrix Composites

A. *Longitudinal Strength*

The strength of a material is not a function averaged over the tested section such as the elastic compliance; rather, it is a point function. It can be defined as the average stress on the cross section of a material at its weakest point which causes it to fail. Normally the strength of a material is defined in terms of stress based on the original cross-sectional area (the engineering stress) rather than the instantaneous area. In the case of static tensile stresses, the criterion of failure is simple and refers to the highest or ultimate tensile stress, based on the original area, that the material specimen can support. With high-modulus metal-matrix composites, this failure terminates stage-IV deformation as described in the previous section, and results as the load is increased while the load-carrying ability is decreased by individual fiber fractures.

Figure 5 is an illustration of the growth of fiber cracks and damage

Fig. 5. Fiber cracking prior to composite failure. Broken white filaments are tungsten–boride cores of boron fibers in an aluminum matrix near the edge of a hole.

immediately prior to specimen rupture. This performance in aluminum–boron is typical of metal-matrix composites reinforced with brittle fibers having less elongation to fracture than the metal matrix. As described above, the matrix transfers loads back into the broken fiber by shear stresses on the surface of the fiber leading away from the broken end.

The recent development of the scanning electron microscope has dramatically improved the capabilities for studying fracture in metal-matrix composites. Since the fracture surfaces in composites are usually very irregular and have deep ridges, the outstanding depth of field of the scanning electron microscope is a valuable asset. Studies by Breinan and Kreider (1970), Prewo and Kreider (1972), Olster and Jones (1972), and Hancock and Swanson (1972) have graphically illustrated the potential of the instrument. Scanning electron fractographs of failed tensile specimens of boron–aluminum and BORSIC–titanium are given in Figs. 6 and 7. Notice the contrasting ductile shear of aluminum compared to the glasslike boron failure. The depth of the fracture surface is dependent on the shear strength of the matrix or the fiber–matrix interface, and the 6061 aluminum alloy contrasts sharply with the much stronger Ti–6Al–4V titanium alloy. The boron–

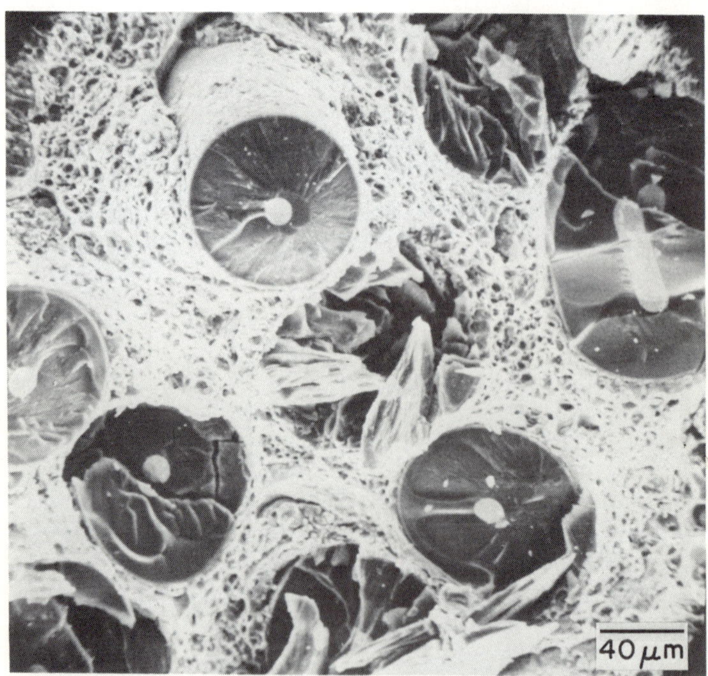

Fig. 6. Fracture surface of boron–aluminum (6061). Fibers are 150 μm in diameter.

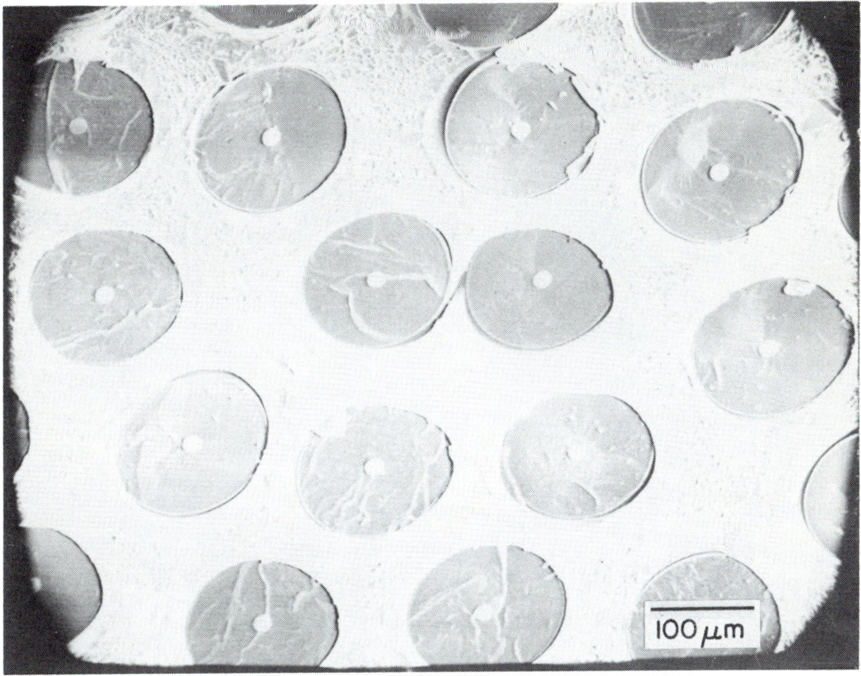

Fig. 7. Fracture surface of BORSIC–titanium (Ti–6Al–4V). Fibers are 150 μm in diameter.

aluminum fracture surface (Fig. 6) has deep undulations and dimpling in the ductile aluminum. Severe damage is done to the fibers, which are tightly bonded to the matrix. The titanium–BORSIC fracture surface in Fig. 7 is characterized by a rather flat surface, with a debonding of the fibers from the titanium. This weak interface plays an important role in the fracture behavior of the composite and gives the material good fracture toughness.

In addition, higher-magnification scanning electron fractographs are particularly useful in determining fracture mechanisms in the fibers. Boron fracture surfaces are shown in Figs. 8 and 9. Figure 8 illustrates a typical high-strength fiber fracture where the initiation site is in the tungsten-boride core. In Fig. 9a and b, boron-fiber fractures are presented that illustrate surface-initiated failures that may have been generated by damage to the fiber surface. In these fracture surfaces, one can find the fracture-initiation site by following the fracture propagation lines back to their point of convergence.

A simple expression has been used by McDaniels *et al.* (1963) to relate composite strength to constituent properties:

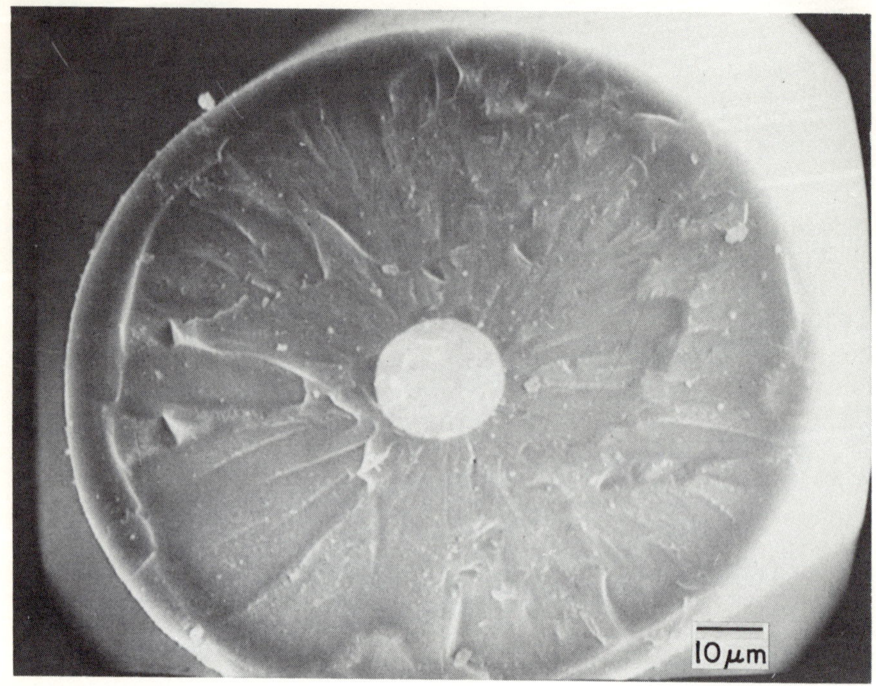

Fig. 8. Fracture surface of 4-mil boron fiber with flaw in tungsten–boride core.

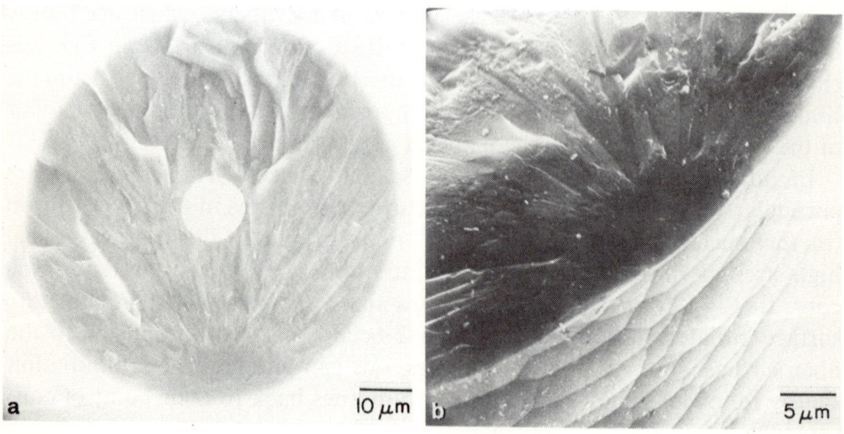

Fig. 9. Fracture of 4-mil boron fiber with surface flaw.

$$\sigma_C{}^* = \sigma_F V_F + \sigma_M V_M \tag{12}$$

where $\sigma_C{}^*$ is the ultimate strength of the composite expressed as a stress based on the original area, σ_F is the average stress on all the fibers, and σ_M is the average stress on the matrix at failure. V_F and V_M are the volume fractions of fibers and matrix. If there is no porosity or third phase, $V_F + V_M = 1$.

If all the fibers have approximately the same strength and the remaining matrix cannot sustain the load at fiber failure, σ_F can be equated to the average strength of the fibers and σ_M can be interpreted as the stress on the matrix at a strain equal to the fiber strain at failure. This case has been studied by McDaniels *et al.* (1963a) in a system consisting of tungsten fibers in a copper matrix. The strength of the composite can be plotted as a function of the fiber volume fraction and is given in Fig. 10. It can be seen that Eq. (12) holds for fiber volume fractions greater than $V_F{}^*$, the critical volume fraction. At lower volume fractions, the matrix can support the load after all fibers have failed, and the above assumptions do not hold. This case is not normally important for structural engineering composites, since fiber reinforcement is not being achieved with respect to strength, which is usually a prime consideration in the choice of the fiber.

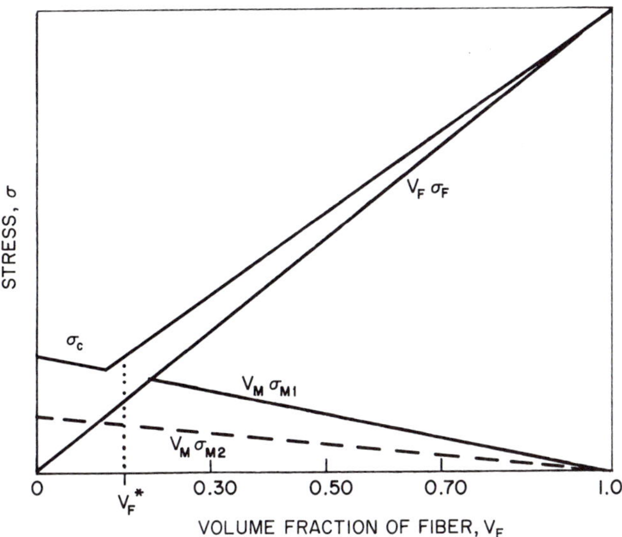

Fig. 10. Rule-of-mixtures strength for a brittle, strong fiber with a ductile matrix. Composite strength $\sigma_C = V_M\sigma_{M1}$ (for $V_F < V_F{}^*$) or $\sigma_C = V_F\sigma_F + V_M\sigma_{M2}$ (for $V_F \geq V_F{}^*$), where σ_{M1} is the ultimate tensile strength of the matrix, σ_{M2} is the matrix stress at the fiber-rupture strain, and $V_F{}^*$ is the critical volume fraction of fiber.

The effective strength of the fibers that should be used in Eq. (12) is not as simple to determine in the case of brittle fibers having a significant range of tensile strengths. Although the average fiber strength can be used when the fibers have similar strengths, this average strength does not predict composite ultimate tensile strength in the case of brittle fibers such as boron. For example, Rosen (1964) has determined the composite strength as a function of fiber-strength coefficients of variations for composites with test gauge lengths equal to the critical stress-transfer length. A plot of this type is given in Fig. 11. Note that a coefficient of variation of 20% would lower the composite strength about 35% according to this theory.

Three significant modifications are generated when weak fibers fracture. First, the cross-sectional strength at the location of the broken fiber is lowered by the loss in strength of that fiber. Second, a static stress concentration around the crack generated by the broken fiber can lower the effective strength. Third, the dynamic stress wave generated as the broken fiber unloads can shock the composite, thereby lowering the instantaneous load-carrying ability of that cross section.

The first problem, that of the composite load-carrying ability in the presence of broken fibers, is also related to the critical load-transfer length of the fiber in the matrix as described in Eq. (11). At the exact location of the filament break, the strength is decreased by the load-carrying ability of the broken fiber. At distances of up to the critical load-transfer length, the load-carrying ability is diminished by the fiber strength less the load that has been transferred back into the broken ends by shear stress on the matrix. If the load-transfer length were infinite—for example, when no shear stresses can be supported between fiber and matrix—the strength of the composite is equal to the dry-bundle strength of the fibers. This fiber strength can be determined by testing a bundle of fibers for tensile strength or by testing each individual fiber and calculating the maximum load the group of fibers

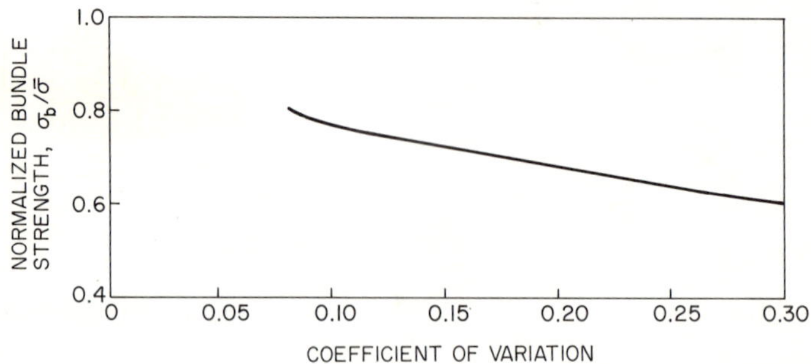

Fig. 11. Bundle strength as a function of variation of fiber strength. (After Corten, 1967.)

would support. The determination can be made by multiplying the stress times the number–area product of unbroken fibers at that stress and maximizing the stress.

Analytical treatments of the concept of the bundle strength of fibers and how it relates to the statistical probability of failure in brittle fibers have been given by Corten (1967) and others. This analysis has been based primarily on the Weibull function, which relates the probability distribution function of flaws in a brittle elastic solid to the size of the specimen. The reader is referred to Volume 2 of this treatise, entitled "Micromechanics," since this subject will not be treated in detail in this volume.

The stress-concentration effect around broken fiber ends can also lower the effective strength of the composite. One of the important assets of composite materials is that a crack propagating normal to the applied tensile load can be stopped at the fiber–matrix interface, because the maximum stress amplitude at the crack tip in the matrix is approximately equal to the ultimate tensile strength of the matrix, and this stress is low compared to the fiber-breaking stresses. For example, in aluminum–boron, the crack-tip stress as it propagates through the aluminum may be up to 350 MN/m^2 (50,000 psi), but the local strength of fiber would be more nearly 4.20 GN/m^2 (600,000 psi). This method of crack-blunting is depicted in Fig. 12. Therefore, stress concentrations around crack tips do not generate unstable crack growth in that system. In the system titanium–aluminum oxide, however, where the relative fiber–matrix strength is more nearly two to one, this crack-tip stress concentration may seriously embrittle or weaken the composite.

Although the crack tip itself does not seriously weaken the aluminum–boron composite, local stress concentrations are significant. At the location

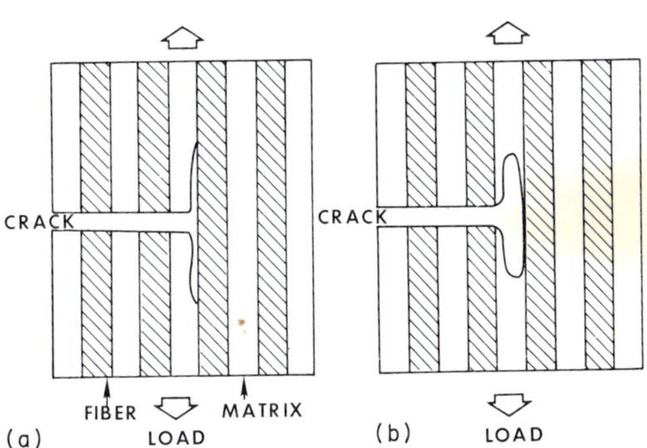

Fig. 12. Crack-blunting in composites: (a) interfacial splitting; (b) matrix shear deformation and splitting.

of the fiber break, the broken fiber ends exert shear stresses on the matrix as they contract. These shear forces are supported primarily by the nearest-neighbor fibers, since the matrix cannot sustain the high loads the broken fiber was carrying. In addition, if the nearest-neighbor fibers are unbroken, little of this local stress is passed on to the more distant fibers. These local added tensile forces on the unbroken fibers can lead to unstable crack growth, since a second adjacent fiber break would lead to even higher local shear-lag forces. This shear lag is depicted in two dimensions in Fig. 13. If these forces were shared equally by the six nearest neighbors and the average fiber stress were 2.8 GN/m^2 (400,000 psi), the local augmentation of tensile stress at the break on the neighboring fibers would be 2.8 GN/m^2 (400,000 psi), or 0.45 GN/m^2 (66,000 psi) per fiber. This augmented stress would be at the peak value at the fiber break and would decrease to zero at a distance of l_C, the shear-transfer length, away from the break. An effect similar to this tensile-stress augmentation has been studied by Rosen in fiber glass.

The third modification on the stress state of the composite when weak fibers break pertains to the shock wave generated by the fiber fracturing. One can calculate the approximate energy release when a 5.6-mil boron fiber

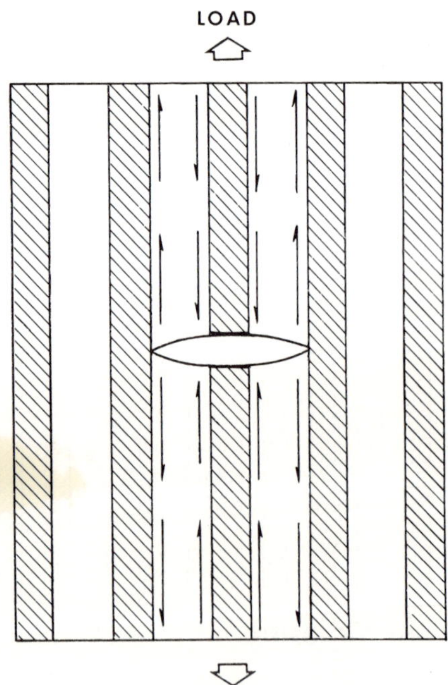

Fig. 13. Two-dimensional crack propagation. Arrows represent shear stresses on the fibers.

fails in a soft aluminum matrix. This energy J is the integral of the elastic strain energy lost from the fiber between two points $l_c/2$ away from the break in both directions. The fracture is often detected as an acoustical report in metal-matrix composites. The dynamic fracture energy is absorbed primarily by the specimen but has not been treated quantitatively in the literature with respect to its effect on the ultimate tensile strength of the composite.

In summary, the relationship between constituent properties and composite strength is much more subtle than the one that relates the elastic moduli, because the strength is a point function rather than an average material constant. Although the rule of mixtures can be applied to composites in which the effective fiber strength can be well predicted, it is not nearly so accurate when brittle reinforcing fibers are used.

B. *Transverse Strength*

The behavior of filament-reinforced composites loaded normal (or transverse) to the reinforcement direction has been predicted by Tsai (1966), Adams and Doner (1967), and Chen and Lin (1968), among others. The transverse properties are not as simply derived as the longitudinal, but using certain assumptions, the modulus and strength properties can be predicted. To predict the macroscopic modulus values, Tsai (1966) assumed that (1) both constituents are linearly elastic up to their failure stresses, (2) interfacial bonding is perfect, and (3) fibers are arranged in a regular array. By using a finite element equal to a quadrant of one fiber zone, the governing differential equations of displacement are generated. From these equations, the transverse stiffness E_{22} of the composite can be derived in addition to the transverse plane Poisson ratio:

$$\nu_2 = (E_{22}/E_{11})\nu_1 \tag{13}$$

In Fig. 14, the ratio of the composite transverse modulus to the matrix modulus is plotted as a function of the volume fraction of fiber and the ratio of the fiber modulus to the matrix modulus for a square array. This graph indicates the powerful effect on transverse modulus of the reinforcing fiber in a metal matrix. For example, at 60 v/o reinforcement of boron in aluminum, the composite transverse modulus is nearly three times that of the matrix.

The transverse strength of composites is not typified by this reinforcement behavior. In this case, the composite fails at the weakest cross-sectional surface. In addition, because of the severe constraint placed on the matrix by the fibers, the strain to failure of the composite is much less than that of unconstrained matrix material. Strength predictions for transversely rein-

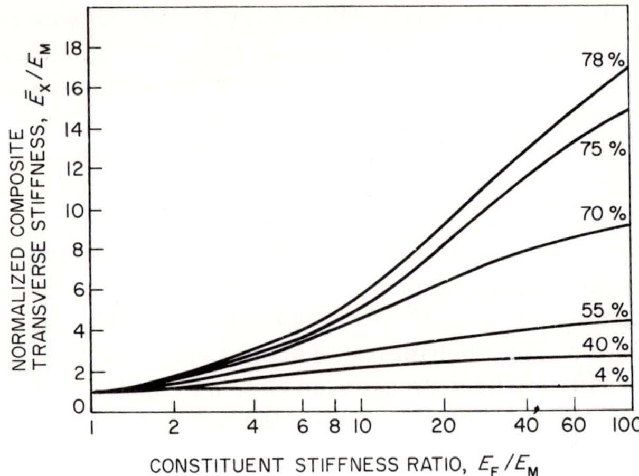

Fig. 14. Normalized composite transverse stiffness for circular filaments in a square array.

forced metal-matrix composites have been presented by Adams and Doner (1969) and by Chen and Lin (1969). As was the case with the modulus, a finite-element construction is used to calculate the stiffness coefficients; this stiffness matrix is then inverted to obtain an influence matrix that gives nodal displacements as a function of external loads. From these nodal displacements, the stress coefficients are calculated. By assuming a regular (square or hexagonal) array of fibers one can use the von Mises–Henchy criterion for yielding at points of maximum dilational strain energy. From these postulations, Chen and Lin calculate the transverse strength of the composite by the following equation:

$$\sigma_T = \sigma_M \bar{\sigma}_C / (U_{max})^{1/2} \tag{14}$$

where σ_M is the matrix strength, σ_C the average stress on the composite, and U_{max} the maximum normalized distortional energy. They evaluated this expression by testing boron plus 6061 aluminum alloy composites; the results are depicted in Fig. 15.

V. Types of Reinforcements for Metal-Matrix Composites

Although the subsequent chapters will discuss specific reinforcing materials for given composite systems, some general statements with respect to

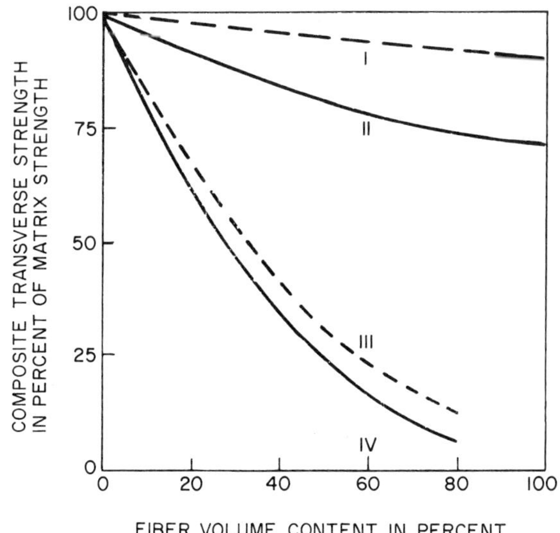

Fig. 15. Theoretical transverse strength of metal-matrix composites: I,III, hexagonal array; I,II, bonded interface; II,IV, square array; III,IV, free interface. (After Chen and Lin, 1969.)

reinforcing materials are considered useful. Filamentary reinforcement has been found to make possible the most effective reinforcement of metal-matrix systems and is the subject of this section. These reinforcement filaments are considered for their worth in combination with three classes of structural engineering metals: low-temperature alloys such as aluminum alloys; intermediate-temperature alloys such as titanium; and high-temperature alloys such as nickel-based superalloys or columbium alloys. Although the requirements of the reinforcement change as the matrix alloy is changed, several desirable properties are nearly universal:

1. High strength. Strength in the fiber is necessary primarily for composite strength, but it also simplifies fabrication and handling.
2. High modulus. This property is important with metal matrices in order that the fiber will attract the load without extensive matrix plastic flow. The comparatively low modulus of fiber glass has made it much less useful than certain other fibers as a reinforcing filament for metal matrices.
3. Ease of fabrication and cost. These engineering considerations are very important if significant structural applications are to be developed.
4. Good chemical stability. This property requirement is quite sensitive to the matrix alloy selected, since the system will have specific fabrication and environmental requirements. However, stability in air and resistance to reactions with matrix materials is important in all fibers.

5. Size and shape of filament. The larger-diameter, circular filaments are to be preferred for metal-matrix composites for solid-state fabrication techniques. These filaments are far simpler to incorporate in a metal-matrix composite by plastic flow, and with less surface are less reactive in liquid-metal fabrication techniques.

6. Reproducibility or consistency of properties. This consideration is always extremely important with brittle or high-strength materials. In many cases, as described earlier, the composite strength depends on the bundle strength, which is a function of the strength distribution of the fibers rather than the mode or peak strengths.

7. Resistance to damage or abrasion. Several brittle fibers are particularly sensitive to moisture exposure (for example, E-glass) or abrasion (for example, Al_2O_3 or SiC). These properties interfere with standard composite handling techniques. High residual compressive stresses as in the case of boron fibers protect the fiber from surface-sensitive degradation.

A summary of the most important reinforcing filaments and their properties is presented in Table I.

High-strength reinforcing wires, rocket wire (steel), molybdenum, and tungsten are particularly useful as reinforcements because of their high strength. Higher strengths combined with good ductility can be obtained in the wire form than any other form of the metal alloys. These wires also have excellent high-temperature creep properties: AFC-77 has a strength of 2.8 GN/m^2 (400,000 psi) at 600°C, TZM molybdenum alloy has a strength of 1.0 GN/m^2 (140,000 psi) at 1100°C, and thoriated tungsten wire (0.005 in. diam) has a strength of 1.8 GN/m^2 (270,000 psi) at 1100°C. However, these wires do not have a high modulus-to-density ratios of the other filaments. In contrast, beryllium wire has an excellent modulus-to-density ratio, as given in Table I. However, its high cost has limited the usefulness of drawn beryllium wire and has led to the incorporation of other forms of beryllium reinforcement.

Both E-glass and S-glass fibers have excellent strength-to-density ratios and low cost, and are perhaps the most important reinforcing fiber for resins. Their low modulus and their chemical reactivity have made fiber glass much less useful in the reinforcement of metals.

Aluminum-oxide fiber is produced by withdrawing a seed crystal from a melt. This fiber, typically 250 μm in diameter, is a single crystal and has very high strength. The fiber is sensitive to abrasion, however, and is expensive. The sapphire fiber is being considered primarily for high-temperature reinforcement.

Boron fiber offers the best combination of properties for the reinforcement of aluminum alloys and magnesium alloys. This fiber is produced by chemical-vapor deposition of boron from boron-trichloride gas on a

TABLE I

Typical Properties of Commercially Available Reinforcement Filaments

Filament material	Diameter (μm)	Manufacturing technique	Average strength (GN/m^2)	Average strength (ksi)	Density (g/cm^3)	Modulus (GN/m^2)
Boron	100–150	Chemical vapor deposition	34	500	2.6	400
SiC-coated boron	100–150	Chemical vapor deposition	31	450	2.7	400
SiC	100	Chemical vapor deposition	27	400	3.5	400
Carbon monofilament	70	Chemical vapor deposition	20	300	1.9	150
B_4C	70–100	Chemical vapor deposition	24	350	2.7	400
Boron on carbon	100	Chemical vapor deposition	24	350	2.2	
Graphite HS	7[a]	Pyrolysis	27	400	1.75	250
Graphite HM	7[a]	Pyrolysis	20	300	1.95	400
Al_2O_3	250	Melt withdrawal	24	350	4.0	250
S-glass	7[a]	Melt extrusion	41	600	2.5	80
Beryllium	100–250	Wire drawing	13	200	1.8	250
Tungsten	150–250	Wire drawing	27	400	19.2	400
"Rocket Wire" AFC-77	50–100	Wire drawing	41	600	7.9	180

[a] 10^4 fil per tow.

tungsten substrate at approximately 1200°C. In the BCl_3 process, a tungsten wire 12 μm in diameter is normally drawn through a gas-tight mercury seal, which also acts as a dc electrode. Almost all production fiber is produced in a vertical-tube reactor with the tungsten pay-off spool at the top and a boron take-up spool beyond the bottom mercury electrode seal. A gas mixture of boron trichloride and hydrogen is reacted along the hot substrate tungsten wire to coat the boron. In order to obtain the high-strength amorphous structure, the temperature is kept below 1250°C. Crystalline boron, formed at 1400°C, has a considerably lower strength (~200,000 psi) but can be formed much faster. The tungsten substrate is to be preferred because of its reproducible properties, high strength, low cost, and chemical purity; however, monofilament carbon and various metallic filaments have been used. The structure of the boron appears amorphous on an X-ray pattern and is homogeneous. The X-ray pattern consists of four broad halos at spacings of 4.3, 2.5, 1.7, and 1.4 Å. However, Gillespie (1966) and Otte and Lipsitt (1966) believed that the boron was crystalline but that the grain size was less than 50 Å. They have detected β-rhombohedral reflections on fine powder fragments. However, the fiber has a very high residual compressive stress state on the surface, which makes the fiber insensitive to surface microscratches or microcorrosion.

Boron has excellent modulus-to-density and strength-to-density ratios, chemical compatibility with solid aluminum and liquid magnesium, large diameter, reproducible properties, and a competitive price for some applications.

Mechanical properties of boron fibers have been reported by numerous investigators. Modulus values for tension and shear reported by Talley (1959), Lasday and Talley (1966), and Herring and Krishna (1966) are approximately 58×10^6 and 25×10^6 psi, respectively. Salkind and Patarini (1967) have reported the fatigue properties and Ellison and Boone (1967) the creep and stress rupture values at 650 and 815°C. The coefficient of thermal expansion was found to be 4.8×10^{-6}/°C in the range 0–300°C by Wawner (1967). High-temperature strength and retention of strength after high-temperature exposures have been reported by Veltri and Galasso (1968); the latter is presented in Fig. 16. Silicon-carbide-coated boron is also included in Fig. 16

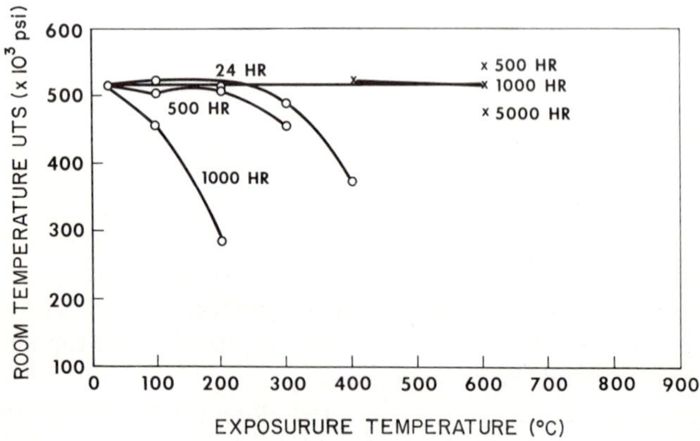

Fig. 16. Average room-temperature ultimate tensile strength (UTS) of filaments after being heated in air: ○, uncoated boron; ×, SiC-coated boron (0.15-mil coating).

for comparison. The silicon carbide coating (1–2 μm thick) is placed on the boron to protect it from oxidation or retard metal reactions during fabrication or high-temperature exposure in application.

Silicon-carbide fiber and boron-carbide fiber have been produced in experimental quantities. These fibers are produced by chemical-vapor deposition on a hot substrate of tungsten or carbon in a way very similar to that used for making boron. Methyldichlorosilane plus hydrogen has been the most useful mixture for producing silicon carbide, and mixtures of methane plus boron trichloride have been used for boron carbide. These deposits are crystalline, and the fibers are sensitive to surface abrasion. The B_4C and SiC crystalline structures are more creep resistant than boron fiber, and the fibers are considered primarily for high-temperature reinforcement.

Graphite fibers or yarn have excellent modulus-to-density and strength-to-density ratios and are available in a range of properties. The fiber is manufactured using a polyacrylonitrile or rayon precursor yarn, pyrolyzing the yarn to carbon, and then stretching at high temperature to graphitize the carbon. The final elastic modulus is generally a function of the severity of the high-temperature graphitizing step, and modulus values of 240–500 GN/m^2 (35–75 $\times$ 10^6 psi) are generally available. The highest modulus is generated at above 2500°C. A typical yarn of graphite made from Orlon contains 10,000 filaments of approximately 7-μm diam. Although fiber strengths of 2.8 GN/m^2 (400,000 psi) are available, problems of composite fabrication due to the reactivity of the fiber to molten metals have limited the usefulness of this fiber for metal-matrix reinforcement. Composites made with this fiber will be treated in detail in Chapter 5.

Other fibers, such as monofilament graphite, are available on an experimental basis and will be discussed later. The technology of whisker-reinforced composites is treated as a separate chapter.

VI. Compatibility of Composite Constituents

Composite materials by their nature include two or more dissimilar phases. These two phases have to be compatible with each other both physically and chemically if a synergistic coupling of the constituents is to take place. With metal-matrix composites, the physical compatibility problems of reinforcing a metal matrix with a lamina or filament often are associated with the material constants relating dilatation to pressure (stress) or thermal changes. The chemical compatibility problems relate primarily to interfacial bonding, interfacial chemical reactions, and environmental chemical reactions during composite fabrication.

The physical compatibility requirements dictate that the matrix should have sufficient ductility and strength in order to transfer the imposed structural loads to the reinforcing members evenly and without strong discontinuities. In addition, local stresses in the matrix due to flaws or dislocation movement should not cause high local stress intensities on the fiber. The matrix should perform in many ways like a soloist's accompanist, smoothing out the rough parts, covering for incidental flaws, but never detracting from the prime assets of the featured member of the team. For many applications, the mechanical properties of the matrix should include high ductility and compliance.

A very significant physical property relationship between the reinforcement and matrix is the thermal coefficient of expansion. Since the matrix is normally the more ductile material, it is preferred that the matrix have the higher coefficient of thermal expansion. This relates to the fact that the phase having the higher expansion coefficient is stressed in tension upon cooling from the normally high fabrication temperature. The reinforcement, a brittle material, is nearly always stronger in compression than it is in tension. This guideline does not hold with very-low-modulus matrices such as resins combined with extremely fine reinforcements such as graphite, where internal fiber buckling is a problem. With the higher-yield-strength matrices such as titanium, it is important that the disparity of thermal expansion coefficients be not too great, since high residual thermal stresses are generally to be avoided.

Chemical compatibility is a more complex problem. Two general types of composites are considered in this volume; *in situ* composites in which the two phases are in thermodynamic equilibrium at the fabrication temperature, and artificial composites in which the chemical kinetics of degrading reactions between the two phases are sufficiently slow to permit compatibility. The former case is typified by the eutectic composites, which are solidified under equilibrium conditions. With the eutectics, the chemical potentials of the phases are equal and specific surface-energy effects are minimized. These composites can have stability problems at temperatures away from the fabrication temperature if phase transformations or changes in concentration are significant. In addition, agglomeration can be a problem with respect to the minimization of total-surface free energy.

The chemical compatibility problem is much more serious with non-equilibrium composites. Problems can occur in bonding the two phases together, as is the case with graphite fibers, where fiber wetting and bonding are very difficult. Also, the chemical reactivity of the fiber with the environment can present serious problems in fabrication. Degrading reactions on high-strength, brittle fibers include stress corrosion and oxidation. Thermal shock can also degrade the fibers.

Probably the most important chemical compatibility problem relates to the direct reaction between the fiber and matrix. With low-temperature metal-matrix composites such as boron–aluminum, the chemical-reaction problem is averted by keeping the fabrication temperature as low as possible. High pressures on a low-creep-strength matrix permit the lower temperatures to be used while still obtaining good consolidation and bonding. Some systems such as boron–magnesium or copper–tungsten can be handled using melt infiltration, since the two phases do not react and are mutually insoluble.

With high-temperature composites, however, several considerations are important with respect to chemical compatibility. These considerations pertain to: (1) ΔF, the free energy of reacting the two phases; (2) μ, the chemical potential; (3) τ, the surface energy; (4) D_G, the grain-boundary diffusion coefficient; and other special diffusion effects.

The first consideration, ΔF, is significant not only during fabrication but also during high-temperature application. The free-energy change of reacting the fiber and matrix represents the driving force for that reaction, and at elevated temperatures the amplitude of this driving force becomes very important. The materials engineer who is designing a high-temperature composite system should determine the free-energy changes for the possible reactions in a proposed system. Systems such as a columbium alloy plus silicon-carbide fibers have very high degradation-reaction driving forces.

The second consideration for chemical compatibility pertains to the chemical potential of each of the elements present in the constituents. The chemical potential relates to the concentration of an element relative to the saturation of that element in each phase. For example, if a composite is to be made with a nickel alloy plus aluminum oxide, then the chemical potential of aluminum (and of oxygen, etc.) should be the same in both phases to prevent aluminum diffusion. Uneven chemical potentials in the constituent phases often lead to interfacial instability and degradation of fiber properties.

The surface energy of a two-phase mixture can be excessively high and therefore lead to interfacial instability. This problem has been significant in whisker-reinforced composites and will be discussed in Chapter 4. On the other hand, with the composite system tungsten carbide plus cobalt, the surface-energy relationship is beneficial and leads to faceted tungsten carbide particles that have excellent cutting edges.

Secondary diffusion effects controlled by grain-boundary or surface diffusion constants can often drastically modify the relationship between two constituent phases of a composite system. A significant problem encountered in the nickel-alloy plus tungsten-fiber composite relates to the reduction of the recrystallization temperature of the tungsten wire caused by nickel diffusion into the tungsten grain boundaries. Other secondary diffusion effects have been identified as liquid-metal embrittlement and hydrogen embrittlement. The latter effect can occur when the interstitial hydrogen concentration in one phase of the composite is high and can endanger the second phase.

The chemical compatibility of each of the metal-matrix composite systems of this volume will be treated in later chapters. This compatibility is of great importance during fabrication and application of the structural composite.

VII. Systems Not Featured in This Volume

The intent of this volume is to provide a solid review of several of the most prominent metal-matrix structural composite systems. The systems that have been chosen for the chapters appear to be those for which engineering structural application is being seriously considered. Little emphasis has been placed on model systems where the intent of the research was more to study composite behavior than to develop a competitive engineering material. In addition, only high-temperature reinforced superalloys will be discussed in depth. These systems appear to have the greatest promise for application. Composite systems such as whisker-reinforced aluminum and silver and eutectic composites of aluminum are treated in by Levitt (1970).

Several other composite systems have received lesser interest to date, and research in these systems is proceeding at a slower pace. In order to familiarize the reader with some of this work, a brief review of the research performed on a number of these composite systems is presented in this section.

Aluminum-matrix composites have been reinforced with fiber glass, beryllium, high-strength steel wire, silicon carbide, and numerous types of whiskers. Work on the silica-fiber-reinforced aluminum alloys was presented by Cratchley and Baker (1964). Fabrication was accomplished by a high-speed aluminum-coating operation on the fibers from a melt, followed by hot pressing of the coated wires. Composites were made with approximately 50 v/o fiber, and strengths of up to 0.85 GN/m^2 (130,000 psi) were attained. It was found that the strength of the composite was very sensitive to hot-pressing parameters, and of course no modulus improvement was made compared to the matrix alloy. Little work has been carried on with this system recently because of the general superiority of the aluminum–boron system and because of the serious compatibility problem between the fiber and matrix.

Aluminum–beryllium composites have been reported by Toy (1968). Fabrication can be performed using hot-press diffusion bonding of beryllium wire plus aluminum foil. Excellent mechanical properties were attained for both specific modulus and specific strength using wire with strengths in the range of 1.25 GN/m^2 (180,000 psi). Both fatigue and creep strength were evaluated and also were characteristic of the strong fibers. The extremely high cost of the beryllium wire in the small diameters that yield the high strengths, however, has retarded the incorporation of this composite system in significant structural applications.

Aluminum alloys reinforced with high-strength steel have been investigated by Sumner (1967) and others. Composites were fabricated using 2024 aluminum alloy plus NS-355 stainless-steel wire (Fe + 0.10–0.18 C, 4–5 Ni, 2–3 Mo, 15–16 Cr) by hot pressing at 475°C at 100 MN/m^2 (14,000 psi). These composites had 25 v/o fiber and had a room-temperature axial strength of 1.2 GN/m^2 (170,000 psi). More recent composites using 50 v/o of AFC-77 wire have been made at United Aircraft with tensile strengths in excess of 1.7 GN/m^2 (250,000 psi). The properties of this composite system can be predicted using the rule-of-mixtures calculations from constituent properties as described earlier, since the fibers are of uniform strength. This system permits the utilization of the outstanding strength and toughness properties of hard-drawn-steel wires. AFC-77 wire of 0.02 cm diam has a room-temperature strength of 3.8 GN/m^2 (550,000 psi) with 2% elongation to fracture, and a strength of 2.8 GN/m^2 (400,000 psi) at 500°C. Aluminum–steel laminae have also been used as a secondary reinforcement for aluminum–boron composites in order to improve the impact strength and off-axis strength.

Boron fibers have been used to reinforce numerous metal alloys, including magnesium and lead. Studies on the continuous-casting fabrication technique on the magnesium–boron system have been reported. High-volume-fraction composites were fabricated without fiber degradation, and excellent tensile strengths can be achieved. The technique includes the continuous casting of a 15–40 filament preform, followed by diffusion bonding or a remelt-bonding operation to fabricate the structural shape. High-strength composites have also been made by plasma-spraying magnesium onto drum-wound layers of boron, followed by hot-press diffusion bonding as reported by Abrams *et al.* (1971). This composite system has excellent strength-to-density and modulus-to-density ratios and should find many applications in lightweight, highly stressed structures.

The strength and modulus of lead can be dramatically increased by additions of low (5 v/o) volume fractions of boron. These composites were produced at United Aircraft by hot-press diffusion bonding of lead foils and drum-wound boron layers. Properties measured to date indicate ultimate-tensile-strength improvements of from 18 MN/m^2 (2500 psi) for pure lead to over 140 MN/m^2 (20,000 psi) for the boron composite with 5 v/o (1.2 w/o) fiber. Similar improvements are anticipated in fatigue and creep strength.

Other metal-matrix composite systems studied have included composites with aluminum, copper, titanium, iron, cobalt, nickel, and tungsten matrices using boron-carbide or silicon-carbide fiber or steel, tantalum, molybdenum, beryllium, or tungsten wire. These systems have been discussed by Galasso (1970).

References

Abrams, E. F., Davies, L. G., Powers, W. M., and Shaver, R. G. (1971). *In* "Space Shuttle Materials," p 39. Soc. Aerosp. Mater. and Proc. Eng., Azusa, California.

Adams, D., and Doner, D. R. (1969). *J. Compos. Mater.* **3**, 368.

Breinan, E. M., and Kreider, K. G. (1970). *Trans. AIME* **1**, 93.

Chen, P. E. and Lin, J. M. (1968). ASM Tech. Rep. P9-55.2.

Chen, P. E., and Lin, J. M. (1969). *Mater. Res. Stds.* **2**, 29.

Cohen, W., and Ishai, O. (1967). *J. Compos. Mater.* **1**, 390.

Corton, H. (1967). *In* "Modern Composite Materials" (L. J. Broutman and R. H. Krock, eds.). Addison-Wesley, Reading, Massachusetts.

Cratchley, D., and Baker, A. A. (1964). *Metallurgia* **69**, 153.

Daniels, H. E. (1945). *Proc. Roy. Soc. (London)* **183**, 405.

Dow, N. F., and Rosen, B. W. (1964). *J. Appl. Mech.*

Ellison, E. G., and Boone, D. H. (1967). *J. Less Common Metals* **13**, 103.

Galasso, S. F. (1970). "High Modulus Fibers and Composites." Gordon and Breach, New York.

Gillespie, J. S. (1966). *J. Amer. Ceram. Soc.* **88:11**, 2423.

Griffith, A. A. (1921). *Phil. Trans. Roy. Soc.* **A221**, 163.

Hancock, J. R., and Swanson, G. D. (1972). Composite Materials: Testing and Design, ASTM STP 497, Philadelphia, Pennsylvania.

Herring, H. W., and Krishna, V. G. (1966). NASA TM-X-1246.

Kelly, A., and Davies, G. J. (1965). *Met. Rev.* **10**, 37, 1–77.

Kelly, A., and Tyson, W. R. (1965). *Int. Mater. Conf., 2nd, Berkeley, California* (V. F. Zachay, ed.), p. 578. Wiley, New York.

Lasday, A., and Talley, C. (1966). *Nat. Symp. Soc. Aero Mater. Proc. Eng., 10th.*

Levitt, A. (1970) "Whisker Technology." Wiley, New York.

McAdam, G. D. (1951). *J. Iron. Steel Inst.* **168**, 346.

McDaniels, D. L., and Signorelli, R. A. (1966). Stress-Rupture Properties of Tungsten Wire, NASA TND 3467.

McDaniels, D. L., Jech, R. W., and Weeton, J. W. (1959). *Proc. Sagamore Ordnance Mater. Res. Conf., 6th* MET 661–601, Syracuse Univ. Res. Inst., p. 116.

McDaniels, D. L., Jech, R. W., and Weeton, J. W. (1960). *Met. Prog.* **78** (6), 118–121.

McDaniels, D. L., Jech, R. W., and Weeton, J. W. (1963a). Stress–Strain Behavior of Tungsten-Fiber-Reinforced Copper Composites, NASA TND 1881.

McDaniels, D. L., Jech, R. W., and Weeton, J. W. (1963b). *Trans. AIME* **233**, 636.

Olster, E. F., and Jones, R. C. (1972). Composite Materials: Testing and Design, ASTM STP 497, Philadelphia, Pennsylvania.

Otte, H. M., and Lipsitt, H. A. (1966). *Phys. Status Solidi* **13**, 439.

Piehler, H. R. (1965). *Trans. AIME* **233**, 12.

Rosen, B. W., Dow, N. F., and Hashin, Z. (1964). Mechanical Properties of Fibrous Composites, NASA CR-31.

Salkind, M., and Patarini, V. (1967). *Trans. AIME* **239**, 1268.
Schuster, D. N., and Scala, U. (1968). *AIAA J.* **6**, 527.
Sumner, W. (1967). *Advan. Struct. Compos. SAMPE* **12**, AC 19.
Talley, C. P. (1959). *J. Appl. Phys.* **30**, 1114.
Toy, A. (1966). *J. Mater.* **3**, 43.
Tsai, S. (1966). Air Force Rep. AFML-TR-66-149.
Veltri, R., and Galasso, F. (1968). *Nature* (*London*) **220**, 781.
Wawner, F. E. (1967). "Modern Composite Materials," p. 244. Addison-Wesley, Reading, Massachusetts.

2

Laminated-Metal Composites

E. S. WRIGHT and ALBERT P. LEVITT

Army Materials and Mechanics Research Center
Watertown, Massachusetts

I. Introduction

A laminated-metal composite consists of two or more layers or laminae of different metals completely bonded to each other so as to yield a composite material whose properties differ from and are more desirable than those of its constituents. By careful selection of these layers, laminated-metal composites may be designed to have outstanding properties for a wide variety of special applications, which may require resistance to wear, corrosion, and impact or enhanced thermal or electrical characteristics.

Metal laminates have a long history of usefulness. Dietz (1969) and Smith (1960) have presented some fascinating examples of the early use of metal laminates. The advantage of using a very hard cutting edge backed up by a more ductile, impact-resistant steel in swords was recognized as early as the Viking period, as exemplified by the famous Viking sword. The Merovingian sword (Lorraine, 6th century), German armor (15th century), Indonesian hand weapons (15th century), and Japanese swords (16th century) attest to the intuitive understanding and appreciation of metal laminates by artisans in Europe and the Near and Far East.

As late as 1900, laminated-steel shotgun barrels formed by forge-welding twisted bars of partially decarburized steel were considered superior to homogeneous steel barrels of that day.

Composite metal laminates provide a means for tailoring a composite to desired specifications. The desired properties may be one or more of the following:

a. corrosion resistance
b. surface hardness
c. wear resistance
d. impact resistance
e. toughness
f. strength
g. improved heat-transfer characteristics
h. improved electrical properties
i. improved magnetic properties
j. controlled deformation (thermal expansion) in response to temperature change
k. lower cost
l. appearance
m. flexibility
n. availability
o. formability

The purpose of this chapter is to provide an overview of the broad and rapidly growing field of metal laminates. To this end, this chapter discusses in turn the principal metal-laminate fabrication methods, the properties of metal laminates, materials engineering of laminated-metal composites for specific applications, and a summary and assessment of this field.

The laminate constituents may be selected so that one or more desired properties listed above may be achieved. For example, copper-clad stainless steel is an excellent roofing material because each constituent enhances the properties of the whole laminate. The copper provides the desired appearance and workability, while the stainless steel contributes strength and reduces the requirement for the more expensive copper. Furthermore, the lower thermal conductivity of the stainless steel as compared to copper improves the solderability of the laminate. On the other hand, copper cladding on stainless steel improves its heat-transfer characteristics when used in cooling applications.

Cupronickel cladding on copper provides a corrosion-resistant, low-cost, attractive appearance to coinage; and the difference in thermal expansion of the two metals in a thermostat provides a controlled, reproducible deformation in response to temperature change. These and other applications, to be described in more detail in Section IV of this chapter, illustrate the remarkable improvement in properties and performance that can be obtained through the proper use of metal laminates.

II. Metal-Laminate Fabrication Methods

Commercial metal laminates have been fabricated by a variety of techniques, the most common of which are roll bonding, coextrusion, explosive welding, and brazing. Although the details of these methods are largely proprietary in nature, the methods will be described briefly to familiarize the reader with the general concepts involved.

A. Roll Bonding

In the roll-bonding process, the metals are bonded by mill rolling under heat and pressure so that an integral bond is formed over the entire interfacial area. The possible combinations of clad and base metal are extremely broad. Table I illustrates the very wide range of feasible metal-laminate combinations. Cladding thickness generally ranges from 2.5 to 20% of total laminate thickness, depending on the application. Since the specific rolling techniques and conditions (temperature, rolling pressure) depend on the properties (melting point, ductility, thickness ratio) of the metals composing the laminate, it would be impossible to describe them all here. The general principles of roll bonding may be illustrated by describing the procedures

TABLE I

Composite Metal Combinations[a,b,c]

Cladding	Aluminum and alloys	Beryllium copper	Brass	Bronze	Copper	Cupro-Nickel	Gold	Gold-Nickel alloy	Indium	Fe Fe-Ni alloys	Magnesium	Molybdenum	Nickel	Ni-Fe alloys	Ni-Cu alloys	Ni-Cr-Fe alloys	Nickel-silver	Platinum	Silver	Stainless steels	Steel, carbon	Tantalum	Titanium	Tungsten	Zinc	Zirconium
Aluminum	x	x	x	x	x		x		x	x	x	x	x	x				x	x		x	x	x	x	x	
Aluminum alloys	x				x					x		x		x					x		x				x	
Bismuth alloys					x																					
Brass	x																				x					
Bronze					x																					
Cadmium	x				x					x									x							
Calcium													x						x		x					
Copper and alloys	x									x	x	x	x	x	x	x			x	x	x	x				
Gold and alloys	x	x	x	x	x					x		x	x	x			x	x	x	x		x	x	x		
Indium	x				x					x		x	x	x								x	x			
Fe/ Fe-Ni low exp alloys	x		x	x	x					x																
Lead and alloys	x		x	x	x					x		x	x	x					x		x	x				
Magnesium										x											x					
Nickel and alloys	x		x		x							x				x				x	x			x		
Palladium and alloys		x	x	x	x	x						x	x	x	x	x	x			x	x	x				x
Platinum and alloys		x	x	x		x				x		x	x	x	x	x	x		x	x	x	x				x
Silver and alloys	x	x	x	x	x	x				x	x	x	x	x	x	x	x			x	x	x	x	x	x	x
Stainless steels	x				x																x					
Steel, low carbon	x				x					x																
Tantalum					x										x											
Tin and alloys	x	x	x	x	x	x	x			x		x	x	x			x		x	x	x	x	x			
Titanium					x															x	x					
Zinc	x							x																		

[a] x denotes feasible combinations.
[b] Ronan (1970).
[c] Courtesy of *Stamping/Diemaking*.

used by Lukens Steel Company to fabricate clad-steel laminates such as stainless-steel-clad carbon steel. After the correct thickness ratio of cladding to backing steel has been selected to provide the final cladding proportion, the surfaces to be bonded are carefully cleaned. The high-alloy cladding layers (except nickel) are then nickel plated to promote bonding to the backing metal. The nickel also prevents diffusion of carbon from the backing metal into the cladding. The side of the cladding that is not to be bonded is then coated with a refractory (nonmelting) parting compound. The backing metal is made slightly wider and longer than the cladding to allow for expansion of the core during rolling and to permit the insertion of spacer bars around the perimeter of the core. Then the layers are assembled so that the steel backing plates are on the outside and the cladding inserts are between them such that the parting compound prevents bonding of the cladding inserts to each other during rolling. Steel spacer bars are then welded into the groove formed around the perimeter of the pack, thereby sealing and protecting the interior of the pack from the atmosphere during subsequent heating. A schematic cross section of the pack prior to rolling is shown in Fig. 1.

The pack is then heated in a soaking pit to the temperature appropriate to the materials in the cladding pack (Fig. 2), and rolled to the final thickness (Fig. 3). The combination of uniform high pressure produced by the rolling mill and the high temperature of the pack serve to forge-weld the components together into an integral, metallurgically bonded composite. A metallograph (Fig. 4) of the interface between stainless steel (Multimet), cladding nickel plating, and carbon-steel backing metal shows the sound metallurgical bond obtained. After rolling, the welded edges on all four sides are removed by shearing or flame cutting. The resulting single-clad sheet can then be heat treated to further improve its properties, descaled in a sodium-hydride bath (a 700°F bath of 1.2–1.7% sodium hydride in molten sodium hydroxide), cooled in a water quench, brightened with a sulfuric-acid bath, and, in the case of stainless cladding, passivated in a nitric–hydrofluoric-acid bath. After leveling in a roller leveler, the clad plates are ready for shipment.

B. Coextrusion

Coextrusion is particularly well adapted to the formation of long, continuous wire, rod, and flat rectangular shapes. Yans *et al.* (1962) have shown

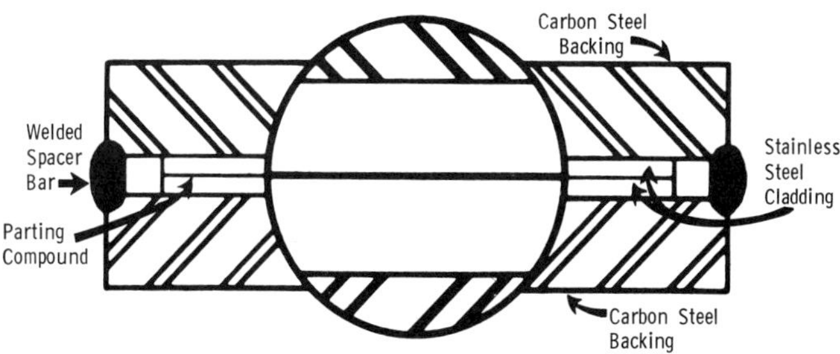

Fig. 1. Schematic cross section of the single-clad pack prior to rolling. (By permission of Lukens Steel Company, 1960)

Fig. 2. Heating of clad pack prior to rolling. (By permission of Lukens Steel Company, 1960)

Fig. 3. Rolling of clad pack. (By permission of Lukens Steel Company, 1960)

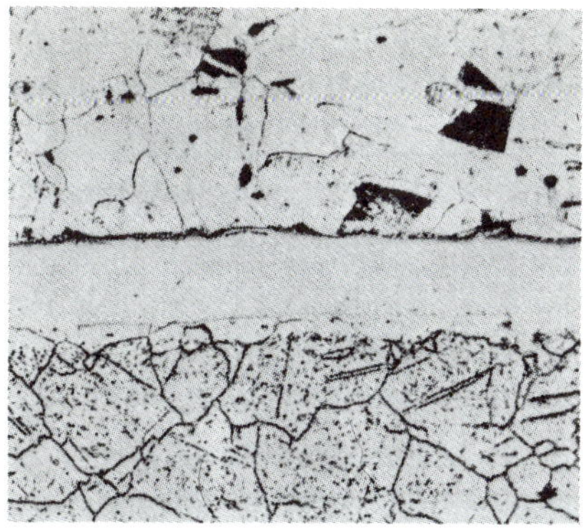

Fig. 4. Metallograph of interface between stainless-steel cladding and carbon steel backing. (70 ×) (By permission of Lukens Steel Company, 1960)

that this technique is especially useful for forming seamless clad nuclear fuel elements. In this process, the metals to be clad are assembled as an extrusion billet as shown schematically in Fig. 5. The extrusion pressure, temperature, and amount of reduction depend on the materials being coextruded. The primary bonding mechanism is pressure diffusion bonding, in which clean metal surfaces are brought into intimate contact under pressure. Yans *et al.* have provided the useful relationships between components of billets and components of round rods and tubes extruded from round billets shown in Table II.

Cable extrusion is a modification of the coextrusion process described above. In this process, the cable or core to be clad is fed through a central

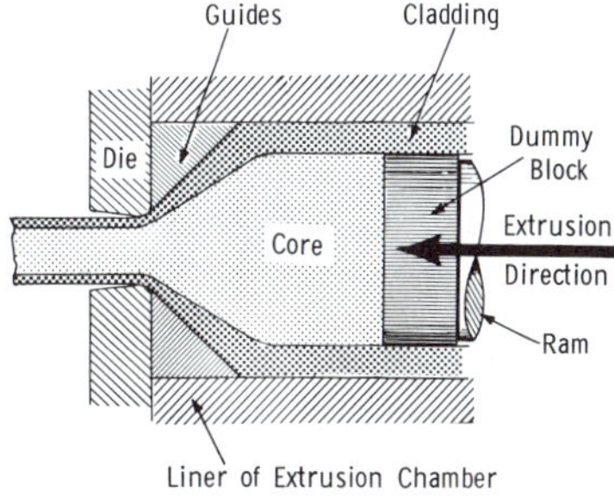

Fig. 5. Schematic of coextrusion process. [After Yans *et al.* (1962) by permission of Wiley (Interscience), New York]

TABLE II

Useful Relations between Components of Billets and Components of Round Rods and Tubes Extruded from Round Billets[a,b]

1. *CSA* of component in billet = *CSA* of same component after extrusion × *R*

2. $$\frac{CSA \text{ of component in billet}}{\text{total } CSA \text{ of billet}} = \frac{CSA \text{ of same component after extrusion}}{\text{total } CSA \text{ of extrusion}}$$

3. For solid rods only:

   a. $$\frac{CSA \text{ of billet core}}{\text{total } CSA \text{ of billet}} = \left(\frac{\text{diameter of billet core}}{\text{total diameter of billet}}\right)^2 = \left(\frac{\text{diameter of core in extrusion}}{\text{total diameter of extrusion}}\right)^2$$

   b. Any cross-sectional dimension in billet $= \text{same dimension in rod} \times \sqrt{R}$

[a] From Yans *et al.* (1962) by permission of Wiley (Interscience), New York.
[b] *CSA* is cross-sectional area, *R* is reduction in area of billet.

guide or mandrel into a chamber where the heated cladding is extruded over it and through the die opening. Copper-clad aluminum wire and aluminum- and lead-clad cables are examples of composites made by this process. The double-ram cable-extrusion technique is schematically illustrated in Fig. 6.

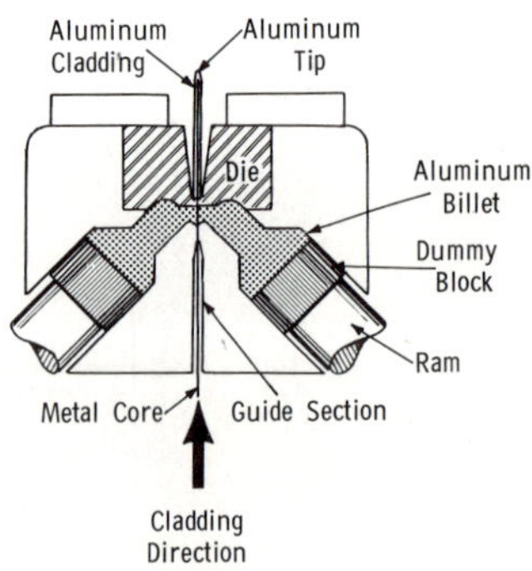

Fig. 6. Schematic of cable-extrusion process. [After Yans *et al.* (1962) by permission of Wiley (Interscience), New York]

C. *Explosive Welding*

Explosive welding or explosive bonding is a relatively new method for producing metal laminates. Crossland (1971), Crossland and Williams (1970), and Wright and Bayce (1964) have presented detailed reviews of this method, which permits the rapid economical bonding of dissimilar metals, many of which could not be bonded otherwise because of their widely different properties. Basically this process involves the formation of a high-velocity metal jet through removal of a surface layer of one of the metals to be joined. As this jet shoots across the surface of the other metal to be joined, it also removes a thin surface layer from it. The resulting clean surfaces are then forced together by the explosive pressure to form a sound metallurgical bond. Thus, this process satisfies the two basic conditions for joining any two metals: (1) the surfaces must be clean (the jet does the cleaning), and (2) the clean surfaces must be brought into intimate contact (the explosive pressure does this). Bonding by this method is not limited by large differences in melting point or plastic properties of the metals to be joined.

Two dissimilar metals can be readily bonded using the parallel-plate standoff technique schematically illustrated in Fig. 7. A rubber or plastic buffer layer serves to protect the flyer plate from the explosive which, when detonated at one end of the flyer plate, causes it to strike the stationary plate with an impact pressure much greater than the yield strengths of the metals to be joined. If the velocity of the collision point is less than the velocity of sound in the two metals, a metal jet is formed at the lower surface of the flyer plate, which scours the interfacial surfaces. The explosive pressure then bonds them together.

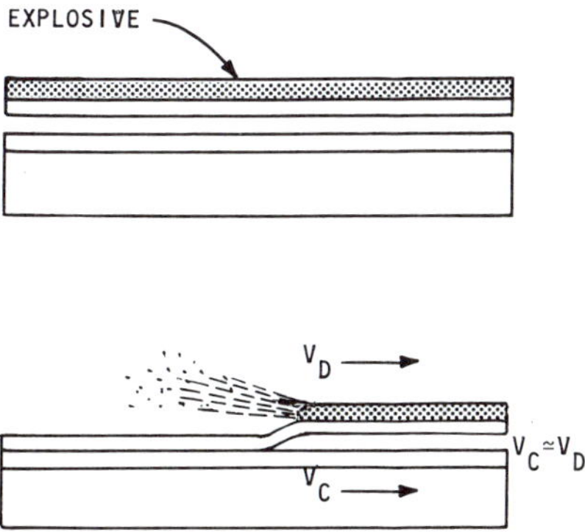

Fig. 7. Schematic of explosive-welding setup with parallel standoff. V_D is the detonation velocity, V_C is the velocity of the collision point. (By permission of Stanford Research Institute)

Since the detonation velocities of most explosives (7–8000 m/sec) is significantly higher than the velocity of sound in the materials to be joined (6000 m/sec), the parallel standoff arrangement makes it difficult to satisfy the requirement that the collision point be subsonic. The angular-standoff configuration (Fig. 8) circumvents this problem because the velocity of the collision point is a function of the plate velocity and the initial standoff angle, but only indirectly dependent on the detonation velocity of the explosive.

Wright and Bayce (1964) have shown that multiple-layered laminates can be sucessfully bonded by explosive welding using the arrangement shown in Fig. 9. Chemically cleaned zinc and aluminum sheets 3.0 × 6.0 × 0.020 in. were stacked in alternate layers with spacers along their edges producing a gap of 0.005 in. between each sheet. The driver plate was then accelerated by an explosive charge so that (1) it struck the stack with a pressure (1.4×10^6 psi) that was several times the yield strength of the materials, and (2) the velocity with which the region of impact traversed the surface of the pack (3500 m/sec) was less than the velocity of sound in zinc or aluminum. Under these conditions, jetting and melting with severe plastic deformation occurs at the point of impact, producing intimate contact and bonding of freshly exposed metal surfaces. The driver-plate momentum forces each sheet of the stack against the succeeding layer such that bonding occurs at each interface. This continues until the impact or collision angle decreases to the level at which jetting no longer occurs at the impact point. In this way, up to 200 sheets of foil have been joined in a single operation.

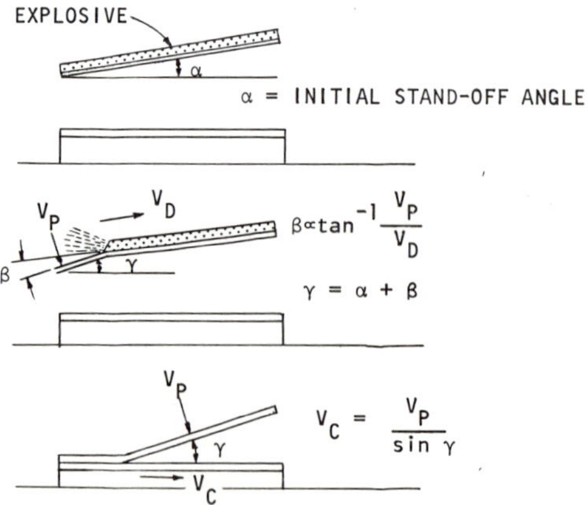

Fig. 8. Schematic of angular-standoff arrangement for explosive welding. (By permission of Stanford Research Institute)

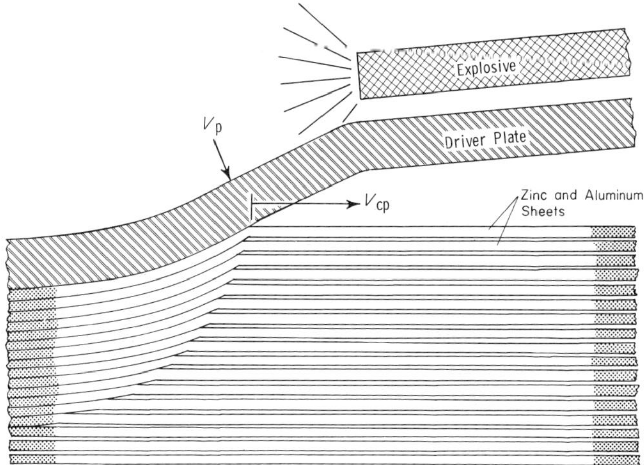

Fig. 9. Schematic of arrangement for explosive welding of multiple-layered laminates.

The excellent welds, speed, and versatility of this method have made it practical and economical. Its capability for bonding dissimilar metals that could not otherwise be bonded further enhances its possibilities. Explosive welding is especially well suited for cladding tubing, for tube–plate welding, and for cladding of wear surfaces in machinery, engine components, and nuclear-reactor and chemical-process equipment. Figure 10 illustrates two methods for cladding tubing, based on the work of Blazynski and Dara (1971). The arrangement for explosively cladding the inside of a tube is shown in Fig. 10a. Here the explosive charge forces the cladding radially outward until it bonds with the outer tubing. Figure 10b depicts the method used to clad the outside of a tube by imploding the cladding onto it. Table III indicates the many combinations of metals that have been sucessfully bonded by this method.

D. Brazing

Brazing basically involves the use of a wetting liquid-metal phase which, on solidification, bonds the two laminae together. When the brazing material has a melting point that is lower than that of the two surfaces to be bonded, the technique is known as high-temperature brazing. The system is heated to a temperature between the melting point of the brazing alloy and that of the laminae to be bonded. The brazing alloy may be in the form of foil or

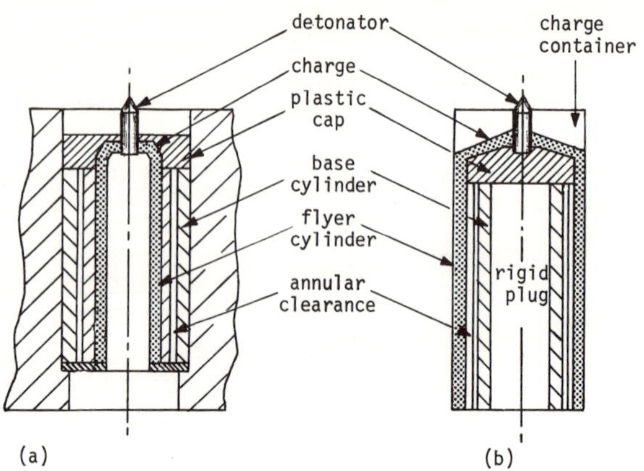

Fig. 10. Setup for cladding cylinders: (a) explosive system; (b) implosive system. [After Blazynski and Dara (1971) by permission of Univ. of Denver]

wire and may be used with or without a flux. When the brazing alloy is used to form a low-melting eutectic between the metal layers, the process is known as eutectic brazing. This method has been used to bond titanium to steel, using a copper–silver eutectic alloy without a flux in a high vacuum.

III. Properties of Laminated-Metal Composites

The use of a metal laminate in any engineering application is advantageous only if it offers improvement in some property or combination of properties at a cost lower than that of a monolithic structure. Properties to be considered in the selection of utilization of metal laminates include mechanical properties such as elasticity, strength, ductility, and fracture toughness, or physical and chemical properties such as density, thermal and electrical conductivity, and chemical reactivity or corrosion resistance.

A. *Elasticity*

The effective elastic modulus or specific stiffness (stiffness-to-density ratio) of a metal laminate depends not only on the elastic moduli, density, and volume fractions of the individual metals comprising the laminate but also on their arrangement and the manner of loading of the laminated structure.

TABLE III

Metal Combinations Bonded by Explosive Cladding[a,d]

METALS	ZINC	PALLADIUM ALLOY	TD NICKEL	TUNGSTEN	NICHROME	MAGNESIUM	MOLYBDENUM	COLUMBIUM AND Cb ALLOYS	PLATINUM	SILVER AND SILVER ALLOYS	GOLD ALLOYS	TANTALUM	HAYNES STELLITE ALLOY 68 c	HASTELLOY ALLOY X c	HASTELLOY ALLOYS BCF c	ZIRCONIUM AND ZIRCALLOYS	TITANIUM AND TI ALLOYS 6Al-4V	NICKEL AND NICKEL ALLOYS b	BRONZE	CUPRO-NICKEL	BRASS	COPPER	ALUMINUM AND Al ALLOYS	MARAGING STEEL	HADFIELD STEEL	STAINLESS STEEL, 200 SERIES	STAINLESS STEEL, 300 SERIES	STAINLESS STEEL, FERRITIC	ALLOY STEEL, AISI 4340	ALLOY STEEL, AISI 4130	LOW ALLOY STEEL, ASTM-387	LOW ALLOY STEEL, ASTM-302	LOW ALLOY STEEL, ASTM-204	MEDIUM C STEEL, ASTM-212	MEDIUM C STEEL, ASTM-201	MEDIUM C STEEL, ASTM-285	LOW C STEEL, AISI 10041	LEAD
LOW C STEEL, AISI 1004 to 1020	●					●	●			●		●						●		●	●	●	●			●	●	●									●	●
MEDIUM C STEEL, ASTM A-285						●	●			●		●	●	●	●	●	●	●	●	●	●	●	●			●	●	●								●		
MEDIUM C STEEL, AST						●	●	●		●		●	●	●	●	●	●	●	●	●	●	●	●			●	●								●			
MEDIUM C STEEL, ASTM A-212						●	●	●		●		●	●	●	●	●	●	●	●	●	●	●	●			●	●							●				
LOW ALLOY STEEL, ASTM A-204														●	●	●	●	●		●	●	●	●			●	●						●					
LOW ALLOY STEEL, ASTM A-302														●	●		●	●		●	●	●	●			●	●					●						
LOW ALLOY STEEL, ASTM A-387														●	●		●	●					●			●	●	●			●							
ALLOY STEEL, AISI 4130																											●			●								
ALLOY STEEL, AISI 4340																											●		●									
STAINLESS STEEL, FERRITIC																												●										
STAINLESS STEEL, 300 SERIES			●				●	●					●				●	●			●	●	●	●			●											
STAINLESS STEEL, 200 SERIES																						●	●			●												
HADFIELD STEEL																							●															
MARAGING STEEL																																						
ALUMINUM AND Al ALLOYS																	●					●	●															
COPPER							●					●					●					●																
BRASS																		●			●																	
CUPRO-NICKEL																				●																		
BRONZE																			●																			
NICKEL AND NICKEL ALLOYS b									●		●			●			●	●																				
TITANIUM AND TI ALLOYS 6Al-4V			●					●									●																					
ZIRCONIUM AND ZIRCALLOYS																●																						
HASTELLOY ALLOYS BCF c															●																							
HASTELLOY ALLOY X c			●											●																								
HAYNES STELLITE ALLOY 68 c																																						
TANTALUM												●																										
GOLD ALLOYS																																						
SILVER AND SILVERALLOYS										●																												
PLATINUM									●																													
COLUMBIUM AND Cb ALLOYS		●					●	●																														
MOLYBDENUM					●		●																															
MAGNESIUM						●																																
NICHROME					●																																	
TUNGSTEN				●																																		
TD NICKEL			●																																			
PALLADIUM ALLOY																																						
ZINC	●																																					

[a] A blank space means bonding of that combination has not been attempted. It does not mean those metals cannot be explosion bonded.

[b] Includes Inconel, *Monel,* and Incoloy, registered trademarks of International Nickel Company.

[c] Registered trademark of Union Carbide Corporation.

[d] Crossland and Williams (1970) after Holzman and Cowan (1965), by permission of Welding Research Council.

The elastic behavior of a laminated metal composite under uniaxial loading within the plane of the composite, i.e., in any direction parallel to the laminae, is analogous to that of a uniaxial fiber-reinforced composite stressed parallel to the fiber direction, and can be readily predicted on a rule-of-mixtures basis. Thus, the effective Young's modulus of a laminar composite is given by

$$E = E_1 f_1 + E_2 f_2 + \cdots + E_n f_n \tag{1}$$

where $E_1, E_2, \ldots, E_n$ are the Young's moduli and $f_1, f_2, \ldots, f_n$ are the volume fractions of the component materials.

Similarly, stiffness under bending loading *within* the plane of the composite can be calculated by rule-of-mixtures equations.

In *transverse* bending or torsional loading, however, rule-of-mixtures behavior is approximated for multiple-layered composites only if there is a large number of laminae and the distribution of high- and low-modulus materials is uniform across the thickness of the composite.

In such a composite the stiffness of a rectangular beam or plate composed of many alternate layers of thin laminae of two materials, loaded as shown in Fig. 11a, would be similar to that of a homogeneous material:

$$M/(1/r) = EI = Ebh^3/12 \tag{2}$$

where M is the applied bending moment, r the radius of curvature of the loaded beam, and b and h the base and height dimensions of the beam. The effective elastic modulus E is the same as the rule-of-mixtures modulus computed for uniaxial loading.

If the distribution of the two materials is not uniform throughout the cross section (Fig. 11b), the uniaxial rule-of-mixtures modulus cannot be used because stiffness in bending is determined predominantly by the outer

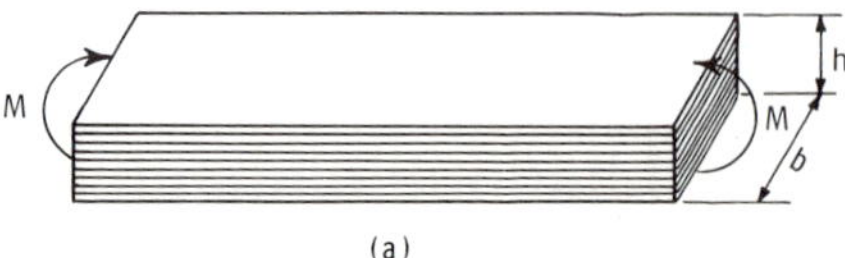

(a)

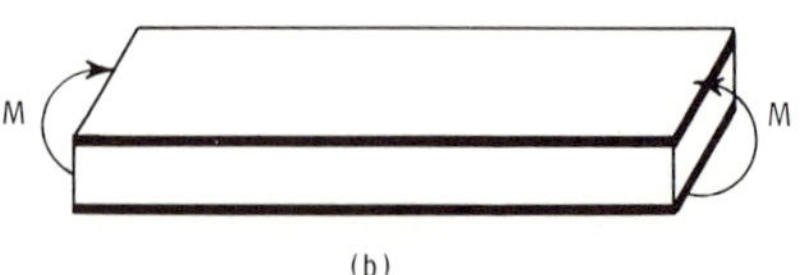

(b)

Fig. 11. Schematic diagram of (a) uniform and (b) nonuniform laminates loaded in transverse bending.

layers, the contribution of each lamina being proportional to the square of its distance from the neutral axis. Stiffness of such a beam is given by

$$M/(1/r) = (b/6) \int Ex^2 \, dx \tag{3}$$

where x is the distance from the neutral axis. If the thicknesses of the laminae are small in comparison with the overall thickness of the composite, the stiffness of the laminate can be approximated by

$$M/(1/r) = (b/12)\,(E_1 t_1 x_1{}^2 + E_2 t_2 x_2{}^2 + \cdots + E_n t_n x_n{}^2) \tag{4}$$

where $E_1, E_2, \ldots, E_n$ are the Young's moduli of the individual laminae, $t_1, t_2, \ldots, t_n$ their thicknesses, and $x_1, x_2, \ldots, x_n$ their distances from the neutral axis. Thus a small amount of high-modulus material can be applied to the outer surfaces of a low-modulus core to produce a laminate having relatively high stiffness in the bending mode.

Figure 12 illustrates the increase in specific bending stiffness (effective bending modulus-to-density ratio) that can be achieved by cladding an aluminum alloy with thin layers of stainless steel, even though the steel is considerably more dense than aluminum.

B. *Yield Strength and Plasticity*

Yielding under uniaxial tension occurs in one of the components of a laminated composite when the stress in that component exceeds its yield strength. Inasmuch as the stress in each component is not necessarily the same as the average stress in the composite, it is more convenient to consider a critical *yield* strain criterion given by

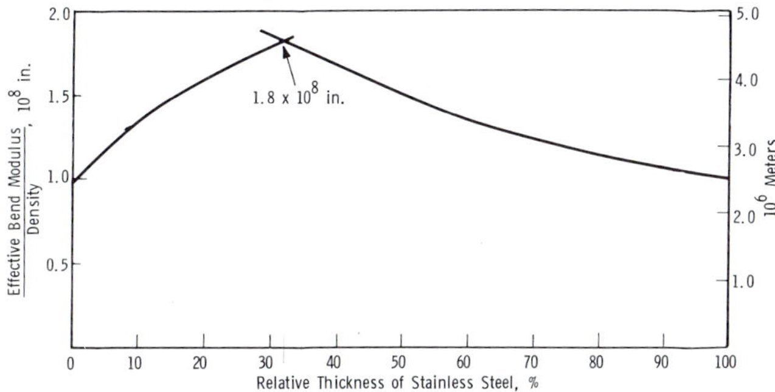

Fig. 12. Effective specific bending stiffness of stainless-steel-clad aluminum. (By permission of Charles Pfizer and Co., Inc., 1971)

$$\varepsilon_{y_1} = \sigma_{y_1}/E_1 \tag{5}$$

where σ_{y_1} and E_1 represent the yield strength and Young's modulus of the particular component of the laminate. Thus, yielding of a laminate occurs when the strain in the laminate exceeds the yield strain of any of the component laminae. Thus, the criterion for yielding is

$$\varepsilon_c > \varepsilon_{y_1}, \varepsilon_{y_2}, \cdots, \text{or } \varepsilon_{y_n} \tag{6}$$

In terms of stress this criterion is

$$\sigma_c/E_c > \sigma_{y_1}/E_1, \qquad \sigma_{y_2}/E_2, \cdots, \quad \text{or} \quad \sigma_{y_n}/E_n \tag{7}$$

It is interesting to note that in a hypothetical laminate consisting of steel ($\sigma_y = 80 \times 10^3$ psi, $E = 30 \times 10^6$ psi) and an aluminum alloy ($\sigma_{y_s} = 30 \times 10^3$ psi, $E = 10 \times 10^6$ psi), the stronger steel laminae would yield first.

The predicted yield strength of such a composite laminate is thus give by

$$\sigma_{y_c} = \sigma_{y_*} E_c/E_* = (\sigma_{y_*}/E_*)(E_1 f + E_2 f_2 + \cdots + E_n f_n) \tag{8}$$

where σ_{y_*} and E_* are the yield strength and Young's modulus of the component laminae having the lowest σ_y/E_c ratio.

A typical engineering stress–strain curve for laminates composed of alternating layers of a mild steel and 60:40 lead–tin solder is shown in Fig. 13. Similar curves for the individual components are shown for comparison. The observed Young's-modulus and yield-strength values (Table IV) agree reasonably well with rule-of-mixture values computed according to Eqs. (1) and (8).

Mechanical properties reported by the manufacturer for stainless-steel–aluminum laminates consisting of a 3003 aluminum core clad on both sides with type 304 stainless steel are shown in Table V and Fig. 14.

Similar data for roll-bonded steel–copper composites have been reported by Hawkins and Wright (1971), who found good agreement between their experimental results and values predicted on the basis of the rule of mixtures using an equal-strain hypothesis (Fig. 15). They also found that the plastic behavior (i.e., the shape of the stress–strain curve after yielding) could also be predicted up to the point of plastic instability or necking in one of the components. It was found that the laminates followed the classical work-hardening equation, $\sigma = K\varepsilon^n$, and that the values of K and n could be synthesized either graphically from the stress–strain curves of the components using the equal-strain rule-of-mixtures approach or analytically from their individual work-hardening equations using the relationships

$$K = K_{Cu} t_{Cu} + K_{st} t_{st} \tag{9}$$

and

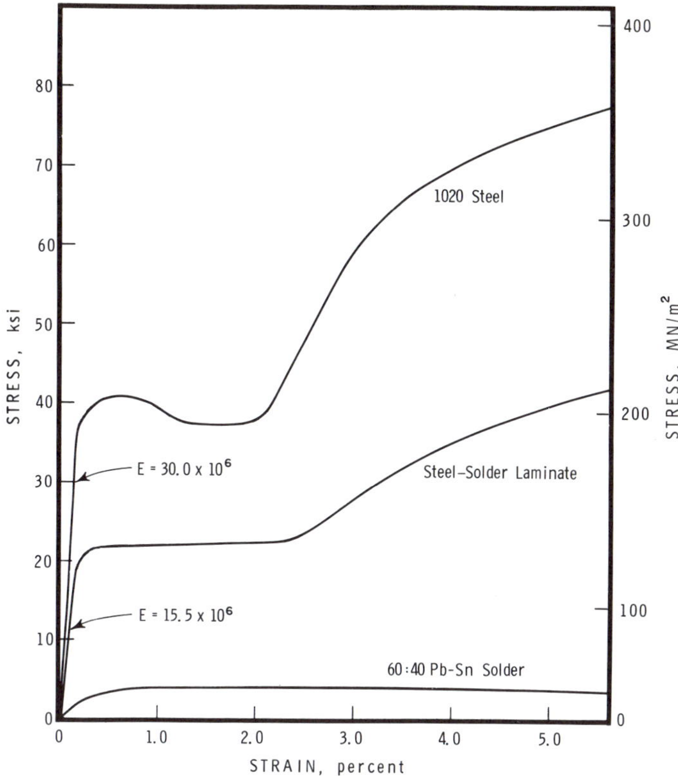

Fig. 13. Stress–strain curves for mild steel, 60 : 40 Pb–Sn solder, and a 50 : 50 steel–solder laminate.

$$n = \log(\sigma_{Cu} t_{Cu} + \sigma_{st} t_{st}) - \log(K_{Cu} t_{Cu} = K_{st} t_{st}) \tag{10}$$

where the subscripts Cu and st are used to denote the coefficients, flow stress, and thickness of the copper and steel components, respectively. Hawkins and Wright also showed that drawing and stretch-forming stresses and punch loads can be predicted for two-layer and three-layer laminates. Qualitative predictions of drawability, formability, and earing tendencies could be made, but exact values of limiting draw ratios or Erickson values could not be calculated.

C. Fracture Toughness

A number of experimental studies have demonstrated that metal laminates having bond strengths or intermediate layers weaker than the major con-

TABLE IV

Young's Modulus and Yield Strength of Mild-Steel—Solder Laminates

Composition (%)		Predicted				Observed			
		Young's modulus		Yield strength		Young's modulus		Yield strength	
Steel	Solder	GPa[a]	(10^6 psi)	MPa[b]	(10^3 psi)	GPa[a]	(10^6 psi)	MPa[b]	(10^3 psi)
100	0	—	—	—	—	207	(30.0)	276	(40.0)
90	10	190	(27.2)	250	(36.3)	187	(27.1)	255	(37.0)
75	25	160	(23.0)	210	(30.7)	158	(22.8)	215	(31.2)
50	50	110	(16.0)	147	(21.3)	107	(15.5)	144	(20.8)
0	100	—	—	—	—	14	(2.0)	128	(4.0)

[a] Gigapascals. [b] Megapascals.

TABLE V

Mechanical Properties of Stainless-Steel—Aluminum Laminates[a,b]

Thickness (in.)	Al	Yield strength		Tensile strength		Elongation (%)	Young's Modulus	
		MPa	(10^3 psi)	MPa	(10^3 psi)		GPa	(10^6 psi)
0.050	40	330	(47.6)	470	(68.1)	34	150	(21.8)
0.060	50	282	(40.7)	410	(59.4)	34	137	(19.8)
0.070	57	247	(35.7)	368	(53.2)	33	127	(18.4)
0.080	63	222	(32.1)	336	(48.6)	33	120	(17.4)
0.090	67	201	(29.1)	310	(44.9)	32	104	(16.5)
0.110	73	172	(24.9)	274	(39.7)	32	106	(15.4)
0.125	76	156	(22.6)	254	(36.8)	32	102	(14.7)
Aluminum	100	41	(6.0)	110	(16.0)	30	69	(10.0)
Stainless steel	0	505	(73.0)	690	(100.0)	35	200	(29.0)

[a] 3003 aluminum core, 0.015-in. 304 stainless-steel faces.
[b] Data supplied by Charles Pfizer and Co., Inc. (1971).

stituents of the composite possess better resistance to crack propagation than a monolithic material or strongly bonded laminate. Arnold (1960) pointed out the advantages of laminating steel sheet from the standpoint of improvement in toughness, and Bluhm (1961) discussed this effect in terms of a bimodal fracture model. Two principal mechanisms were identified by Embury *et al.* (1967). In a crack-arrester configuration (Fig. 16a) in which the crack-propagation direction is perpendicular to the laminae, improved fracture resistance is attributed to delamination at an interface ahead of the

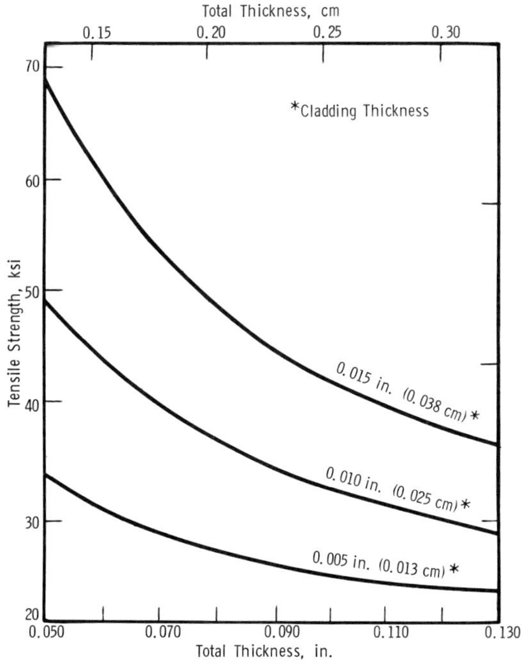

Fig. 14. Tensile strength of stainless-steel–aluminum laminates. (By permission of Charles Pfizer Co., Inc., 1971)

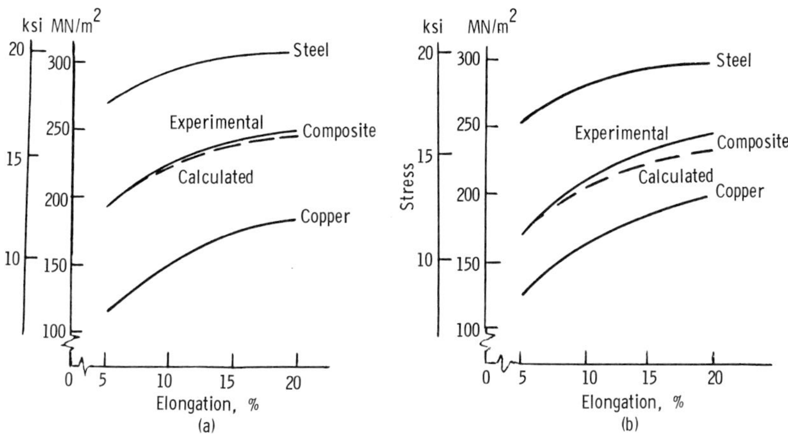

Fig. 15. Comparison of calculated and experimental stress–strain curves for roll-bonded composites: (a) 50% Cu, 50% steel two-layer laminate; (b) 33% Cu, 34% steel, 33% Cu three-layer laminate. [From Hawkins and Wright (1971), by permission of Inst. of Metals]

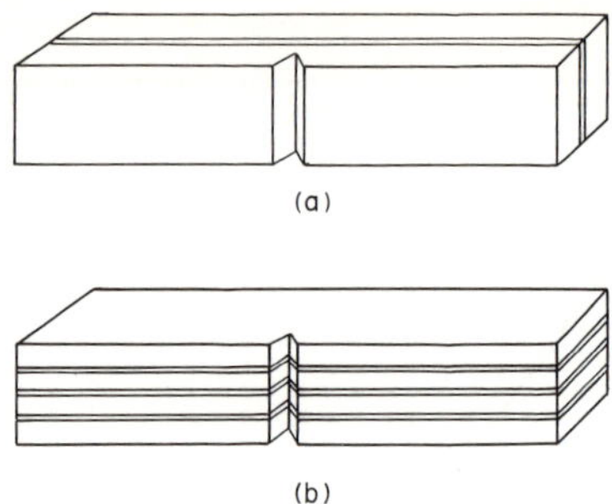

Fig. 16. Schematic diagram of (a) crack-arrester and (b) crack-divider laminates. [From Embury *et al.* (1967), by permission of AIME]

propagating crack, and consequent blunting of the crack and relief of the triaxial tensile stress state at its tip. Further propagation of fracture requires reinitiation of the crack, which absorbs considerably more energy than continued propagation of an existing crack. In many cases reinitiation does not occur, because further deformation of the laminate is accommodated by plastic flow rather than fracture.

In crack-divider laminates (Fig. 16b), the improvement in resistance to crack propagation is also attributed to delamination ahead of the crack tip, but the proposed mechanism is simply one of dividing the laminate into a number of thin layers in the vicinity of the crack tip, thus relieving the third-dimensional plastic constraint. Since yielding depends upon exceeding a critical shear stress rather than a critical maximum tensile stress, relief of the tensile stress in the third direction lowers the principal tensile stress required to produce yielding, thus encouraging plastic flow in lieu of cleavage. The difference between the fracture toughness of the weakly bonded laminate and that of a monolithic specimen or a strongly bonded laminate is essentially equivalent to the difference between the plane stress and plane strain behavior of the major component.

The results of Embury *et al.* (1967) demonstrated the striking decrease in ductile–brittle transition temperature that can be achieved in mild steel by the incorporation of a single crack-arresting lamina of soft solder or copper (Figs. 17 and 18). Their results also showed that the ductile–brittle transition temperature of mild-steel crack-divider laminates decreases gradually with decreasing thickness of the individual steel laminae (Fig. 19).

Wright (1968) has shown that similar effects can be achieved in high-strength steels, even though the brittle-fracture mode is not one of cleavage. Charpy impact test results for monolithic and solder-laminated AISI 4130 steel at a yield-strength level of 170 ksi indicate an increase in fracture energy

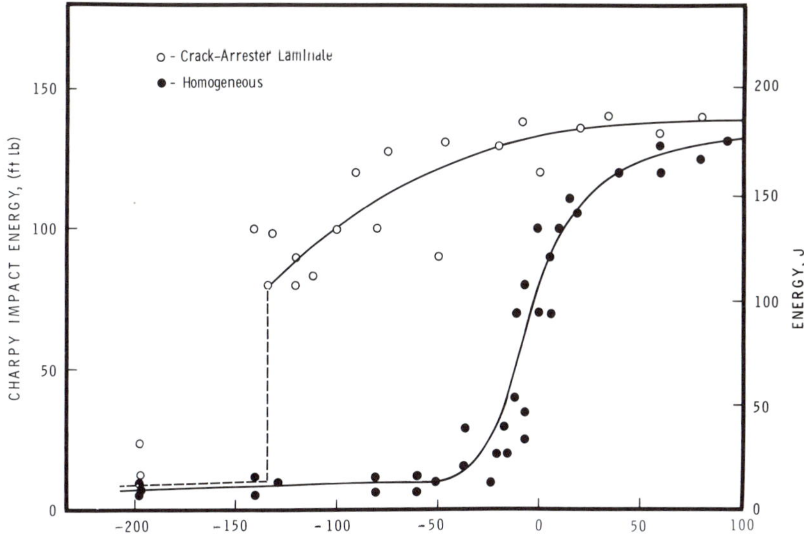

Fig. 17. Charpy impact energy vs. temperature for homogeneous mild steel and soft-solder–steel crack-arrester laminates. [From Embury *et al.* (1967), by permission of AIME]

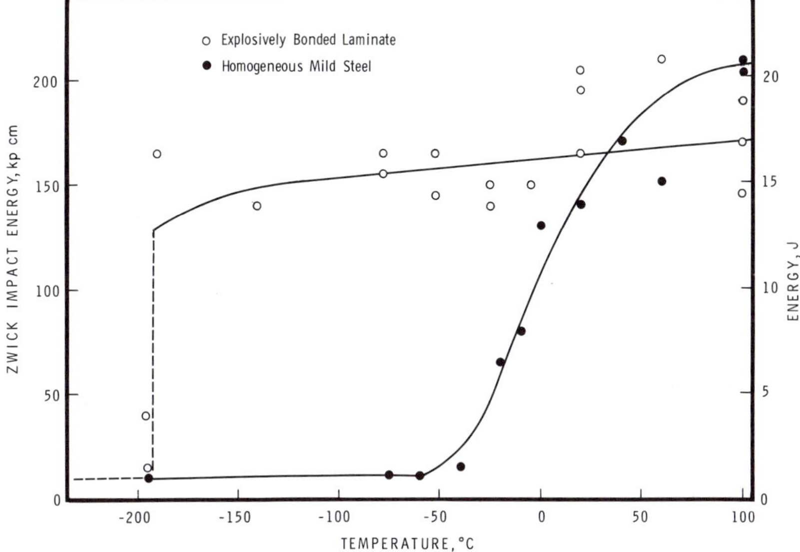

Fig. 18. Zwick impact energy vs. temperature for homogeneous mild steel and explosively bonded steel–copper crack-arrester laminates. [From Embury *et al.* (1967), by permission of AIME]

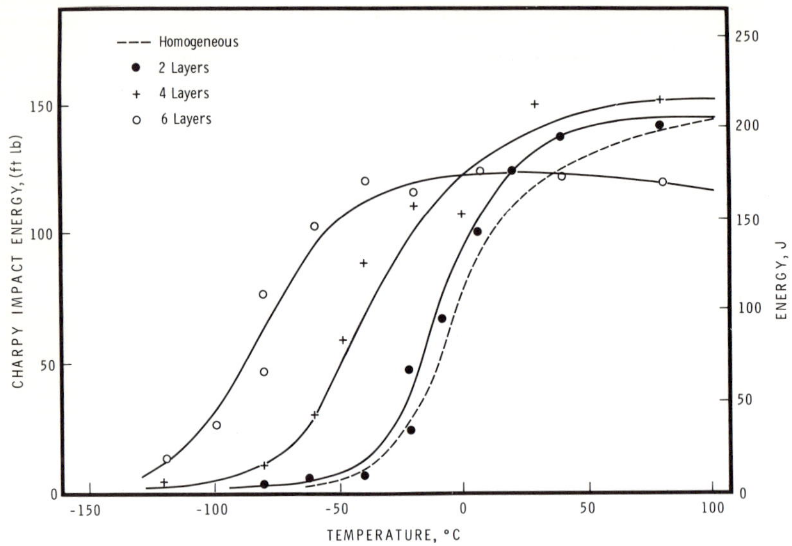

Fig. 19. Charpy impact energy vs. temperature for homogeneous mild steel and silver-brazed steel crack-divider laminates. [From Embury *et al.* (1967), by permission of AIME]

ranging from 200 to 1200% over the entire range of test temperatures from +100° to −196°F. (Fig. 20).

Almond *et al.* (1969) summarized a number of papers on fracture of metal laminates and also conducted a quantitative examination of factors controlling crack propagation in crack-arrester laminates. Dynamic load deflection curves for mild-steel laminates bonded with copper, solder, epoxy, and rubber were determined by means of an instrumented impact test. Initial fracture loads and postfracture loads were found to agree with values predicted on the basis of simple geometrical considerations and known material properties (Fig. 21).

Leichter (1966) showed that increases in Charpy fracture energy of up to 100% could be obtained in brazed crack-divider laminates of maraging steels and Ti–5A–2.5Sn alloy (Fig. 22). Kaufman (1967) observed similar increases in K_{Ic} values, from 34.3 ksi in.$^{1/2}$ to 74.4 ksi in.$^{1/2}$, in epoxy-bonded 7075 aluminum alloy composites.

Almond *et al.* (1969) studied the fracture behavior of solid-wall and laminated-wall pressurized steel cylinders containing through-wall simulated cracks. It was found that the solid-wall vessels failed in a brittle manner at all temperatures below −5°C, with crack-branching and fragmentation below −68°C, while the laminated vessels showed stable crack growth and

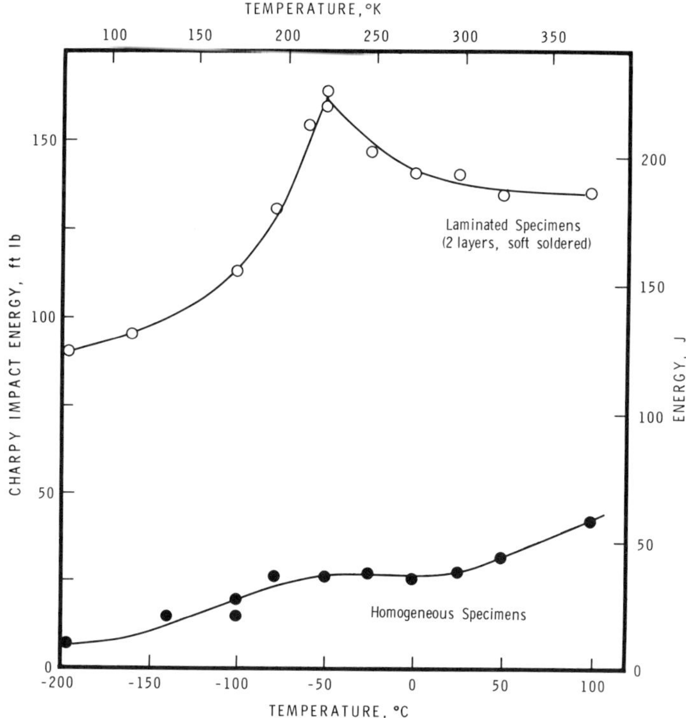

Fig. 20. Charpy impact energy vs. temperature for homogeneous AISI 4340 steel and soldered 4340 steel crack-arrester laminates. (From Wright, 1968)

ductile behavior at $-75°C$ and above (Fig. 23). Initial rapid crack growth was exhibited by a laminated-wall vessel at $-105°C$, but the crack did not propagate to the ends of the vessel and no branching or fragmentation was observed. Similar behavior would be expected for bonded laminates having bond strengths weak enough to permit delaminations at or ahead of the crack tip.

It is thus readily apparent that lamination can be used as an effective means of controlling fracture toughness and avoiding catastrophic brittle fractures. It is considered particularly applicable to pressurized systems where lamination would facilitate the use of leak-before-burst design concepts even in thick-walled pressure vessels. It is obvious that the crack-arrester geometry exists in a pressure vessel containing a crack that has not yet perforated the vessel wall, but once perforation occurs, the crack-divider geometry exists and catastrophic failure is unlikely.

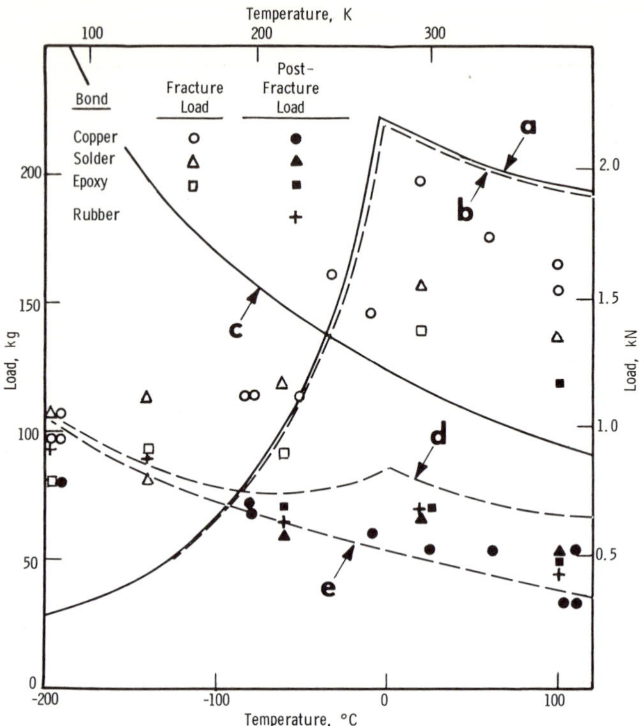

Fig. 21. Fracture loads vs. temperature for mild-steel specimens and crack-arrester laminates with different bonds. Data points are from instrumented impact tests; curves are: (a) fracture loads of homogeneous specimens, (b) predicted fracture load of strongly bonded laminates, (c) general yield loads of an unnotched specimen, (d) predicted fracture loads of weakly bonded laminates, (e) predicted postfracture loads of laminates. [From Almond *et al.* (1968), by permission of ASTM]

A third mechanism for fracture control in laminates containing, as a major constituent, laminae of plastically anistropic metals having an insufficient number of operative deformation systems to permit ductile behavior in a polycrystalline aggregate has been described by Wright (1967).

General yielding of a polycrystalline material occurs when the dislocation pile-up stress in grains that are favorably oriented for slip (i.e., those having available slip systems aligned near the maximum-shear-stress direction) exceeds the minimum stress necessary to cause slip in neighboring grains and thus propagate slip across the entire specimen. This stress is characteristically higher in anistropic metals than in isotropic metals because of fewer available slip systems and the consequent lower probability of slip

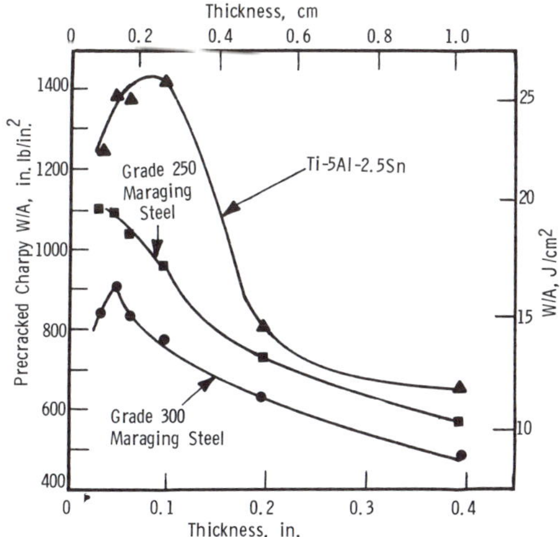

Fig. 22. Variations of the fracture toughness of Ti–5Al–2.5Sn and 18% Ni maraging steels with thickness at room temperature. [From Leichter (1966), by permission of *J. Spacecraft*]

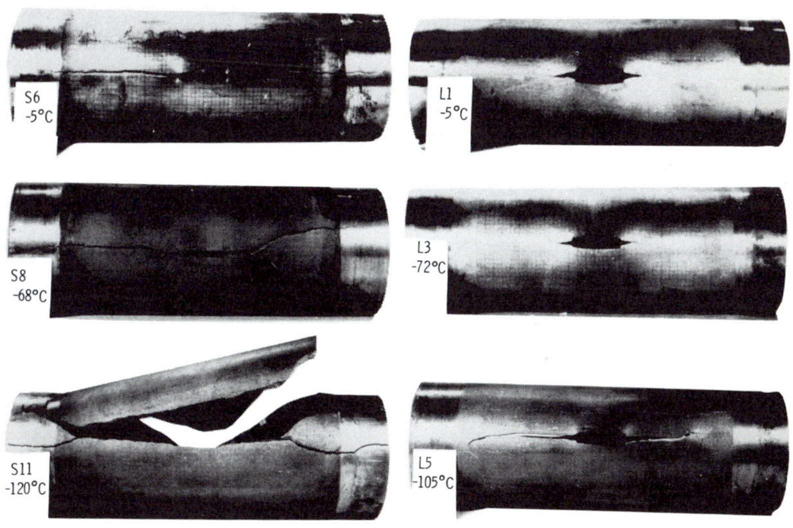

Fig. 23. Fracture behavior of solid-wall (left) and laminated-wall (right) pressurized steel cylinders. [From Almond *et al.* (1969), by permission of *JISI*]

systems being favorably oriented in the unyielded grains for propagation of yielding.

In terms of the yield equation of Petch (1953), it can be shown that the yield stress of a polycrystalline anistropic material should be higher in relation to the critical shear stress for activation of dislocation sources than that of an isotropic material, since

$$\tau_y = m\tau_0 + m^2_{\tau_c} r^{1/2} d^{-1/2} \tag{11}$$

where τ_y is the tensile yield stress, τ_0 the friction stress on unlocked dislocations, τ_c the critical shear stress to activate dislocation sources, d the average grain diameter, and m a constant reflecting the ratio of applied stress to the resolved shear stress on favored slip systems. Sachs (1928) calculated the orientation factor for face-centered cubic structures to be 2.2 by a statistical averaging of this ratio for collection of randomly oriented free crystals. Armstrong *et al.* (1962) calculated the equivalent value for close-packed hexagonal metals that exhibit only basal slip (such as beryllium or zinc at low temperatures) to be 6.5. The increase in yield or flow stress that results from the higher orientation factor thus increases the stress available for fracture initiation and propagation.

Interposing layers of an isotropic metal between anisotropic metal layers that have been recrystallized to produce layers of single-grain thickness appears to reduce the tendency for brittle behavior of the anisotropic material by providing suitably oriented slip systems for yield propagation (Fig. 24).

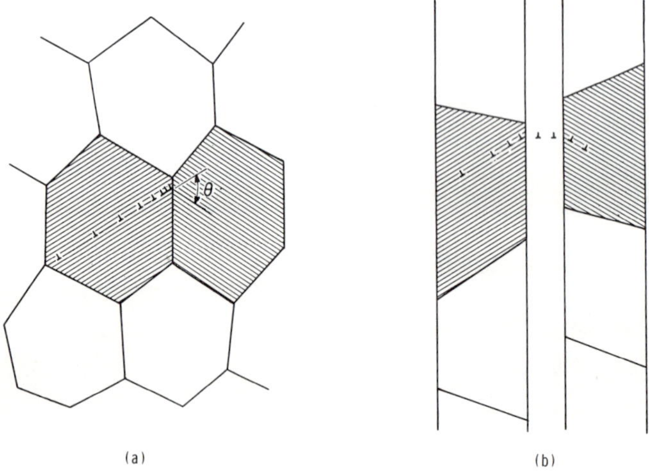

Fig. 24. Proposed mechanism for prevention of brittle fracture in laminates containing brittle anisotropic lamina. (From Wright, 1967)

This mechanism for fracture control is illustrated by a comparison of the stress–strain behavior of monolithic zinc and aluminum specimens with that of zinc–aluminum laminates at $-196°C$ (Fig. 25). At this temperature polycrystalline zinc specimens fail brittlely before general yield by the initiation of Stroh microcracks at grain boundaries, followed by prompt propagation of fracture. Aluminum deforms readily, but at a low stress. The zinc–aluminum laminates exhibit general yielding and considerable plasticity at a relatively high stress level. Failure eventually occurs, but only after twins develop in the zinc laminae, thus providing zinc–zinc boundaries to serve as sites for Stroh crack nucleation. Such a concept offers a method of producing ductile beryllium composites, but a practical process for producing the required laminates containing beryllium laminae of single-grain thickness needs to be developed.

D. Stress-Corrosion Cracking

Most high-strength materials are highly sensitive to stress-corrosion cracking in mild corrodents and at low stresses. Even materials that possess good general corrosion resistance, such as 18% maraging steels, are not immune. Stavros and Paxton (1970) have shown that 250-grade maraging steel, which has a K_{Ic} value of over 100 ksi in.$^{1/2}$, has a K_{Iscc} value of only 20 ksi in.$^{1/2}$ in distilled water. Floreen *et al.* (1971) evaluated the stress-corrosion cracking behavior of 18 Ni (280) maraging steel and 18 Ni (280) maraging steel–Armco iron laminates in aerated 3.5% NaCl solutions. The homogeneous material had a K_{Ic} of 60 ksi in.$^{1/2}$ in air and a K_{Iscc} of 15–20 ksi in.$^{1/2}$ in the salt solution. Time to failure at a level of 30 ksi in.$^{1/2}$ was approximately 20 hr. The crack-arrester laminates showed no failures within 240 hr at K levels of 30 or 80 ksi in.$^{1/2}$. Crack-divider laminates containing only 8, 9, and 24% Armco iron gave failure times of 63, 63, and 80, respectively.

E. Electrical Conductivity

Metal laminates are frequently used for applications such as corrosion-resistant wire and cable, heating elements, switch-contact springs, and circuit-breaker thermoelements, all of which involve the conduction of electricity. In clad wire or strip or in laminar composites in which the conduction direction is parallel to the laminae, the conductivity of the laminate can be readily calculated on a rule-of-mixtures basis in which

$$k_C = k_1 f_1 + k_2 f_2 + \cdots + k_n f_n \tag{12}$$

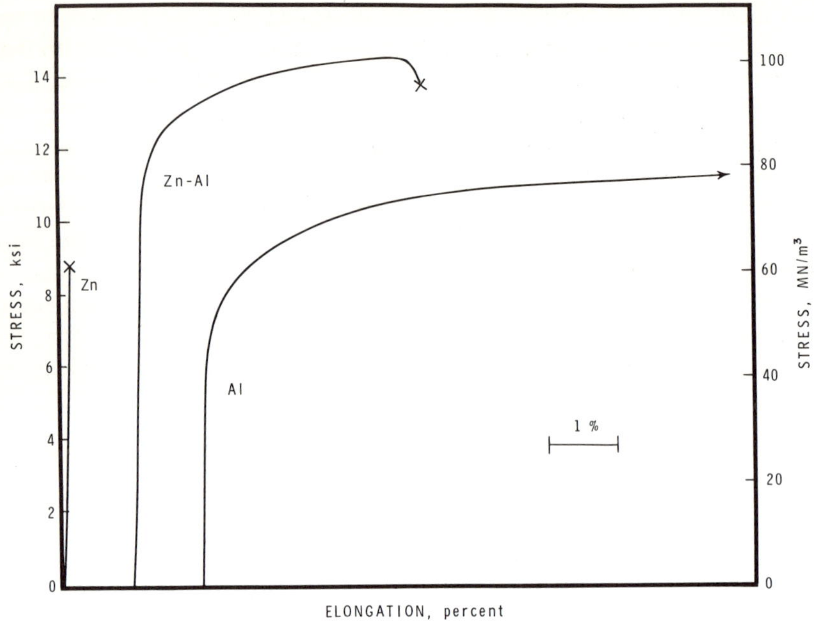

Fig. 25. Stress vs. elongation curves for zinc, aluminum, and zinc–aluminum laminate tensile specimens tested at $-196°C$. [From Almond *et al.* (1968), by permission of ASTM]

where $k_1, k_2, \ldots, k_n$ are the electrical conductivities of the individual components and $f_1, f_2, \ldots, f_n$ are their respective volume fractions. A plot of the conductivity of stainless-steel-clad copper composites as function of thickness is shown in Fig. 26.

Since resistivity is the reciprocal of conductivity, the effective resistivity of a metal laminate parallel to the laminae is given by

$$1/R_C = f_1/R_1 + f_2/R_2 + \cdots + f_n/R_n \tag{13}$$

where $R_1, R_2, \ldots, R_n$ are the resistivities of the component laminae.

F. Thermal Conductivity

A major usage of metal laminates is for applications in which improved heat flow is desired in directions parallel to the sheet surface, even though the surface layers must be made of materials of low thermal conductivity because of corrosion or wear resistance, strength, or other considerations.

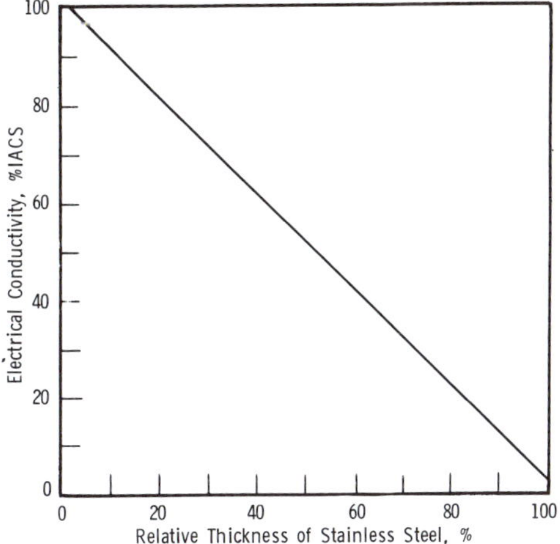

Fig. 26. Electrical conductivity of stainless-steel–copper laminates. (By permission of D. E. Makepeace Division, Engelhard Industries, 1971)

By incorporating layers of high-conductivity materials within the laminate, high conductivity in directions parallel to the sheet surface can be achieved. Typical applications are cooking utensils, laboratory ware, heat exchangers, and chemical process vessels.

The relationships governing thermal conduction in laminates *parallel* to the laminae are similar to those governing electrical conduction:

$$K_C = K_1 f_1 + K_2 f_2 + \cdots + K_n f_n \tag{14}$$

where $K_1, K_2, \ldots, K_n$ are the thermal conductivities of the component laminae. A plot of thermal conductivity of stainless-steel–aluminum laminates is shown in Fig. 27.

In some applications such as heat exchangers, thermal conduction *perpendicular* to the surface is of paramount importance. Thermal conductivity in this direction is given by

$$1/K_C = f_1/K_1 + f_2/K_2 + \cdots f_n/K_n \tag{15}$$

provided there is a perfect thermal bond between layers and there are no low-conductivity layers that have formed by surface contamination or intermetallic diffusion. Such layers would obviously decrease thermal conductivity and would need to be accounted for by the addition of appropriate terms to the conductivity equation.

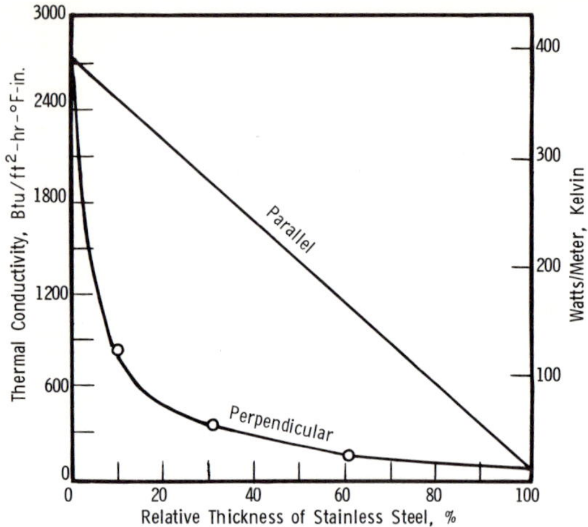

Fig. 27. Thermal conductivity of stainless-steel–copper laminates. (By permission of D. E. Makepeace Division, Engelhard Industries, 1971)

G. Thermal Flexivity

Another important application of metal laminates is bimetallic (or trimetallic) thermostat elements. The basic quantity used to describe the thermal activity or sensitivity of thermostat elements is *flexivity*, defined by an ASTM Standard (1965) as "the change in curvature of the longitudinal centerline of the specimen per unit temperature change for unit thickness," or, in terms of Fig. 28,

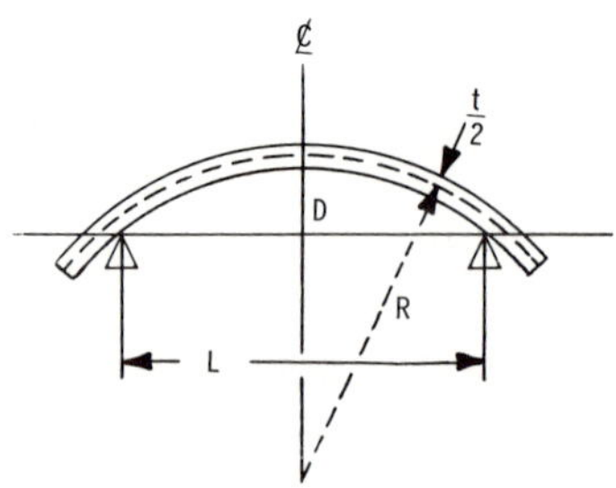

Fig. 28. Relation of radius of curvature to thickness in thermostat laminates. (From ASTM Standard B106-65T, by permission of ASTM)

$$F = \frac{(1/R_2 - 1/R_1)t}{T_2 - T_1} \tag{16}$$

where $1/R_1$ and $1/R_2$ are the curvatures at temperatures T_1 and T_2, respectively, and t is the total thickness of the element. Flexivity, in turn, depends upon a number of material properties and geometrical factors. The change in curvature of a two-layer laminate with change in temperature can be calculated by means of an equation developed by Timoshenko (1925):

$$\frac{1}{R_2} - \frac{1}{R_1} = \frac{6(X_2 - X_1)(T_2 - T_1)(1 + m)^2}{h[3(1 + m)^2 + (1 + mn)(m^2 + 1/mn)]} \tag{17}$$

where $m = t_1/t_2$, the thickness ratio of the two components, $n = E_1/E_2$, the ratio of their elastic moduli, X_1 and X_2 are their respective linear expansion coefficients, T_1 and T_2 the initial and final temperatures, and $1/R_1$ and $1/R_2$ the initial and final curvatures.

The actual behavior of real thermostat elements is usually more complex because the expansion coefficients of alloys used in thermostats are frequently not constant over the temperature range of use, and real elements may contain more than two laminae. The design of thermostat elements is thus usually done either empirically or by means of a computer, as described by Ornstein (1965).

The expansion of three typical thermostat component alloys as a function of temperature is shown in Fig. 29. Also shown are plots of the expansion-difference curves (which determine flexivities) of laminates composed of the high-expansion alloy versus standard Invar, 36Ni–64Fe (curve Bl), and versus 40Ni–60Fe (curve B2). It can be seen that the element containing Invar would have slightly higher flexivity up to 500°F, but the one containing 40Ni–60Fe would have a slightly longer near-linear range.

H. Corrosion and Erosion Resistance

The ability to obtain good corrosion resistance concomitant with other desired properties such as strength, thermal or electrical conductivity, or reduced cost, accounts for a large portion of the modern uses of metal laminates, particularly in the fields of architecture, chemical engineering, housewares, and nuclear power engineering. Thus, stainless-steel-clad cooking utensils or chemical process vessels have corrosion resistance equal to that of homogeneous stainless-steel vessels, but may have superior heat transmission and be cheaper. The resistance of metal laminates to corrosion, erosion, or wear is largely determined by the properties of the surface layers.

Exceptions to this rule occur because of galvanic corrosion reactions between the cladding and the exposed core if unprotected edges, holes, or

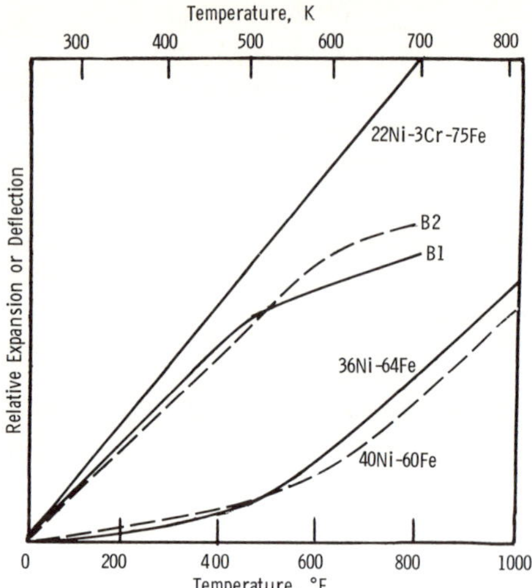

Fig. 29. Expansion of thermostat alloys and resultant deflection of thermostat elements vs. temperature.

cladding defects are exposed to the corroding medium. This can seriously reduce the observed corrosion resistance of the surface of the composite in the vicinity of the exposed edge, or it can cause rapid corrosion of the core and apparent delamination of the composite.

If the surface layers of the composite are strongly anodic with respect to the interior layers, the surface may corrode sacrificially, leaving the core material cathodically protected against corrosion even when it might otherwise be highly susceptible. Advantage is taken of this effect in many clad products, most notably in Alclad aluminum alloy sheet, in which a strong structural alloy is clad on one or both sides with either commercially pure aluminum or an aluminum alloy that is anodic to the core. This makes possible the use of aluminum in many applications where its use would be prevented by tendencies toward pitting corrosion, which would lead to perforation of the material or to serious reduction in strength and possible failure from applied loads. Figure 30 shows cross sections of 6063 T5 and Alclad 6063 T5 pipes that had been buried in moist soil for seven years. The surface cladding on the Alclad pipe effectively prevented the penetration of corrosion into the 6063 core alloy. A similar effect is depicted graphically in Fig. 31 for 5050 and Alclad 5050 tubing exposed to a saline environment. Figure 32 demonstrates how Alclad coatings prevented the deterioration of tensile strength of 2024-T3 aluminum alloy sheets exposed to sea water.

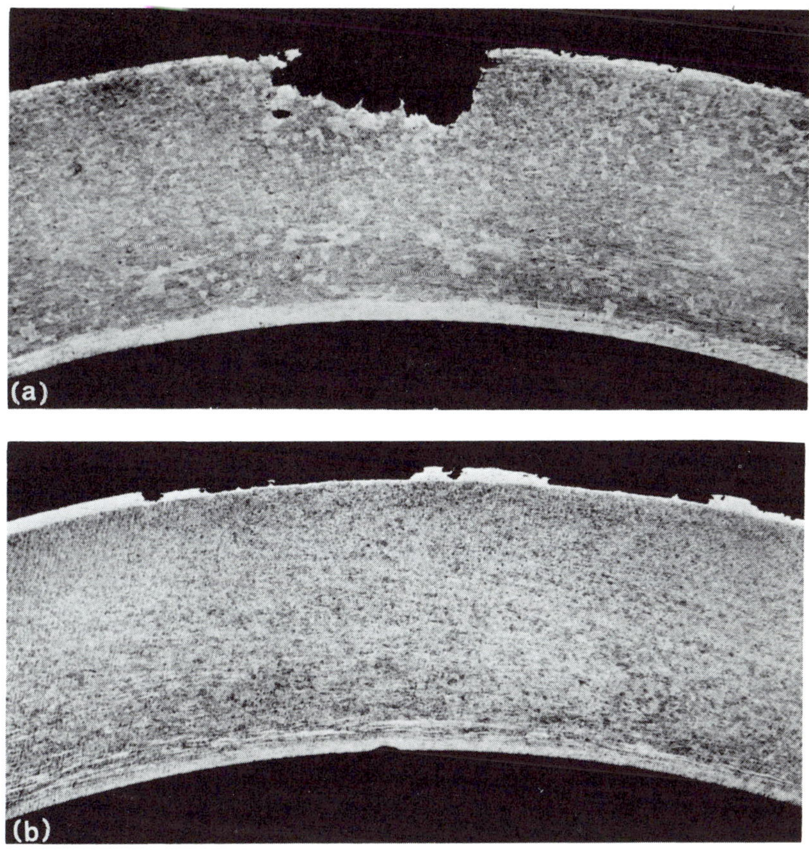

Fig. 30. Cross sections of (a) 6063-T5 and (b) Alclad 6063-T5 pipes after seven years' exposure to moist soil (15 ×). [From Brown (1969), by permission of MIT Press]

A number of nuclear reactors have used metal laminates as fuel elements and/or control rods. Fuel elements are composed of a core alloy that contains a fissile material such as ^{235}U, ^{233}Th, or ^{239}Pu, surrounded by a cladding alloy that is resistant to corrosion by the coolant. Fast reactors—i.e., those that employ high-energy neutrons—can use aluminum or stainless-steel cladding, but thermal reactors—those using relatively slow or low-energy neutrons—must use a cladding material such as a zirconium alloy that has a low cross section for neutron capture but a high thermalizing or "slowing-down" power. Control rods, on the other hand, usually have a core containing elements such as hafnium, cadmium, or boron, which have

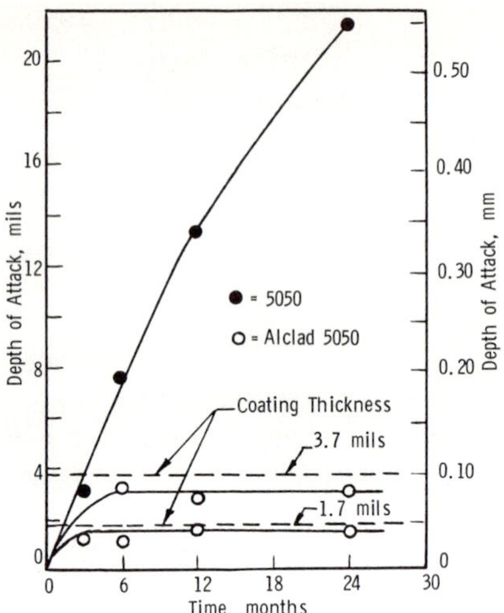

Fig. 31. Depth of attack of 5050 and Alclad 5050 tubing by a saline environment. [From Brown (1965), by permission of MIT Press]

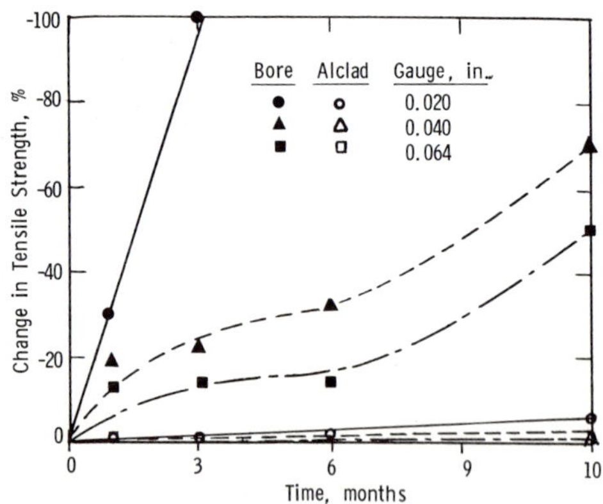

Fig. 32. Change in tensile strength of bare and Alclad 2024-T3 sheets resulting from exposure to sea water. [From Brown (1969), by permission of MIT Press]

high cross sections for neutron capture and can, when inserted into the reactor, absorb neutrons to reduce the operating power level of the reactor. Here again a corrosion-resistant cladding may be required to protect or contain the control-rod core material.

A notable example of a nuclear reactor that had laminated fuel elements was used in the first nuclear submarine, the Nautilus. The fuel elements were composed of a uranium–zirconium alloy, clad with Zircaloy-2, a Zr–1.45Sn–0.15Fe–0.1Cr alloy. The coolant was pressurized water at 345°C.

IV. Engineering Criteria for Selection and Application of Metal Laminates

The following examples illustrate the materials engineering considerations used in the advantageous application of metal laminates. In each case, in addition to the laminate's mechanical and physical properties, its fabricability, which involves all those characteristics required to bring the laminate into its final shape and condition, including its ability to be bent, worked, and joined to itself and to dissimilar metals, plays a key role.

A. Aircraft Structures

Aircraft structures and skins must be strong, lightweight, corrosion resistant, and fatigue resistant. These favorable properties are embodied in an aluminum laminate known as Alclad. The term Alclad, originated by the Aluminum Company of America, is now used by the Aluminum Association to identify a class of metallurgically bonded metal laminates having an aluminum-alloy coating that electrochemically protects a core alloy. By proper selection of coating and core, the laminate may be designed to have a range of properties. The core alloy provides the desired mechanical properties, while the coating provides long-term corrosion protection. Alclad sheet such as Alclad 7079-T6 has been, and continues to be, widely used for the skins of aircraft because of its light weight, dependability, and corrosion resistance. Figure 33 presents the cross section of Alclad 2024-T3 sheet, 0.032 in. thick. The excellent metallurgical bond between the 1110 aluminum coating and the 2024-T3 core is clearly visible. The total coating thickness is 5% of the laminate thickness. Coating thickness in other Alclad products ranges from 2.5 to 15% of the laminate thickness. For a detailed treatment of Alclad products, the reader is referred to Brown (1969).

B. Anodes for Aluminum Potlines

In the production of primary aluminum by electrolysis, consumable carbon anodes are used to carry electric current to the cryolite bath. These anodes are connected to steel brackets, which in turn are bolted to aluminum

Fig. 33. Cross section of Alclad 2024-T3 sheets, 0.032 in. thick. Cladding is 1100 aluminum. (125 ×) (By permission of Aluminum Company of America)

rods. The mechanical joint between the aluminum and the steel was unsatisfactory because of its variable electrical resistance and its tendency toward rapid deterioration due to corrosion and/or arcing. This condition was corrected by using aluminum–steel transition joints. The aluminum rod could then be reliably welded to the aluminum side of the couple, and the steel bracket could be similarly joined to the steel side of the couple. The bolted assembly is compared with its welded counterpart in Fig. 34. This transition joint made for a much improved, longer-lasting, reliable, anode assembly having low, uniform, electrical resistance.

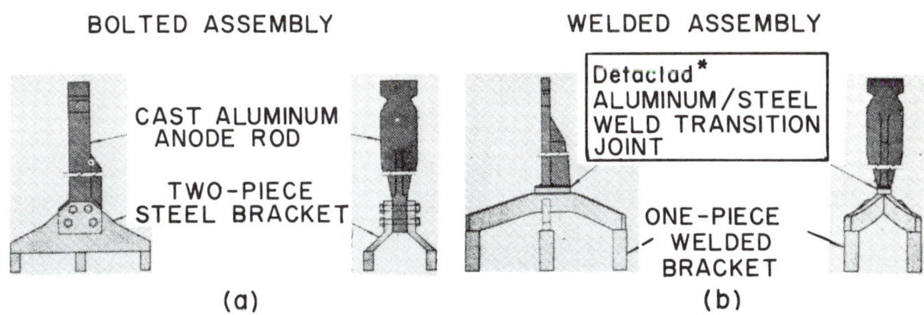

Fig. 34. Comparison of bolted anode assembly (a) with improved welded anode assembly using Detaclad aluminum–steel weld transition joint (b). (By permission of *Light Metal Age*, 1969)

C. *Architectural and Building-Trade Applications*

1. *Roofing and Related Applications*

Copper-clad stainless steel is a particularly attractive laminate for architectural and building-trade applications. The commercial laminate is fully annealed and generally consists of a 10% thickness of copper, metallurgically bonded to both sides of an 80% core of type-404 stainless steel. The sound metallurgical bond attainable in this system is shown in Fig. 35. It is lighter, stronger, stiffer, and about 10–15% less expensive than solid copper; yet it provides the warm beauty, fabricability, and durability of solid copper. It is readily formed by using conventional sheet-metal tools and procedures, and is easier to solder than solid copper because the lower thermal conductivity of the stainless-steel core keeps the heat of the soldering iron at the joint. Thus, smaller, cooler soldering irons may be used, with resulting savings in soldering time and cost. Copper-clad stainless steel has been effectively and economically used in roofing (Fig. 36), as fascia, on mansards, and as face sheets for wall panels, flashings and rain drainage (gutters and downspouts), siding, window and door frames and hardware, and roof ventilators. Table VI shows the superior mechanical properties of copper-clad stainless steel as compared to the 0.020-in.-thick copper commonly used in architectural applications.

The light weight, ease of handling, and corrosion resistance of Alclad sheet, described in Section IV.A, give it wide applications as industrial siding and roofing panels. The missile-assembly building at Cape Kennedy, which is the world's largest building, is completely covered with Alclad 3004-H16.

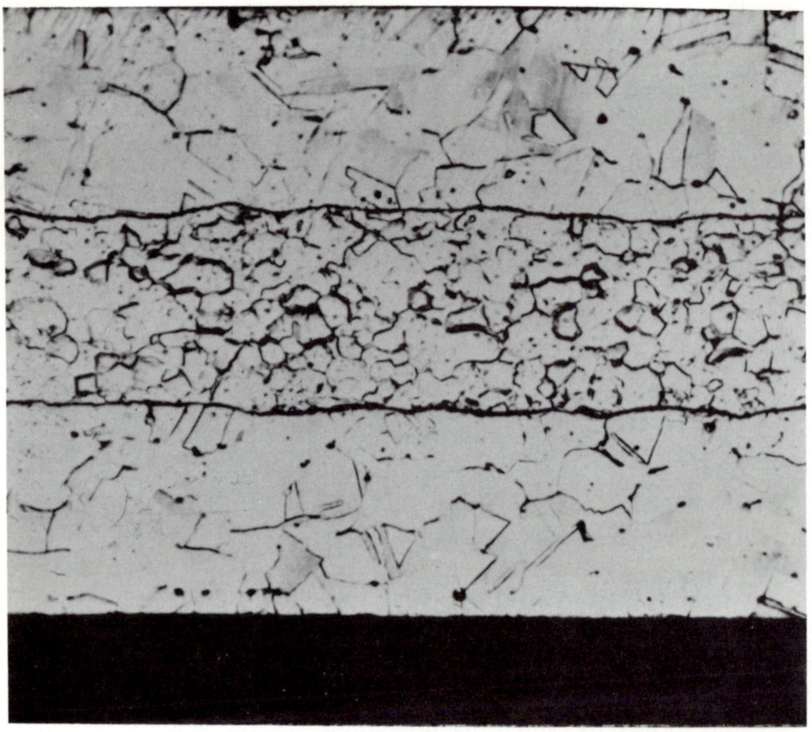

Fig. 35. Photomicrograph of typical TiGuard Cu–SS–Cu bond (280 ×). (By permission of Texas Instruments Inc., 1971)

2. Wiring

Copper-clad aluminum wire has recently become a serious contender in the building-wire field because it performs as well as pure copper, although it is lighter and less expensive. This composite eliminates the contact and creep problems associated with all-aluminum wire and also greatly reduces the amount of expensive copper required as compared to solid copper wire. The clad conductor shown in Fig. 37, which consists of 30 w/o copper and 70 w/o aluminum, provides users a 12–15% cost savings over solid-copper cable.

Fig. 36. TiGuard copper-clad stainless-steel roofing. (By permission of Texas Instruments Inc., 1971)

TABLE VI

Comparison of Copper and Copper-Clad Stainless Steel (TiGuard®)[a]

Property	Copper	Copper-clad stainless steel
Nominal temper	Soft	Soft
Standard thickness (in.)	0.020	0.015
Relative cost (copper = 100)	100	75
Yield strength (1000 psi)	11	35
Tensile strength (1000 psi)	35	63
Elongation (% in 2 in.)	36	30
Weight/sq. ft (lb)	1.000	0.628
Thermal expansion coefficient ($\times 10^{-6}$ in./in.-°F)	9.8	6.1
Thermal conductivity (Btu/hr-ft^2-ft-°F)	226	37

[a] By permission of Texas Instruments Incorporated.

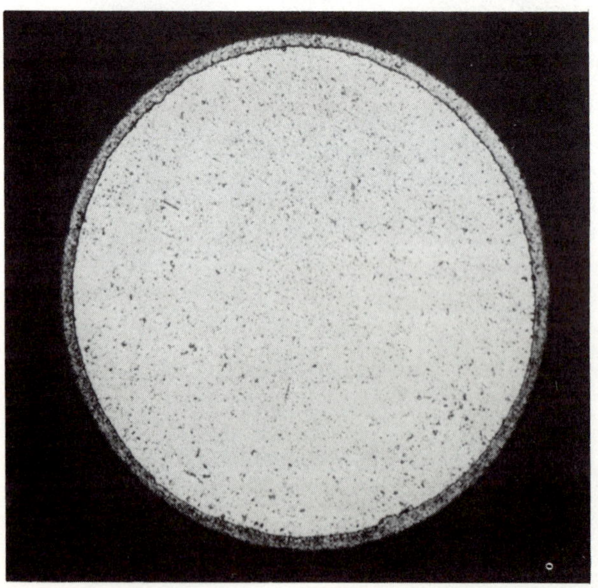

Fig. 37. Cross section of copper-clad aluminum wire. (By permission of Texas Instruments Inc., 1971)

D. Chemical Process Equipment

Titanium has become the metal of choice for chemical process equipment because of its superior corrosion resistance, erosion resistance, and anti-fouling properties. However, the use of solid titanium pressure vessels was seriously limited by their prohibitively high cost of two to four times that of the equivalent stainless steel. The advent of titanium-clad steel has greatly broadened the use of this metal in chemical process equipment because of the cost savings produced by replacing monolithic titanium by a titanium cladding. The inner titanium liner provides the desired corrosion resistance, while the outer layer of steel provides the needed strength.

Titanium has been successfully bonded to steel by brazing, roll bonding, and explosive bonding. Brazing is accomplished by the technique described earlier in Section II.D. Roll bonding is performed by mill rolling under an inert gas at forge-weld temperature (~1020°C). Titanium may also be explosively bonded to steel by the method described in Section II.C, using a separation of at least 0.025 mm between the titanium and the steel. Since these three cladding methods subject the titanium and steel to significantly different temperature, pressure, and time environments, the resulting interfacial bonds are significantly different. Figures 38–40 show the bond lines and fractures in the interfacial bonds produced by brazing, roll bonding, and explosive bonding, respectively. A continuous phase of a brittle intermetallic compound was evident at the braze-bonded interface. Some coarsening of the titanium grain size was also produced by exposure to the brazing temperature. Similarly, there was titanium-grain coarsening in the roll-bonded cladding. The carbon diffusion into the interface from the steel also formed a thin layer of titanium carbide at the interface.

The explosively bonded interface (Fig. 40) produced the characteristic interfacial ripples, due to the extensive plastic deformation caused by the explosive pressure. A minimum of brittle intermetallic compounds was formed because of the very short time the interface was exposed to elevated temperature. Thus, the bond in the explosively clad material is superior to that in braze-bonded and roll-bonded material in strength and toughness. These characteristics make the explosively bonded system the preferred material for pressure-vessel construction.

Figure 41 shows an early titanium-clad steel tank fabricated from a plate produced by the Detaclad† explosion-bonding process. The key problem in fabricating this tank was forming it and achieving good welds in the laminate. This tank was one of several used for the manufacture of aromatic acids.

† Detaclad is a registered trademark of E. I. DuPont de Nemours & Co., Inc.

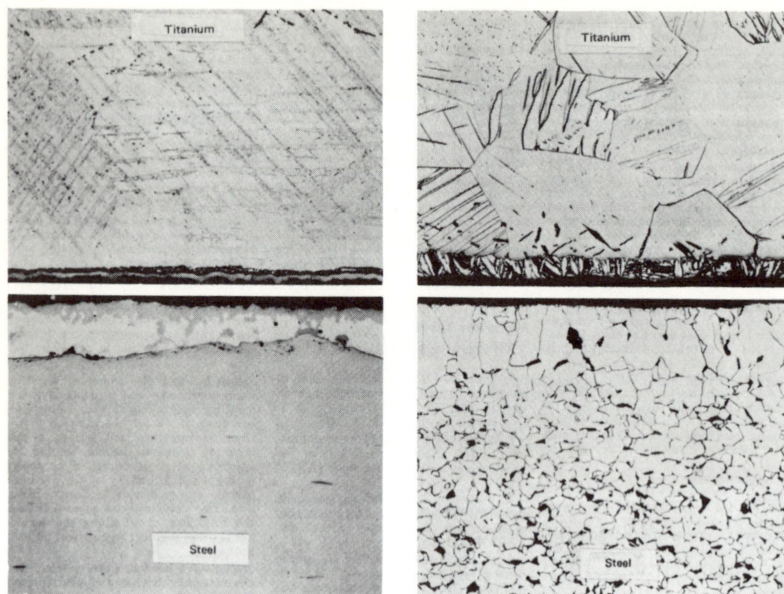

Fig. 38. Bond-line fracture of braze-bonded titanium-clad steel plate (250×). [From Sticha (1969), by permission of *Materials Protection*]

Fig. 39. Bond-line fracture of roll-bonded titanium-clad steel (250×). [From Sticha (1969), by permission of *Materials Protection*]

Vessel dimensions were 7.0 m long × 2.4 m diam. This vessel clearly demonstrated that titanium clad plate could be cut, formed, and welded under shop conditions. The weld that was successfully used in this tank is shown in Fig. 42.

Subsequent advances in clad fabrication have further established the value and utility of titanium-clad steel in chemical process equipment. These advances include: (1) simplification of welding procedures and replacement of the costly silver inlay with less expensive copper; (2) fabrication of larger and wider clad plates (up to 200 ft^2), with attendant reduction of total costs by 25–40%; (3) development of clean-room practices that ensure high-quality welds; (4) development of cladding specifications that ensure uniformly high-quality material that meets pertinent ASME vessel code requirements; (5) improvement in clad manufacturing methods that have resulted in reduced product costs; and (6) the cladding of titanium to high-strength steel has yielded a material that is outstanding for high-pressure, moderately-high-temperature process reactors. The vessel shown in Fig. 43 is made of 0.48-cm titanium clad to 2.5-cm high-strength SSS-100 steel.

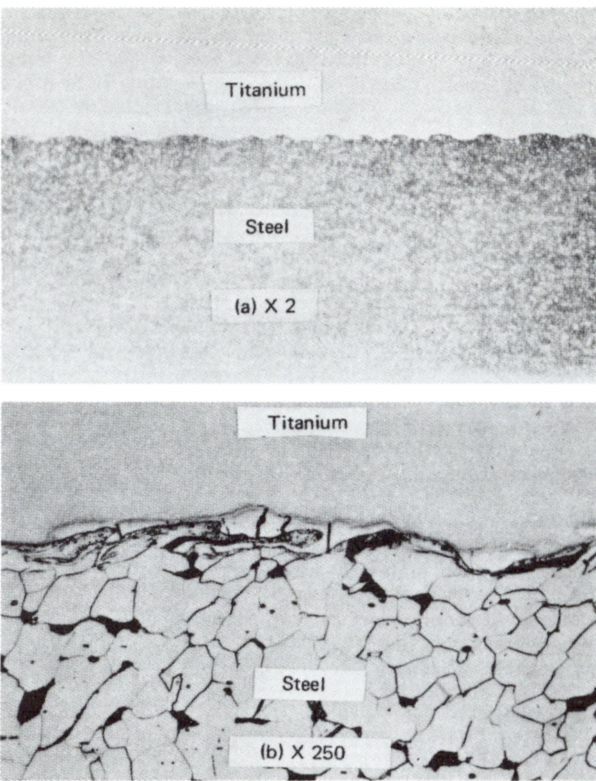

Fig. 40. Interface of explosion-bonded titanium-clad steel: (a) 2 ×, (b) 250 ×. (By permission of National Association of Corrosion Engineers)

It operates at a pressure of 1000 lb/in.^{2} and a temperature of 235°C to produce organic intermediates by reacting organic compounds with hydrochloric and sulfuric acids. In addition to the compatability of the titanium with an acid environment, the need for high heat transfer through the vessel wall was met by the use of thin, high-strength steel integrally clad to titanium. This vessel is in service at the Chamber Works of the DuPont Company, Pennsville, New Jersey. It was fabricated by the Wyatt Division of U.S. Industries, Dallas, Texas.

Titanium-clad steel has also been widely adopted for use in shell and tube heat exchangers because of the excellent corrosion resistance and antifouling properties of titanium.

Fig. 41. Titanium-clad carbon-steel vessel (2.0-mm commercially pure titanium on $\frac{7}{8}$-in. SA-156 grade 70 steel). [From Williams (1970), by permission of *Chemical Engineering Progress*]

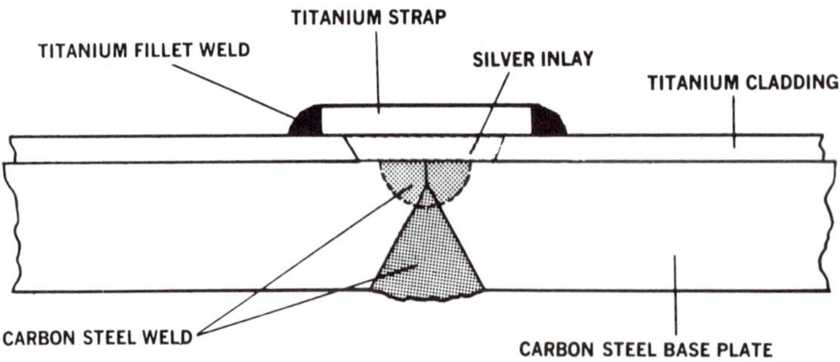

Fig. 42. Double-V silver-inlay batten-strap weld joint design used in early titanium-clad steel vessels. [From Williams (1970), by permission of *Chemical Engineering Progress*]

E. Clad-Metal Bullet Jackets

In order to reduce costs and eliminate the possibility of overdependence on scarce copper for ammunition, the U.S. Army embarked on a program aimed at substituting a copper-alloy-clad steel bullet jacket for the traditional

Fig. 43. Titanium-clad alloy-steel pressure vessel (0.48-cm commercially pure titanium on 2.5-cm SSS-100 quenched and tempered steel). [From Williams (1970), by permission of *Chemical Engineering Progress*]

copper-alloy bullet jacket. Under an Army contract, Texas Instruments successfully demonstrated that steel bullet jackets clad with gilding metal (90% Cu, 10% Zn) could be economically fabricated. The ballistic performance of the laminated bullet is superior to that of its unclad counterpart, with a 72% saving in copper and with assured raw-material availability. The sectioned 0.30-caliber bullet in Fig. 44 shows the gilding-metal-clad steel jacket in cross section. The jacket consists of gilding metal–steel–gilding metal in the proportions by volume of 15:80:5. The inner layer of gilding metal is used to facilitate drawing the jacket. The bullet core is lead. Figure 45 shows the excellent metallurgical bond between the gilding metal and steel. This development illustrates the Army's continuing efforts to improve its weaponry while at the same time reducing costs and conserving critical materials.

F. Coinage

Metal laminates, first introduced into U.S. coinage in 1965 to conserve scarce and costly silver, have clearly proven their value and utility. Prior to the acceptance of cupronickel-clad copper coinage, every known material, including all metals, ceramics, glass, plastics, wood and fabric, was tried, but only clad metals met all of the following requirements:

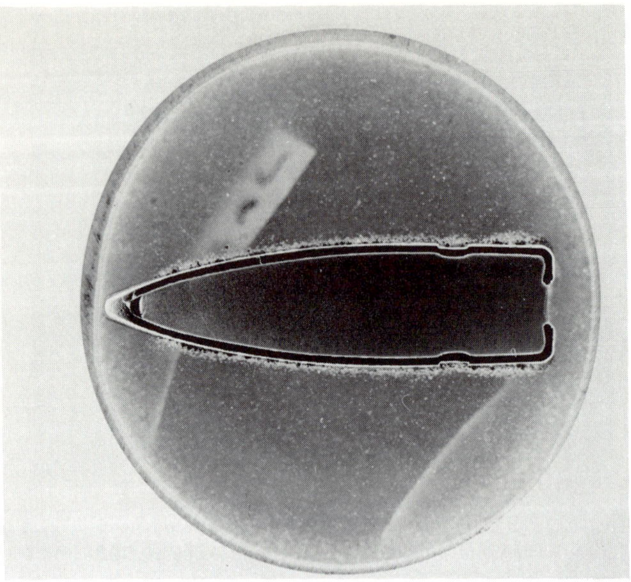

Fig. 44. Cross section of a 30-caliber bullet showing laminated jacket. From the outside the laminate consists of gilding metal–steel–gilding metal in the proportions by volume of 15 : 80 : 5. (3.5 ×) (By permission of Army Materials and Mechanics Research Center)

a. availability
b. corrosion resistance
c. formability
d. appearance
e. abrasion resistance
f. rho density
g. reclaimable scrap

All of these requirements are self explanatory except the rho density factor, which is the property needed to permit the operation of coin discriminators in vending machines. The material also must be capable of being used interchangeably with nonclad coinage in vending machines.

Today, with the exception of special commemorative coins, all United States primary and subsidiary coins (dollars, half dollars, quarters, and dimes) are metal laminates consisting of a layer of copper sandwiched between two thinner layers of cupronickel alloy (75% copper, 25% nickel) in the proportions by volume $\frac{1}{6}:\frac{2}{3}:\frac{1}{6}$. The outer layers of cupronickel alloy provide the required appearance, abrasion resistance, and corrosion resistance, while the inner copper layer contributes to the required density. Figure 46 shows the excellent metallurgical bond developed between the outer layer and the core. A major saving in scarce silver has been realized by the use of this laminate, whose constituents are readily available. This laminate is also easily fabricated and formed and readily reclaimable as scrap.

Fig. 45. Cross section of laminated steel jacket of a 30-caliber bullet showing the excellent metallurgical bond between gilding metal and steel. The laminate consists of gilding metal–steel–gilding metal in the proportions by volume of 15:80:5.(300 ×) (By permission of Army Materials and Mechanics Research Center)

G. Dual-Hardness Armor Plate

Dual-hardness armor is another example of the use of laminates to achieve what cannot be done with monolithic metal. In this application a very-high-hardness (60 Rockwell C) steel is metallurgically bonded to a tougher, more ductile, softer (50 Rockwell C), backing steel. The very hard facing serves to break up the steel core of an armor-piercing projectile, while the tough backing holds the facing together and absorbs the deformation caused by the projectile impact without cracking. Dual-hardness armor substantially reduces the weight required to defeat small-arms projectiles as compared to the standard rolled homogeneous steel armor. This metal laminate is made by

Fig. 46. Cupronickel-clad copper coinage strip. (50 ×) (By permission of Texas Instruments, Inc.)

grinding two plates, seam-welding the periphery, and evacuating the interfacial region through an evacuation tube, which is then sealed off. The composite pack is then heated in a slabbing furnace to about 2100°F, and cross-rolled to accomplish the metallurgical bonding, with an accompanying thickness reduction of slightly more than 50%. It is then final-rolled to the desired gage, sheared to size, and annealed.

Two methods have been used to produce the desired hardness levels. The first method, known as ausforming, requires an additional 50% total thickness reduction in the temperature range 1100–1400°F, followed by a water-spray quench. This laminate is in the fully hardened condition when it leaves the mill. The difference in composition between the two layers is such that after ausforming, the frontal layer is harder than the more ductile backing layer. The second method utilizes conventional quench-and-temper heat-treating procedures after the bonding operation. Each method produces equivalent hardness in each layer. The ausformed composite requires the use of a composition having carbide-forming elements such as chromium, molybdenum, and vanadium to obtain the strengthening response during thermal–mechanical working. Slightly lower carbon content can be used in both the frontal and rear plates than with the heat-treated product. The heat-treated composite uses a minimum alloy content to obtain hardenability, and a greater latitude is available in composition selection so that the transformation temperatures of the two compositions can be matched to maximum flatness.

The Army Materials and Mechanics Research Center (AMMRC) has been a prime mover in recent advancement of dual-hardness armor technology. Because the ausformed product is in the final hardened condition when it leaves the mill, it is less amenable to subsequent fabrication. Therefore, in recent years it has been generally replaced by the heat-treatable composite, which can be readily fabricated into the desired shape and then hardened.

The composition of the armor constitutents is shown in Table VII. Macroscopic views of the armor components (before and after bonding) are shown in Fig. 47. The metallurgical bond achieved by hot rolling is so good (Fig. 48) that there is a gradual change in texture in the interfacial zone with no visible, well-defined, localized interface. Experiments at AMMRC by Anctil and Kula (1969) on fatigue crack growth in dual-hardness armor and the unbonded constituents under the conditions shown in Fig. 49 indicate that the bonded composite is superior in its resistance to fatigue crack growth (Fig. 50). This characteristic makes this laminate attractive for use as structural armor in Army helicopters.

TABLE VII

Dual-Hardness Armor Composition[a]

	Ni	Co	C	Fe
Front	9	4	0.45	86.55
Back	9	4	0.25	86.75

[a] By permission of Army Materials and Mechanics Research Center.

Although, in the early stages of development, problems of surface decarburization, edge cracking in sectional plates, and quench cracking after hardening were encountered, these have been solved, with the result that dual-hardness armor is a commercial product. Abbott and Manganello (1972) have recently presented a detailed analysis of the metallurgical factors affecting the ballistic behavior of laminar and monolithic armor.

H. Marine Structures

The use of aluminum superstructures joined to steel decking is very attractive in marine applications. The lightweight aluminum provides

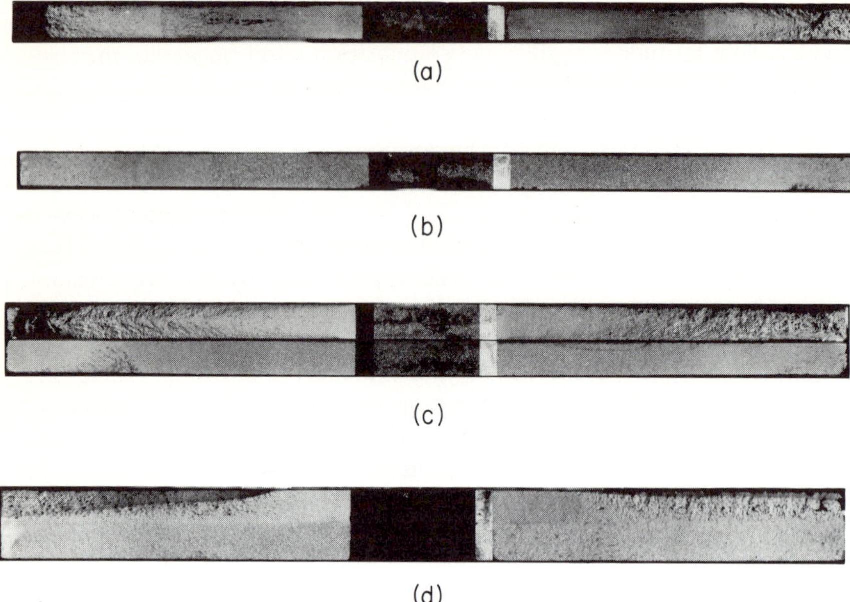

Fig. 47. Dual-hardness steel composite fracture surfaces (2 ×): (a) back-up component (0.3% C); (b) frontal component (0.5% C); (c) composite unbonded; (d) composite bonded. (By permission of Army Materials and Mechanics Research Center)

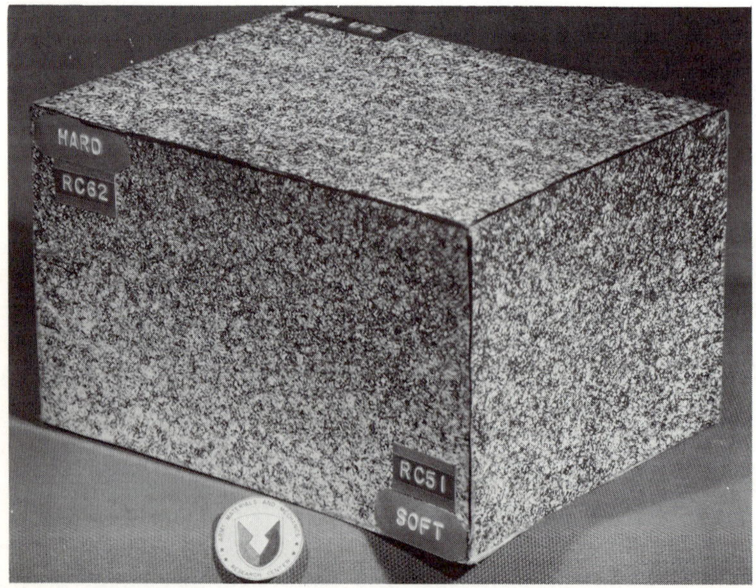

Fig. 48. Typical microstructure of heat-treatable dual-hardness steel armor (250 ×). (By permission of Army Materials and Mechanics Research Center)

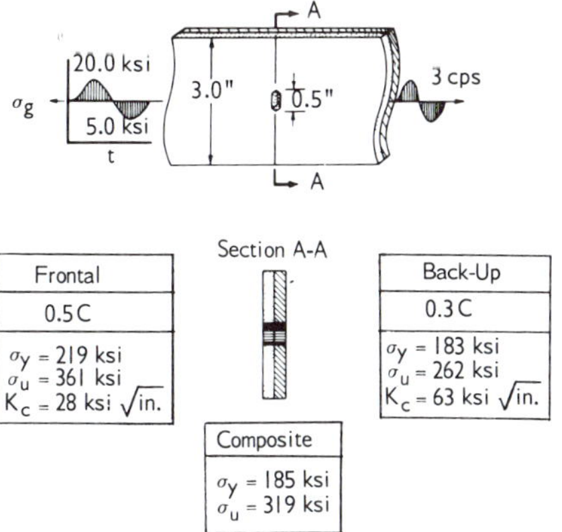

Fig. 49. Properties of dual-hardness armor compared with those of its constituents. (By permission of Army Materials and Mechanics Research Center)

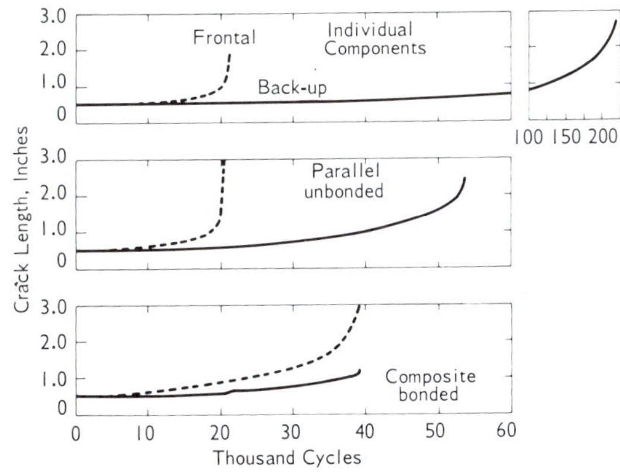

Fig. 50. Comparison of fatigue crack growth in unbonded constituents and bonded dual-hardness composite. (By permission of Army Materials and Mechanics Research Center)

corrosion resistance and weight savings, while the steel provides hull strength and stability. Before the advent of aluminum-clad steel, the aluminum was mechanically joined to the steel. This type of joint was very unsatisfactory because of severe galvanic corrosion. Despite the most careful painting and insulation, ship motion would cause chipping of the paint or insulation, metal-to-metal contact, and severe galvanic corrosion.

The development of aluminum-clad steel offered a way of circumventing this problem, by the use of an aluminum–steel transition joint that would permit joining the aluminum superstructure to the aluminum side of the joint and steel decking to the steel side of the joint by conventional welding techniques.

Detacouple† aluminum–steel transition joints have been found to be very effective for marine structures. The explosive-bonding process provides a strong, ductile joint. The best joint was obtained using a three-layered laminate consisting of 0.6-cm 5456 aluminum clad to 0.9-cm 1100 aluminum, which is clad to 1.8-cm A516 grade 55 steel. "Splash–spray" corrosion tests indicated surprisingly good corrosion resistance, which is apparently due to the formation of hydrated aluminum oxide at the interface. Since this compound occupies a larger volume than the aluminum used up in the oxidation reaction, it fills the crack in the interface, thereby reducing further corrosion. Corrosion resistance of the material is summarized in Table VIII. The mechanical properties of this transition joint are presented in Table IX. Figure 51 illustrates the impact resistance of the aluminum–steel interfacial bond. The clad izod impact specimen failed preferentially in the 1100 aluminum rather than in the interfacial zone. The fatigue resistance, shown in Table X, is much greater than that of equivalent mechanical joints and is better than that of aluminum welds joining them to adjoining aluminum

TABLE VIII

Corrosion Resistance of Detacouple Transition Joint[a]

Test and Specimen Description	Specimen Condition	Exposure Duration	Depth of Penetration*** (in.)
Transition joint (3 in. x 1 in. x 1⅜ in.) welded to 3-in. x 7-in. x ¼-in. 5456 aluminum and mild steel panels. Splash-spray test, Wrightsville Beach, North Carolina, begun 11/10/67.	Unpainted.	3 months	0.027
		12 months	0.033
		27 months	0.042
	Completely painted.	12 months	None
		34 months	None
	Aluminum panel unpainted; steel panel and transition joint painted.	12 months	None
		34 months	None
16-in. x 1-in.-wide transition strip. Welded on stern area of S.S. American Legion behind stack area.*	Unpainted.	12 months	0.033
	Primed with zinc chromate.	12 months	None
	Painted.	12 months	None
Continuous 5% salt-spray tests.	Unpainted.	1,000 hr**	0.060
(ASTM B-117-57T)	Painted.	1,000 hr	None

*Owned by United States Lines.
**Sixteen hours of testing is considered comparable to one year exposure in the Detroit, Michigan environment.
***Maximum depth of corrosion penetration at the bond zone.

[a] By permission of E. I. DuPont de Nemours & Co., Inc.

† Registered trademark of E. I. DuPont de Nemours & Co., Inc.

TABLE IX

Mechanical Properties of Detacouple Transition Joint[a]

Property	Condition	Value (psi) Typical	Minimum
Ultimate tensile strength	As clad	20,000	15,000
	As welded	14,000	11,000**
Ultimate shear strength (Shear Test conducted per SA263-264 or 265)	As clad	13,000	10,000
	As welded	11,500	9,000**
	Thermal cycled*	13,000	
Impact Resistance			
a. Charpy "Keyhole"	Test temperature −50°F	45-75 ft-lb	
b. Charpy "V-Notch"	Test temperature +15°F	60-130 ft-lb	
c. Dropweight (ASTM E-208)		−15°F NDT	

*Cycled 2500 times between 500°F and 80°F water before testing.
**Peak welding temperatures in excess of 600°F may lower this value.

[a] By permission of E. I. DuPont de Nemours & Co., Inc.

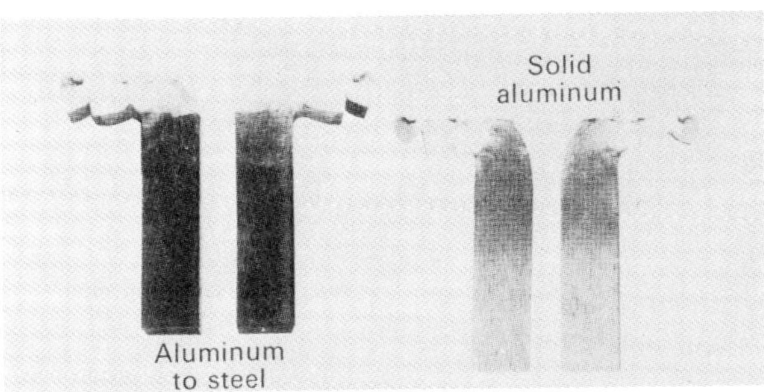

Fig. 51. Izod impact test. Impact resistance of the Detacouple transition-joint bond zone is greater than that of the aluminum itself. (By permission of E. I. DuPont de Nemours & Co., Inc.)

structures. Figure 52 shows ship structural sections using this transition joint.

Figure 53 shows how the bolted connection in (a) might be repaired using a Detacouple transition joint as in (b), or replaced using the design shown in (c).

TABLE X
Fatigue Resistance of Detacouple Transition Joint[a]
(Krouse Double Direct Stress Testing Machine)

Specimen Type	Stress in Web** Compression (psi)	Stress in Web** Tension (psi)	Cycles to Failure	Comments
A. Recommended Design Ratio (Transition joint 4 times as thick as Al and steel webs welded to it).	15,000	5,000	395,000	All failed in heat-affected zone of 5456 Al weld.
	15,000	1,000	721,500	
	10,000	3,000	1,267,400	
B. Typical mechanical connection* (14 CRES ⅜-in. rivets/ft).	15,000	5,000	31,600	Rivet fractured.
	15,000	5,000	63,300	

*Samples fabricated by commercial shipyard using standard production technique.
**In all cases, webs consist of 5456-H321 Al and HY-80 steel, ¼-in. thick.

[a] By permission of E. I. DuPont de Nemours & Co., Inc.

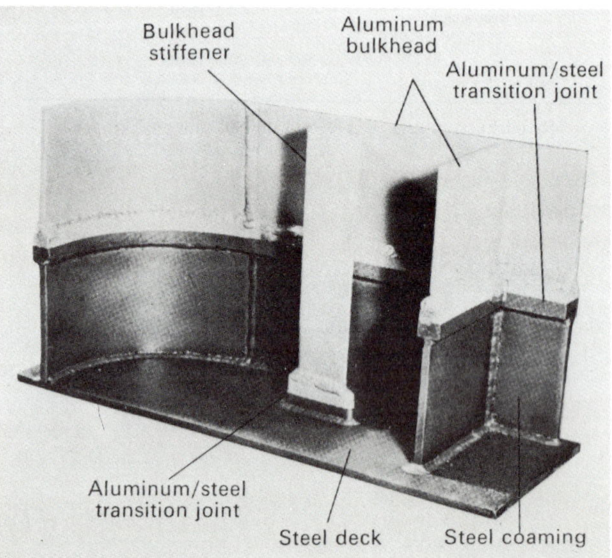

Fig. 52. Ship structure mock-up showing aluminum superstructure welded to steel decking using Detacouple transition joint. Examples of bends, butt welds, T's, and pads are shown. (By permission of E. I. DuPont de Nemours & Co., Inc.)

I. Nuclear-Reactor Fuel Elements

The cost and limited availability of fossil fuels in the face of continually increasing worldwide demand for energy has caused a major shift to the use

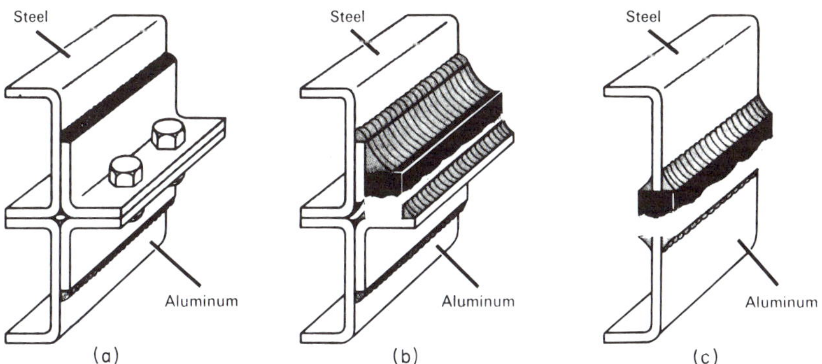

Fig. 53. The bolted connection (a) might be field repaired using a Detacouple transition joint by design (b) and completely replaced on new structures by design (c). Note that the spacing between the outlet flanges is not changed by either design. Users of Detacouple are able to design specific joint configurations to meet their detailed engineering requirements. (By permission of E. I. DuPont de Nemours & Co., Inc.)

of nuclear power in the industrialized nations. In this power system, the heat generated by the fission of radioactive nuclear fuels is absorbed by a working fluid such as water, which is in turn converted to steam to drive a turbogenerator. The nuclear fuel elements, which contain the nuclear fuel, must be clad with a nonfissionable material to prevent corrosion, distortion, and the loss of radioactive particles to the coolant. Nuclear fuel elements have been clad with a variety of metals, including aluminum, stainless steel, magnesium and its alloys, zirconium and its alloys, graphite, nickel, beryllium, niobium, and vanadium. Of these, the four major claddings are aluminum, zirconium, magnesium, and stainless steel. The choice of cladding depends upon its nuclear properties (i.e., its ability to maintain the desired neutron energy spectrum), its chemical and physical compatibility with the nuclear fuel, its corrosion resistance, and its mechanical properties. The cladding must have sufficient creep strength to resist distortion due to the pressure of gases generated during atomic fission. Since methods for cladding nuclear fuel elements have been described in detail by Yans *et al.* (1962), they will not be discussed here. The objective of this section is to point out the basic requirement for metal laminates in nuclear reactors and to present some examples of clad nuclear fuel elements. Figure 54 shows schematically a tubular fuel-element assembly used in cladding by drawing.

Figure 55 shows four billet configurations that have been used in the cladding of flat rectangular fuel elements by rolling. The billet consists of a fuel-element core surrounded by cladding. The rolling operation reduces the billet to the desired dimensions while metallurgically bonding the cladding to the fuel element. Fuel elements with roll-bonded claddings have

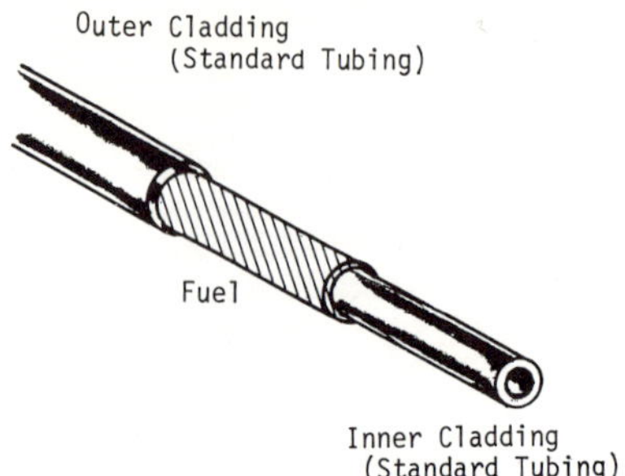

Fig. 54. Tubular fuel-element assembly used in cladding by drawing. Fuel mixture: uranium oxide–aluminum powder or uranium oxide–stainless-steel powder. [From Yans *et al.* (1962), by permission of Wiley (Interscience), New York]

been widely used in research and power reactors. Research reactors such as the Materials Testing Reactor (MTR) and the Engineering Testing Reactor (ETR) have used aluminum–uranium alloys clad with aluminum. Power reactors such as the experimental boiling-water reactor (EBWR) and the stationary medium power-plant reactor (SM-1) have used a uranium–zirconium–niobium alloy clad with zircaloy-2 and a dispersion of UO_2 plus B_4C particles in a 302B stainless-steel matrix clad with 304L stainless steel, respectively (see Abrahamson and Vallee, 1974).

J. Thermostats

1. Principle of Operation

Thermostats represent an extensive, growing, and indispensable application of metal laminates. Basically, a thermostat consists of bonded layers of two or more metals or other materials in strip or sheet form having different coefficients of expansion such that the composite material undergoes a change in curvature when its temperature is changed.

2. Fabrication

The metal layers are generally bonded by hot or cold rolling, hot pressing, or casting the lower-melting alloy on the solid higher-melting layer. The method used depends on the properties of the component and the application. After bonding the metal layers, subsequent manufacturing operations include slitting, flattening, fabricating, and heat treating. The last procedure is required to relieve and control the stresses introduced during fabrication, so that thermostat calibration and performance are not affected. Since a

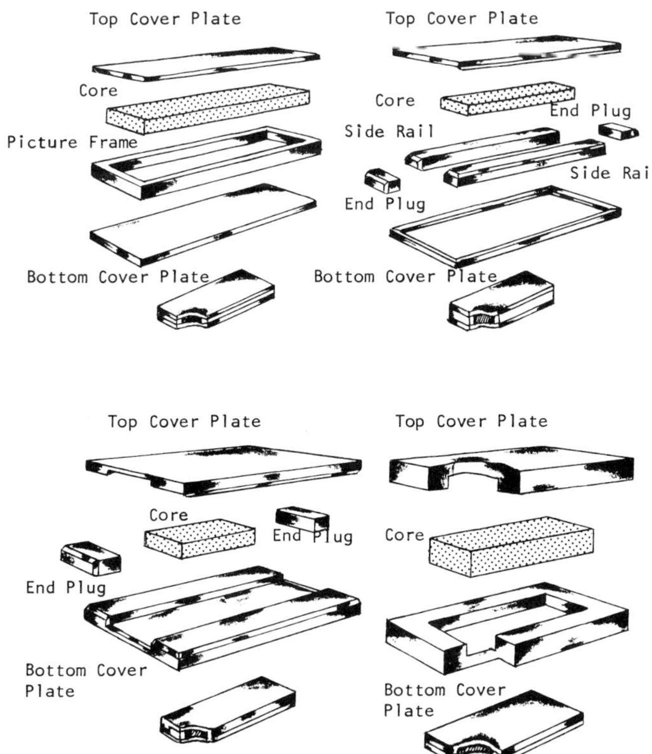

Fig. 55. Four variations of the picture-frame, core, and cover-plate rolling Billet. [From Yans *et al.* (1962), by permission of Wiley (Interscience), New York]

comprehensive discussion of the theory and practice of thermostats has recently been presented by Savolainen and Sears (1969), the purpose of this section is to emphasize the essential and widespread role of metal laminates in temperature measurement and control and to indicate the basic applications of thermostat metals.

3. Basic Applications

The five basic applications of thermostat metals are:

(1) direct temperature indication, as in a dial thermometer;

(2) indication of other characteristics involving temperature change indirectly, as in a watt-demand meter where current flow is indicated by heat produced;

(3) direct temperature control, as in a room thermostat;

(4) control of other characteristics, as in a circuit breaker where current flow is controlled by heat produced;

(5) temperature compensation, as in an automatic choke in a carburetor.

Table XI, which lists a number of applications, illustrates the indispensable role of thermostats in many types of equipment.

4. *Selection of Thermostat Components*

A. SELECTION FACTORS FOR OPTIMUM PERFORMANCE. The factors shown in Table XII must be considered in selecting thermostat components

TABLE XI

Typical Thermostat Applications[a]

Air dryers	Fluorescent-lamp ballast protectors	Popcorn machines
Air heaters	Gasoline-gauge indicators	Portable electric tools
Air valves	Gas meters	Radiator shutters
Alarm devices	Gas safety pilots	Radiosondes
Altimeters	Generator cutouts	Ranges, electric and gas
Aquarium heaters	Glue pots	Recording thermometers
Automatic chokes	Hat stretchers	Refrigerators
Automotive exhaust heat controls	Heating pads	Relays, overload
Automatic transmission	Humidifiers	Relays, signal
Blueprint machines	Incubators	Room thermometers
Bread-wrapping machines	Instruments, electric	Room thermostats
Candy mixers	Instruments, testing	Scales
Carburetor temperature regulators	Ironing machines	Shock absorbers
Chicken brooders	Irons, electric	Signal devices
Cigar lighters	Laboratory ovens	Sign flashers
Circuit breakers	Lamps, electric	Soldering irons
Cord sets	Lamps, therapeutic	Speedometers
Current regulators	Light flashers	Stack controls
Damper controls	Lighting systems	Starting devices
Demand meters	Machine tools	Steam radiators
Dental furnaces	Motor protection	Steam traps
Dental sterilizers	Motor starters	Time switches
Draft controls	Necktie pressers	Toasters, electric
Electric-light plants	Oil-burner controls	Transformer temperature indicators
Electric meters	Oil gages	Type-metal pots
Electric-motor protectors	Oil purifiers	Voltage regulators
Electric sheets	Ovens, electric and gas	Waffle irons
Fans	Percolators	Water heaters, electric and gas
Fire alarms	Photomounting machines	Windshield defrosters

[a] By permission of Texas Instruments, Incorporated.

TABLE XII

Factors in Selecting Thermostat Components

Factor	Requirement
1. Thermal expansivity	Must have required linearity or nonlinearity. Must be reversible. Must match components with highest and lowest expansivity to provide maximum thermal response or activity.
2. Modulus of elasticity	High-modulus materials preferred because of their ability to do more work.
3. Tensile and creep strength proportional limit	These factors should be high so that the thermostat can resist the high mechanical and thermal stresses induced in the components in high-temperature, heavy-duty service.

The requirements listed in the table indicate that elemental metals or solid-solution alloys that are stable in the desired temperature range are best suited as thermostat metals.

B. PROPERTIES OF THERMOSTAT METALS. High-expansivity components include brass, monel, austenitic nickel–chromium-iron, and nickel–manganese–iron alloys, pure nickel–molybdenum–iron alloys, and high-manganese–nickel–copper alloys. These are bonded to low-expansivity components, which are usually of the Invar group (nickel–iron alloys).

When thermostats are used in corrosive environments, there is always the possibility of severe corrosive attack due to the development of destructive galvanic couples produced between dissimilar metal components. Great care in selection of thermostat components must be taken to eliminate or minimize this effect. Fortunately, there are a number of thermostats whose components are corrosion resistant and do not form a damaging galvanic couple. In some cases, corrosion protection is obtained with tin and lead dipped coatings or, less effectively, by electroplating with nickel, chromium, cadmium, or zinc.

V. Summary and Assessment

The concept of laminating similar or dissimilar metals together to produce a composite having properties superior to its constituents is an old, useful, and important one. By proper selection of the constituent laminae, materials can be made with properties tailored to a great number and variety of desired applications. For example, the wear resistance, stiffness, strength, and light weight of stainless-steel-clad aluminum make it very useful in aircraft and rapid-transit applications; the same advantages plus controlled thermal conductivity of copper-clad stainless steel give it a key position in architecture, building construction, and cookware; corrosion resistance combined with strength, toughness, and reduced weight make titanium-clad steel particularly attractive for chemical process equipment; and cupronickel-clad copper provides coinage with the necessary attributes of appearance, wear resistance, weight, and fabricability. Similarly, thermostat metals can be fabricated to have precisely controlled deformation in response to temperature in or out of corrosive environments. Furthermore, these composite properties can be achieved, in many cases, with great savings in costly and scarce metals, a factor of increasing importance as growing demands are made of the Earth's limited resources.

The use of laminates of similar metals is also important because of their enhanced toughness and resistance to catastraphic failure when used in pressure vessels and other structures.

The availability of a number of well-established methods for fabricating sound, well-bonded laminates, such as roll bonding, coextrusion, explosive welding, and brazing, provides a basis for selecting the optimum fabrication process for a given laminate system.

A. *Technology Gaps*

The primary technology gaps involve the technical and cost limitations in currently available laminate fabrication methods. These are briefly described below:

1. Roll Bonding

As noted earlier, when two dissimilar metals are bonded together, internal stresses are built up because of differences in thermal expansion coefficients. Their magnitude may be such that the composite is seriously distorted. Extensive experience, together with careful selection and treatment of the metals, can minimize, if not eliminate distortion.

The high cost of roll bonding large steel plates to form dual-hardness armor has been a primary barrier to the wider use of this laminate. Current studies on the fabrication of this armor by electroslag remelting offer the potential of reducing its cost by 50% by eliminating the costly fabrication procedures, such as edge-welding the plates together, evacuating the pack, and rolling, that must be carried out in conventional roll bonding. The electroslag-remelting method produces the laminate by electrically remelting an ingot of each of the armor constituents into adjacent molds, which are then bonded together with molten metal of one of the constituents used as a third electrode. The slag formed during the remelting protects and cleans the underlying molten metal of impurities. The feasibility of this approach has been demonstrated by Philco-Ford. Work is currently underway to bring this method into commercial application.

In the case of thermostat metals, precise control over material dimensions, rolling pressure, temperature, and time are required to produce new laminates whose deformation in response to temperature change is precise and controlled.

2. Explosive Welding

Explosive welding is limited to the fabrication of flat plates. With the exception of concentric cylinders, contoured shapes cannot be made because (1) a constant stand-off distance cannot be maintained, and (2) the critical constant angle between the two layers cannot be maintained, since the angle between layers must change as the contour changes. Therefore, flat plates must be fabricated first, followed by secondary fabrication of the laminate into the desired contour. Fabrication of metal laminates into contoured shapes also presents difficulties because one must deal with dissimilar metals having different deformation characteristics. The development of a method for the simultaneous bonding of a laminate and its fabrication into a contoured shape by explosive welding would greatly advance the use of metal laminates in industrial and military applications.

B. Outlook

The outlook for metal laminates is one of continued, accelerating growth, involving traditional applications such as coinage, thermostats, building construction, electronic equipment and electrical wiring, jewelry, printing, pressure vessels, automobiles, rapid transit, ships, and aircraft. Strong growth projections for each of these areas can be made, although actual market forecasts are beyond the scope of this chapter.

The driving forces behind the use of metal laminates continue to be the cost savings, improved properties or performance, and conservation of expensive and precious metals that can be realized through their use. The rapid advances in industrial and defense technology, together with the ever growing need for worldwide pollution control will require the broadened use of metal laminates to conserve precious metal resources. Efforts to control automotive pollution will be greatly enhanced by the development of clad catalysts, which greatly reduce the need for platinum, palladium, and other scarce metals that are necessary for conversion of automotive exhausts to harmless gases. Metal laminates have become indispensable elements of our technological society.

References

Abbott, K. H., and Manganello, S. J. (1972). Metallurgical Factors Affecting the Ballistic Behavior of Steel Targets, AMMRC TR 72-31, September.

Abrahamson, S. G., and Vallee, R. (1974). In this treatise, Vol. 3.

Almond, E. A., Embury, J. D., and Wright, E. S. (1968), Interfaces in Composites, ASTM STP 452.

Almond, E. A., Petch, N. J., Wraith, A. E., and Wright, E. S. (1969). *J. Iron Steel Inst.* **207**, 1319–1323.

Anctil, A. A., and Kula, E. B. (1969). Fatigue Crack Propagation in Armor Steels, AMMRC TR 69–25, November.

Armstrong, R. W., Codd, I., Douthwaite, R. M., and Petch, N. J. (1962). *Phil. Mag.* **7**, 45.

Arnold, S. V. (1960). Toughness of Steel Sheet: The Advantage of Laminating. Watertown Arsenal Lab. Tech. Rep. # WAL-TR-834.21/2.

ASTM Standard B106–65T (1965). ASTM, Philadelphia, Pennsylvania.

Blazynski, T. Z., and Dara, A. R. (1971). *Proc. Int. Conf. Center High Energy Forming, 3rd* p. 2.4.2.

Bluhm, J. I., (1961). *Proc. ASTM* **61**.

Brown, R. H. (1969). "Composite Engineering Laminates" (A. Dietz, ed.) Chapter 11. MIT Press, Cambridge, Massachusetts.

Crossland, B. (1971). Review of the State of the Art in Explosive Welding. Report to Commission 4 of the Int. Inst. of Welding, Rep. No. 567.

Crossland, B., and Williams, J. D. (1970). *Met. Rev.* **15**.

Dietz, A. G. H. ed. (1969). "Composite Engineering Laminates." MIT Press, Cambridge, Massachusetts.

Engelhard Industries Brochure (1971). D. E. Makepeace Div., Econ-o-clad Clad Metal Strip.

Embury, J. S., Petch, N. S., Wraith, A. E., and Wright, E. S. (1967). *Trans. TMS AIME* **245**, 2529–2536.

Floreen *et al.* (1971). *Corrosion* **27** (12), 519–524.

Hawkins, R., and Wright, J. C. (1971). *T. Inst. Metals* **99**, 357–371.

Holzman, A. H. and Cowan, C. G. (1965). Bonding of Metals with Explosives. Welding Res. Council Bull. 104.

Kaufman, J. G. (1967). *J. Basic Eng.* **89**, 503.

Leichter, H. L. (1966). *J. Spacecraft* **3**, No. 7, 113.

Lukens Steel Co. (1960). "Fabrication of Lukens Clad Steels." Coatesville, Pennsylvania.

McKenny, C. R., and Barker, J. G. (1971). *Mar. Techn.*

Ornstein, J. L., (1965). *Mater. Design Eng.*

Petch, N. (1953). *JISI* **174**, 25–36.

Pfizer Metals and Composite Products (1971). Stainless Clad Aluminum. Tech. Rep. No. TR-130.

Rochenbach, D. J. (1969). "Design and Fabrication of Welded Anode Assemblies for a Prebaked Aluminum Potline," *Light Metal Age.*

Ronan, J. T., (1970). "Composite Metals." *Stamping/Diemaking.*

Sachs, G., (1928). *Ver Deut.–Ing.* **72**, 734.

Salvaluiren, U. U., and Sears, R. M. (1969). "Composite Engineering Laminates" (A. G. H. Dietz, ed.), Chapter 10. MIT Press, Cambridge Massachusetts.

Sears, R. M. (1963). *Mater. Res. Std.* **3**, 982–986.

Smith, C. S. (1960). "A History of Metallography." Univ. of Chicago Press, Chicago, Illinois.

Stavros, A. S., and Paxton, H. W. (1970). *Corrosion* **27**, 519–524.

Sticha, E. A. (1969). *Mater. Protect.* **8**, No. 10.

Texas Instruments, Inc. (1971). T. I. Rep. 51-70-30, Final Rep. under Contract No. DAAA-25-69-CO462, R. V. Barone, Project Manager.

Timoshenko, S. (1925). *J. Opt. Soc. Amer.* **2**, 233.

Wright, E. S. (1967). Deformation and Fracture of Plastically Anisotropic Laminates. Univ. of Newcastle-upon-Tyne, Brittle Fracture Study Group Ann. Rep.

Wright, E. S. (1968). Fracture of Laminated Composites. Stanford Res. Inst. Annu. IRSD Rep.

Wright, E. S., and Bayce, A. E. (1964). *Proc. NATO Advan. Study Inst. High Energy Rate Working of Metals, Oslo.*

Yans, F. M., Loewenstein, P., and Greenspan, J. (1962). "Nuclear Reactor Fuel Elements" (A. R. Kaufmann, ed.), Chapter 12. Wiley (Interscience), New York.

3

Directionally Solidified Eutectic Superalloys

E. R. THOMPSON and F. D. LEMKEY

United Aircraft Research Laboratories
East Hartford, Connecticut

I. Introduction

Since Sorby's pioneering applications of the light microscope to the study of polished sections of meteorites and metals in 1863, scientists have recognized the existence of microduplex structures. Little attention was paid to this feature until it was demonstrated that a solid could be produced wherein the phases take regular form, such as parallel fibers or lamellae of one phase in a matrix of another over extended distances. This development was accomplished as a result of many detailed studies on the manner in which certain alloys solidify or transform and followed, almost one-hundred years later, Sorby's period of "unencumbered research" at his home in Sheffield, England.

During the past decade, metallurgists cooperating with nature in the controlled solidification of eutectic alloys have formed new kinds of materials consisting of aligned interpenetrating phases. From the directional freezing of eutectics, anisotropic microduplex microstructures have been produced *in situ*, and one name applied to this class of materials has indeed been *in situ* composites. The potential of this unique approach to the production of composite materials in terms of special physical and mechanical properties was recognized in the invention by R. W. Kraft (1964). A thorough description of these early investigations involving fibrous and lamellar systems, Cu–Cr, Al–Al_3Ni, and Al–$CuAl_2$, has been given in previous reviews (Hertzberg, 1964, 1967; Salkind *et al.*, 1970; Lemkey, 1970).

These early investigations identified several important characteristics of aligned eutectics that have been responsible for the continued interest in their development. These features include the presence of a high-strength constituent, good bonding between phases, strength retention at temperatures close to the eutectic melting point, and excellent high-temperature phase stability. It is the attractive high-temperature properties of these directional *in situ* structures that this chapter will principally consider. For a more thorough account of the physical metallurgy and microstructure of eutectic grains, the reader is directed to previous reviews by Tiller (1958), Scheil (1959), Chadwick (1963), Kerr and Winegard (1966), and Hogan *et al.* (1971).

The first evidence that a controlled eutectic structure could behave as a classical reinforced composite was found by Lemkey *et al.* (1965) and Hertzberg *et al.* (1965) in their studies of the Al–Al_3Ni system. This work was followed by the evaluation of other eutectic systems possessing a still higher volume fraction of the reinforcing phase; for example, the Nb–Nb_2C eutectic was shown by Lemkey and Salkind (1967) to be stronger and more creep resistant than commercially available niobium alloys. In 1966 exciting indications of the potential use of controlled eutectic alloys at elevated temperature came with the discovery by Lemkey and Thompson of carbide-

and intermetallic-compound-strengthened nickel and cobalt eutectics. These alloys were both stronger and more creep resistant than the so-called "superalloys" based on the same metals. Because of these superior high-temperature characteristics we have termed these high-strength cobalt- and nickel-based eutectics "eutectic superalloys."

Today, eutectic superalloys are a special and sophisticated class of materials. Unlike conventional superalloys, they are anisotropic and derive their remarkable strength and creep resistance from controlled solidification. Currently, complex milticomponent alloys with matrices compositionally similar to classical superalloys have resulted in an even greater spectrum of improved, reproducible, and predictable properties combining resistance to both corrosion and creep. However, as a class of materials they are only now beginning to emerge from the laboratory stage of development. They show great attractiveness for gas-turbine usage and aerospace application where high service temperature is a pacing requirement.

The purpose of this chapter is to review the current technology in eutectic superalloys of nickel and cobalt and to present experimental data from systems that typify the behavior of these materials. Much of the discussion will be subdivided into two categories, in keeping with the micromorphology of normal eutectics†: (1) lamella and (2) rod or whisker. The information presented is not intended to be fully comprehensive but rather to point out the general mechanical behavior of these materials and trends in their development.

II. Systems Approach

An examination of available metallic elements in terms of their respective melting points, mechanical properties, densities, and oxidation behavior will demonstrate that only a few of these are suitable as a matrix of an alloy for structural service at temperatures exceeding 1100°C. Based on the mechanical demands alone, current nickel and cobalt superalloys do not possess mechanisms for further strengthening at service temperatures a few hundred degrees or less below their incipient melting points, which is required to satisfy the projected needs of the 70s as discussed by Tien (1972). Microstructures of the γ'-strengthened superalloys are often adversely affected at such temperatures (e.g., re-solution of the precipitate). Solid-solution strengthening is rather ineffective at these temperatures, and the high rates of diffusion permit dislocation climb, which brings about an increased ease of plastic deformation in dispersion-strengthened alloys. In addition to the

† For a definition of this term, see Section II.A.

high-temperature mechanical characteristics, corrosion resistance and fabricability are important criteria to be considered in the design of an improved heat-resistant material.

Of the available elements, nickel and cobalt are perhaps the most practical choices, since they have served as the basis for currently used alloys. In addition to these, it should be noted that iron, chromium, and the precious metals also possess the high melting point required and may be the basis for future eutectic research. Since the number of high-melting metals that are practically useful is limited, eutectics based on intermetallic and covalent compounds have been considered and certainly will be considered further in the future. As a group, these compounds have properties that are markedly different from their elemental components. They are normally brittle materials with high hardnesses and may have melting points substantially higher than those of the parent metals.

A. *Binary, Pseudobinary, and Ternary Eutectics*

Information is now available on the microstructure of almost every binary and pseudobinary nickel- and cobalt-based eutectic where the equilibrium phase diagram exists. Over 150 directionally solidified eutectic alloys have been tabulated and the microstructures and crystallographic relationships listed in the review of Hogan *et al.* (1971). The intelligent selection of a binary or pseudobinary eutectic as a possible elevated-temperature structural material is not altogether easy, however. The mechanical behavior of eutectic composites is such that their strength is strongly dependent on the volume fraction and properties of the aligned reinforcement, with the continuous metal matrix serving as a load-transfer medium and contributing toughness. Thus, knowledge of the volume fractions of the phases is desirable. This can, in most cases, be obtained from the lever rule, provided the phase diagram is complete. Provided the surface energy between the two solid eutectic phases is isotropic, a minimum-surface-energy criterion predicts the rod form when the volume fraction of the minor phase is less than about 32% and the lamellar form when it is greater, as pointed out by Hunt and Chilton (1962). The relative merits of lamellar and rodlike microstructures for a reinforced composite are debatable. Generally the fibers within a eutectic composite will withstand a higher elastic strain before deformation or fracture and are therefore most effective in terms of reinforcing the ductile matrix. However, since this microstructural form is usually accomplished in the lower-volume-fraction systems, their strengths may still be less than the lamella-reinforced eutectic composites. For nearly equal volume fractions, mechanical considerations would probably favor the rodlike microstructure. On the other hand, the relative thermal stability of the two microstructural forms would favor the lamellar microstructure.

Prediction of the microstructural form that the reinforcement will assume in any eutectic is still uncertain. Although there are some exceptions to the rules set forth, the best classification of eutectic microstructures is the one based on the growth characteristics of the component phases proposed by Hunt and Jackson (1968). The component characteristic is a term related to the component's latent heat of fusion divided by its absolute melting temperature, i.e., the entropy of fusion. If the entropy of fusion of a phase is less than $2R$, where R is the gas constant, the solid–liquid interface was predicted to be nonfaceted atomically. Metals and most alloys fall within this group. For materials having an entropy of fusion greater than $2R$, the interface was predicted to be atomically smooth or crystallographically faceted. The metalloids, carbides, and certain compounds fall in this group. Thus, binary eutectics have been conveniently divided into three groups: nonfaceted–nonfaceted, nonfaceted–faceted, and faceted–faceted, assuming that each component will freeze in the same manner during coupled eutectic growth as it does independently. Examples of the first group, these being defined as normal eutectics, constitute the majority of the systems presented in Table I and include Ni–Cr, Ni–W, NiAl–Cr, etc. Nonfaceted–faceted systems, which have shown a surprisingly large range of coupled growth, include the refractory-metal-monocarbide and chromium-carbide (Cr_7C_3) systems with nickel and cobalt, as discussed by Lemkey and Thompson (1966).

Having selected a system from volume-fraction and growth characteristics of the reinforcing phase, aligned periodic eutectic structures containing no primary phases must be produced. In the absence of impurities, simple binary nonfaceted–nonfaceted eutectics permit the growth of aligned microstructures over a wide range of growth conditions, and their steady-state growth is rather well understood, although the theoretical model is not entirely complete, as shown by Jordan and Hunt (1971). The growth of faceted–nonfaceted systems, in contrast, is only qualitatively understood.

During the steady-state directional growth of eutectics, coupled plane-front growth may locally break down. This causes growth to a nonplanar liquid–solid interface and results in a colony or cellular microstructure. This breakdown has been associated with constitutional supercooling at the interface due to impurities. It was shown by Tiller *et al.* (1953) that the criterion for the onset of constitutional supercooling and interface stability for single-phase dilute alloys could be expressed as

$$G/R \leq mC_0(1 - k)/Dk \qquad (1)$$

where G is the thermal gradient in the liquid, R the freezing velocity, D the liquid diffusion coefficient, C_0 the impurity concentration, k the partition coefficient, and m the slope of the liquidus line. For a eutectic, an expression

TABLE I

Summary of Room-Temperature Mechanical and Certain Physical Properties of Aligned Nickel and Cobalt Alloys

System A–B	v/o B	Melting point (°C)	Density (g/cm^3)	E (GN/m^2)	Tensile strength (MN/m^2)	Elongation (%)
Ni–NiBe	38–40	1157	—	215	918	9.0
Ni–Ni_3Nb	26	1270	8.8	—	745	12.4
Ni–Cr	23	1345	8.0	—	718	29.8
Ni–NiMo	50	1315	9.5	—	1250	<1
Ni–Ni_3Ti	29	$\sim$1300	8.2	—	650	<1
Ni–W	6	1500	—	—	830	45
Ni–TiC	5.5	1307	—	—	—	—
Ni–HfC	15–28	1260	—	—	—	—
Ni–NbC	11	1328	8.8	—	890	9.5
Ni–TaC	$\sim$10	—	—	—	—	—
Ni–Cr–NbC	11	1320 range	—	—	—	—
Ni,Co,Cr,Al–TaC	$\sim$9	—	8.8	—	1650	$\sim$5
Ni_3Al–Ni_3Nb	44	1280	8.44	242	1240	0.8
Ni_3Al–Ni_3Nb	32	1280 range	—	—	1230	2.0
Ni_3Al–Ni_3Nb	$\sim$32	1270–1285 range	8.5	—	1130	29
Ni–Ni_3Al–Ni_3Nb	—	1270	—	—	1140	2.3
Ni_3Al–Ni_3Ta	$\sim$65	$\sim$1360	10.8	—	930	<1
Ni–Ni_3Al–Ni_3Ta	—	$\sim$1360	—	—	1060	5
Ni_3Al–Ni_7Zr_2	42	1192	—	—	—	—
Ni_3Al–Mo	26	1306	8.18	138	1120	21

NiAl–Cr (rod)	34	1450–1455 range	6.4	182	1240	<1
$(Ni,Cr)–(Cr,Ni)_7C_3$	30	1305 range	—	200–290	685–960	2–11
Co–CoAl	35	1400	—	172	500–585	~ 6
Co–CoBe	23	1120	—	—	—	—
$Co–Co_3Nb$	~ 50	1235	—	—	—	—
$Co–Co_2Ta$	35	1276	—	—	—	—
$Co–Co_7W_6$	23	1480	—	—	750	<1
Co–TiC	16	1360	—	—	—	—
Co–HfC	15	—	—	—	—	—
Co–VC	20	—	—	—	—	—
Co–NbC	12	1365	8.8	—	1030	2
Co–TaC	16	1402	9.1	222	1035	11.8
Co–Cr–NbC	12	1340 range	—	—	1280	<2
Co–Cr–TaC	~ 9	1360 range	9.0	210	1035–1160	16–20
$(Co,Cr)–(Cr,Co)_7C_3$	30	1300 range	8.0	296	1280	1.5
$(Co,Cr,Al)–(Cr,Co)_7C_3$	28	1295 range	7.8	283	1730–2011	2.5–1.0
$(Co,Cr)–(Cr,Co)_{23}C_6$	~ 40	1340 range	7.91	~ 276	1200	0.96

similar to this may be derived, and it has been shown by Kraft and Albright (1961) that the planar-to-colony transition in impure eutectics has a functional dependence on G, R, and C_0 similar to the above expression. Thus, provided the ratio of gradient to rate of solidification is above a certain limit, impure eutectics may be successfully aligned. For example, in processing the Ni_3Al–Ni_3Nb pseudobinary eutectic, the critical ratio of G/R with its relatively low impurity content was found to be 23°C hr/cm^2 by Thompson *et al.* (1970a). The aligned lamellar microstructure of this eutectic containing 44 v/o Ni_3Nb, grown at 2 cm/hr under a gradient in the liquid of ~70°C/cm, is shown in Fig. 1. The (Co,Cr)–$(Cr,Co)_7C_3$ pseudobinary eutectic composition, which contained a much higher level of impurity derived from the use of foundry-grade elements, exhibited a critical G/R ratio of ~ 7°C hr/cm^2 in the study by Thompson and Lemkey (1970).

The amount of systematic investigation of ternary nickel and cobalt eutectic alloys, in contrast to the binary and pseudobinary alloys, is very small. The ternary Ni_3Al–Ni_3Ti–Ni_2TiAl eutectic, examined by Thompson and Lemkey (1969), and more recently the Ni–Ni_3Al–Ni_3Nb eutectic, examined by Kraft and Thompson (1972), indicate an increased complexity in structure. The microstructure of the latter system grown at 2 cm/hr under a gradient in the liquid of ~ 70°C/cm is shown in Fig. 2. The growth range of these alloys, which exhibit invariant equilibria, was found to be equivalent to the respective binary eutectics of similar impurity content.

B. Off-Eutectic Coupled Growth

Cooperative growth to a planar solid–liquid interface has been demonstrated by Mollard and Flemings (1967) and Jordan and Hunt (1971) under special growth conditions with alloys displaced from the equilibrium eutectic composition. Stable plane-front growth was favored by a low growth rate R, a steep gradient G, and the essential absence of convective mixing. The resultant composite structures were either lamellar or rodlike, again depending on the volume fraction. Primary-phase dendritic growth was suppressed during unidirectional growth of hypo- or hypereutectic melts when the temperature gradient ahead of the interface was sufficiently steep to prevent the hyper- or hypoeutectic liquid from becoming supercooled. It was shown that for each alloy composition there exists a critical value of G/R above which stable planar growth is possible and below which the interface is unstable and dendritic growth of the primary phase occurs. The critical condition employed analytically was a derivative form of the interface stability criterion [Eq.(1)] where the ratio of thermal gradient to growth rate should equal or exceed the product of the liquidus slope and the difference between the eutectic and starting composition divided by the liquid diffusion coefficient. Directional solidification of an off-eutectic alloy in the pseudo

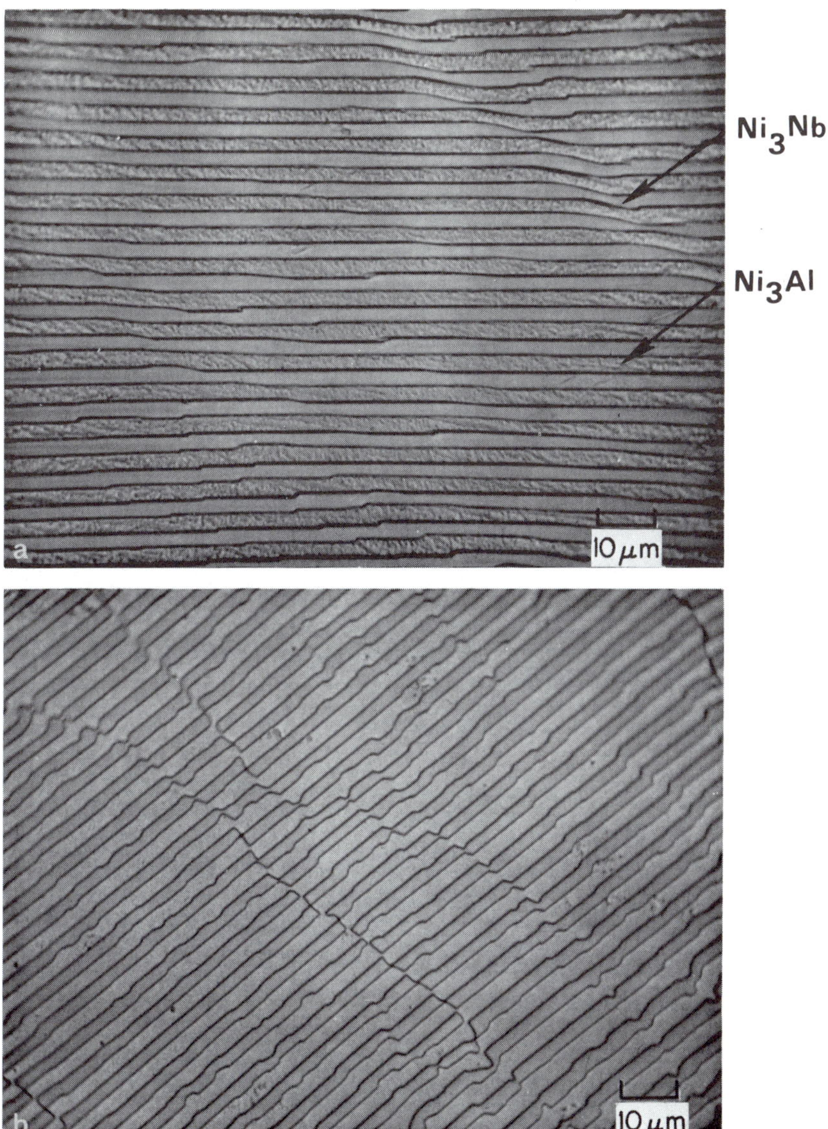

Fig. 1. Microstructures of unidirectionally solidified Ni_3Al–Ni_3Nb eutectic: (a) longitudinal, (b) transverse.

binary Ni_3Al–Ni_3Nb whose composition is depicted in Fig. 3 was recently investigated by Kraft and Thompson (1972). A significant change in the volume fraction was achieved, specifically, a 12 v/o reduction in Ni_3Nb, as

Fig. 2. Transverse microstructure of Ni–Ni_3Al–Ni_3Nb ternary eutectic.

shown in Fig. 4. However, more stringent growth conditions, i.e., steeper temperature gradient and/or slower freezing rate, were required than for the pseudobinary γ'–δ.

An additional disadvantage of this technique is that while the volume fraction of the phase may be altered considerably, the composition of the phases cannot be adjusted but remains the same as that found in the eutectic phases. This would limit compositional changes of the matrix phase, as might be required, for instance, for the optimization of corrosion resistance.

C. Monovariant Eutectics

Another approach to varying the volume fraction of the phases present that permits variation in their composition is to consider the biphase

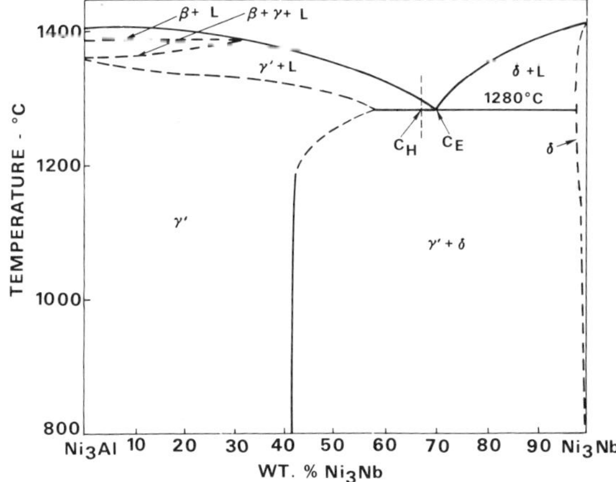

Fig. 3. Equilibrium diagram of Ni_3Al–Ni_3Nb system.

solidification of monovariant ternary alloys, as described by Thompson and Lemkey (1970). This approach is similar to that considered above for an impure binary eutectic, but attention here is given to true ternary alloys rather than dilute impurities in binary alloys. Again the onset of supercooling and the resulting instability of the liquid–solid interface could be expressed by modifying the constitutional supercooling equation [Eq. (1)] as follows:

$$G/R \leq m_T C_0(1 - k^c_{\alpha\beta})/Dk^c_{\alpha\beta} \tag{2}$$

where m_T is the slope of the liquidus trough, C_0 the third-element concentration, D the diffusion coefficient of third element in the liquid, and $k^c_{\alpha\beta}$ the weighted average of the distribution coefficient of c for the α and β phases.

The ternary phase diagram containing the constitutional terms employed above is illustrated in Fig. 5. Consider the manner in which steady-state solidification of an alloy e_1, C_0, on the eutectic liquidus trough e–e_2 would proceed. As freezing begins, solids of composition a_1 and b_1 will be in equilibrium with liquid of composition e_1. As the solids form, the impurity c will build up ahead of the interface until it reaches a composition e_2, $C_0/k^c_{\alpha\beta}$, and the crystallizing solids have the same impurity concentration as the original liquid. In general, the impurity concentration in each of the two solid phases will be different. Each phase will reject impurity according to the value of its individual solid–liquid distribution coefficient k^c_α and k^c_β (i.e., a_2/e_2 and b_2/e_2, respectively). The average solid–liquid distribution coefficient will be $k^c_{\alpha\beta} = e_1/e_2$. $k^c_{\alpha\beta}$ may approach unity when the impurity raises the melting point of one phase $k^c_\alpha > 1$ and lowers the melting point of the other $k^c_\beta < 1$. From the modified constitutional supercooling criterion,

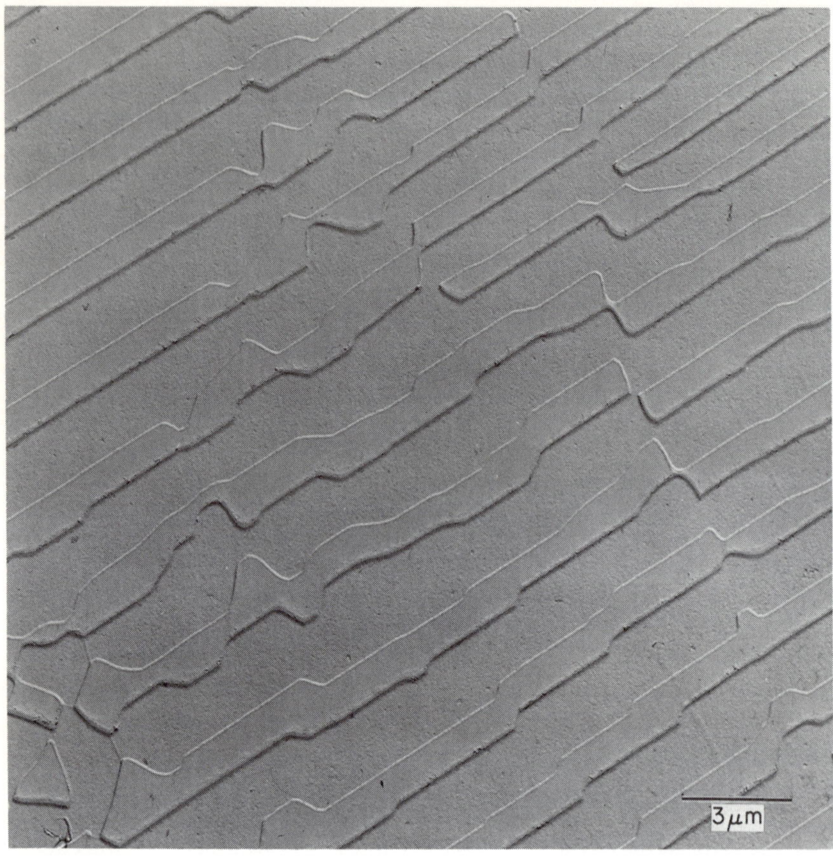

Fig. 4. Electron micrograph of center section of bar of hypoeutectic γ'–δ(32 v/o Ni_3Nb).

a planar solid–liquid interface would be predicted over the widest range of growth conditions when m_T approaches zero (nearly horizontal slope) and $k^c_{\alpha\beta}$ approaches unity.

The freezing of Co–Cr–C and NiAl–Cr alloys was shown qualitatively to fit this expression by Thompson and Lemkey (1970) and Lemkey (1973a), respectively. Recently, monovariant alloys of Ni,Cr–Ni_3Nb and Ni,Al–Ni_3Nb have been unidirectionally solidified by Lemkey (1973b) over the wide range of composition shown in Fig. 6, and the plane-front–cellular transitions generally follow the criterion cited in Eq. (2). From microstructural evidence as exemplified in Figs. 7 and 8, experience has been gained that suggests that coupled growth over a range of composition is possible for

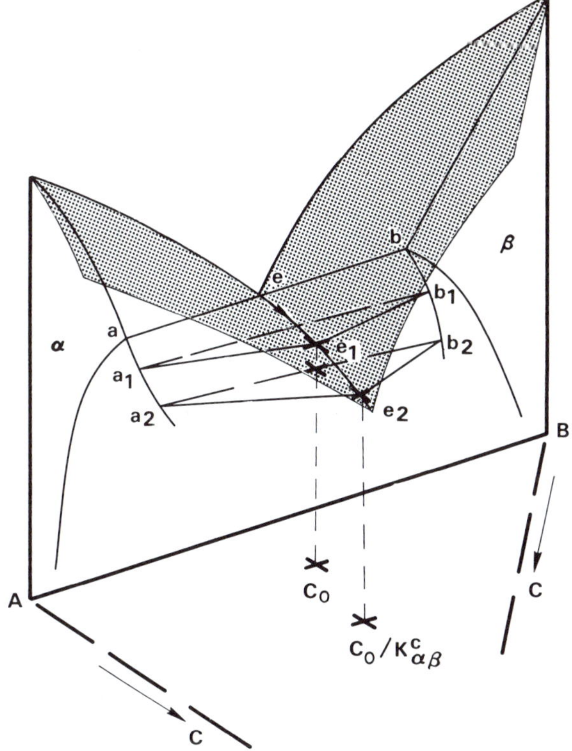

Fig. 5. Schematic of the freezing of a monovariant alloy e_1.

numerous nickel- and cobalt-based alloys existing on shallow monovariant eutectic troughs without the stringent conditions that are usually required in the growth of off-eutectic alloys. Using the monovariant eutectic approach, for example, the composition of the matrix phase in a cobalt alloy was adjusted so that a chromium content corresponding to the maximum oxidation resistance was obtained in the study by Thompson and Lemkey (1970). Further, the optimum amount of γ' can also be selected with alloy heat treatment to maximize matrix strength, as demonstrated by Lemkey (1973b) for the $Ni,Ni_3Al–Ni_3Nb$ system.

D. *Multivariant, Multicomponent Eutectics*

Just as today's nickel and cobalt superalloys are complex partly because of the adjustments in composition demanded by the physical, mechanical, and

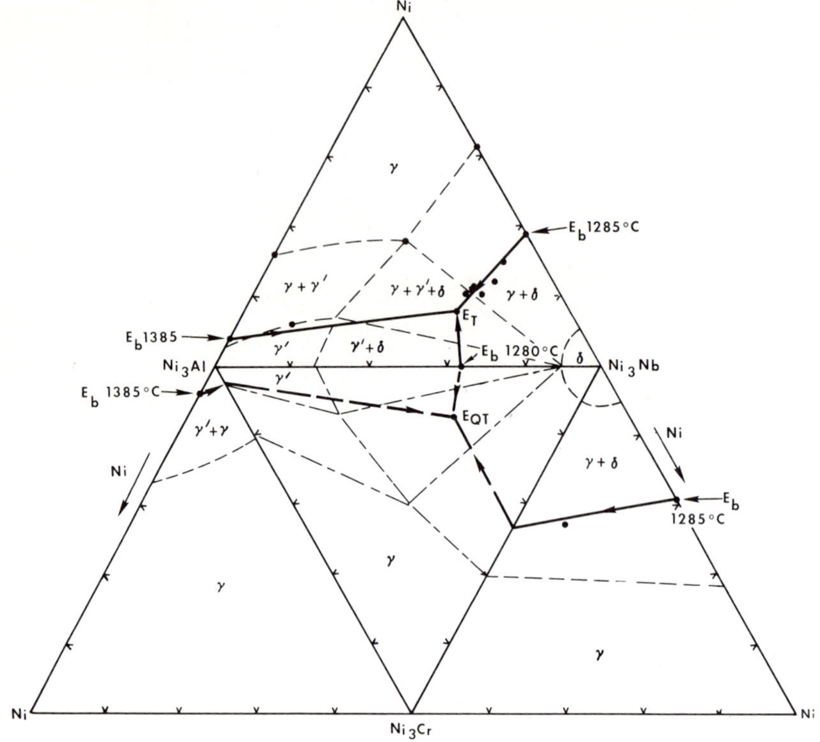

Fig. 6. The liquidus surface and 1200°C phase fields in the nickel-rich corner of the Ni–Nb–Cr–Al diagram: —— liquidus trough, --- 1200°C, –·– hypothetical.

chemical criteria, eutectic alloys have also evolved in their complexity. Thus, although monovariant eutectics offer a unique capability of changing chemistry and volume percent along the eutectic trough, a still greater degree of freedom in composition is sometimes required. In particular, aligned biphase structures have been produced from compositions that are thermodynamically multivariant, rather than invariant or monovariant as in the binary or ternary systems discussed previously.

An example of this approach is described for the nickel-rich quaternary system shown in Fig. 9, for the sake of convenience and ease of graphical illustration, although we have shown it to apply also with higher-order systems. For the quaternary system, the reaction that produces the desired anisotropic biphase structure is that of the simultaneous separation of two solids from the liquid. Figure 9 shows a polythermal projection of the quater-

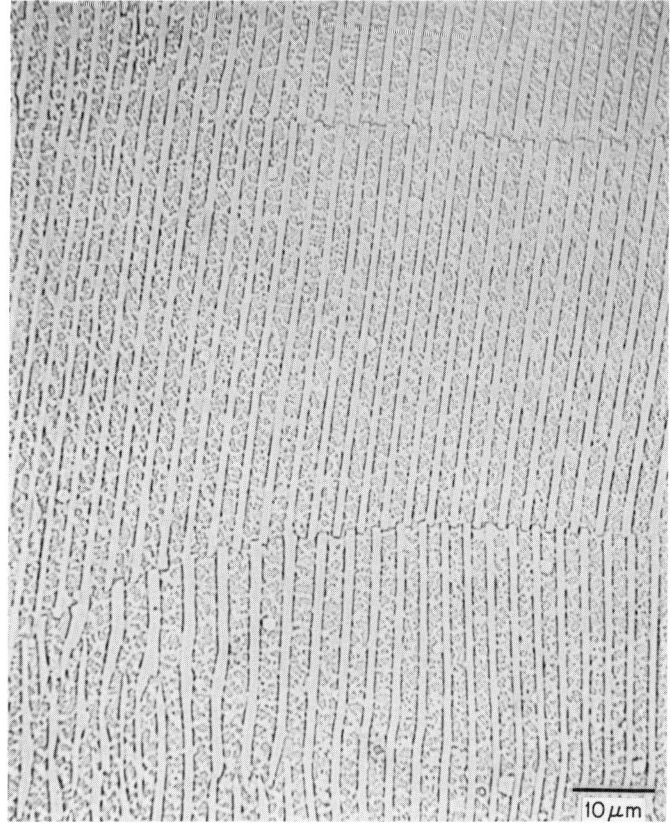

Fig. 7. Transverse microstructure of Ni,Ni_3Al–Ni_3Nb monovariant eutectic.

nary system Ni–Al–Nb–Cr. The faces of the tetrahedron show polythermal projections of the ternary systems Ni–Al–Nb, Ni–Cr–Nb, and Ni–Cr–Al. The growth of the binary eutectic Ni–Ni_3Nb and the monovariant eutectic alloys between it and the two ternary eutectics, $\gamma + \beta + \delta$ and $\gamma + \gamma' + \beta$, have been considered independently above. The liquidus surface, excluding the invariant binary and ternary eutectic as end points, defines the alloy compositions that melt over a temperature interval and that allow the achievement of an aligned biphase composite. An electron photomicrograph taken of a bivariant composition consisting of a γ' precipitation-hardened nichrome matrix reinforced by Ni_3Nb lamellae is shown in Fig. 10.

In terms of the phase rule, at a fixed pressure, two independent variables must be selected in a quaternary system to define the state of the system

Fig. 8. Transverse microstructure of Co–15Cr–8.45Nb–1.05C, growth parallel to [100] Co, 1 cm/hr.

wherein two solid phases are in equilibrium with the liquid. Concentration of two of the components, or one of the components and a temperature, may be used to map the liquidus surface. In a quinary system three independent variables for the reaction $L \longrightarrow \alpha + \beta$ describe a liquidus volume, while the nonisothermal reaction of the type $L \longrightarrow \alpha + \beta + \gamma$ would be located on the liquidus surface.

The ability to form anisotropic biphase structures in quinary and higher-order systems has been demonstrated in the (Co,Cr,Ni,Al)–$(Cr,Co,Ni)_7C_3$ system (Fig. 11), the Ni,Cr,Ni_3Al–NbC system (Fig. 12), and the (Co,Cr, Ni,Al)–$(Cr,Co,Ni)_{23}C_6$ system (Fig. 13). Growth occurred by macroscopically plane front solidification, and again the criteria for stability may be based on constitutional supercooling arguments where the ratio of gradient to rate should equal or exceed the ratio of the alloy melting range to the effective diffusion coefficient of atoms in the liquid. The stability criterion is expressed

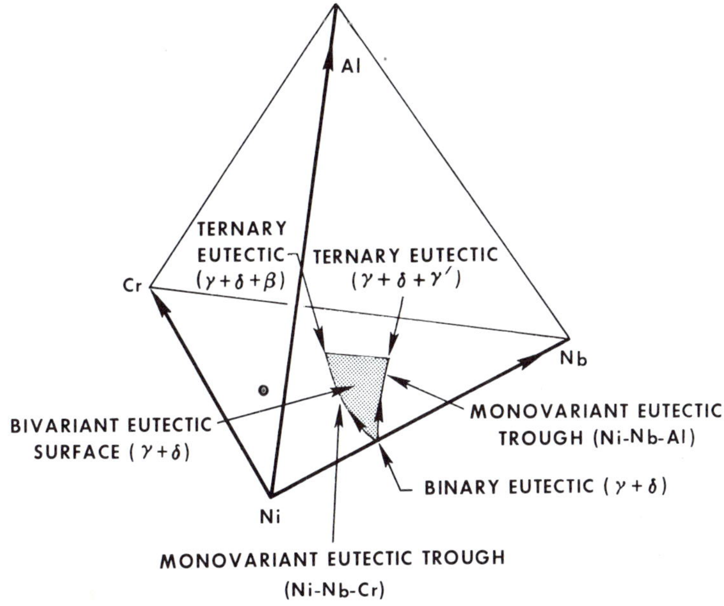

Fig. 9. Polythermal projection showing bivariant eutectic surface wherein $L \rightleftharpoons \gamma + \delta$.

in terms of melting range rather than the normally used parameters of solute distribution coefficient and liquidus slope because such parameters are not definable for multicomponent alloys. Using this multivariant approach to eutectic alloy design, which was first demonstrated by Thompson and Lemkey (1972), considerable compositional latitude is provided to afford not only reasonable high-temperature strength but also a balance of other selective criteria.

III. Alloy Preparation Techniques

The laboratory-scale production of aligned eutectics of nickel and cobalt is at present part science and part art. Each experimental setup is unique to the investigator, and no standard technique has emerged. In general, however, vertical solidification (upward interface movement) techniques appear favored because more symmetrical heat losses and a quiescent melt results. Heating is normally achieved by induction or with large thermal mass resistance furnaces.

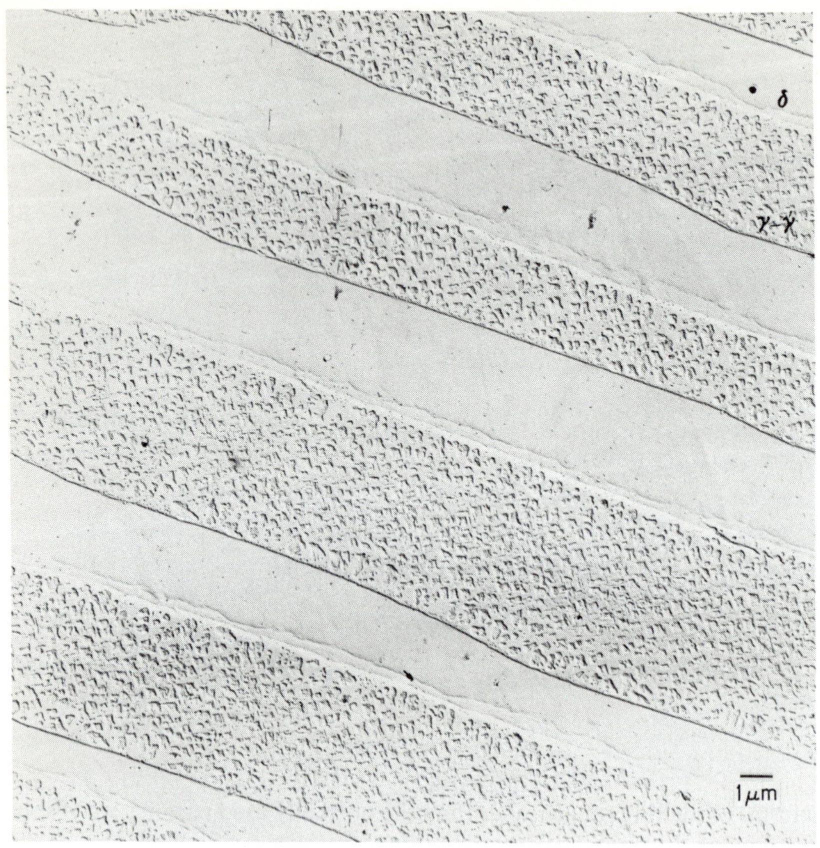

Fig. 10. $Ni_3Al(\gamma')$-precipitation-hardened and Ni_3Nb (δ)-reinforced nichrome (γ) eutectic.

A. Bridgman Technique†

Directional freezing has been normally accomplished in our laboratory by mold withdrawal from the large graphite resistance furnaces pictured in Fig. 14. Nickel and cobalt eutectics, contained in Al_2O_3 molds, have been supported in graphite sleeves or water-cooled rods on a movable cooled pedestal. Withdrawal rates between 0.5 and 100 cm/hr have been used, while melt temperatures have been typically 1600°C. This system has produced

† Bridgman (1925).

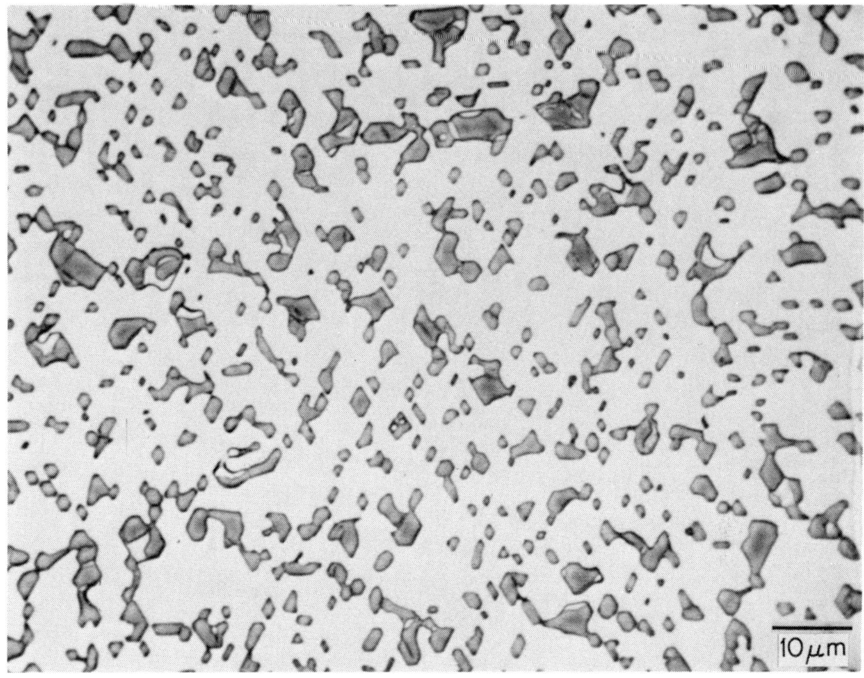

Fig. 11. Transverse microstructure of Co–35Cr–3.2Ni–2.5Al–2.15C.

thermal gradients in the liquid during directional freezing in the range from 50 to 70°C/cm. The temperature of these furnaces is controlled automatically by a two-color (ratio) optical pyrometer whose electronic servo loop monitors a 35-kva power supply regulated by a saturable-core reactor. The thermal mass of these resistance furnaces has provided the stability required for the uniform motion of the liquid–solid interface.

Vertical Bridgman controlled-growth has also been accomplished at United Aircraft in a high-gradient apparatus shown schematically in Fig. 15. In this setup a known mass of alloy (typically 200 gm), contained in a cylindrical alumina tube nominally 1.2 cm in diameter that is either open or closed at one end, is positioned within the induction coil, water spray ring, and constant-water-level tank. With the spray on, the alloy is melted by a radiating graphite sleeve that is inductively heated. Power requirements are established by experience. Controlled freezing commences upon the withdrawal of the Al_2O_3 mold through the water spray ring. In this setup thermal gradients in the liquid of approximately 350°C/cm may be achieved. In this

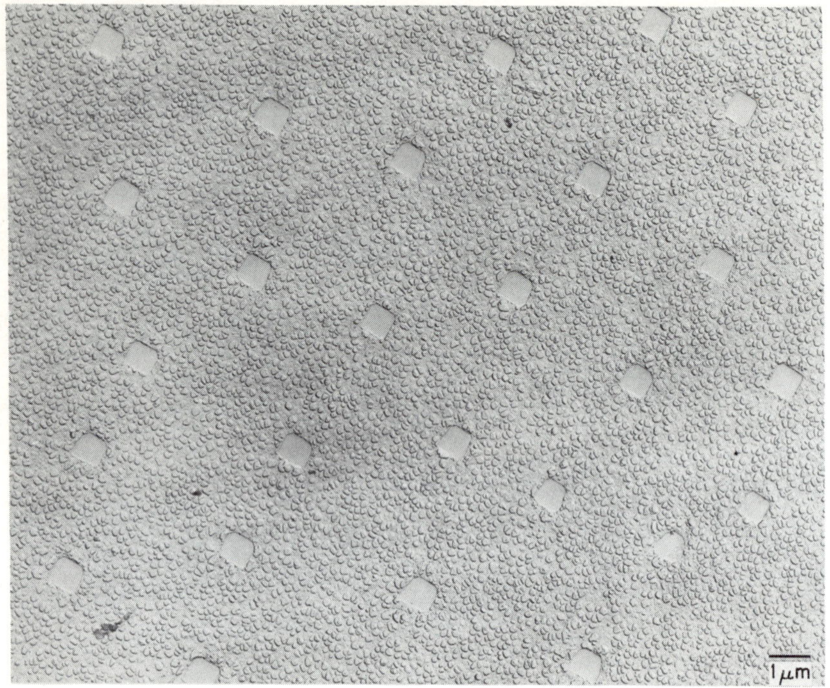

Fig. 12. Transverse microstructure of Ni–10Cr–5Al–8.9Nb–1.1C; growth rate 1 cm/hr.

technique, excluding small end-affected regions, the rate of freezing is equal to the velocity of mold withdrawal over the 15–25 cm of specimen length.

B. Zone Melting

A small molten zone may also be passed along the length of bar to achieve unidirectional growth. Both electron-beam and concentrated induction heating have been successfully applied to nickel and cobalt eutectics. No particular preference has been given for the direction of movement of the zone. The rod diameter is usually 3–6 mm, and the alloy grown is generally several inches in length.

Usually one half of the specimen—the half having the freezing solid–liquid interface—is rotated to overcome thermal asymmetry. Rotation also aids in mixing in the molten zone, causing the effective distribution coefficient to approach the equilibrium coefficient, thereby producing more effective solute segregation.

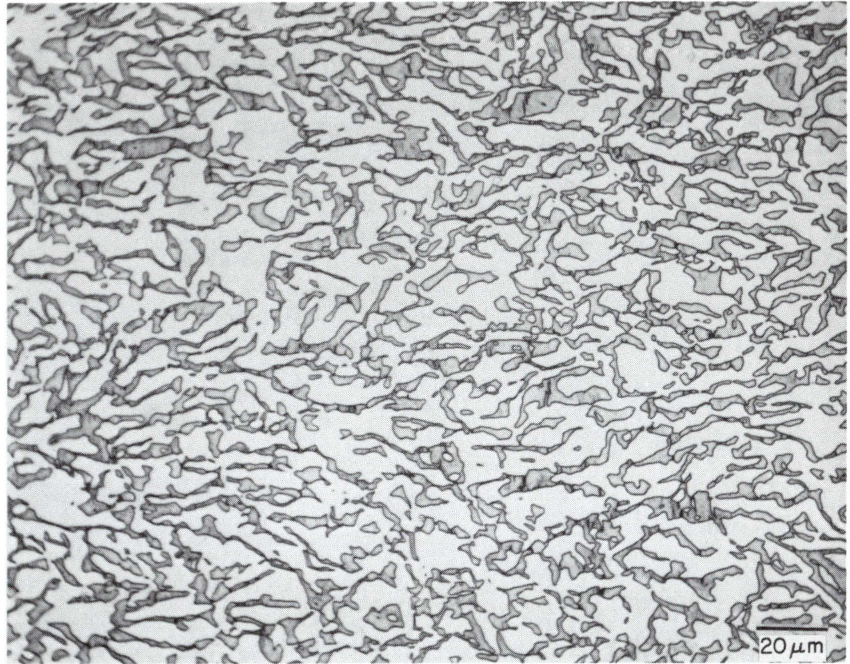

Fig. 13. Longitudinal microstructure of Co–47.0Cr–2.0Ni–1.0Al–1.8C [(Co,Cr,Ni,Al)–$(Cr,Co,Ni)_{23}C_6$] directionally solidified at 2 cm/hr.

In induction heating, good temperature control is achieved by keeping the power to the system constant rather than using temperature controllers. Constant power to the system is maintained by placing a constant-voltage transformer in the supply circuit of the electronic generators or by keeping constant the voltage produced by a motor–generator set.

Induction zone melting within a mold has been used by Bibring *et al.* (1972) in the production of alloys of nickel and cobalt reinforced by aligned refractory monocarbides. Gradients in the liquid as high as 160°C/cm have been reported with this technique. Steeper thermal gradients in the range of 500°C/cm have been obtained with electron-beam float zone melting.

IV. Mechanical Characteristics

The properties of eutectics have indicated that their behavior is generally that of a matrix containing dispersed second-phase particles, elongated in

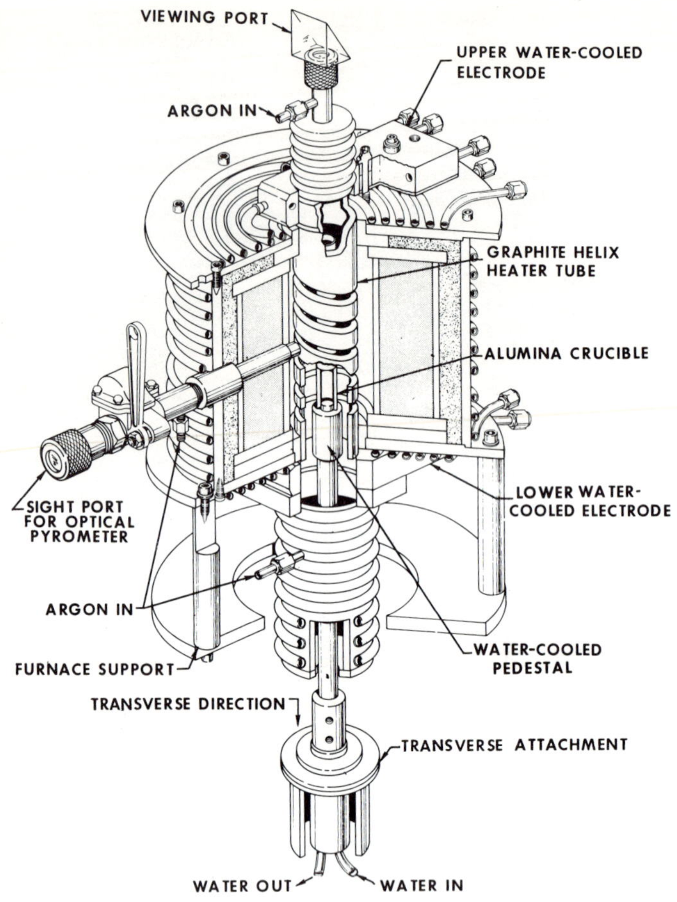

Fig. 14. Graphite directional-solidification furnace.

the direction of growth, with near-infinite aspect ratios. In the following dis-discussions, this will generally be assumed, although justification may be given in certain instances. The fact that the dispersed phase within a eutectic grain is virtually continuous in the direction of growth is acceptable from a solidification viewpoint if allowances for fiber–lamella terminations and branching are made.

A. Elastic Properties

The columnar grain of a directionally solidified eutectic consists of interpenetrating single crystals of the eutectic phases, usually exhibiting

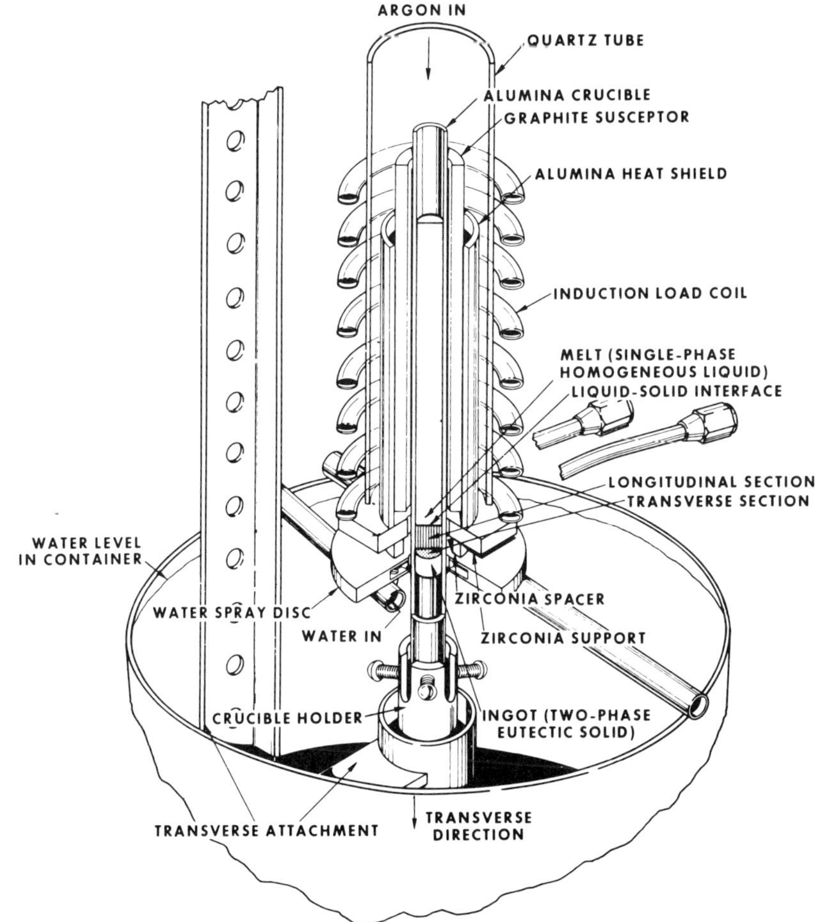

Fig. 15. High-thermal-gradient directional-solidification apparatus.

specific growth directions and crystallographic relationships. Since the microstructure of most eutectic composites is an assemblage of these columnar grains randomly oriented about the growth direction, the material can be considered as inhomogeneous but elastically isotropic in the transverse plane. The problem of specifying elastic constants is similar to that for the case of directionally solidified superalloys, as treated by Wells (1967).

The elastic properties of a eutectic composite may be expressed in terms of the same five constants, which are independent in the artificial composites, namely:

E_{11}: Young's modulus parallel to the direction of phase alignment;
E_{22}: the transverse Young's modulus;
G_{12}: the first shear modulus;
ν_{12}: the Poisson ratio of transverse contraction to longitudinal extension;
ν_{23}: the Poisson ratio in tension normal to the phase alignment.

Of these constants, the longitudinal Young's modulus E_{11} is most often measured; using the rule-of-mixtures equation, the contribution of each phase may be understood. Values for this elastic constant are included for the various eutectic composites in Table I.

As the elastic properties of the individual phases are examined to assess their relative contribution to the composite elastic coefficients, the elastic anisotropy of the phases will be of concern. Unfortunately, the elastic coefficients of the phases are often unknown.

B. Tensile Properties

1. Fibrous Eutectics

A. COMPOSITE STRENGTHENING. The longitudinal tensile properties of eutectic composites may be best understood by comparing the observed tensile behavior with that predicted from a simple rule-of-mixtures approach, i.e.,

$$\sigma_c = V_m\sigma_m + (1 - V_m)\sigma_r \tag{3}$$

where an isostrain assumption is made, V_m is the matrix volume fraction, σ_m is the matrix stress at a specific value of strain, and σ_r is the reinforcing-phase stress at this level of strain.

The room-temperature stress–strain behavior of a representative fiber-reinforced eutectic alloy from the study by Lemkey and Thompson (1971) is plotted in Fig. 16. The microstructure of the eutectic composite consists of 11 v/o of niobium carbide within a nickel matrix. In Fig. 16, the stress–strain curve of an alloy approximating the matrix phase is also given so that the degree of reinforcement is made apparent and so a comparison may be made with a stress predicted by Eq. (3). The fibrous carbide has been found to exhibit whiskerlike properties and to be elastic to failure at room temperature. At a composite strain of 1.5%, the weighted summation of matrix and fiber ($0.015E_f$) stresses predicts a composite stress of 680 MN/m^2, which agrees closely with the observed stress and demonstrates the usefulness of Eq. (3) for estimations.

The ductile behavior exhibited by the Ni–NbC eutectic composite is

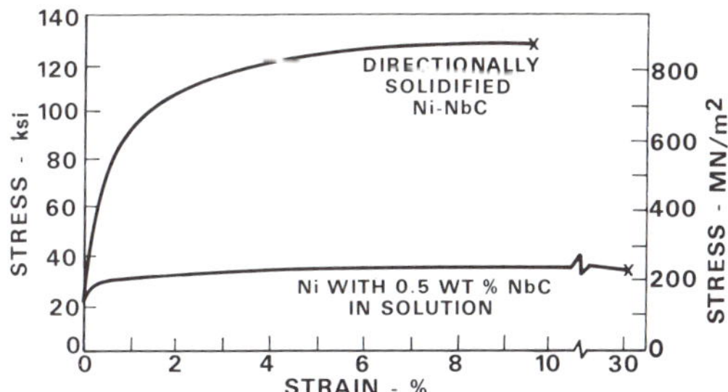

Fig. 16. Room-temperature stress–strain behavior of the Ni–NbC eutectic composite and an alloy approximating the eutectic matrix phase.

atypical of fibrous composites in general. It has been found to be a characteristic of the rather low-volume-fraction, refractory-monocarbide-reinforced alloys with nickel-, cobalt-, and iron-based matrices, however. In these systems, the fibers begin to fail at elastic strains of approximately 2 %, but the matrix surrounding the fibers is sufficiently tough and is present in sufficient volume to prevent the propagation of the crack through the matrix or the stress overload of adjacent fibers. The carbide fibers progressively decrease in length, leading to the segmented fibers shown in Fig. 17.

In Fig. 18 are plotted the room-temperature stress–strain curves of two monovariant eutectic alloys that consist of fibrous $(Cr,Ni)_7C_3$ within nickel–chromium solid-solution matrices. The microstructures of these directionally solidified eutectics are similar to that of the complex, M_7C_3-reinforced alloy shown in Fig. 11, except that the carbide volume fractions are less. The alloy containing 40 w/o chromium has a matrix with more chromium in solution (~ 30 % compared to ~ 25 %) and a lower carbide volume fraction. Although the alloys are compositionally and microstructurally similar, their properties are very different. Microstructures of sections from the fractured specimens are displayed in Fig. 19. In one case the fibrous phase is found to exhibit multiple fractures, while in the second, fiber fractures are relatively infrequent.

If a comparison is made at a strain of 1 % between the stresses observed in the higher-strength eutectic composite of Fig. 18 and that anticipated from the rule of mixtures for continuous elastic fibers in a ductile matrix, the observed strength is appreciably less. The reason for the discrepancy in this system, which was absent in the niobium-carbide-reinforced eutectic, is traced to the wider variation and lower values of fracture strength of the carbides. This reasoning is also useful in understanding the lower strength and ductile behavior of the alloy containing 40 w/o chromium. The presence of the fibers can effect a high rate of matrix work-hardening so that a matrix

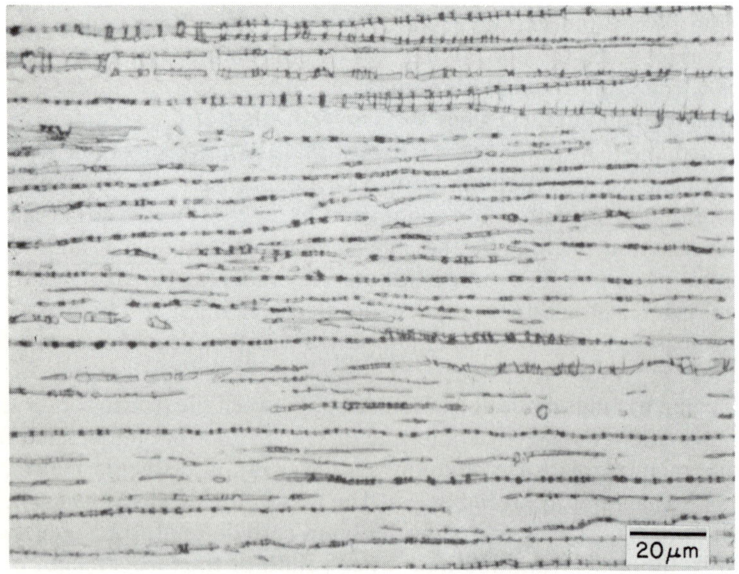

Fig. 17. Whisker breakup in Ni–NbC at room temperature.

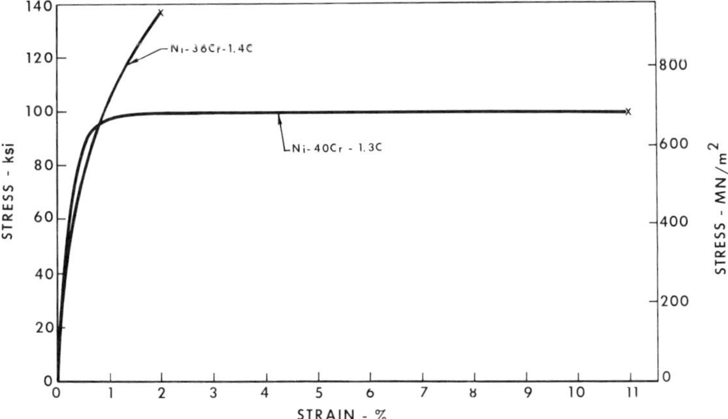

Fig. 18. Room-temperature stress–strain curve for (Ni,Cr)–$(Cr,Ni)_7C_3$ composites.

strength of approximately 690 MN/m^2 may be expected at the point of composite yield. Since the fibers begin failing at strains on the order of 0.5 %, the composite stress–strain curve approaches a region of no work-hardening. Fiber breakup continues as the composite is deformed to fracture, but true fiber strengthening is not realized because of the high matrix strength and the low effective strength and volume fraction of the carbide.

B. MATRIX STRENGTHENING. By suitable alloying, eutectics can be developed that have matrices hardened significantly by solid-solution effects or by phases that are precipitated in the solid state. Although this has been accomplished in both types of alloy discussed above, a very interesting demonstration of combining matrix strengthening with fiber reinforcing has been given by Bibring *et al.* (1972) in their study of tantalum-carbide-reinforced alloys. As shown in Fig. 20, the tensile properties of a heat-treatable Ni–20Co–10Cr–3Al–TaC eutectic composite (whose microstructure is similar to the alloy pictured in Fig. 12) have been examined after a solutioning treatment and after an aging treatment. The aging promotes the precipitation of an ordered γ' phase, based on Ni_3Al. This is the same approach that is commonly used to harden nickel-based superalloys. The composite, following aging, is stronger by 138 MN/m^2. Thus, a successful combination of fiber and precipitation strengthening has been accomplished and results in a fiber-reinforced, nickel-based superalloy.

2. *Lamellar Eutectics*

A. COMPOSITE STRENGTHENING. As in the case of fibrous eutectics, the presence of a stiffer and more elastic phase aligned within a ductile matrix

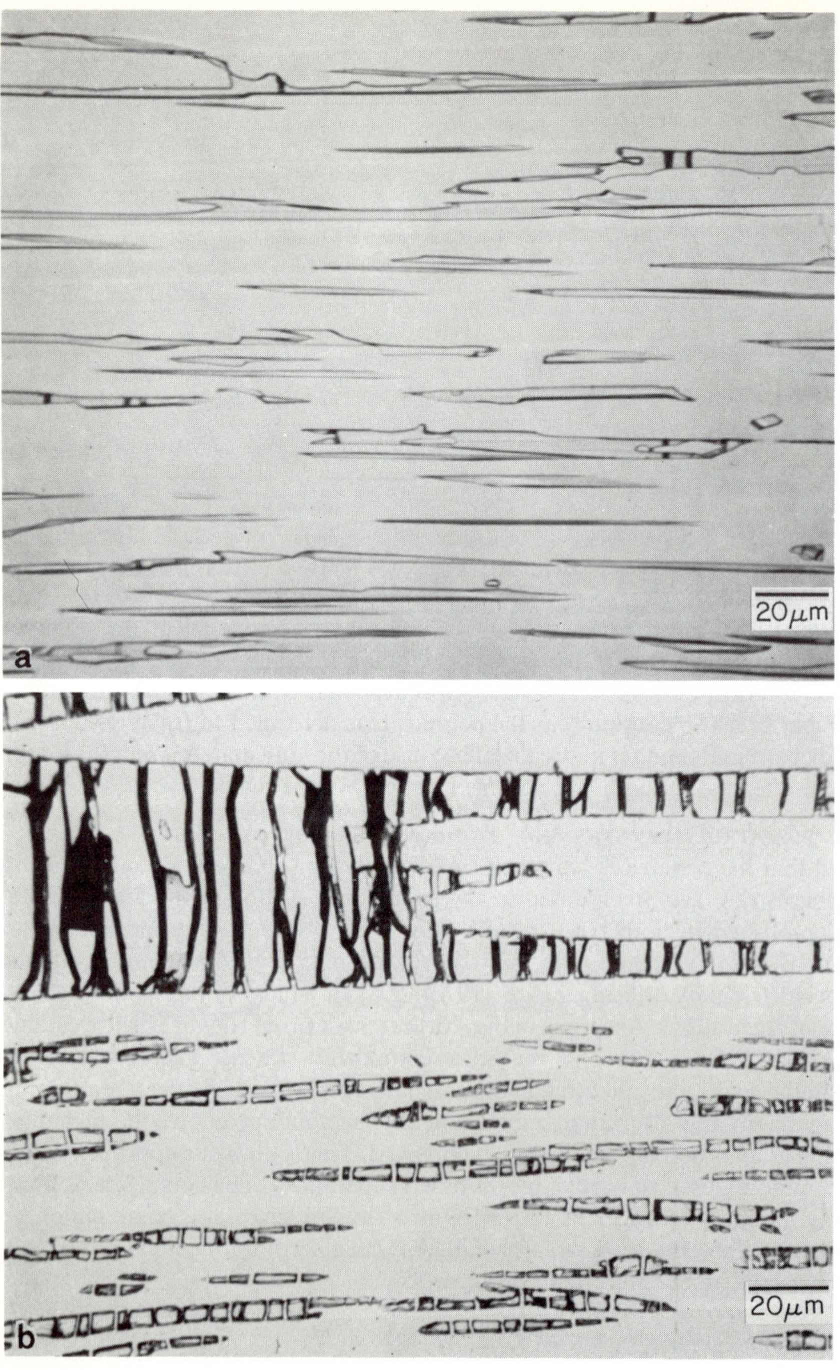

Fig. 19. Microstructures of (Ni,Cr)–$(Cr,Ni)_7C_3$ monovariant eutectic alloys in regions removed from tensile fracture. (a) Ni–36Cr–1.4C; (b) Ni–40Cr–1.3C.

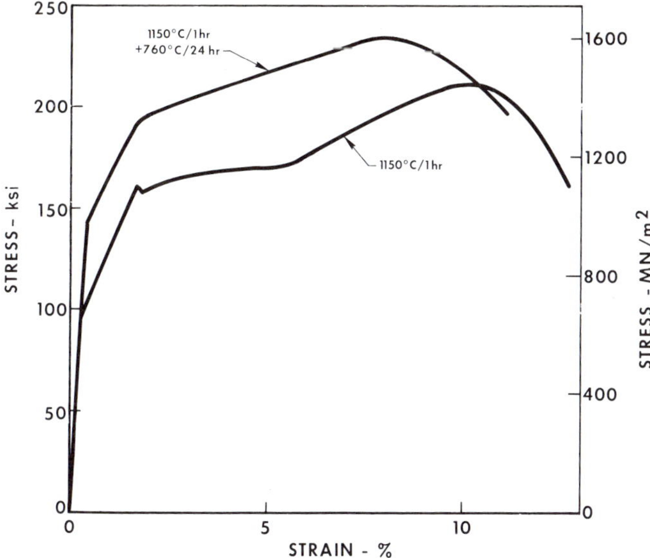

Fig. 20. Room-temperature stress–strain behavior of heat-treatable Ni–20Co–10Cr–3Al–TaC eutectic composites (after Bibring *et al.*, 1972).

permits us to consider the strength of the lamellar eutectics in light of the rule-of-mixtures prediction.

The room-temperature stress–strain behaviors of two trinickel-niobium-reinforced lamellar eutectics are plotted as a function of solidification rate in Figs. 21 and 22. In these eutectics, namely, Ni(γ)–Ni_3Nb(δ) and Ni_3Al(γ')–Ni_3Nb(δ), the volume fractions of the δ phase are 32 and 44 v/o, respectively, and the growth orientation of the compound is the same. The compound does not act as a fully elastic phase in either composite, but instead displays a twinning mode of deformation, as shown by Hoover and Hertzberg (1971a) and Thompson *et al.* (1970a). The microstructure of the Ni_3Al–Ni_3Nb eutectic composite grown at 0.5 cm/hr and fractured in tension at room temperature is displayed in Fig. 23. The δ phase is twinned in the section away from the fracture surface and contains twins and cracks on twin boundaries in the region near the fracture surface.

Although composite strengthening is shown in the present examples of Figs. 21 and 22, as was the case in the lamellar eutectic studies of Kossowsky *et al.* (1969) and Cline and Stein (1969) wherein the eutectic phases were ductile, an understanding of their tensile properties requires more than the simple rule-of-mixtures theory. This need is made obvious when the effect of an increased solidification rate (which decreases lamellar spacing) on

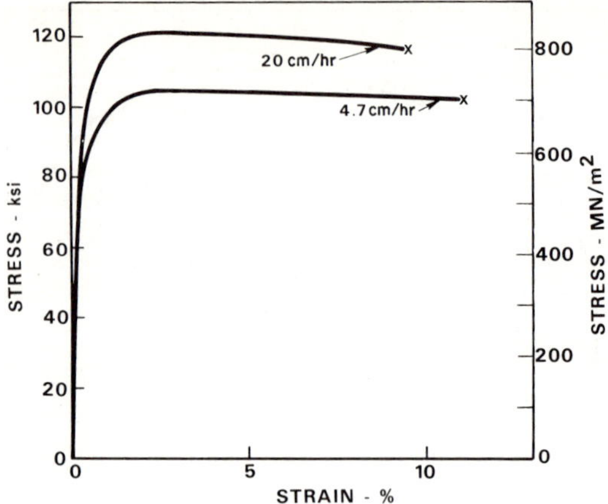

Fig. 21. Room-temperature stress–strain behavior of Ni–Ni_3Nb eutectic composites.

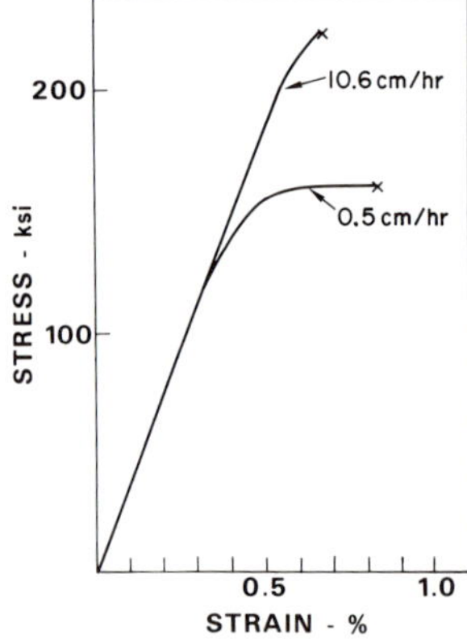

Fig. 22. Room-temperature stress–strain behavior of Ni_3Al–Ni_3Nb eutectic composites.

tensile characteristics is considered. The changes in strength of the alloys have been brought about without modifying either the volume fraction or strength of the eutectic phases.

B. MATRIX STRENGTHENING. Alloying of the lamellar eutectics can be accomplished in like manner to that of the fibrous eutectics, and may be attempted as a means of strengthening the lamellar reinforcing phase as well as the matrix. At present, most attempts have been aimed at affecting the matrix of the eutectic composite, however; an example similar to that cited for the fibrous eutectics will be given.

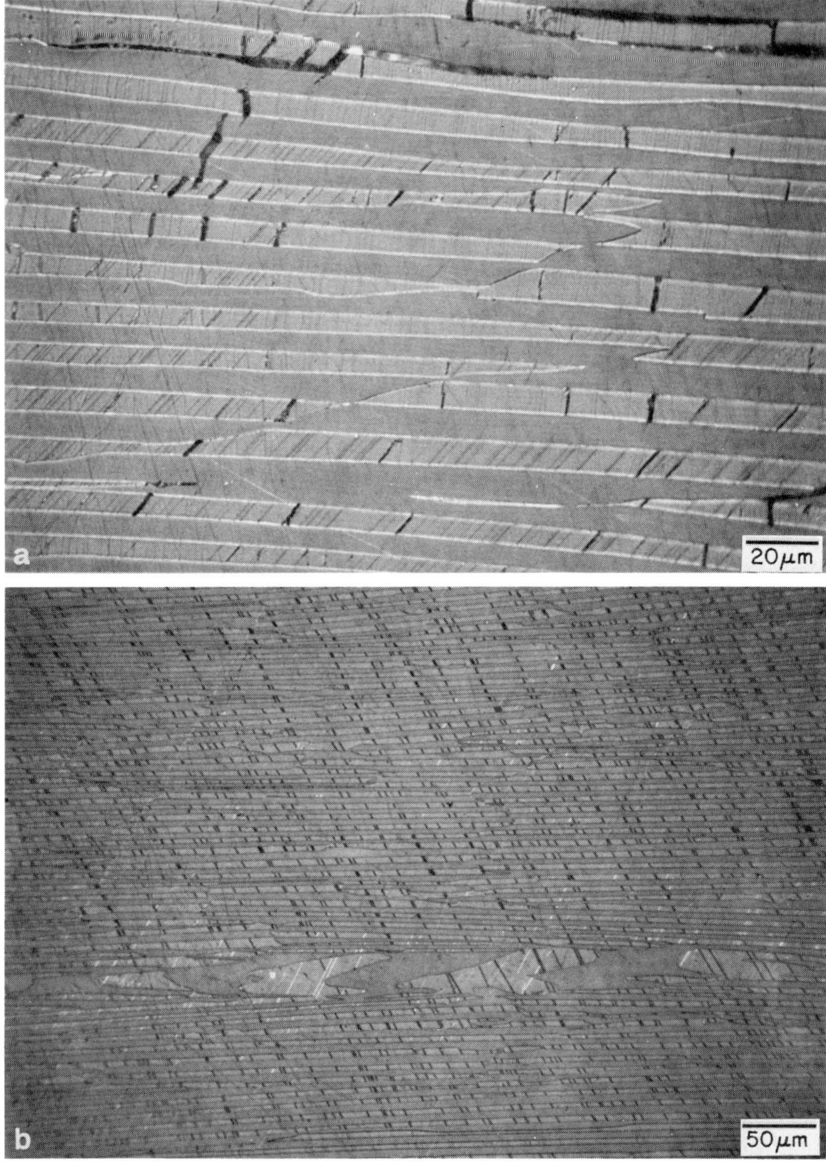

Fig. 23. Microstructure of Ni_3Al–Ni_3Nb eutectic (0.5 cm/hr) fractured in tension at 24°C. (a) Near fracture surface, (b) gauge section far removed from fracture.

As shown earlier (Fig. 21), the ultimate strength of the Ni(γ)–Ni_3Nb(δ) eutectic grown at 4.7 cm/hr is approximately 725 MN/m^2. In Fig. 24 is shown the stress–strain behavior of this eutectic modified by an addition of 2.5 w/o aluminum. The alloy retains a high ductility and exhibits a strength of almost 1100 MN/m^2. The presence of this amount of aluminum causes the precipitation of a phase based on Ni_3Al within the nickel matrix as the ingot is cooled from its solidification temperature. As was the case with the fibrous-eutectic example, this demonstrates the successful combination of composite and precipitation strengthening.

C. INTERLAMELLAR SPACING INFLUENCE. Lamellar eutectics have been shown by Shaw (1967), Cline and Stein (1968), and Thompson *et al.* (1970a) to be strengthened as the interlamellar spacing is reduced. This effect is responsible for the differences in the stress–strain curves of the eutectic composites of Figs. 21 and 22. Fibrous eutectics are also subject to strengthening by microstructural refinement, but in tension the effects reported have not been as marked as in the lamellar composites.

In the lamellar systems, the yield strength has been successfully correlated with the spacing according to a Hall–Petch equation,

$$\sigma = \sigma^* + k\lambda^{-1/2}$$

where $\sigma^*$is a frictional stress, k is a constant, and λ is the spacing. Eutectoid steel, consisting of lamellae of ferrite and cementite, has also been found by Embury *et al.* (1966) to follow this relation and to exhibit remarkable strength as a consequence of pearlite refinement. In the case of directionally solidified

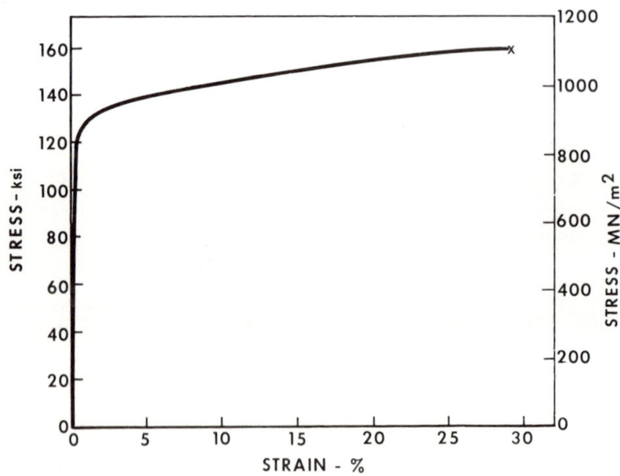

Fig. 24. Room-temperature stress–strain behavior of Ni-2.5Al–21.5Nb eutectic composite.

eutectics, however, the microstructural stability and the columnar grains permit such strengthening to contribute to high- as well as low-temperature strength.

An example of high-temperature strengthening accomplished in a eutectic composite by a reduction of the interlamellar spacing is given in Fig. 25. By reducing the spacing in the Ni_3Al–Ni_3Nb eutectic by a factor of 5, the 1093°C tensile strength, measured at a strain rate of 0.01 min^{-1}, has been doubled.

3. *Comparisons*

In terms of their possible engineering usefulness, the properties of eutectic superalloys at high temperature are more important than the ambient properties. In Fig. 26 the temperature dependences of the ultimate strengths of the fibrous eutectic superalloys are compared with one another as well as two high-strength, cast nickel-based superalloys. The comparison superalloys are directionally solidified Mar-M200 and conventionally cast TRW–NASA VIA. In the temperature range from 760 to 980°C, the superalloys have higher tensile strengths than any of the fiber-reinforced eutectic composites yet developed. Above 1010°C, the eutectic consisting of Ni_3Al fibers within a Ni_3Ta matrix (which is the stronger phase) has superior strength.

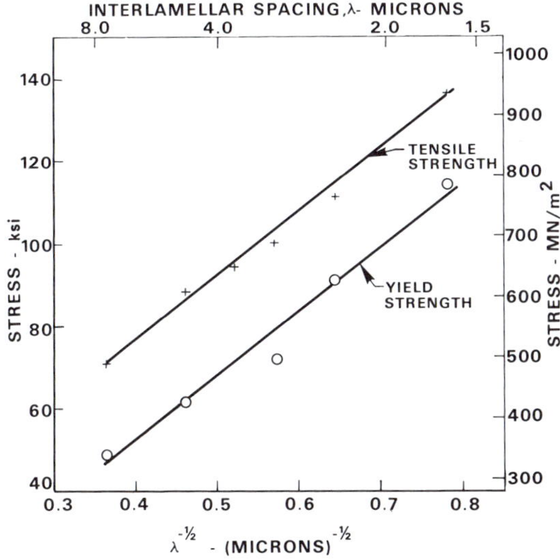

Fig. 25. 1093°C yield and tensile strengths of Ni_3Al–Ni_3Nb eutectic as a function of (interlamellar spacing)$^{-1/2}$.

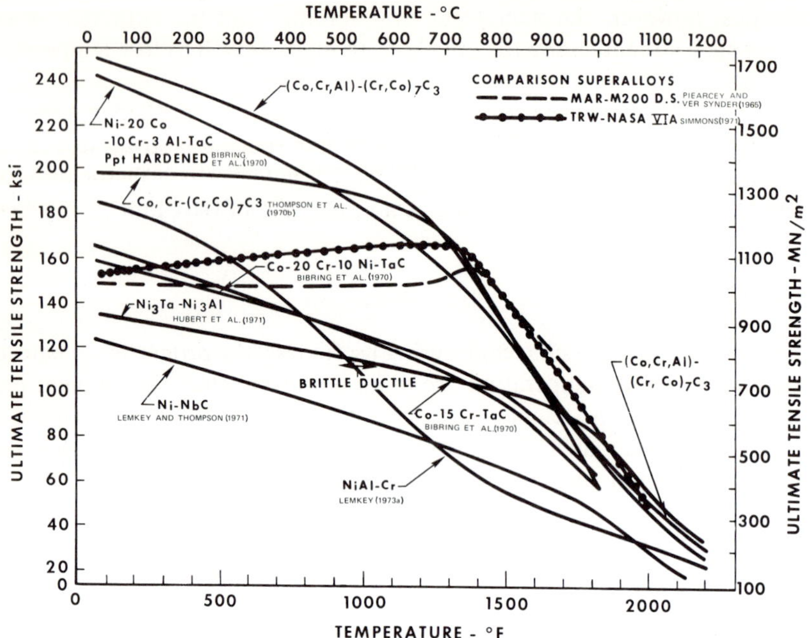

Fig. 26. Temperature dependence of tensile strengths of fibrous eutectics.

The strengths of the nickel-based superalloys are again used for comparison in Fig. 27 where the ultimate tensile strengths of the lamellar eutectics are plotted as a function of temperature. In the temperature region of 760°C where the γ' strengthening is so effective, the strengths of the superalloys are shown to at least equal those of the eutectics. Above 816°C, the strength of the Ni_3Al–Ni_3Nb eutectic is greater, and its superiority increases with increasing temperature.

The phase-reinforced eutectic superalloys are expected to maintain their strengths much closer to their melting points than conventional superalloys, which lose their strengths because of re-solutioning of the phases responsible for hardening. Furthermore, as an assemblage of columnar grains, at high temperatures the directionally solidified eutectic superalloys benefit not only from the aligned phases but also the absence of transverse grain boundaries, which would otherwise weaken the material.

C. Creep Properties

In a creep environment, microstructural perfection of the eutectic composites assumes an even greater importance than in tension. Defects that are

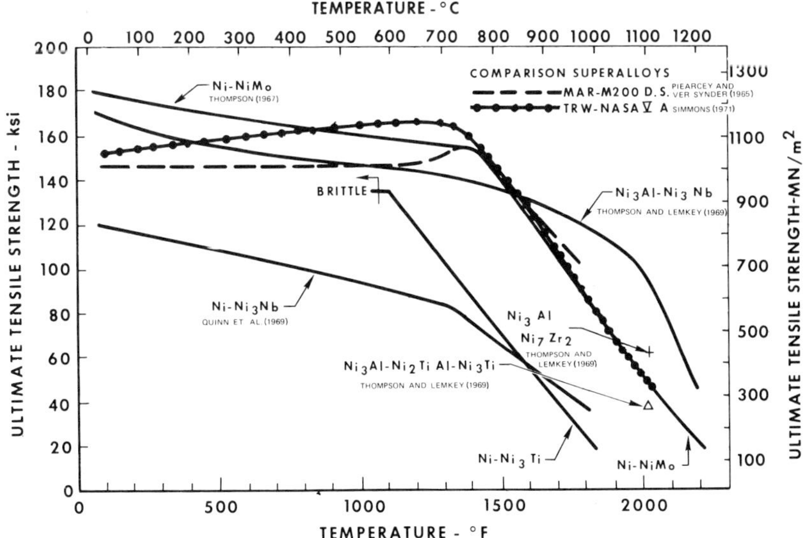

Fig. 27. Temperature dependence of tensile strengths of lamellar eutectics.

of no consequence in tension can be the source of premature creep-rupture, as shown by Breinan *et al.* (1972a).

Nevertheless, the anticipation of good creep resistance of the high-temperature eutectic composites, as suggested by their thermally stable microstructures and retention of tensile strength at high temperatures, has been borne out in numerous systems.

1. Fibrous Eutectics

The most comprehensive evaluation of the creep behavior of a fibrous eutectic superalloy has been made of the $(Cr,Co)_7C_3$-reinforced (Co,Cr) eutectic alloy by Thompson *et al.* (1970b). This alloy does not represent the most creep-resistant fibrous eutectic that has been studied, however. These are the monocarbide-reinforced alloys with complex matrices and the "inside-out" alloy Ni_3Ta–Ni_3Al (in which the matrix is the more creep-resistant phase) developed by Hubert *et al.* (1971).

A similar explanation to that proposed by Thompson *et al.* (1970b) in the paper concerning the chromium-carbide-reinforced alloy may be used to develop an understanding of the creep behavior of the monocarbide-reinforced alloys. In order to simplify, the pseudobinary NbC-reinforced nickel alloy will be considered. The 1093°C 100-hr rupture stress for this eutectic composite has been reported by Lemkey and Thompson (1971) to be approximately 55 MN/m^2. Upon loading to this stress, which is below yielding, the stress is shared between the constituents, the ratio of their stresses being equal to the ratio of their elastic moduli. At this temperature, the ratio of fiber modulus to matrix modulus is taken to be 4 to 1, and the fiber and matrix stresses are then 164 and 41 MN/m^2, respectively. The matrix cannot withstand a stress of this magnitude without rupture in a short time. The matrix therefore relaxes by creep, and stress is transferred to the carbide phase. It may be concluded that the creep resistance of the eutectic composite must be controlled by the strong fibrous phase, which functions as if it were continuous in length. This should not be interpreted to mean that the matrix is unimportant in the development of a creep-resistant eutectic composite. The more creep resistant the matrix, the less the load transferred to the reinforcing phase. Ways of strengthening the matrix of the fibrous eutectics include those mentioned in the previous discussion of tensile properties and, without compositional modification, reducing the interfiber spacing, which acts to constrain matrix flow, thereby rendering it more creep resistant, as shown by Breinan *et al.* (1972a).

2. Lamellar Eutectics

The high-temperature lamellar eutectics whose properties have been evaluated in a creep environment have generally consisted of rather high volume fractions ($>$30 v/o) of an intermetallic reinforcing phase within a less creep-resistant matrix. Similar arguments to those given above for the behavior of the fibrous eutectics are apropos. Because of the higher volume fraction, however, less stress-bearing capability is required of the reinforcing phase in this case in order to develop a material with superior creep resistance.

Two studies have been reported in which the creep process in the lamellar eutectics has been examined. In both, the steady-state creep rate $\dot{\varepsilon}$ was evaluated by the following equation:

$$\dot{\varepsilon} \approx k\sigma^n \exp(-Q/RT)$$

where k is a constant based primarily on material properties and Q is the creep-activation energy. By measurements of the creep rate at two temperatures for a material at constant stress and structure, the creep-activation energy was determined. In the case of the $Ni_3Al(\gamma')$–$Ni_3Nb(\delta)$ eutectic alloy, the activation energy for creep of the composite was found by Thompson *et al.* (1972) to be approximately 150 kcal/mole. The good comparison between this measured value and that reported by Kornilov *et al.* (1963) for single-phase Ni_3Nb (157 kcal/mole) was interpreted by the authors as an indication that the creep process in the eutectic composite was controlled by creep of the δ phase. In the second study, the Ni–Cr lamellar eutectic was reported by Kossowsky (1970a) to have an activation energy of 80 kcal/mole. It was suggested that this indicated that the steady-state creep deformation was diffusion controlled; this conclusion was reached from activation-energy estimation for diffusion in the nickel- and chromium-rich phases, namely, 70 and 75 kcal/mole, respectively. Such a measurement is also consistent with a diffusional process in the more creep-resistant chromium-rich phase being rate controlling. We prefer this interpretation.

3. Comparisons

A comparison of the rupture properties of the fiber-reinforced eutectic composites is made in the Larson–Miller parametric plot of Fig. 28. In the figure are also included comparison curves for two high-strength nickel-based superalloys, directionally solidified MAR M-200 and TRW–NASA VIA. The nickel-based superalloys are superior at the lower temperatures, corresponding to Larson–Miller parameters less than 46×10^3. Above this value, the Ni_3Ta–Ni_3Al eutectic and, above 50×10^3, several carbide-reinforced alloys are more rupture resistant than the nickel-based superalloys. At a stress of 345 MN/m^2, the temperature for 1000-hr rupture would be 890 and 865°C for the Ni_3Ta–Ni_3Al eutectic composite and the superalloys,

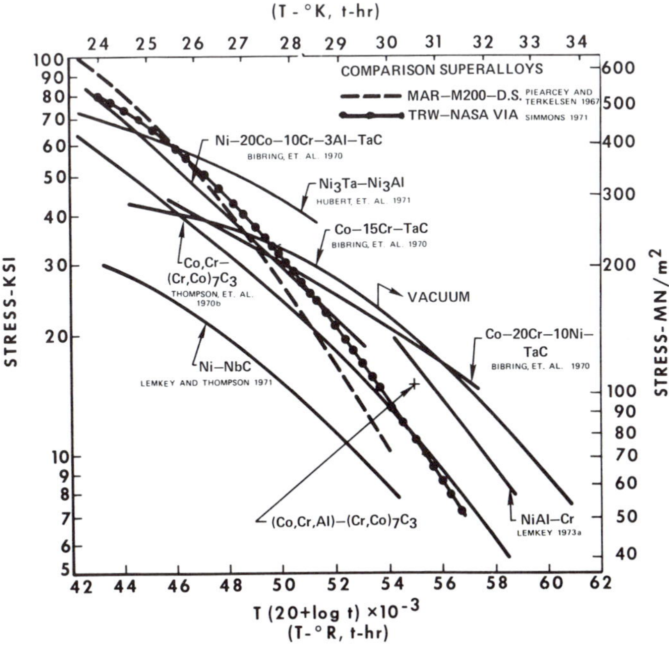

Fig. 28. Larson–Miller parameter curves for rupture of fibrous eutectics.

respectively. At a stress of 138 MN/m^2, the nickel-stabilized, tantalum-carbide-reinforced cobalt alloy would rupture in 1000 hr at 1055°C, while TRW–NASA VIA would do so at 992°C. The improvement noted in the aluminum-containing (Co,Cr)–$(Cr,Co)_7C_3$ eutectic over that without aluminum is a consequence of matrix strengthening accomplished by a faster growth rate, which produces a reduced interfiber spacing. The point given is for the alloy grown at 50 cm/hr. When the alloy is grown at 2 cm/hr, its rupture life is comparable to that shown for the (Co,Cr)–$(Cr,Co)_7C_3$ eutectic composite.

A similar comparison to that given for the fibrous eutectics is made for the lamellar eutectic composites and the superalloys in Fig. 29. The Ni_3Al–Ni_3Nb lamellar eutectic, grown at 2 cm/hr, is shown to be superior over the entire temperature range represented in this plot. At stresses of 345 and 138 MN/m^2 the temperatures for 1000-hr rupture are 945° and 1055°C. This lamellar eutectic therefore exhibits an advantage of 80° and 63°C at these stresses over the best superalloys. Furthermore, as shown by Thompson *et al.* (1972), the Ni_3Al–Ni_3Nb eutectic may be provided with additional creep resistance by plane-front growth at faster rates. For example, in this alloy grown at 4.2 cm/hr, a 345-MN/m^2 stress requires a temperature of 975°C to cause rupture in 1000 hr.

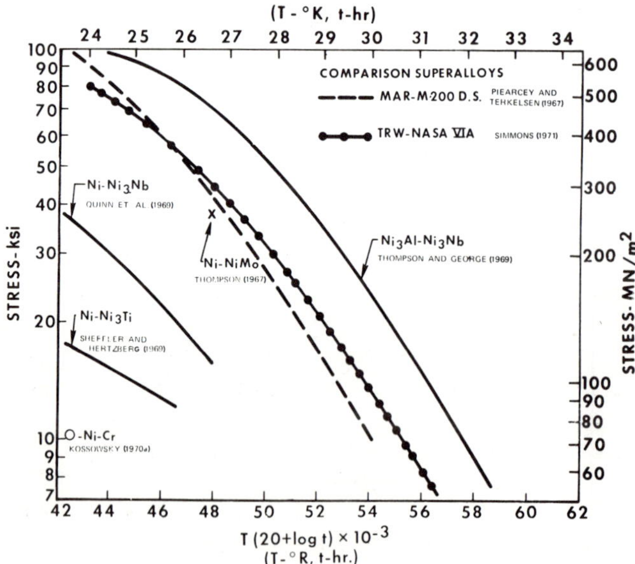

Fig. 29. Larson–Miller parameter curves for rupture of lamellar eutectics.

These eutectic superalloys are extremely attractive candidates for blades in advanced gas turbine engines because of allowable temperature increases or stress increases at a specific temperature and the attendant turbine performance improvements that have been indicated as discussed by Thompson and George (1969).

D. Fatigue Properties

In order to aid the materials designer in his development of fatigue-resistant materials, Grosskreutz (1969) has cited helpful guidelines, these being factors that produce (1) high static yield stress that is stable under cyclic loading, (2) crack-initiation suppression, and (3) reduction in crack-growth rates. These criteria may be used to enlighten observations that have been made on the fatigue properties of eutectic composites. There are indications that composites in general and eutectics in particular have characteristics that favorably influence items 1 and 3.

Eutectic composites by virtue of their well-bonded, high-modulus reinforcing phases provide a stable and high yield stress, thereby contributing to the development of a fatigue-resistant material. Once initiated, cracks moving through the microstructures of some eutectic composites may be stopped or diverted by interfacial debonding or longitudinal splitting of the reinforcing phase. A classic example of such behavior is shown in the microstructure of a Ni_3Al–Ni_3Nb eutectic specimen fatigued at ± 930 MN/m^2 (Fig. 30).

The most comprehensive study of the fatigue failure mechanism in a high-temperature eutectic composite (Ni–Ni_3Nb) has been reported by Hoover and Hertzberg (1971b). On the basis of their room-temperature evaluations, it was concluded that under high-stress, low-cycle fatigue conditions the fracture of the Ni_3Nb lamellae in advance of the crack front controlled the fatigue behavior, while the low-stress, high-cycle resistance was controlled by the matrix. This work suggests that similar behavior may be expected in other eutectic composite systems.

The endurance limits of a number of fiber- and lamella-reinforced eutectic composites are listed in Table II. For comparison, data for Udimet 700 and directionally solidified Mar-M200 are also included. The eutectic composites in general display excellent fatigue resistance, with endurance limits on the order of 60 % of their tensile strengths.

Figure 31 depicts a high-temperature fatigue failure of the Ni_3Al–Ni_3Nb lamellar eutectic. The specimen shows the periodic upheaval of the oxide on the maximum-stress surface. Oxide extends into the material beneath these areas. Failure occurred when a notch produced by the deformation and

TABLE II

Fatigue Properties of High-Temperature Materials

Material	Temperature (°C)	Endurance limit 10^7 cycles (MN/m^2)	Tensile strength (MN/m^2)	Type of test	Reference
Nickel superalloys					
U 700 (wrought)	24	345	~1380	Rotating bending	*Aerospace Structural Metals Handbook*
	704	380	~1100	Reversed bending	
	927	318	485	Reversed bending	
MAR-M200 (directionally solidified)	24	276	1000	Fluctuating tension, 34.5 MN/m^2 minimum	Gell and Leverant (1968)
	760	420	1030		Leverant and Gell (1969)
	927	276	760		Leverant and Gell (1969)
Lamellar Eutectics					
Ni–Cr	24	585	~830	Fluctuating tension	Kossowsky (1970b)
	760	310	~380		
$Ni–Ni_3Nb$	24	415	745	Notched specimen, fluctuating tension, 13.8 MN/m^2 minimum	Hoover and Hertzberg (1971b)
$Ni_3Al–Ni_3Nb$	24	690	1170	Fluctuating tension, $R = 0.1$	Thompson *et al.* (1970a)
$Ni_3Al–Ni_3Nb$	871	430	860	Reversed bending	Thompson *et al.* (1971)

Fibrous Eutectics					
Co–15Cr–TaC	24	600	1150	Rotating bending	Bibring (1972)
	800	400	680	Fluctuating tension, 20 MN/m^2 minimum	Bibring (1972)
Co–20Cr–10Ni–NbC	24	650	1090	Rotating bending	Bibring (1972)
	800	500	700	Fluctuating tension, 20 MN/m^2 minimum	Bibring (1972)
Co–20Cr–10Ni–NbC	24	650	1030	Rotating bending	Bibring (1972)
	800	500	—	Fluctuating tension, 20 MN/m^2 minimum	Bibring (1972)
Ni–20Co–10Cr–3Al–TaC	24	450	1480	Rotating bending	Bibring (1972)
	800	600	~900	Fluctuating tension, 20 MN/m^2 minimum	Bibring (1972)
(Co,Cr,Al)–$(Cr,Co)_7C_3$	24	620	1730	Fluctuating tension, $R = 0.1$	

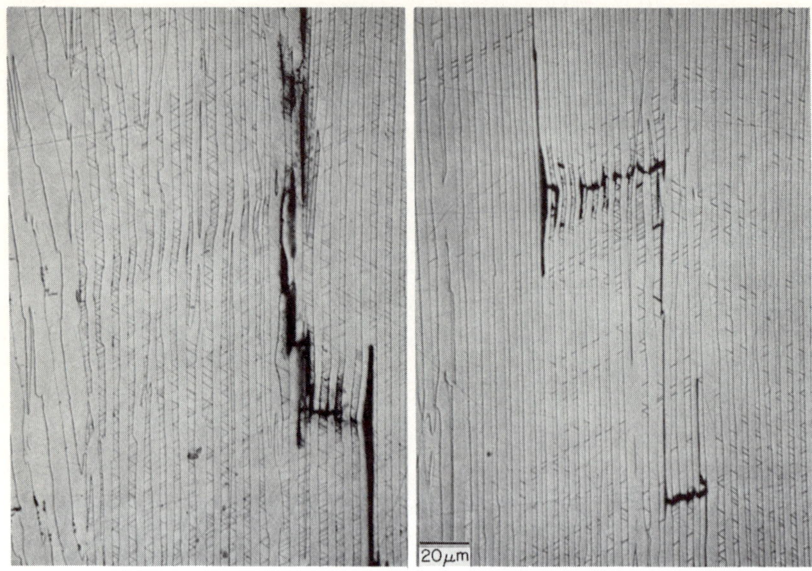

Fig. 30. Nonpropagating cracks generated in fatigue specimen of Ni_3Al–Ni_3Nb tested at ± 930 MN/m^2 (± 135 ksi).

oxidation became critical. In such a case, surface protection against oxidation should benefit the fatigue properties.

In the case of two eutectics we have studied, namely, Ni_3Al–Ni_3Nb and (Co,Cr,Al)–$(Cr,Co)_7C_3$, the ability to coax the specimens to higher fatigue strengths has been demonstrated. The mechanism for coaxing in two-phase composite materials is not yet understood, and it remains to be shown whether this is a general phenomenon.

E. Toughness

Based on the work of Cooper and Kelly (1967), as the interfiber or interlamellar spacing is reduced, we might expect a corresponding reduction in the toughness of the eutectic composites. This would result from the increased constraint of the reinforcing phase, which would act to limit the plastic-zone size in the ductile matrix. In the one case where this effect was evaluated, a reduction in toughness was found by Thompson (1971) to accompany the reduction in interfiber spacing.

In systems that have a high volume fraction of brittle phase, toughness must be a source of concern. There are means, however, to achieve useful fracture toughness in such systems if a method of crack blunting or diversion

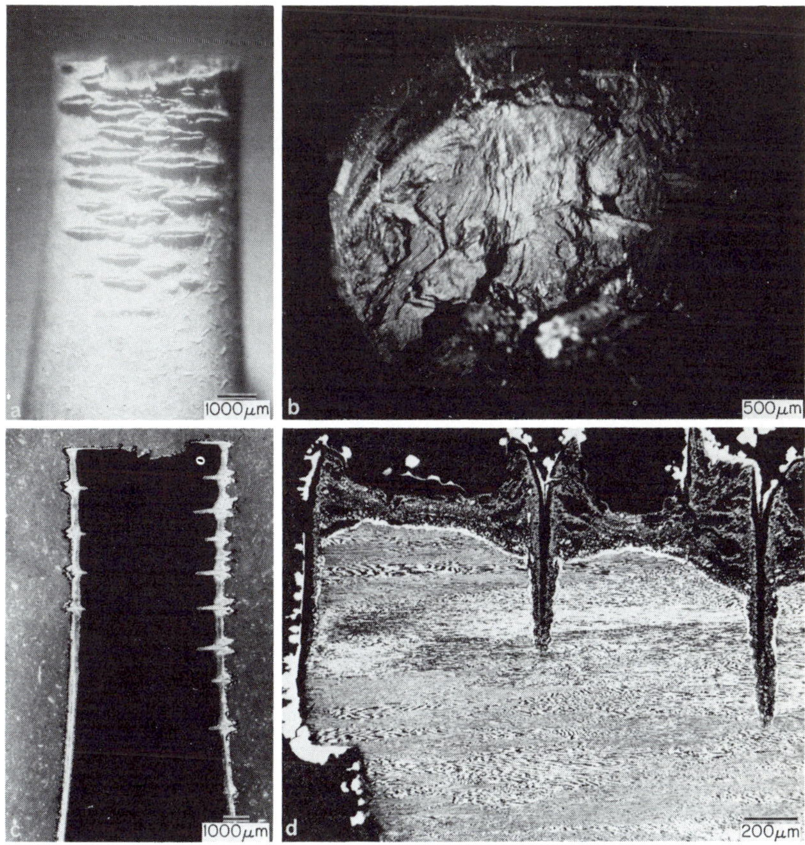

Fig. 31. Fatigue specimen fractured at 871°C, ± 404 MN/m^2 (± 58.5 ksi). (a) Specimen surface, (b) fracture surface, (c) longitudinal macrostructure, (d) longitudinal microstructure.

is available. This may arise in the case of debondable interfaces, which may open as a crack approaches. Perhaps the best example of this has been shown by Thompson and George (1969) in the Ni_3Al–Ni_3Nb lamellar eutectic. The effects of temperature on the impact toughness and the appearance of fractured specimens are shown in Fig. 32. Crack diversion and blunting have been accomplished by longitudinal splitting of lamellar interfaces and grain boundaries.

The toughness observed by Thompson (1971) in the (Co,Cr)–$(Cr,Co)_7C_3$ eutectic is believed to be reasonably representative of systems containing a high volume fraction (~ 30 v/o) of a brittle fibrous phase within a relatively ductile matrix. It was shown that the work to fracture of this system was

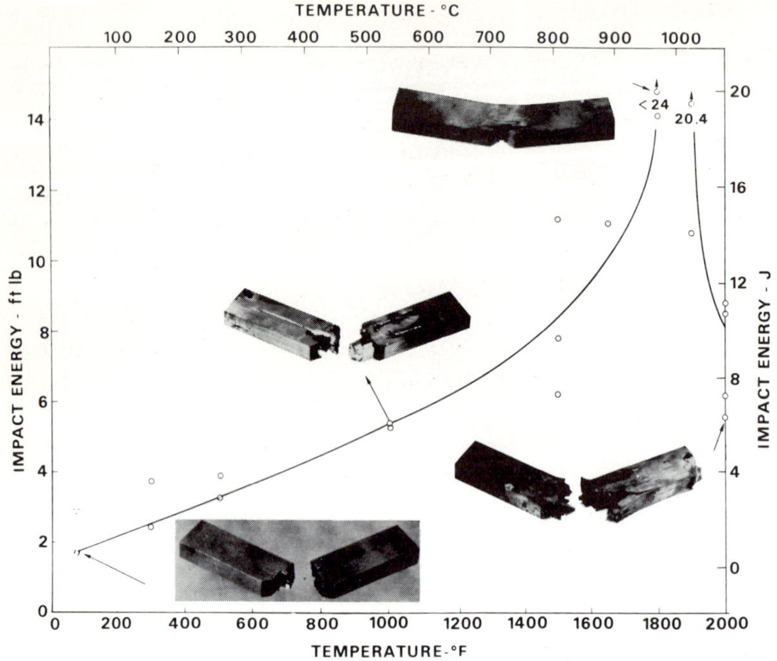

Fig. 32. Charpy impact behavior as a function of temperature. ○ Ni_3Al–Ni_3Nb, ▽ MAR-M200, □ MAR-M302.

sufficiently high to allow the material to be considered as useful for elevated-temperature applications. This indicates that as long as the high-strength fibers are dispersed within a tough matrix, the longitudinal toughness of fibrous eutectics may be structurally acceptable.

Bibring (1971) has shown that the low-volume fibrous eutectic systems ($V_F \approx 10$ v/o), consisting of the fibrous refractory monocarbides within nickel- and cobalt-based matrices, may possess exceptional low-temperature toughness. On the niobium-carbide-reinforced nickel eutectic, we have measured impact strengths of 55 and 27 J for smooth and notched Izod specimens. This greatly exceeds the impact toughness of the cast nickel-based superalloys.

F. Mechanical Anisotropy

Data concerning the nonaxial mechanical properties of the directionally solidified eutectic superalloys are meager. In this section we shall summarize these observations.

1. Tension

The (Co,Cr)–$(Cr,Co)_7C_3$ eutectic alloy has been studied by Thompson *et al.* (1970b) in tension normal and at 45 deg to the growth direction at room and elevated temperatures. The strengths of these orientations are plotted as a function of temperature and are compared with the longitudinal orientation in Fig. 33. The nonaxial strengths are considerably less than that when the fibers are parallel to the stress direction. At temperatures below 1000°C, the off-axis ductilities are adversely affected by cracks that form in the carbides and propagate along their length.

The Co–20Cr–10Ni eutectic alloy reinforced with tantalum-carbide fibers has been reported by Bibring (1972) to have the following properties. At room temperature the transverse strength is ~ 900 MN/m², which is 85% of the longitudinal strength, with a tensile elongation of 4.5%, which is 15% of the longitudinal. At 800°C, the transverse strength is ~ 345 MN/m² (50% of the longitudinal), with a transverse elongation of ~ 7%, which is similar to that longitudinally.

2. Creep-rupture

The only off-axis creep measurements of which we are aware are from our unpublished study on the $(Cr,Co)_7C_3$-fiber-reinforced (Co,Cr) eutectic alloy

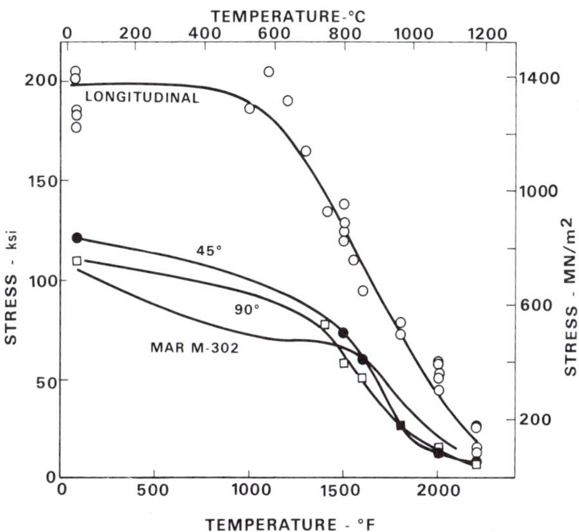

Fig. 33. Longitudinal, transverse, and 45-deg tensile strengths of (Co,Cr)–$(Cr,Co)_7C_3$ eutectic alloy. Curve for MAR-M302 is included for comparison.

in the transverse orientation. The stresses to cause rupture in 100 hr were found to be approximately 165, 103, 69, and 31 MN/m^2 at 760°, 871°, 982°, and 1093°C, respectively. The comparable longitudinal 100-hr rupture strengths are 497, 304, 172, and 90 MN/m^2. The transverse elongations to failure at these temperatures were about 20%.

3. *Fatigue*

The transversely oriented tantalum-carbide-fiber-reinforced Co–Cr–Ni alloy has been evaluated by Bibring (1972) at room temperature and 800°C. The transverse room-temperature endurance limit of approximately 450 MN/m^2 was 70% of that of the longitudinal orientation; the 800°C fatigue-endurance limit of approximately 248 MN/m^2 was 50% of the longitudinal.

4. *Toughness*

The toughness of the (Co,Cr)–$(Cr,Co)_7C_3$ eutectic has been evaluated by Thompson (1971) in the orientations shown in Fig. 34. In the B and C orientations the fracture involved the longitudinal splitting of the carbide fibers. The toughness of the three orientations increased with an increase in interfiber spacing (decreased solidification rate). The works to fracture of the A, B, and C orientations were on the order of 1.47, 0.63, and 0.36 J/cm^2, respectively. Corresponding Charpy V-notch impact strengths were 7.05, 1.35, and 1.15 J.

V. Environmental Effects

A. *Thermal Fatigue*

In the application of a material as a turbine airfoil, the designer must be concerned with the reponse of the material to fatigue arising from restrained expansion and contraction of thin sections during thermal excursions. The high longitudinal fatigue strength of the eutectic composites suggests that they should resist transverse cracking of edges, which is a common failure mode in monolithic materials. Longitudinal cracks that result from thermally induced shear stresses may occur, however, and normal thermal-fatigue sensitivity should be evaluated.

Another effect that is absent or unimportant in the thermal cycling of monolithic materials must be considered by the developer of eutectic composites, namely, the internal stress that results as a consequence of the differing thermal expansion coefficients of the eutectic phases. Those stresses may be estimated from an elastic analysis and are proportional to the product of the

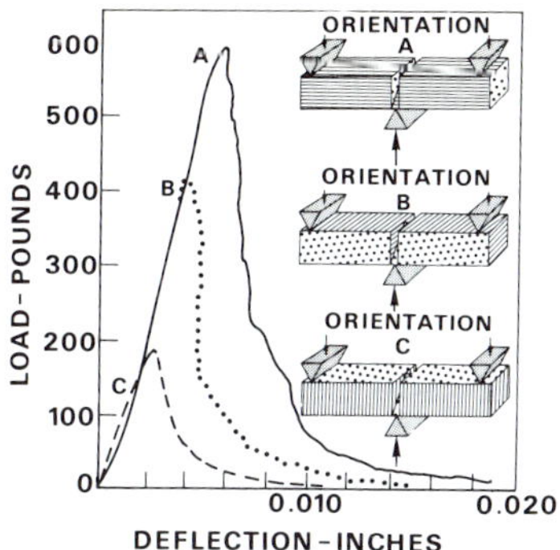

Fig. 34. Load–deflection curves for different orientations of the (Co,Cr)–$(Cr,Co)_7C_3$ eutectic alloy.

expansion mismatch and the temperature interval, $\Delta\alpha\ \Delta T$, which has been termed by Lazlo (1943) the strain potential of tessellation. The residual stress may exceed the yield stress of the ductile phase and may cause plastic strains sufficiently large to lead to cyclic thermal-fatigue damage, as discussed by Hoffman (1970). In addition to the thermal-expansion effect, phase transformations, as encountered in unstabilized iron- or cobalt-based alloys, can also affect the residual stress state.

In Table III are listed the approximate thermal expansion coefficients of some of the phases that are included in the eutectic superalloys we have considered. Following Hoffman's (1970) treatment, we shall estimate elastic

TABLE III
Thermal-Expansion Coefficients

Material	Temperature range (°C)	α (10^{-6}/°C)
Nickel and cobalt alloys	20–1100	16–19
$Ni_3(Al,Nb)$	20–1000	11.3
Ni_3Nb	20–1000	~9.7
Ni_3Ta	20–1000	~11
TaC	20–1000	6.6
NbC	20–1000	7.1
TiC	20–1000	7.7
Cr_7C_3	20–800	10.1

stresses produced in a representative fibrous and lamellar eutectic system. In the case of a 10 v/o niobium-carbide-fiber-reinforced nickel alloy, rapid cooling to room temperature from 1100°C causes a matrix stress of

$$\sigma_M = \frac{V_R E_R E_M \,\Delta\alpha\, \Delta T}{V_R E_R + V_M E_M} \approx 310 \text{ MN/m}^2$$

where the elastic moduli of matrix and fiber are taken as 207 and 345 GN/m^2, respectively, and $\Delta\alpha$ is taken as 10×10^{-6}/°C. The corresponding elastic fiber stress is about -2760 MN/m^2 (compressive). These are average stresses, and gradients of strain in the matrix are ignored.

In the case of a 44-v/o Ni_3Nb-lamella-reinforced Ni_3Al alloy, rapid cooling from 1100°C will produce a matrix stress of

$$\sigma_M = \frac{V_R E_R E_M \,\Delta\alpha\, \Delta T}{(V_R E_R + V_M E_R)\,(1 - \nu)} \approx 235 \text{ MN/m}^2$$

where the elastic moduli of the matrix and reinforcement are taken as 186 and 290 GN/m^2, respectively, and $\Delta\alpha$ is taken as 1.6×10^{-6}/°C. This is substantially below the yield strength of the matrix compound, as shown by Thornton *et al.* (1970). The corresponding Ni_3Nb stress is about -304 MN/m^2.

In the fiber-reinforced example, the matrix stress calculated is above the yield stress of some nickel alloys and indicates the possibility of matrix yielding during the quench and matrix fatigue by continued thermal cycling. By reducing the volume fraction of reinforcement and the thermal-expansion mismatch, the matrix stress will be decreased.

The fact that thermal cycling over a wide temperature range may cause damage in eutectic composite systems where an appreciable expansion mismatch is present is demonstrated pictorially in Figs. 35 and 36. In the first case the pseudobinary eutectic Ni–NbC was exposed to approximately 1800 thermal cycles between 400 and 1130°C. The test was carried out in a gas-burner rig. The transverse and longitudinal sections of the material after test show (Fig. 35) that the matrix has recrystallized and that the fibrous phase has been broken. In the second case, an alloy of Co-15Cr–NbC was thermally cycled 1500 times between 400 and 1130°C in a test with heating provided by electrical resistance. Matrix allotropy, as well as thermal-expansion mismatch, produce the thermally fatigued microstructure of Fig. 36. The carbides that were long and needlelike crystals have been broken and pushed about as the matrix has undergone repetitive transformations and formed new grains. The observation of broken carbides in these cases was unexpected because of their high strength, but serves to illustrate the large local stresses that may arise. In more complex monocarbide-reinforced alloys, Breinan

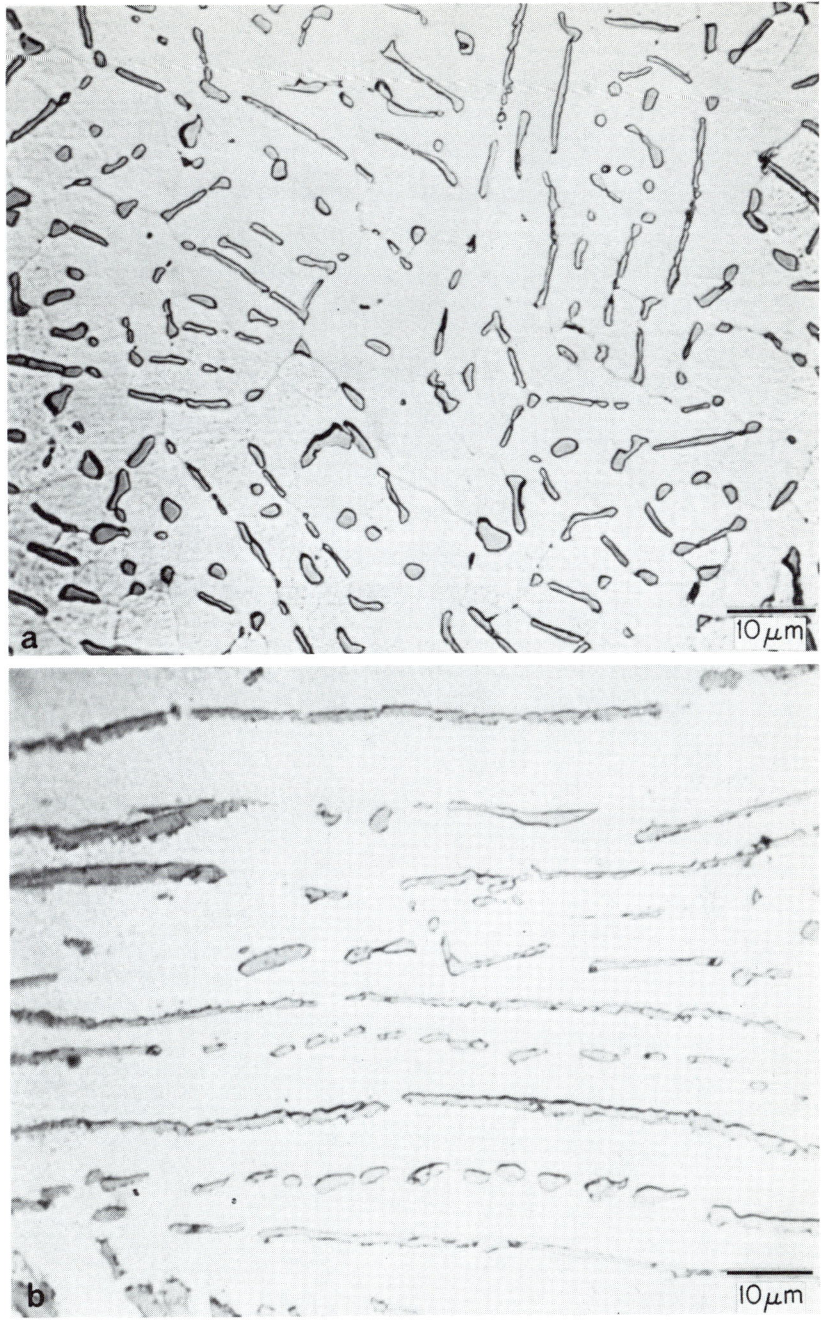

Fig. 35. Microstructure of thermally cycled Ni–NbC. (a) Transverse, (b) longitudinal.

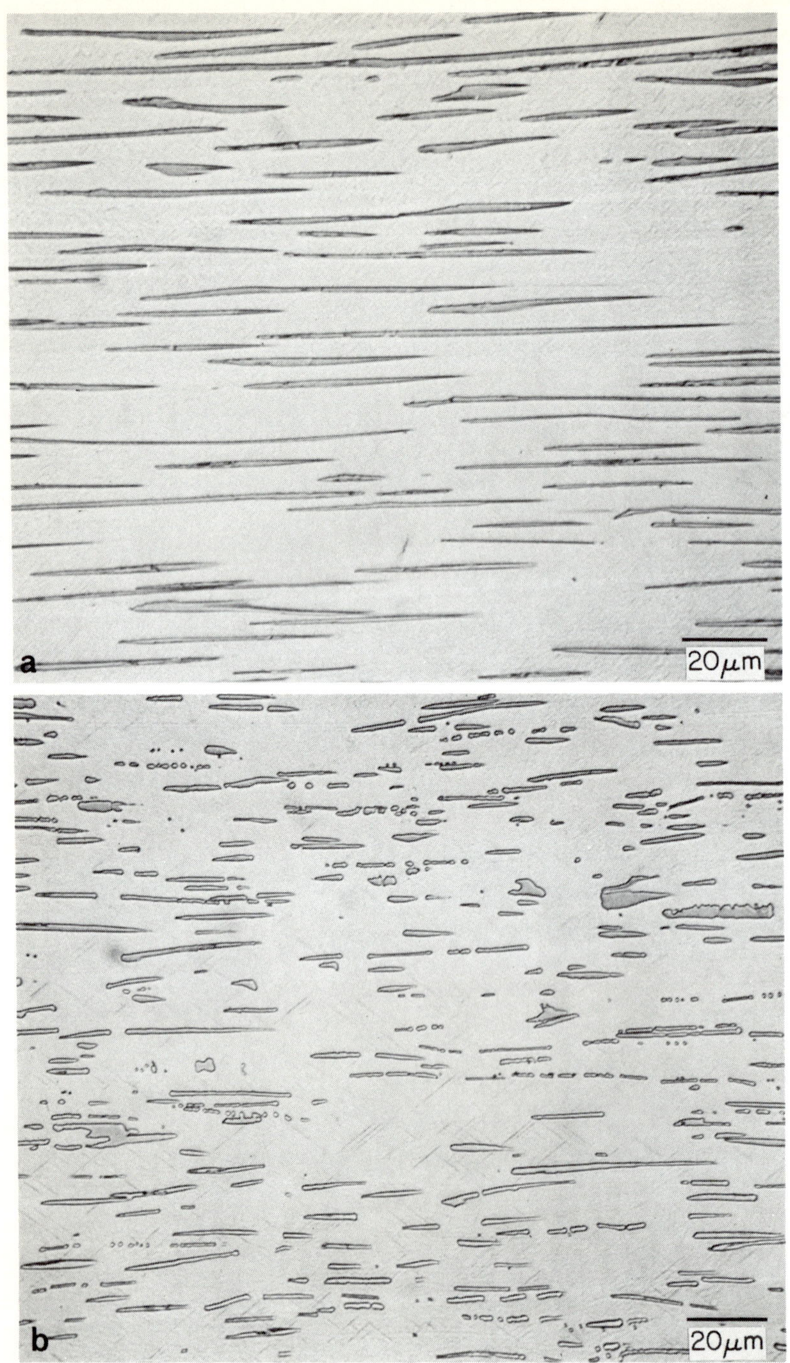

Fig. 36. Mechanical breakup of fibers in Co,Cr–NbC alloy after thermal cycling. (a) Uncycled, as directionally solidified alloy; (b) cycled, alloy after 1500 cycles, 750–2050°F.

et al. (1972b) have also shown that physicochemical instability may play an important role in the damage process.

B. Oxidation

In order to achieve maximum usefulness, the eutectic superalloys that are developed must not only exhibit superior creep, fatigue, and thermal-fatigue properties but must also be resistant to high-temperature oxidation.

The development of oxidation-resistant materials is discussed in a paper by Wallwork (1970). As suggested by this article, the achievement of oxidation resistance in eutectics will depend on the incorporation of aluminum and/or chromium as part of their alloy composition.

The oxidation data that have been reported for the high-temperature eutectic composites studied to date have generally consisted of still-air furnace tests, either static or cyclic. In Fig. 37, static 1093°C oxidation results are given for a number of the eutectic alloys that we have evaluated. In Fig. 38, the 1000°C oxidation results of Bibring *et al.* (1972) for a number of complex eutectic alloys reinforced by a monocarbide are shown. Although some of the high-chromium- and aluminum-containing eutectic alloys have shown good resistance to high-temperature oxidation, only the NiAl–Cr monovariant eutectic is sufficiently resistant so as not to require a protective coating for extended high-temperature use.

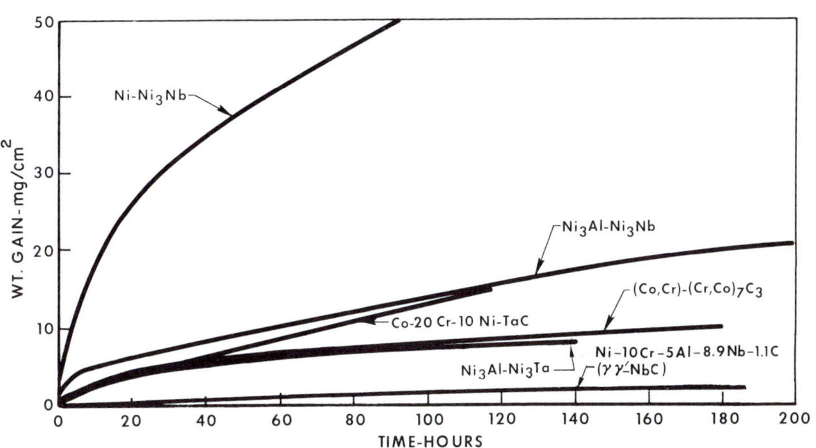

Fig. 37. Interrupted oxidation of eutectic alloys at 1093°C (Ni_3Al–Ni_3Ta after Hubert *et al.*, 1971)

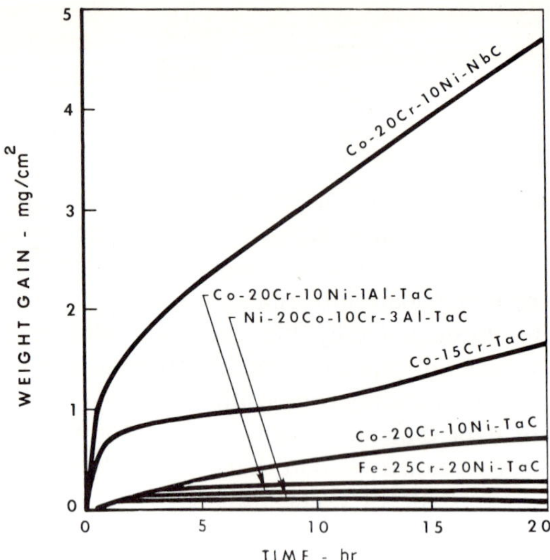

Fig. 38. Calm-air oxidation at 1000°C of various monocarbide-strengthened eutectics (after Bibring *et al.*, 1972).

The protection of superalloys for elevated-temperature use is mainly based on diffusion aluminide coatings, as discussed by Goward (1970). Under oxidizing conditions, a protective oxide of Al_2O_3 is formed. Coatings of this type have been applied to eutectic composites. In Fig. 39 is shown a β(NiAl) diffusion coating on the Ni_3Al–Ni_3Nb lamellar eutectic. Coating of the fibrous-tantalum-carbide-reinforced cobalt eutectic has been reported by Bibring *et al.* (1972) to have been successfully accomplished using a chromizing–aluminizing process. In addition to the diffusion coatings, overlay coatings have been applied to superalloys and have demonstrated superior protection. Their superiority stems from their compositions, which are essentially independent of the alloy and which are optimized to provide the required protection. The overlay coatings have been applied by vacuum-deposition processes and appear to offer a possible solution to the coating of eutectic superalloys.

VI. Manufacture of Components

A. *Directional Solidification in Investment Molds*

Until recently, superalloys have been directionally solidified as airfoil-shaped turbine components in a stationary, ceramic investment mold,

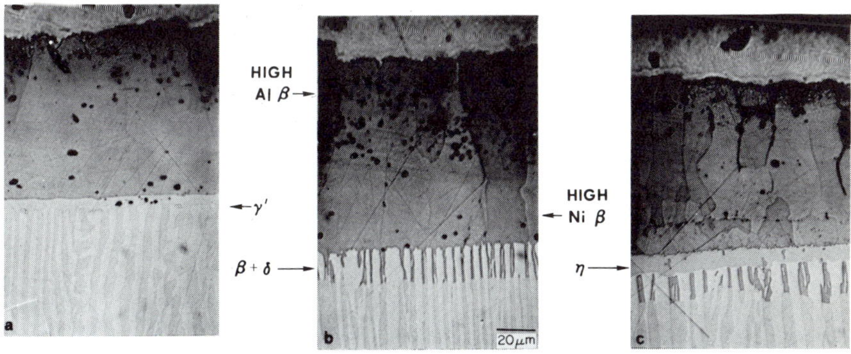

Fig. 39. Effect of (a) low, (b) medium, and (c) high aluminum content in NiAl coating on interaction with Ni_3Al–Ni_3Nb substrate.

following a power-down procedure described by VerSnyder and Shank (1970). The development by Erickson *et al.* (1971) of a mold-translation (withdrawal) method for the directional solidification of superalloys has provided an improved means of increasing the temperature gradient in the freezing liquid and the allowable solidification rate.

Because of the high G/R ratio required for the plane-front growth of most eutectics, the power-down scheme, which is a low-thermal-gradient process, is an unfavorable technique for the directional solidification of eutectic composites. Despite this, the power-down technique has been used with success in the directional solidification of hollow vanes of the (Co,Cr)–$(Cr,Co)_7C_3$ eutectic, as pictured in Fig. 40.

A Bridgman approach, which bears some resemblance to the mold-translation method, wherein a mold is lowered from a stationary furnace, has been successfully used to produce solid turbine blades of the Ni_3Al–Ni_3Nb eutectic (Fig. 41). The mold-translation method of directional solidification appears to offer a possible solution to the growth of eutectic turbine components. We should recognize, however, that further advances in processing technology will be one of the most important keys to the wide-scale applicability of these materials. An improvement in this direction has recently been reported by Gell *et al.* (1974); it involves the immersion of an investment mold into a liquid metal coolant. The technique therefore bears some similarity to the method of controlled growth in the high-gradient apparatus (Fig. 15). The thermal gradient provided by this approach may be twice that of the production withdrawal method. Eutectic airfoils have been directionally solidified in investment molds by liquid-metal immersion.

Fig. 40. Hollow turbine vane, directionally cast from (Co,Cr)–$(Cr,Co)_7C_3$ eutectic composite. (Courtesy of C. Phipps, PWA, Manchester, Connecticut.)

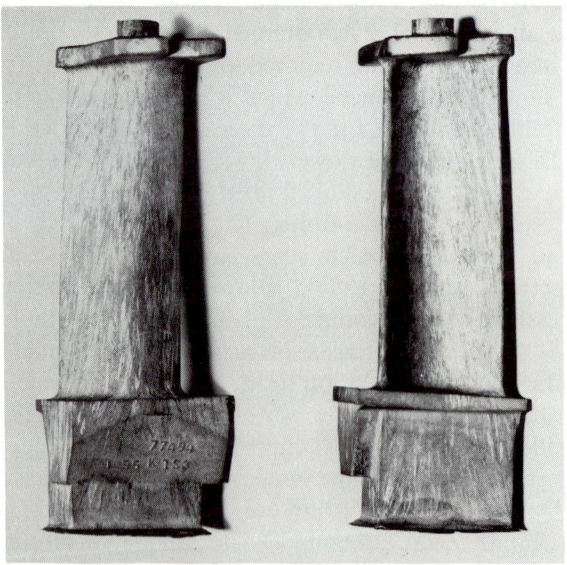

Fig. 41. Solid turbine blade, directionally cast from Ni_3Al–Ni_3Nb eutectic composite. (Courtesy of Dr. Larry Graham, TRW, Cleveland, Ohio.)

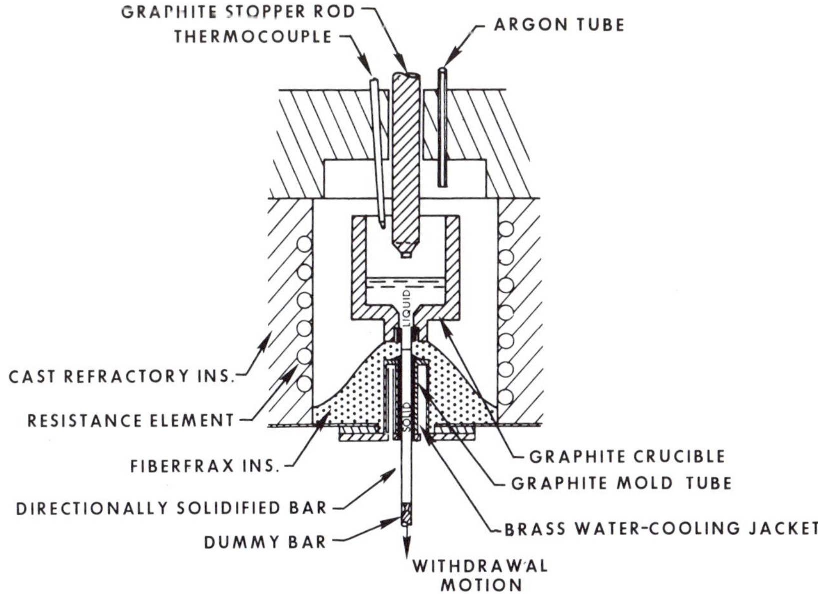

Fig. 42. Schematic diagram of semicontinuous casting apparatus (after Lawson and Kerr, 1971).

B. Semicontinuous Casting

Another possible solution to the problem of producing directionally solidified eutectics in quantity may lie in the area of semicontinuous casting. This type of approach, illustrated in Fig. 42, has been applied by Lawson and Kerr (1971) to the directional solidification of low-melting eutectics. At the rates employed in this study, the eutectics solidified with a cellular liquid–solid interface and would consequently be viewed as marginally successful. However, these results suggest that an approach along these lines may be capable of providing directionally solidified eutectic superalloys at high rates.

Acknowledgment

The authors gratefully acknowledge Dr. A. F. Giamei, who read and criticized portions of the manuscript.

References

Bibring, H. (1971). Aspects Experimentaux des Compartement Mechique des Composites a Fibres Elabores per Solidification Orientee. G.A.M.I. Lecture (in press).

Bibring, H. (1972). Private communication.

Bibring, H., Seibel, G., and Rabinovitch, M. (1972). *Mem. Sci. Rev. Met.*, **69**, 41.

Breinan, E. M., Thompson, E. R., McCarthy, G. P., and Hermann, W. J. (1972a). *Met. Trans.* **3**, 221.

Breinan, E. M., Thompson, E. R., and Lemkey, F. D. (1972b). *Proc. Conf. In Situ Compos.*

Bridgman, P. W. (1925). *Proc. Amer. Acad. Arts. Sci.* **60**, 305.

Chadwick, G. A. (1963). *Prog. Mater. Sci.* **12**, 97.

Cline, H. E., and Stein, D. F. (1969). *Trans. AIME* **345**, 841.

Cooper, G. A., and Kelly, A. (1967). *J. Mech. Phys. Solids* **15**, 279.

Embury, J. D., Keh, A. S., and Fisher, R. M. (1966). *Trans. AIME* **236**, 1252.

Erickson, J. S., Owczarski, W. A., and Curran, P. M. (1971). *Met. Prog.* 58.

Gell, M., and Leverant, G. R. (1968). *Trans. AIME* **242**, 1869.

Gell, M., Sullivan, C. P., and VerSnyder, F. L. (1974). *In* "Solidification Technology" (J. Burke, ed.). Syracuse Univ. Press, Syracuse, New York.

Goward, W. G. (1970). *J. Metals* **22**, 31.

Grosskreutz, J. C. (1969). *Proc. Air Force Conf. Fatigue Fracture Aircraft Struct. Mater*, p 47.

Hertzberg, R. W. (1964). *In* "Fiber Composite Materials," p. 77. ASM, Metals Park, Ohio.

Hertzberg, R. W. (1967). *In* "Modern Composite Materials" (L. J. Broutman and R. H. Krock, eds.), p. 442. Addison-Wesley, Reading, Massachusetts.

Hertzberg, R. W., Lemkey, F. D., and Ford, J. A. (1965). *Trans. AIME* **233**, 342.

Hoffman, C. A. (1970). Effects of Thermal Loading on Composites with Constituents of Differing Thermal Expansion. NASA TN D-5926.

Hogan, L. M., Kraft, R. W., and Lemkey, F. D. (1971). *Advan. Mater. Res.* **5**.

Hoover, W. R., and Hertzberg, R. W. (1971a). *Met. Trans.* **2**, 1283.

Hoover, W. R., and Hertzberg, R. W. (1971b). *Met. Trans.* **2**, 1289.

Hubert, J. C., Jurz, W., and Lux, B. (1971). Etude des Systemes Eutectiques a Base de Nickel Reinforce par des Composes Intermetalliques: Etude Exploratoire des Systemes Ni_3Al–Ni_3Ta et NiAl–NiAlNb. Final Rep. DRME Contract No. 69/537.

Hunt, J. D., and Chilton, J. P. (1962). *J. Inst. Metals* **91**, 338.

Hunt, J. D., and Jackson, K. A. (1968). *TMS AIME* **242**, 1275.

Jordan, R. M., and Hunt, J. D. (1971). *Met. Trans.* **2**, 3401.

Kerr, H. W., and Winegard, W. C. (1966). *J. Metals* 563.

Kornilov, I. I., Shinayew, A. Ya., and Pylayev, Ye. N. (1963). *Russ. Met.* **5**, 67.

Kossowsky, R. (1970a). *Met. Trans.* **1**, 1909.

Kossowsky, R. (1970b). The Ni–Cr Eutectic Alloy System. ARPA Contract No. DAHC 15-67-C-0176.

Kossowsky, R., Johnston, W. C., and Shaw, B. J. (1969). *Trans. AIME* **245**, 1219.

Kraft, R. W. (1964). U.S. Patent 3,124,452.

Kraft, R. W., and Albright, D. L. (1961). *Trans. Met. Soc. AIME* **221**, 95.

Kraft, E. H., and Thompson, E. R. (1972). *Proc. Conf. In Situ Compos.*

Lawson, W. H. S., and Kerr, H. W. (1971). *Met. Trans.* **2**, 2853.

Lazlo, F. (1943). *J. Iron Steel Inst.* **147**, 173.

Lemkey, F. D. (1970). Eutectics Engineering with Microstructure, p 2. Zenith, Univ. Museum, Oxford, England.

Lemkey, F. D. (1973a). Ph.D. Thesis, Oxford Univ.

Lemkey, F. D. (1973b). Final Rep., NASA CR-121038, Contract NAS3-15562.

Lemkey, F. D., and Salkind, M. J. (1967). "Crystal Growth," p 171. Pergamon, Oxford.
Lemkey, F. D., and Thompson, E. R. (1966). U.S. Patents 3,528,808 3,554,817, and 3,564,940.
Lemkey, F. D., and Thompson, E. R. (1971). *Met. Trans.* **2**, 1537.
Lemkey, F. D., Hertzberg, R. W., and Ford, J. A. (1965). *Trans AIME* **233**, 334.
Leverant, G. R., and Gell, M. (1969). *Trans. AIME* **245**, 1167.
Mollard, F. R., and Flemings, M. C. (1967). *TMS AIME* **239**, 1526.
Piearcey, B. J., and Terkelsen, B. E. (1967). *Trans AIME* **239**, 1143–1150.
Piearcey, B. J., and VerSynder, F. L. (1965). A New Development in Gas Turbine Materials—The Properties and Characteristics of PWA 664. PWA Rep. No. 65–007.
Quinn, R. T., Kraft, R. W., and Hertzberg, R. W. (1969). *Trans. ASM* **62**, 38.
Salkind, M. J., Lemkey, F. D., and George, F. D. (1970). *In* "Whisker Technology" (A. Levitt, ed.), Chapter 10. Wiley, New York.
Scheil, E. (1959). *Giesserei* **24**, 1313.
Shaw, B. J. (1967). *Acta Met.* **15**, 1169.
Sheffler, K. D., and Hertzberg, R. W. (1969). *Trans. ASM* **62**, 105.
Simmons, W. F. (1971). Description and Engineering Characteristics of Eleven New High-Temperature Alloys. DMIC Memorandum 255.
Thompson, E. R. (1967). The Structure and Properties of the Ni-NiMo Eutectic Prepared by Unidirectional Solidification, Summary of the Twelfth Refractory Composites Working Group Meeting, Tech. Rep. AFML-TR-67-228, p. 214.
Thompson, E. R. (1971). *J. Compos. Mater.* **5**, 235.
Thompson, E. R., and George, F. D. (1969). Eutectic Superalloys. SAE Paper No. 690689.
Thompson, E. R., and Lemkey, F. D. (1969). *Trans. ASM* **62**, 140.
Thompson, E. R., and Lemkey, F. D. (1970). *Met. Trans.* **1**, 2799.
Thompson, E. R., and Lemkey, F. D. (1972). U.S. Patent 3,671,223.
Thompson, E. R., George, F. D., and Kraft, E. H. (1970a). Investigation to Develop a High-Strength Alloy with Controlled Microstructure. Final Rep. N00019-71-C-0052.
Thompson, E. R., Koss, D. A., and Chesnutt, J. C. (1970b). *Met. Trans.* **1**, 2807.
Thompson, E. R., Kraft, E. H., and George, F. D. (1971). Investigation to Develop a High-Strength Eutectic for Aircraft Engine Use. Final Rep. Contract N00019-71-C-0096.
Thompson, E. R., George, F. D., and Breinan, E. M. (1972). *Proc. Conf. In Situ Compos.*
Thornton, P. H., Davies, R. G., and Johnston, T. L. (1970). *Met. Trans.* **1**, 207.
Tien, J. (1972). *Proc. Int. Conf. "Superalloys—Processing", 2nd* MCIC Rep.
Tiller, W. A. (1958). *In* "Liquid Metals and Solidification." ASM, Cleveland, Ohio.
Tiller, W. A., Jackson, K. A., Rutter, J. W., and Chalmers, B. (1953). *Acta Met.* **1**, 428.
VerSnyder, F. L., and Shank, M. E. (1970). *Mater. Sci. Eng.* **6**, 213.
Wallwork, G. R. (1970). *Int. Conf. Strength Metals Alloys, 2nd*, ASM Paper 17.9.
Wells, C. H. (1967). *ASM Trans. Quart.* **60**, 270.

4

Nickel Alloys Reinforced with α-Al_2O_3 Filaments

R. L. MEHAN

General Electric Company
Research/Development Center
Schenectady, New York

M. J. NOONE

General Electric Company
Space Sciences Laboratory
King of Prussia, Pennsylvania

I. Introduction

One of the most challenging applications for metal-matrix composites is in the blades and vanes of gas-turbine engines. These components operate at high temperatures and at stress levels close to the limits of available metallic alloys, and hence they present highly rewarding goals for composite-materials research. For example, simple calculations, using the rule of mixtures for composite-material behavior, indicate a great potential for the reinforcement of metallic alloys to produce improved materials for turbine components. This potential has provided the impetus for the work discussed in this chapter.

The improved operating efficiency that results from the use of more refractory, higher-strength materials in turbine components has, of course, also spurred the development of improved metallic alloys for these applications. The considerable success achieved by alloy developers has enabled remarkable improvements to be made in the operation of gas turbines during their history of over 30 years. Also, intricate and elaborate improvements in the design of gas turbines, particularly the advent of air cooling of the hot metal components, have enabled turbines to operate with gas temperatures even in excess of the melting point of the uncooled components. During the past decade, alloy development and design improvements have provided progressively more severe goals for composites to achieve in order to compete with, or be more attractive than, metallic components. Progress in composite-materials development over the same period has been considerable in the areas of new or improved reinforcing filaments and filament production methods, the development of practical fabrication processes, and the generation of data on the properties and behavior of composite materials for many applications.

A more recent family of materials that are also being extensively studied for applications in gas turbines is formed by the unidirectional solidification of specific metallic compositions. The resulting microstructure consists of a laminar or a rodlike morphological array of one phase within another: these microstructures have therefore been referred to as *in situ* composites. These materials combine many of the attributes of both conventional alloys and conventional composites (i.e., those formed by incorporating discrete filaments into a chosen matrix) and are currently receiving much attention for high-temperature applications because of the great potential for thermodynamic stability in the eutectic systems; instability and degradation of properties under thermal exposure remains a critical problem in most conventional composite systems. Eutectic composites for elevated-temperature applications are discussed in detail in Chapter 3.

Most of the experimental work in metal-matrix composites during the past decade has been with systems based on aluminum and its alloys as a matrix because of the relative ease of fabrication and the reduced reactivity with available filaments at the low temperatures involved in fabrication and use (below 600°C). For applications such as gas-turbine components, more refractory matrices based on nickel, cobalt, or iron must be employed, and the degree of difficulty in composite fabrication and the possibility for interaction between fibers and matrices increase at these higher fabrication and service temperatures. Similarly, reinforcing filaments with adequate strength and stability at high temperatures are required for these applications: suitable filaments include oxides, carbides, borides, and refractory metals (e.g., Al_2O_3, ZrO_2, SiC, B_4C, TiB_2, Mo, W, etc.).

Nickel has been the base metal for most superalloys developed for gas-turbine service and therefore has also been predominant as a matrix in the development of composites for high-temperature applications. Most of the work described in this chapter employed pure nickel or simple nickel–chromium alloys as matrices. Alloying additives in complex superalloys can cause accentuated attack at filament–matrix interfaces; they also increase the hardness of the matrix, which hampers composite fabrication by diffusion bonding processes. Indeed, many additives are not needed to increase the strength of a composite system, since, by design of the composite, the reinforcing filaments are intended to carry most of the applied loads. The properties of the matrix in a composite should therefore be optimized to provide protection of the reinforcement, ductility and toughness, oxidation resistance, etc., and not necessarily contribute appreciably to the strength of the system.

Single-crystal aluminum oxide (α-Al_2O_3; sapphire) has many properties that make it attractive as a reinforcement for metal matrices at high temperatures. Foremost among these properties are a high elastic modulus, low density (relative to most matrix metals), high strength in filamentary form, high melting point, strength retention at high temperatures, and stability in oxidizing environments. Sapphire whisker crystals were featured among the earliest demonstrations of the effectiveness of filamentary reinforcement of metals at high temperature: the high strength of these crystals stimulated extensive studies directed toward harnessing their potential to reinforce nickel-based matrices.

Sapphire filaments, in the form of whisker crystals, individually fabricated rods, and continuous filaments, have been used in nickel matrices during the past ten years with varying degrees of success in the achievement of useful reinforcement. This work is reviewed in this chapter, together with an assessment of the potential of the system and the current state of the art. The salient points that emerge from this work may be summarized as follows. Large-diameter continuous filaments greatly facilitate composite fabrication and make possible more efficient use of the reinforcement than is possible with discontinuous whisker crystals, despite the higher strength of the latter. The surface of the reinforcement is degraded by chemical interaction with the matrix at high temperature and should be protected by coatings to preserve the strength and hence the effectiveness of the reinforcement. Large differences in expansion coefficient between filament and matrix cause disruption of interfacial bonds during thermal cycling; in extreme cases, fracture of filaments can result from these mechanical interactions. Sapphire undergoes plastic deformation at temperatures within the range where reinforcement of nickel-based matrices is desired, and this limits the degree of reinforcement that the filaments can provide. At the high filament loadings required to provide load-bearing properties that are superior to those of advanced superalloys, fabrication of composites is difficult, and other characteristics of the system, such as poor impact resistance, then detract from the potential of the composite system. In addition, the cost of sapphire filaments suitable for reinforcement remains prohibitively high for many possible applications, despite the considerable progress made recently in the production of continuous filaments.

II. Sapphire Whiskers in Nickel and Nickel Alloys

In the early 1960s, sapphire filaments in the form of whisker crystals became available in sufficient quantities for use in studies of the reinforcement of metals; these whiskers exhibited properties that indicated a high potential for exploitation as reinforcing elements. The early work in the Ni–Al_2O_3 composite system, therefore, was performed using whiskers as the reinforcement, and much basic information concerning fabrication methods, reaction mechanisms, and wetting characteristics of the system was obtained. Much of this early work has been reviewed in detail by Sutton (1970) and Schuster and Scala (1970).

Sapphire whiskers remain among the strongest forms of solids yet discovered. The strength of these whiskers is shown as a function of cross-sectional area in Fig. 1 (Mehan and Herzog, 1970), from which it can be seen that the strength increases with decreasing size, primarily because a smaller surface area allows fewer strength-reducing surface defects to exist. Because of the higher strength of the small-diameter whiskers, and because they were also easier to grow than those of larger diameter, they were used preferentially in the fabrication of composites.

The methods used to fabricate sapphire-whisker-reinforced nickel-based composites included liquid infiltration (Sutton *et al.*, 1965) and powder metallurgy (often preceded by electroplating) (Chorné *et al.*, 1966; Stapley and Beevers, 1969). Although the liquid-infiltration method proved quite successful with matrices of lower melting point such as aluminum (Mehan, 1970) and silver (Sutton *et al.*, 1965), it was found difficult to employ with nickel or the nickel-based matrices of more practical importance. This difficulty arose because sapphire is not spontaneously wetted by liquid metals, and a metallic coating was needed to promote wetting and facilitate fabrica-

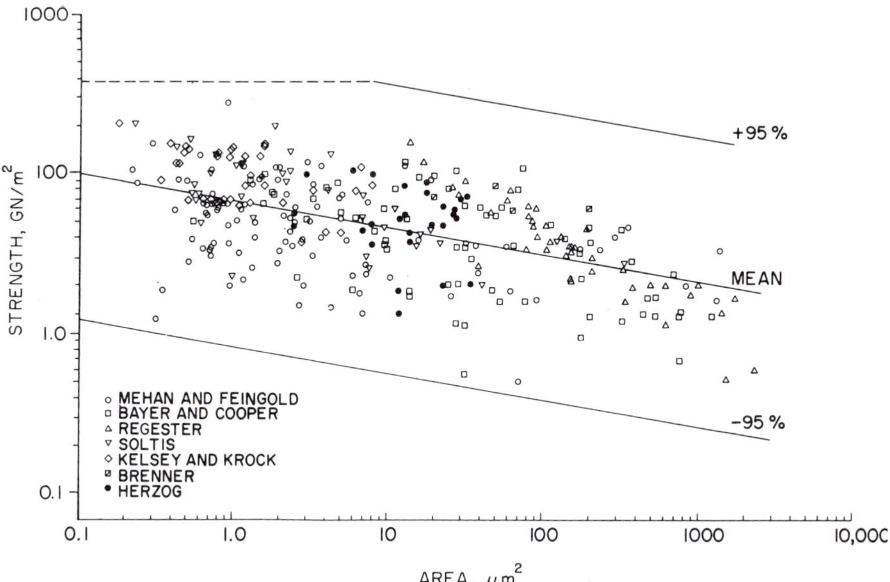

Fig. 1. Compilation of published data on the tensile strength of sapphire whiskers. The lines indicate the mean value and the ±95% confidence limits (Mehan and Herzog, 1970, *from* "Whisker Technology." Wiley, New York, reproduced with permission).

tion. For infiltration of aluminum or silver matrices the sapphire surface was coated with a more refractory metal such as nickel or nichrome (Mehan and Feingold, 1967) to encourage wetting: coatings for this purpose were applied by sputtering processes. For infiltration of nickel alloys, more refractory metals were needed as coatings, but the rate of dissolution of even these materials severely limited the time for the fabrication process.

For fine whiskers the coatings had to be kept thin so that the coating material did not become a significant volume fraction of the reinforcement in the fabricated composite. The significance of coating thickness is illustrated in Fig. 2, where the maximum volume fraction obtainable (assuming rectangular packing) is shown as a function of whisker diameter and coating thickness: for fine whiskers (diameters about 1.0 μm) coatings less than 0.5 μm thick are required in order to attain practically useful volume fractions. Metallic coatings of this thickness, including elementary tungsten and platinum were found to dissolve in liquid nickel or nichrome in a matter of seconds, leading to spontaneous dewetting at the whisker surface (and hence to a poorly fabricated composite) and also to filament-strength degradation (Noone, Feingold, and Sutton, 1969).

Power-metallurgical methods, consisting of mixing matrix-metal powder together with either aligned or random whisker mats, proved disappointing

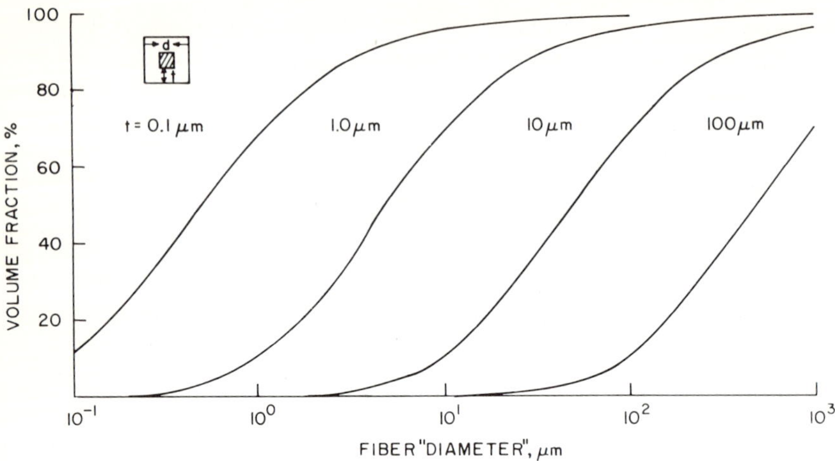

Fig. 2. Effect of coating thickness on the maximum obtainable volume fraction for whiskers of various sizes in whisker-reinforced composites (Noone, Feingold, and Sutton, 1969, *from* ASTM STP 452 with permission).

for composite fabrication. Whisker damage and breakage was an inevitable feature of this process because of poor filament alignment and abrasion between powder and filaments; therefore, no high-strength composites were obtained (Kelsey, 1963). Hot pressing in the presence of a liquid phase has been used, but nonwetting behavior or coating instability then became a problem (Sicka *et al.*, 1964).

The third major fabrication technique used to produce Al_2O_3-whisker–nickel composites consisted of a combination of electroplating nickel onto the whiskers and then consolidating them by hot pressing: the electroplating of nickel was facilitated by previously coating the whiskers with a sputtered metallic film of W or NiCr alloy (Chorné *et al.*, 1968). The plated whiskers were then aligned before consolidation by hot pressing. Substantial reinforcement was achieved in small composites at room temperature; however, at elevated temperature, similar composites were weak because of poor bonding between coating and matrix. Some whisker breakage was also associated with this process.

The best properties obtained with this composite system have not been impressive: at room temperature, maximum strengths of the order of 1170 MN/m^2 were obtained (Chorné *et al.*, 1966). Strengths as high as 621 MN/m^2 at 1000°C were measured, but on very small specimens fabricated by manual alignment of individual whiskers (Chorné *et al.*, 1966). To the authors' knowledge, no substantial reinforcement of nickel and nickel alloys has yet been achieved at temperatures above 1000°C using α-Al_2O_3 whiskers in specimens larger than about 2.5 mm diam. The primary obstacles to successful fabrication using whiskers remain those of whisker breakage, strength degradation by interaction with the matrix, difficulties in alignment of whiskers and achieving useful volume-fraction loading, instability of whisker coatings, and inadequacy of bonding and load transfer between matrix and reinforcement.

The results of an extensive investigation of the reinforcement of nickel with sapphire whiskers over the past six years in England have recently been summarized by Calow and Moore (1972). They also abandoned powder-metallurgical fabrication processes because of whisker fragmentation leading to low-strength composites. To minimize mechanical damage to the whiskers, a nickel matrix was therefore formed by plating directly onto the aligned whiskers by pyrolysis of nickel carbonyl at 170°C; small composites of about 20 v/o whiskers were formed by hot pressing the coated filaments at 1100–1300°C. Tensile strength of these composites at 20°C was about 370 MN/m^2, and the maximum strength at 1100°C was only 48 MN/m^2.

Calow and Moore (1972) attribute the low strength at elevated temperature to the disruption of matrix–filament bonding during heating to the test temperature as a result of the inherent differences in expansion coefficient between matrix and filament. They conclude, therefore, that reinforcement is impossible at high temperatures. This conclusion is supported by the work of Chorné *et al.* (1966), but it is significant to note that in both these cases the whiskers were subjected to extensive degradation by interaction with the matrix metal because no protective filament coatings were used. It is, of course, not unlikely that matrix alloys could be developed with expansion coefficients that closely match that of Al_2O_3 (at least in the direction of the filament length) and thus could alleviate interfacial bond disruption; thermal-expansion mismatch between Al_2O_3 and nickel-based matrices also causes serious damage in large-diameter sapphire filaments, as is discussed in subsequent sections.

Research on whisker reinforcement of metals is currently at a very low level in the United States. The cost of α-Al_2O_3 whiskers of adequate quality for reinforcement remains prohibitively high, and no reliable and economical methods have been developed to process them into aligned mats while removing and discarding defective whiskers and other growth debris. These factors, combined with the coating, reaction, and breakage problems described above, present formidable barriers to the use of sapphire, or indeed other crystals in whisker form, to reinforce metals at high temperatures. It seems likely that if methods were perfected that would enable whiskers to be incorporated into metal matrices and to remain chemically and mechanically stable, their major function could be as a secondary reinforcement; that is, whiskers might be placed at strategic locations such as joints, penetrations, fillets, and in specific components to provide locally enhanced properties.

III. Sapphire Rods as Reinforcement for Nickel and Nickel-Alloy Matrices

A. Properties of Sapphire Rods

Single-crystal sapphire has been available for many years in the form of rods produced by machining and centerless grinding from sapphire boules grown by the Verneuil technique. This material had generally low strength (Watchman and Maxwell, 1959) and was considered to have little potential as a reinforcing filament because of the high cost of production and the short lengths (less than 50 cm) available. However, Morley and Proctor (1962) showed that the strength of sapphire rods was determined by the perfection of their surfaces, not by imperfections associated with their size or inherent in their structure. They showed that large rods (1 mm diam and 5–10 cm long) could be prepared with nearly defect-tree surfaces by a flame-polishing process and that these rods had strengths (measured in bending) comparable to that of sapphire-whisker crystals.

Even in this high-strength form, however, flame-polished sapphire rods could not be considered a practical reinforcing filament, since each short rod was individually prepared, and crystal growth, machining, and polishing were all costly operations. Nevertheless, the ability to produce bulk sapphire in high-strength form greatly facilitated studies of the effects of interactions between sapphire and nickel alloys that were not possible with whisker crystals. The study of flame-polished rods as reinforcing elements in small composites was adopted by the authors as a means of assessing the potential of the Al_2O_3–Ni composite system in a simpler and more straightforward manner than was possible with whiskers. The large size of the rods (relative to whiskers) facilitated handling, application of coatings, performance of controlled heat treatments, and the fabrication of small composites: the rods, although short, were nevertheless "continuous" filaments in the small composites produced in the laboratory (up to 10 cm long).

The use of flame-polished sapphire rods to study the Al_2O_3–Ni-alloy system proved to be a critical phase in the development of this composite material. The strength of individual rods could be determined easily both in the as-produced state and after particular heat treatments, matrix interactions, or fabrication processes: determination of the strength of whisker crystals in similar circumstances was, at best, extremely tedious and often impossible.

Quantitative data on the effects of interfacial bonding reactions, fabrication variables, etc. on the strength of the reinforcing filaments (rods) were obtained that explained much of the previously observed behavior and properties of whisker-reinforced composites (Noone *et al.*, 1969).

Flame-polished sapphire rods were produced for these studies by rotating the centerless-ground rods in a lathe while a small oxyhydrogen flame impinged on the surface and traversed along parallel to the length of the rod. Under the correct conditions, the flame melted a thin layer at the surface of the rod as it traversed along and left in its wake a liquid-smooth, defect-free surface with a slight ripple reflecting the helical path of the flame. Rods polished for composite studies were generally 0.5 mm diam and about 10 cm long. Fig. 3 shows a comparison of the surface smoothness for sapphire rods with various forms of surface finish: (a) centerless ground, (b) commercially flame polished, (c) underpolished in a laboratory rig, (d) flame polished under ideal laboratory conditions (Noone, Feingold, and Sutton, 1969). Full details of the polishing procedure and results for sapphire and ruby (Cr-doped Al_2O_3) material have been described elsewhere (Noone and Heuer, 1972).

It is interesting to note that, despite the simple laboratory arrangement for producing the flame-polished rods, the rate of production of reinforcing filament, in terms of grams of sapphire per hour, exceeded that of laboratory methods for whisker production and that the total cost per pound of filament was competitive with whisker production costs.

The strength of sapphire filaments with varying degrees of surface perfection (ground, flame-polished, etc.) is shown in Fig. 4: strength was measured four-point bending with steel knife edges and a center span and moment arm of 1.27 cm (Noone and Heuer, 1972). No particular care was taken to avoid casual contact between filaments or with the test fixtures, and no noticeable deterioration of the surface (or strength) could be attributed to these contacts. As is well known, the bending test is not a completely satisfactory method of measuring the strength of reinforcing filaments, since only a small volume of material (a line on the surface in the case of four-point test) is subjected to the maximum stress, and the presence of a stress gradient complicates the fracture process. The simplicity of the bending test for this type of filament (short length, high strength, and relatively large diameter), however, was preferable to tensile testing, where gripping and alignment can be difficult. Some direct measurements of tensile strength were made on flame-polished rods using a carefully aligned testing machine with the ends of the rods cemented to rigid anvils with an epoxy cement: the results are shown in Table I. The higher strength of the well-aligned rods is consistent with the bending-strength data, and examination of fracture surfaces showed that in most cases fracture initiated within the bulk of the rod; the surface perfection obtained in

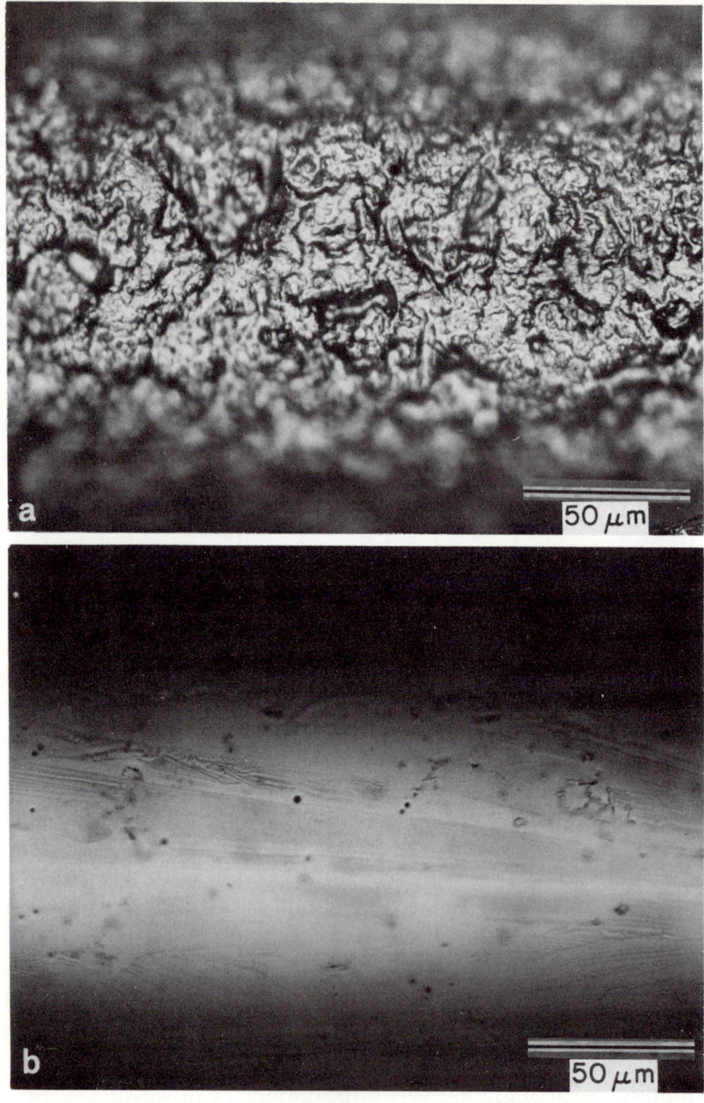

Fig. 3. The surfaces of sapphire filaments resulting from various finishing procedures: (a) centerless ground, (b) commercial flame polish, (c) deliberate laboratory underpolish, and (d) well-polished surface; the ripples reflect the path of the flame during polishing (Noone, Feingold, and Sutton, 1969).

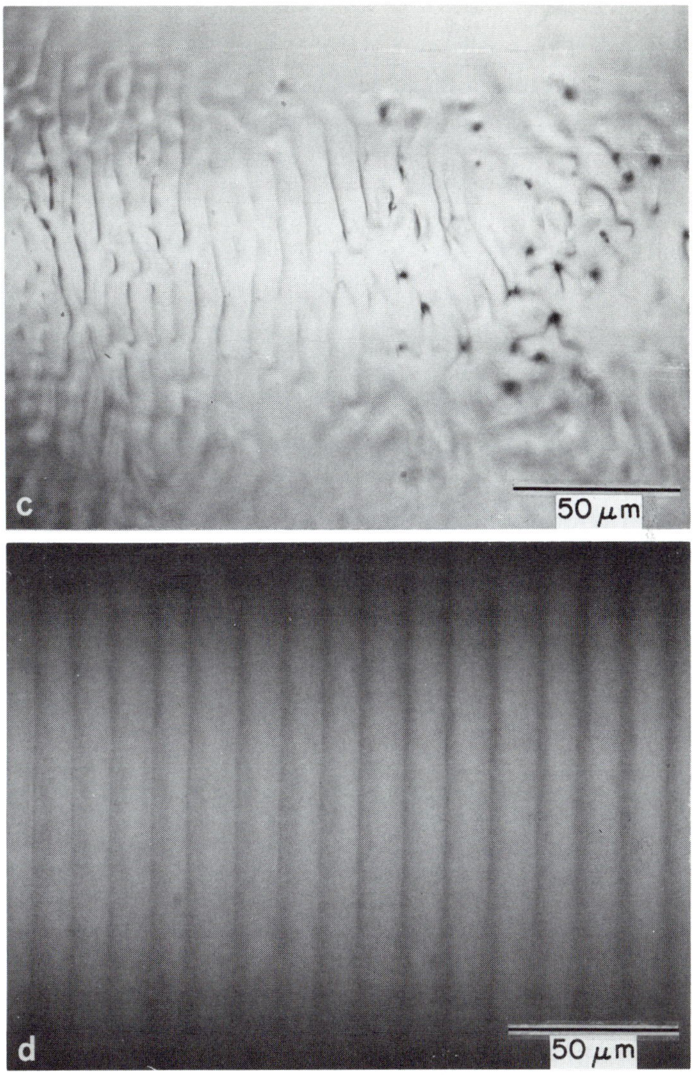

Fig. 3. *continued*

polishing was considered to be greater than the perfection (defect content) of the as-grown bulk material (Mallinder and Proctor, 1966; Noone and Sutton, 1968).

The commercial production of single-crystal sapphire in the form of continuous filaments by Tyco Laboratories (1968) resulted in another major advance toward the practical realization of sapphire-reinforced nickel-alloy

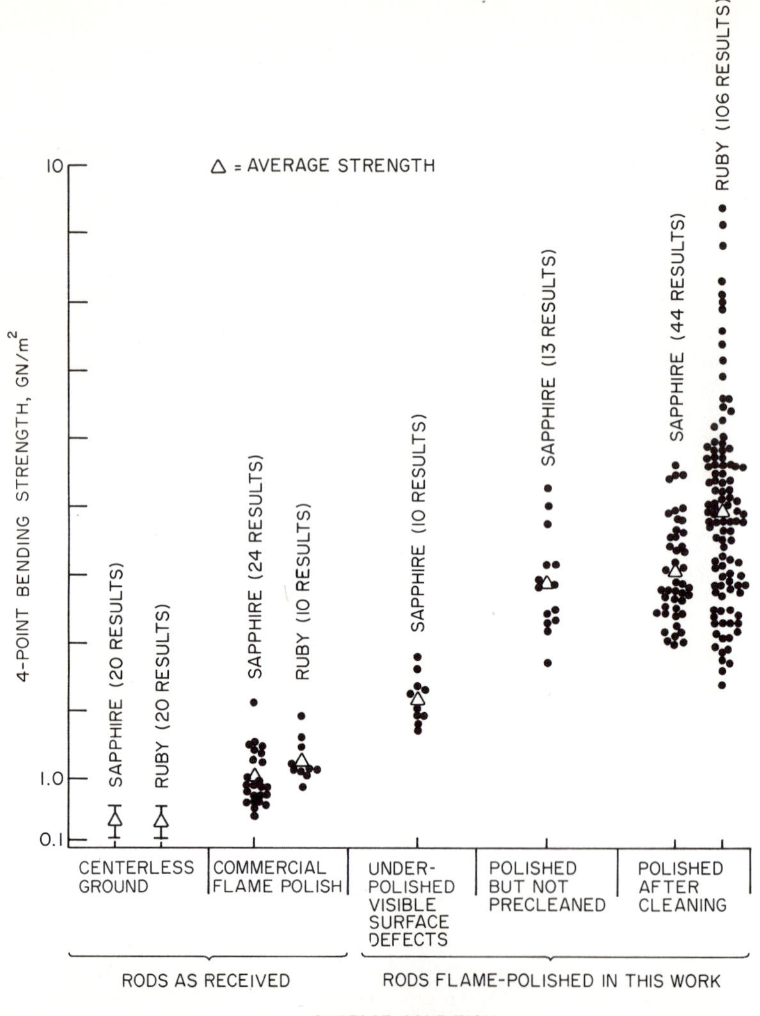

Fig. 4. Bending strength of single-crystal α-Al_2O_3 filaments with various surface conditions (Noone and Heuer, 1972).

composites. These filaments have been used recently by the authors for reinforcement studies, and their behavior has been compared with that of flame-polished Verneuil material. The Tyco filament is produced by an "edge-defined, film-fed growth" (EFG) process (LaBelle, 1971) and has typically a crystallographic orientation with the *c*-axis parallel to the length.

TABLE I

Tensile Strength of Flame-Polished Sapphire Rods[a]

Test	Gauge length (mm)	Strength (MN/m^2)	Alignment
1	26	2860	Good
2	6	Pull-out	Good
3	15	2150	Good
4	26	1930	Good
5	23	3660	Good
6	5	2350	Good
7	28	1890	Good
8	8	1780	Poor
9	20	2500	Poor
10	18	3520	Good
11	10	Pull-out	Good
12	4	3020	Good
13	30	2370	Poor
14	10	3190	Good

[a]Noone and Heuer (1972).

Many cross-sectional geometries can be produced, but filaments are generally circular with a diameter of 0.25 mm. Some interesting reinforcement geometries can be envisioned using ribbons as the reinforcing element.

The strength of the Tyco filament is limited by internal and surface defects, as is that of the other forms of sapphire filament. A feature of this particular growth process, however, is that voids are formed within the material as a result of the volume change as the liquid Al_2O_3 solidifies during filament formation. These voids can be eliminated by careful choice of growth conditions (Pollock, 1972a, b) but at some reduction in growth rate. Many voids apparently form with faceted walls, and their sharp internal notches can be effective as stress raisers. Voids emerging at the surface resulted in low strength in some of the early material used by the authors; however, flame polishing of these filaments produced smooth surfaces such that the strength was then limited only by internal defects. The bending strength of two batches of Tyco filament (obtained in 1969) in the as-grown and flame-polished condition is shown in Fig. 5 (Noone and Heuer, 1972). A flame-polishing stage has since been incorporated into the production process for these filaments with marked improvement in the strength and the uniformity of strength (Pollock, 1972a).

The strength of Tyco sapphire filament in tension, and of larger-diameter (3.2-mm) rod in compression, is shown for two major crystallographic

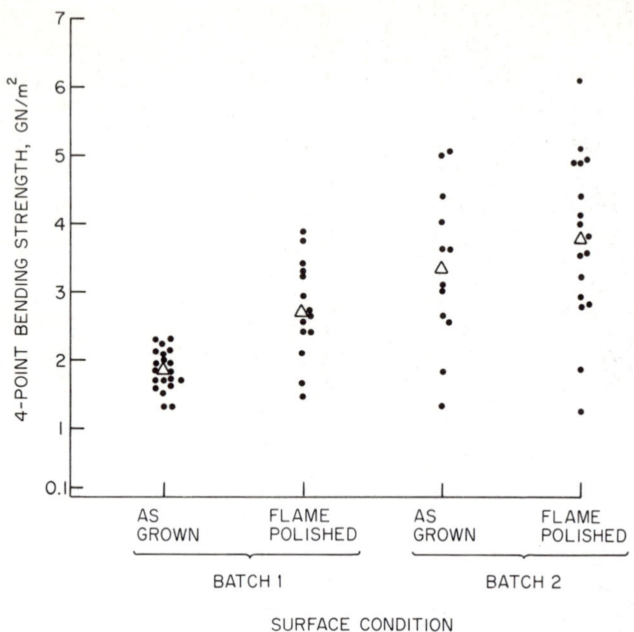

Fig. 5. Bending strength of two batches of Tyco sapphire filaments in the as-grown condition and after flame polishing. △ denotes average strength (Noone and Heuer, 1972).

orientations in Figs. 6 and 7 (Hurley, 1972). The low compressive strength of the filaments at elevated temperature is significant and is no doubt a factor in the difficulties encountered in composite fabrication with these filaments (as discussed in subsequent sections): all the Tyco filament used in the work described in this chapter were *c*-axis filaments, i.e., with <0001> parallel to filament length. Plastic flow can occur in sapphire at elevated temperatures by slip mechanisms and by deformation twinning (Kronberg, 1962; Heuer, 1970). Basal slip occurs readily at temperatures above about 900°C in specimens suitably oriented with respect to the direction of stress. The flame-polished Verneuil crystals typically had an orientation with the basal planes at an angle of about 30 deg to the length of the rod; basal slip was therefore commonly observed in rods stressed in bending (or in tension or compression parallel to the length of the rods) at temperatures above 900°C, as shown in Table II (Noone and Heuer, 1972). Tyco filaments, with the basal planes perpendicular to the filament length, do not deform readily by basal slip when stressed in bending; however, they can deform readily when stressed in compression normal to the filament axis, as in the case during composite

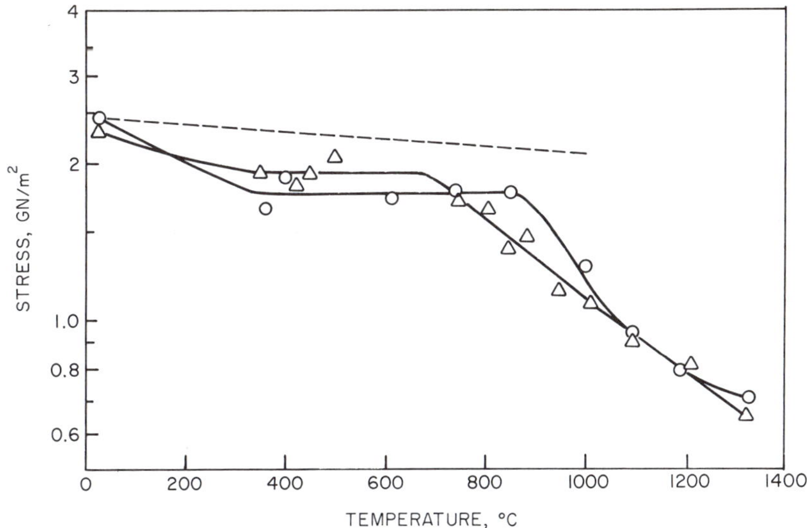

Fig. 6. Tensile strength of Tyco sapphire filaments (0.25 mm diam) with < 0001 > and < 1120 > crystalographic axes parallel to the filament length and stress direction. ○ *c*-axis filament; △ *a*-axis filament; --- calculated values (Hurley, 1972).

fabrication by diffusion bonding for example. As is discussed later, the onset of plastic deformation in sapphire may limit the temperature and stress level at which the various forms (orientations) of filament can be used as effective reinforcement.

For future applications of sapphire as a reinforcement, the continuous Tyco filament has great potential. The production methods are being continually refined and improved (Pollock, 1972b, c), and cost reductions have already been made. It has been claimed that the filament cost could be competitive with that of boron within the foreseeable future (Anon, 1970), which could make it an attractive reinforcement for other matrices at lower temperatures than the current objectives—for example, aluminum alloys (Golland and Beevers, 1972) or titanium alloys (Tressler and Moore, 1970).

B. Nature and Effect of the Nickel–Sapphire Reaction

A reaction between sapphire and nickel or nickel-based alloys takes place at elevated temperatures and is consistent with observations on the stability of Al_2O_3 particles used in dispersion-strengthened alloys (Seybolt, 1968).

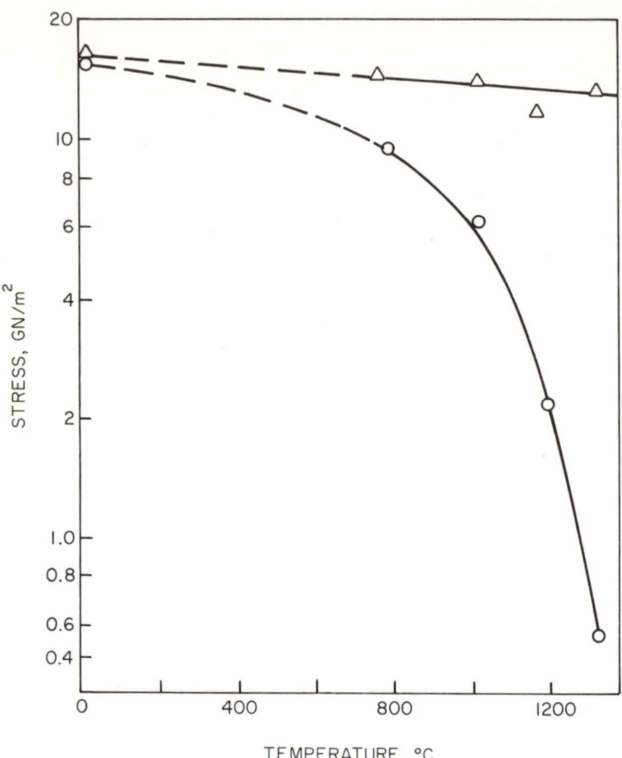

Fig. 7. Compressive strength of short lengths (6.4 mm) of Tyco sapphire rods (3.2 mm diam) with < 0001 > and < 1120 > crystallographic axes parallel to the length and to the stress direction. ○ *c*-axis filament; △ *a*-axis filament (Hurley, 1972).

Unless this reaction removes material uniformly, or forms a uniform reaction product on the filament surface, a reduction in strength would be expected because of local surface roughening. The resulting stress concentration would reduce the filament strength if the roughening were more severe than already existing defects. It is difficult to visualize, for example, reaction short of severe local dissolution that would further reduce the strength of the as-ground sapphire shown in Fig. 3. The better the surface quality of the sapphire filament, the more acute is the sensitivity to strength-reducing surface reactions; very small defects in the near-perfect surfaces of flame-polished sapphire result in severe reductions in the measured strength. The gradual undulations shown in Fig. 3 carry a very low stress concentration, and even thermally etched facets of submicroscopic dimensions that have been observed on flame-polished filaments (Noone and Heuer, 1972) are not severe strength-reducing defects, indeed they are common features of some high-strength whisker crystals (Mehan and Feingold, 1967). The flame-

TABLE II

Four-Point Bending Strength of Flame-Polished Linde Ruby Filaments and As-Ground Tyco Sapphire Filaments[a,b]

	Flame-polished Linde Ruby		As-grown Tyco sapphire[d]	
Temperature (°C)	Strength (MN/m^2)	Slip[c]	Strength (MN/m^2)	Slip
20	4130	0	3310	0
900	2070	0	1430	0
900	2070	0	1120	0
1000	1540	+	1270	0
1000	1490	0	760	0
1100	1120	+	850	0
1100	790	+	550	0

[a] All 0.5 mm diam, tested in air–argon atmosphere at various temperatures.
[b] Noone and Heuer (1972).
[c] + denotes that slip or twins were observed, 0 that they were not, under optical examination at 100 × in polarized light.
[d] Batch 2.

polished filaments were thus ideal for studies of the effects of surface reaction on strength.

The consequences of reactions between Al_2O_3 and metals on the strength of filaments are shown in Table III (Noone, Feingold, and Sutton, 1969): the strength was measured in four-point bending on a small number of filaments as recorded in the table. The filaments were coated with a sputtered layer ($\sim 1\ \mu m$ thick) of the various metals and alloys and then given heat treatments as outlined in Table III. As expected, no reduction in strength resulted from the sputtering operation: a strength of 5520 MN/m^2 was measured on rods after they were sputtered with various coatings. The significant feature of the results in Table III is the marked reduction in strength that followed heating of rods coated with nickel alloys.† After heating, these thin coatings broke up to form a network of metal droplets, as shown in Fig. 8. Droplet formation began on flame-polished surfaces when they were heated at 1000°C but did not occur on commercially polished specimens until they were heated to near the melting point of the alloy, and it is considered that the extreme smoothness of the flame-polished surface contributed to the break-up of the metal layer. The driving force for the formation of droplets is no doubt the tendency to reduce the surface-to-volume ratio of the extremely thin film, and the

† The absence of any strength-degrading reactions between molten Al and Ni–Cr–Fe-coated sapphire (item 5, Table III) was exploited in the fabrication of Al_2O_3-whisker-reinforced Al-matrix composites by Mehan (1970).

TABLE III

Strength of Flame-Polished Verneuil Sapphire Filaments after Receiving Various Coating and Heat Treatments[a]

History of specimens	Mean bending strength (MN/m^2)	Number of results
1. As polished (early work)	4260	4
2. As polished and heated to 120°C for "0" min in H_2	4380	5
3. Ni–Cr–Fe coated, as sputtered	5550	1
4. Ni–Cr–Fe coated, then coating removed in aqua regia	5660	1
5. Ni–Cr–Fe coated and heated to 725°C in molten Al, then metal removed in acids	5840	3
6. Ni–Cr–Fe coated and heated to 1000°C for 18 hr in H_2	1690	4
7. As 6 above, but metal removed in aqua regia before testing	2590	2
8. Ni–Cr–Fe coated and heated to 1450°C for 1 hr in H_2	1410	2
9. Ni–Cr–Fe coated and heated to 1420°C for "0" min in H_2	952	4
10. Ni–Cr–Fe coated and heated to 1000°C for 66 hr in air	1440	4
11. Ni–Cr–Fe coated and heated to 1000°C for 16 hr in air	1480	5
12. Ni–Cr–Fe coated and electroplated with thick Ni coat, then heated to 1000°C for 18 hr in H_2, then metal removed in acids	1240	1
13. As polished (later work)	5760	12
14. Ni–Cr–Fe coated, as sputtered	6100	4
15. NiCr(V) coated and heated to 1000°C for 23 hr in H_2	1570	2
16. NiCr(V) coated and heated to 1000°C for 16 hr in H_2	1630	4
17. NiCr(V) coated and heated to 1450°C for 1 hr in H_2	1580	3
18. NiCr(V) coated and heated to 1420°C for "0" min in H_2	497	9
19. NiCr(V) coated and heated to 1000°C for 16 hr in air	1000	2
20. Ni coated and heated to 1000°C for 16 hr in H_2	1380	2
21. Ni coated and heated to 1450°C for 1 hr in H_2	1640	2
22. Ni coated and heated to 1000°C for 16 hr in air	1360	4
23. Ti coated and heated to 1300°C in carbon-rich atmosphere for 3.5 hr in H_2 to form TiC coating	828	3
24. W coated and heated to 1420°C for "0" min in H_2	3220	13
25. W coated and heated to 1000°C for 16 hr in H_2	2710	7
26. W coated and heated to 1420°C for "O" min in H_2	3220	5

[a] Noone, Feingold, and Sutton (1969).

mechanism is presumably by surface diffusion. Mechanical locking of the layer onto the less-well-polished surface presumably prevented a similar break-up on the commercially polished surface.

The surface of the rod between the droplets (and beneath them when they were removed by dissolution in acid) was appreciably roughened following

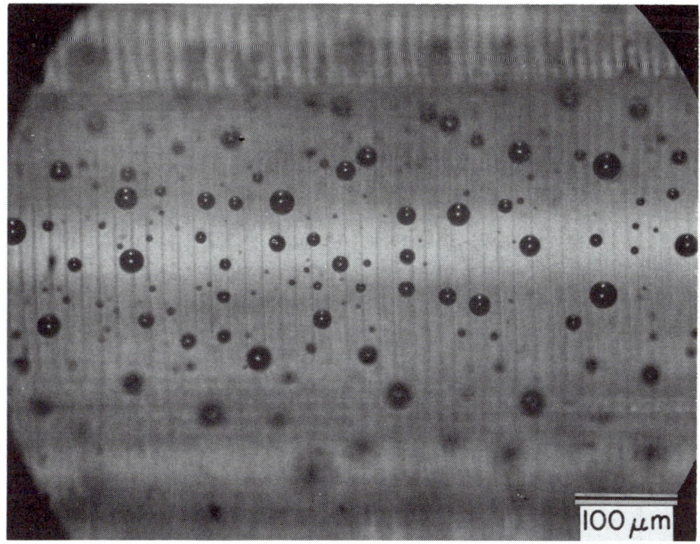

Fig. 8. Metal droplets formed on a flame-polished sapphire surface after heating sputtered NiCr coatings at 1450°C for 1 hr in hydrogen (Noone *et al.*, 1969).

the heat treatment. Figure 9 shows the surface of a rod after the droplet network was dissolved away in aqua regia. The increase in strength after dissolution of the droplets (observed on several occasions) implies either that there was a stress concentration at the root of the drop or that the acid treatment etched out some severe stress raisers from the reaction-product surface layer (Noone and Sutton, 1968).

Isolated droplets formed on the surface of rods or whiskers heated at 1450°C even for "zero" time (rods pushed into the hot zone of a tube furnace as fast as practicable to avoid thermal-shock damage and removed equally fast with no delay in the hot zone). These droplets were nonwetting for all alloys used, as shown in Fig. 10 for nichrome. Heating in air did not result in droplet formation because a glassy oxide film was produced: X-ray examination revealed the presence of NiO and spinel-type phases. In most cases the strength reduction was greater after this treatment than after heating in hydrogen. However, NiO was also detected at the surface of rods heated in "hydrogen," which suggested that there may be some oxygen present since Ni should not reduce Al_2O_3 under these conditions (Glassner, 1959). It was therefore possible that the evaluation of rods with thin sputtered coatings in these circumstances gave unduly pessimistic results, since the formation of

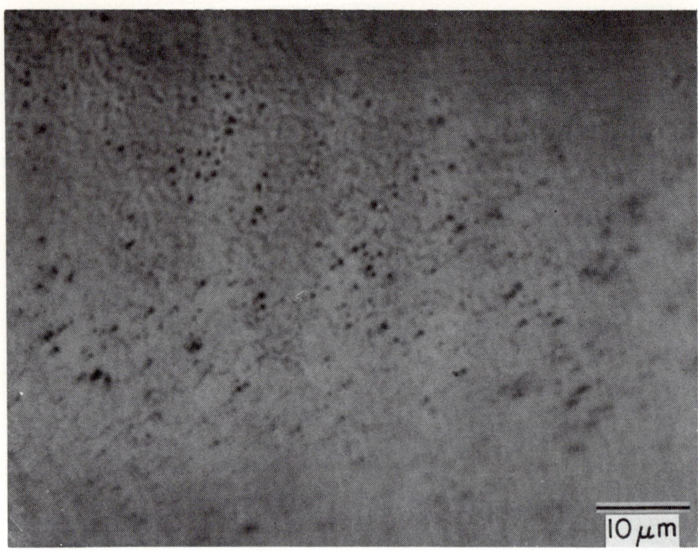

Fig. 9. Surface of a flame-polished sapphire filament following (a) coating with Ni–Cr–Fe by sputtering, (b) heating to 1000°C for 18 hr in hydrogen, and (c) removal of the metal droplets in aqua regia (Noone *et al.*, 1969).

NiO or spinel phases at the sapphire surface would be possible following film break-up. However, other evidence showed that, even without film break-up, the surface was degraded and hence strength was reduced after heating. When a thick (0.5-mm) electroplated Ni coating was applied before heating over a Ni–Cr–Fe sputtered coating (condition 12, Table III) and dissolved away after heating, the strength of the rod was reduced considerably.

The results presented above were corroborated by the examination of rods removed from vacuum hot-pressed composites (Mehan, 1970). The specimen used for these studies is shown in Fig. 11. As-grown Tyco filaments 0.5 mm in diameter were used to produce these specimens; as shown in Fig. 12, the filament strength (measured in bending on extracted filaments) was reduced after heating in contact with the metal matrix (some additional data from sputter-coated and heat-treated filaments are also shown).

The nature of the reaction between sapphire and nickel in both oxidizing and reducing atmospheres has been studied in detail recently (Mehan, 1970). In inert atmospheres, the cause of strength decrease is the formation of small surface imperfections of the order of 1.0 μm in diameter. The size and number of these imperfections increase with increasing temperature and time; photographs of replicas of typical defects are shown in Figs. 13 and 14. Although

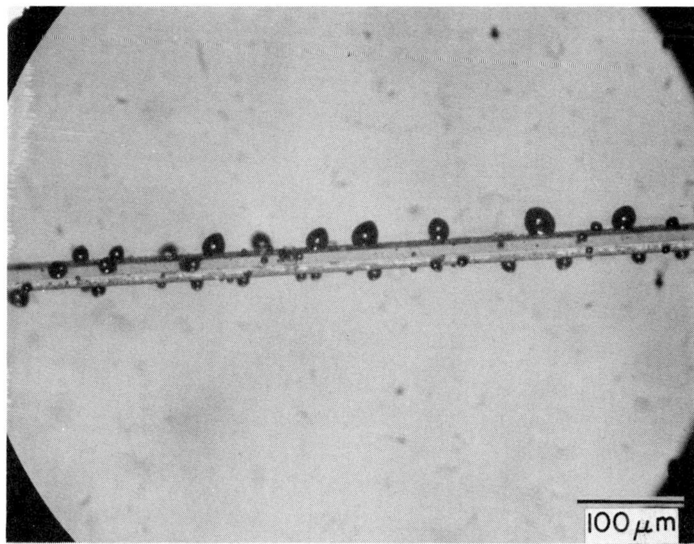

Fig. 10. Metal droplets formed on a sapphire whisker surface after heating sputtered NiCr coatings at 1450°C for 1 hr in hydrogen (Noone *et al.*, 1969).

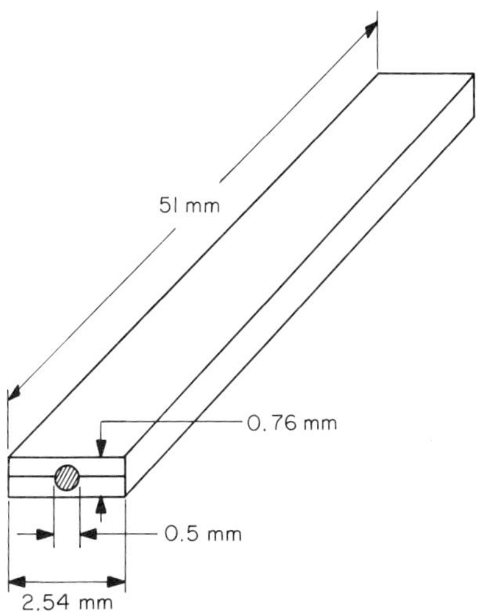

Fig. 11. Test specimen used to investigate sapphire–nickel-alloy reactions (Mehan, 1970).

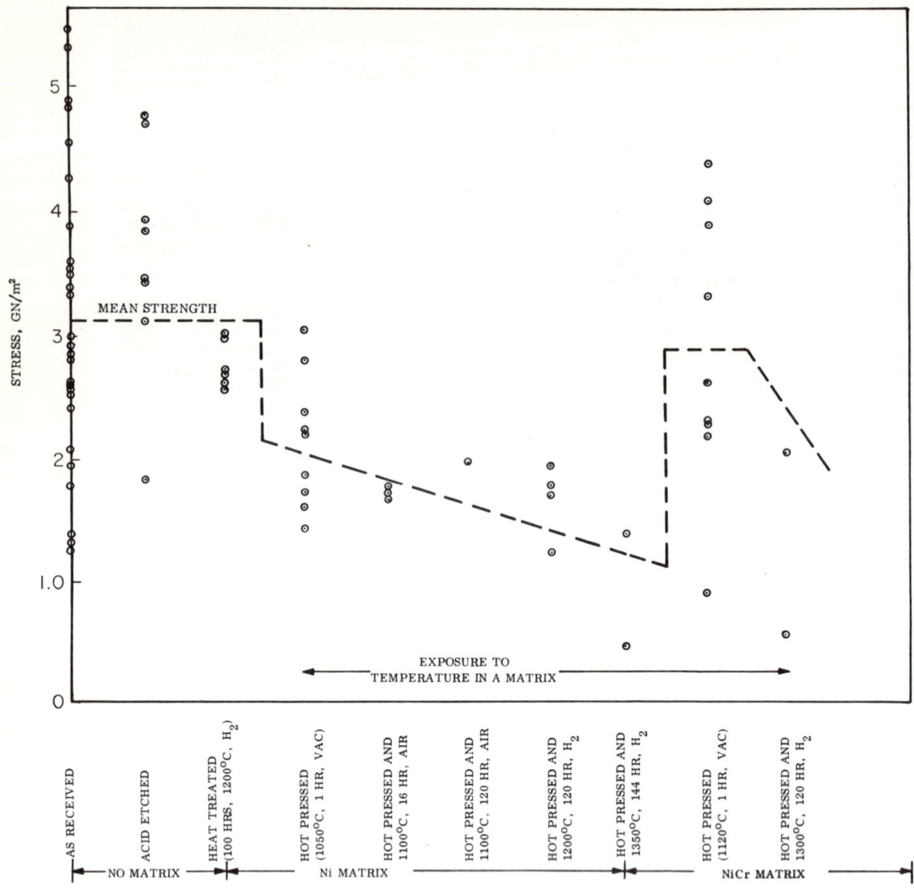

Fig. 12. Bending strength of Tyco sapphire filaments after receiving various heat treatments associated with composite fabrication. Broken line indicates trend (Mehan, 1972).

some replicas were obtained from chemically extracted Tyco sapphire filaments, and hence the surface structure could be affected by the etching operation, a similar structure was observed in mechanically extracted filaments. (Filaments were mechanically extracted by grinding away the surface of a composite until a small length of filament was exposed; the matrix material was then folded back to release the filament.)

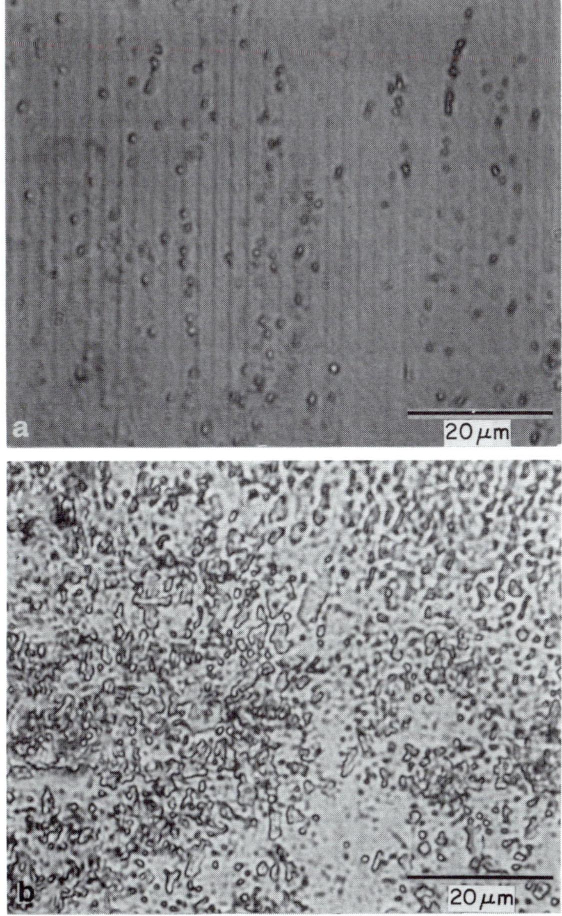

Fig. 13. Surface replicas of Tyco sapphire filaments acid-extracted from nickel matrices after being (a) hot pressed at 1050°C for 1 hr and (b) hot pressed and heat treated at 1200°C for 100 hr in hydrogen (Mehan, 1972).

In addition to observations on the rod surface, microprobe measurements were also conducted on taper sections† of the Ni–Al_2O_3 interface after heat treatments in hydrogen at 1300 and 1350°C. In no case was any evidence obtained to indicate migration of nickel into the filament or of aluminum

† A taper angle of 11.5 deg was used, leading to a tip magnification of 5.

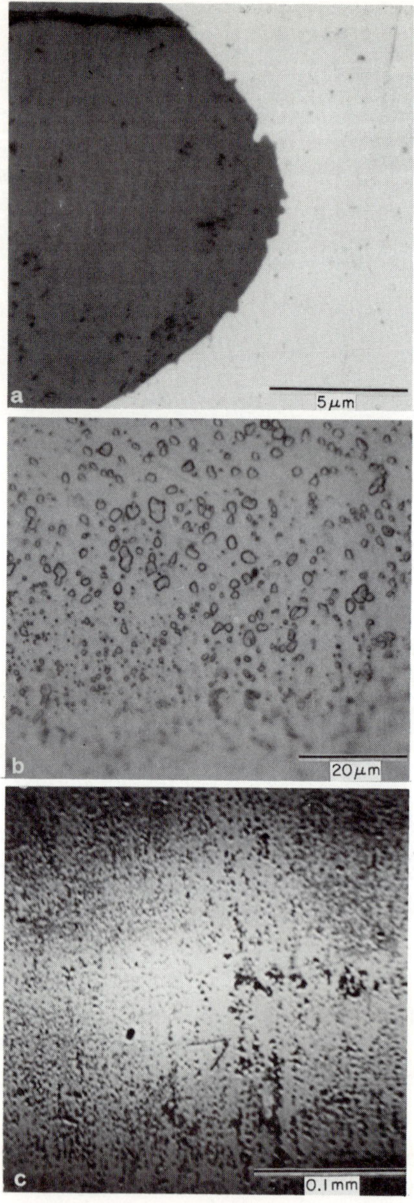

Fig. 14. Tyco sapphire filament after a 100-hr heat treatment in a nickel matrix at 1300°C in hydrogen: (a) taper section, (b) replica of mechanically extracted sapphire filament surface, and (c) electron probe at filament surface in current image mode (Mehan, 1972).

into the matrix. The mechanism for the pit-forming reaction (after Seybolt, 1968), is probably

$$Al_2O_3(m) = 2Al(m) + 3O(m)$$

where (m) refers to the metal matrix. The amount of aluminum and oxygen dissolved is very small, of the order of 0.03% for aluminum in nickel at 1400°C. The amount of aluminum dissolved therefore is too small to be detected by the electron probe, which accounts for the lack of evidence for diffusion previously noted.

Under similar experimental conditions, but with air present, $NiAl_2O_4$ is found at the nickel–sapphire interface, as shown in Fig. 15. The growth of this spinel is governed by a parabolic law of the form $x = K\sqrt{t}$, where x is the reaction-zone thickness and t the time. The activation energy for the reaction, as shown by the Arrhenius plot in Fig. 16, was found to be 61 kcal/mole. Because the mechanism for the formation of both $MgAl_2O_4$ and $NiAl_2O_4$ by solid-state reactions have been shown to be diffusion controlled (Pettit *et al.*, 1966), and because the diffusion coefficient is proportional to x^2/t rather than x/t, the activation energy for diffusion would be twice that calculated from the square-root-of-time plot, or 122 kcal/mole for the above data. This is in good agreement with the value of 115 kcal/mole found by Pettit *et al.* (1966) between 1200° and 1500°C. It is also very close to the activation energy of 116 kcal/mole found for the formation of $MgAl_2O_4$ between 1050° and 1300°C (Bratton, 1969).

The effect of about 18 w/o chromium addition to nickel on the metal–sapphire reaction in an inert atmosphere is a reduction in the amount of pitting as compared to the case of pure nickel. Although it is possible to form pits of similar size to those found in the nickel–sapphire case, the temperature must be increased to about 1350°C. In an oxidizing atmosphere, a reaction zone consisting of mixture of NiO and $NiAl_2O_4$ is formed at the sapphire–nichrome interface, as shown in Fig. 17 (Mehan, 1972). Although strength has not been measured on sapphire rods in this condition, it may be inferred from the surface roughness that the strength will be low.

C. *Effect of Coatings*

Some degree of reaction between sapphire filaments and nickel or nickel–chromium alloys appears to be inevitable at temperatures where reinforcement is required. The interaction reduces the filament strength by creating stress-raising imperfections on the surface, which thereby reduces the reinforcing potential. The exact nature of the matrix–filament interaction is not clear in all circumstances; however, it has been established (Sutton and Feingold, 1966) that reactive-metal constituents in a liquid-nickel matrix

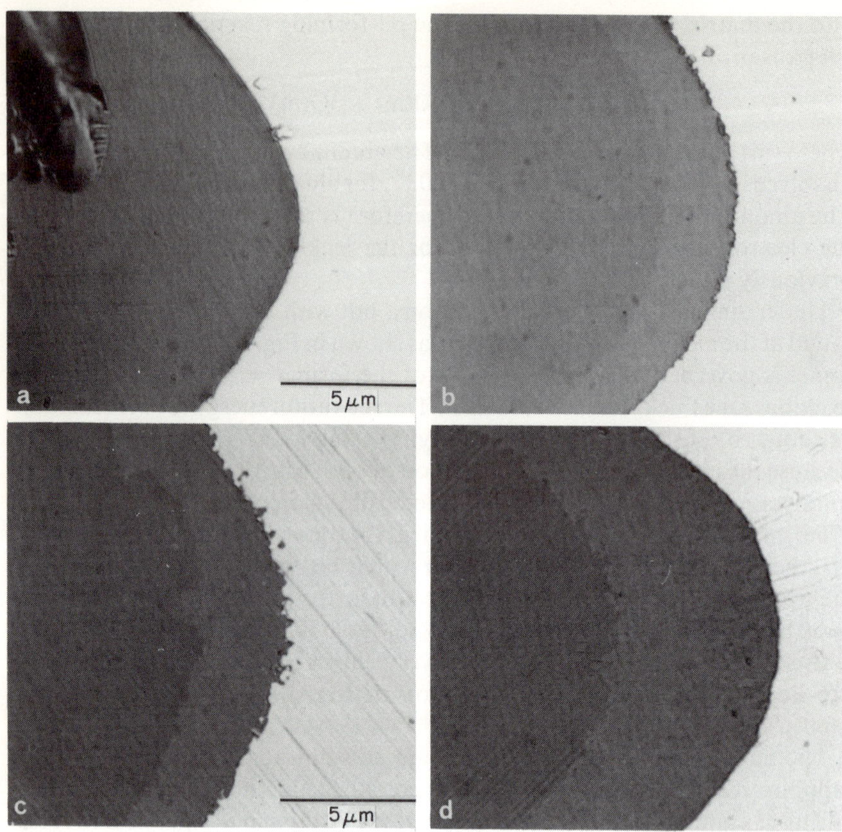

Fig. 15. Growth of $NiAl_2O_4$ at a sapphire–nickel interface after heat treatments in air: (a) 16 hr; (b) 120 hr; (c) 1.6 hr; (d) 40 hr. All micrographs are taper sections (Mehan, 1970).

rapidly segregate at the Al_2O_3 interface and form oxides at the expense of the Al_2O_3; the Al is then released into the matrix. Such reactive metals (those with a more negative free energy of formation of oxide than Al)—for example, Hf and Zr (Glassner, 1959)— can have a significant effect on the nature of the interfacial reaction between nickel and single-crystal Al_2O_3 at concentrations as low as a few parts per million (Sutton and Feingold, 1966). It is clear, therefore, that matrix (and coating) constituents that thermodynamically favor the reduction of Al_2O_3 should be avoided. Of course, it is possible that some degree of reinforcement can be obtained from unprotected and therefore degraded filaments (strength of the order of 1400 MN/m^2) in nickel

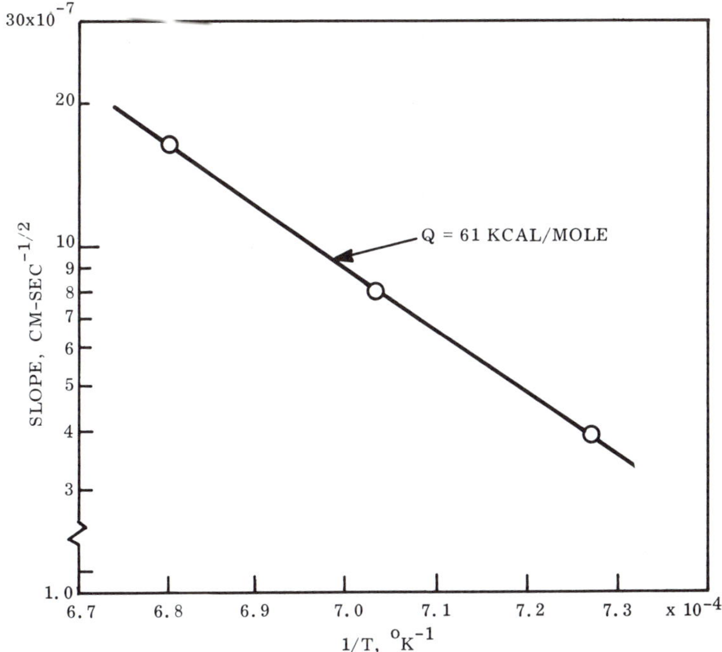

Fig. 16. Rate of $NiAl_2O_4$ formation at a sapphire–nickel interface in air as a function of reciprocal temperature (Mehan, 1972).

alloys. However, for maximum filament strength and utility in the composite it is essential that protective coatings be applied to the filament to prevent, or at least inhibit, interactions with the matrix alloys.

To facilitate fabrication processes, elemental-refractory-metal coatings would be useful; coatings of tungsten or W–26% Re are reasonably stable in contact with nickel at high temperature for short periods of time (Noone *et al.*, 1969). An electron-probe X-ray image of a tungsten–nickel interface after an elevated-temperature heat treatment is shown in Fig. 18a. This may be compared to a less stable material, an iridium–nickel interface, shown in Fig. 18b: iridium was found to be typical of the platinum-group metals. The molybdenum–nickel interface was also examined and showed that molybdenum was much more reactive than tungsten under similar conditions. However, even for tungsten, an exposure of 100 hr at 1200°C resulted in nickel diffusing through the tungsten, and tungsten into the nickel, to form a reaction zone about 100 μm thick. It is thus evident that for long-time stability at elevated temperature, as well as for protecting the sapphire surface during the fabrication process, a coating at least 100 μm thick would be needed. Such

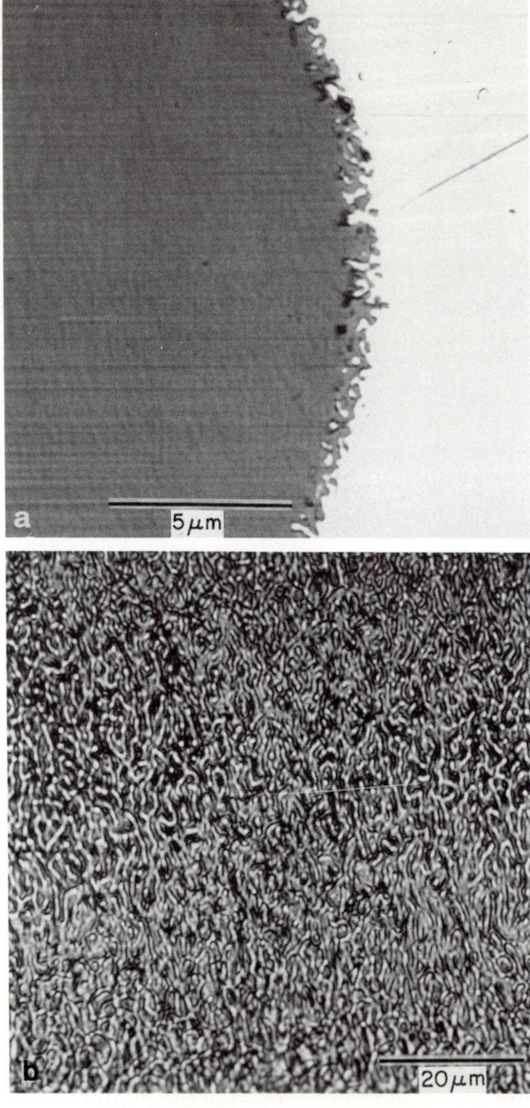

Fig. 17. Interface between a Tyco sapphire filament and a nichrome matrix after a heat treatment of 142 hr in air: (a) taper section, (b) replica of chemically extracted filament surface (Mehan, 1972).

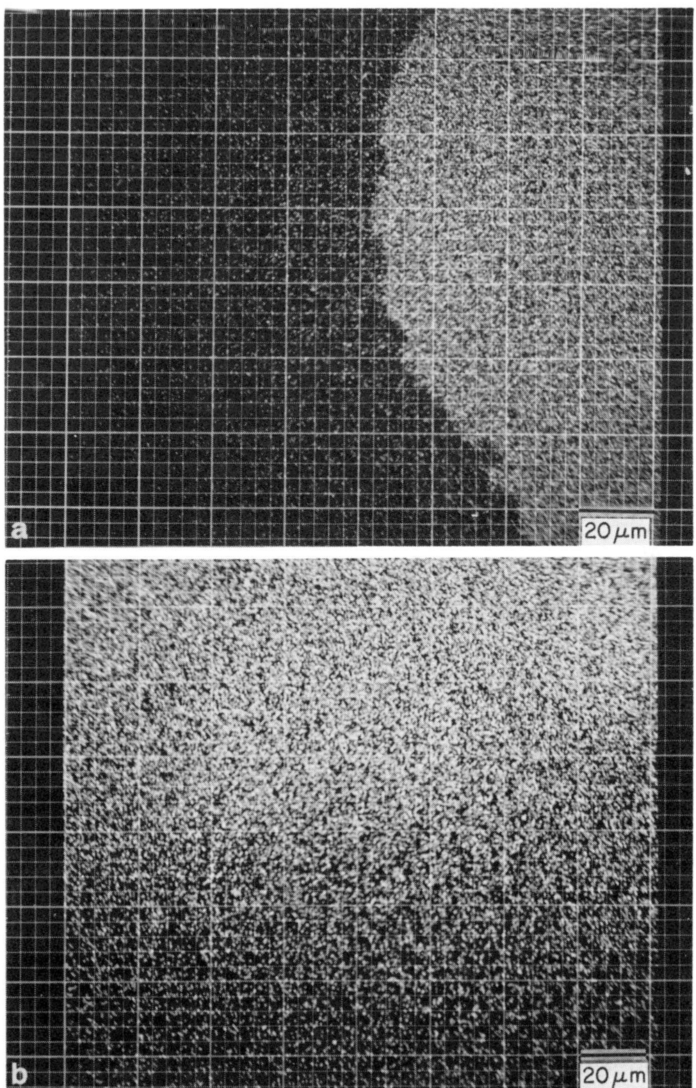

Fig. 18. Electron-probe X-ray images of the interface between (a) W wire and Ni matrix, and (b) Ir wire and Ni matrix, after heating to 1200°C for 100 hr in hydrogen (Noone *et al.*, 1969).

a thick coating of tungsten would appreciably add to the weight of the composite.† Tungsten itself does not degrade the sapphire filament surface (in a reducing atmosphere) as much as elements such as nickel and nichrome, although a degree of strength decrease can be measured, as shown in Table III: the reason for the decrease is not clear.

Studies showed that tungsten was the most satisfactory elemental coating for protection of the filament surface, but even this was not stable enough to allow small thicknesses to be used; therefore, refractory ceramic coatings such as carbides, borides, and oxides were surveyed and evaluated (Chorné *et al.*, 1968). In choosing such coatings, metals that form oxides more stable than Al_2O_3 must be avoided. For example, preliminary experiments using HfC (a highly stable carbide) as a coating were initially found not to degrade sapphire seriously after an elevated-temperature heat treatment. However, when sessile-drop experiments were performed, HfC was found to be unsuitable as a protective coating in the presence of liquid nichrome (Noone, Feingold, and Sutton, 1969). Figure 19 shows the interface after a liquid-nichrome droplet was brought into contact with a sputtered HfC coating on s sapphire disk: the molten alloy appeared to break up the HfC and, more important, resulted in the formation of a second phase at the sapphire interface. This phase was shown to be HfO_2, presumably formed by reduction of the Al_2O_3 by hafnium. Sapphire filaments that were given a similar coating and exposure to liquid nichrome were found to be weakened. Although these results were obtained using the metal in the liquid phase, where diffusion rates are very much more rapid than in the solid state, similar reactions may be expected in the solid state after a sufficiently long exposure to elevated temperature.

Observations of the tendency to form stable oxidic interfacial reaction products led to the conclusion that a logical choice for protective coating should be an oxide with inherently greater stability than Al_2O_3. ThO_2 would be the best choice because of its high negative free energy of formation and its demonstrated stability in nickel alloys, e.g., ThO_2–NiCr. There are, however, some difficulties in handling ThO_2 because of its radioactivity, and there is a considerable density penalty; therefore, work with stable oxide coatings has concentrated on Y_2O_3 (Noone, 1970), which also has high stability.

Y_2O_3 has been routinely deposited onto sapphire rods using rf sputtering techniques (Noone, 1970) that enable coating thicknesses up to 1 μm to be produced in a relatively short time. The degree of protection of the coatings

† A 100-μm-thick tungsten coating on a 0.25-mm-diam sapphire filament would increase the filament weight by a factor of about 12!

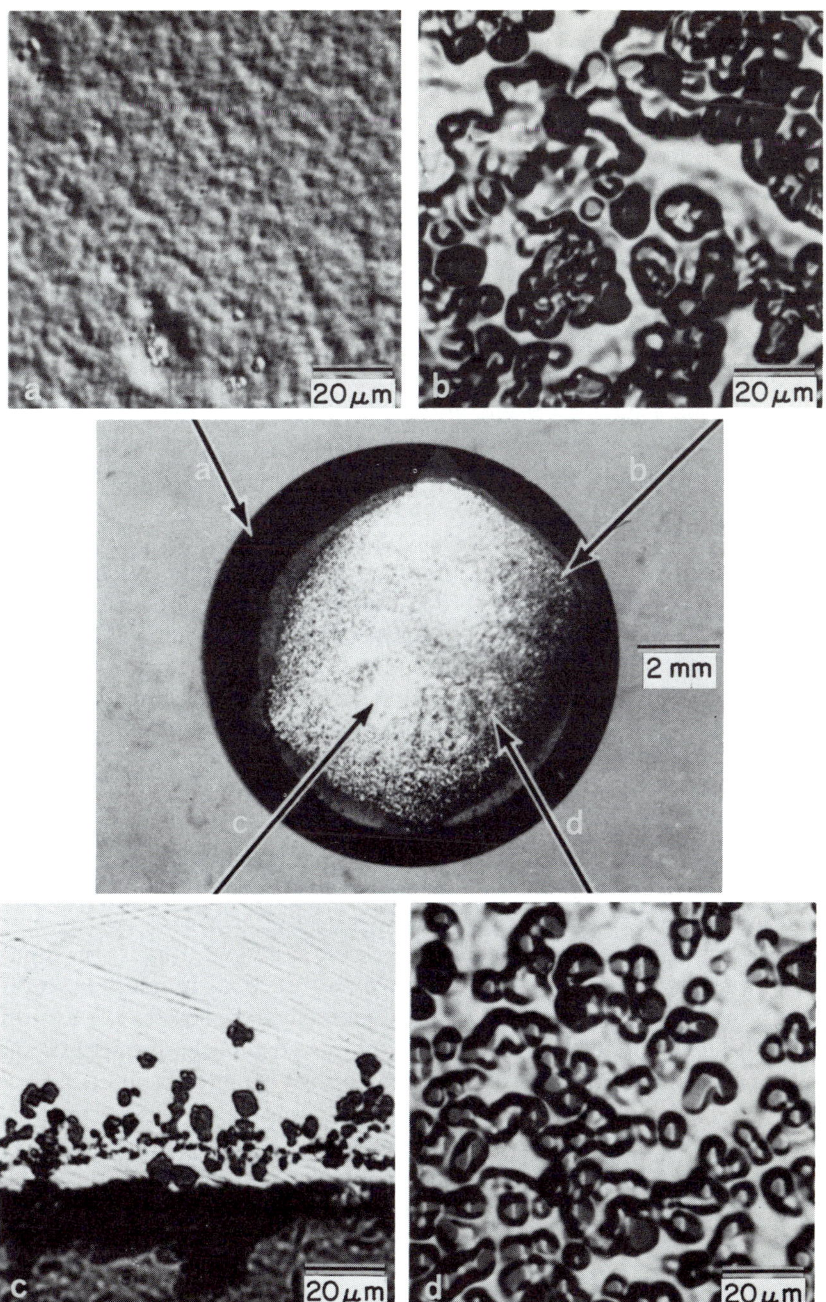

Fig. 19. Structure of a HfC-coated sapphire disk after exposure to liquid 80Ni–20Cr for 9 min in a hydrogen atmosphere: (a) Al_2O_3 surface and HfC after heating; (b) Al_2O_3 surface beneath drop; (c) polished cross section at alloy–disk interface (disk separated) showing reaction product; (d) nichrome surface beneath drop (Noone, Feingold, and Sutton, 1969).

can be assessed from data on the strength of flame-polished ruby filaments† after a variety of heat treatments, as shown in Table IV. The difference in protective behavior between W and Y_2O_3 was not marked in most experiments under similar conditions. For example, as-coated filaments heated to 1200°C for 1 hr in H_2 (items 4 and 7 in the table) had identical strengths (within the scatter bands of each). However, in the presence of NiCr matrix, the superiority of the Y_2O_3 coating became evident (items 5 and 9). The rapid diffusion of tungsten into the matrix during hot pressing accounts for the inferior protection offered by tungsten under these conditions.

Comparison of other data presented in Table IV shows clearly that the deposition of Y_2O_3 can itself, unfortunately, degrade the surface of the filaments (item 6). This is no doubt due to the more energetic nature of the rf sputtering and to the higher temperature reached in this process compared to dc sputtering of tungsten: dc sputtering has no effect on the filament strength (item 3). Further development of optimized techniques for depositing Y_2O_3 (or similar stable oxides) economically and continuously onto high-strength sapphire filaments is needed to ensure freedom from degradation when they are incorporated into metal matrices at high temperature.

A graphic illustration of the ability of Y_2O_3 to protect reactive surfaces in nickel–chromium alloys at elevated temperatures is depicted by Figs. 20 and 21 (Noone, 1970). In this case molybdenum was the substrate rather than sapphire in order to emphasize the effectiveness of Y_2O_3, since uncoated molybdenum (Fig. 20) is vigorously attacked by nichrome at the temperature used (1200°C). Figure 21 shows that the thin coating of Y_2O_3 completely eliminated the reaction between molybdenum and nichrome. Similar encouraging results were observed in the sapphire–Y_2O_3–nichrome system, as will be discussed in the next section.

For a full consideration of coatings in these situations, it must be emphasized that filament coatings must serve other functions in addition to preventing matrix interactions. These functions include the necessity of ensuring adequate filament–matrix bonding to optimize composite properties, and, in the case of liquid-phase processes, promoting wetting to facilitate fabrication. A chemically desirable protective coating (e.g., Y_2O_3) may not completely satisfy all these additional requirements, and duplex coatings have been used in some studies. These duplex coatings have, for example, consisted of thin coatings of tungsten sputtered over Y_2O_3 in order to facilitate further fabrication steps, as will be discussed in the next section.

†Ruby filaments consisting of Al_2O_3 doped with about 0.05% Cr_2O_3 were used in several experiments because of the greater ease in flame polishing (compared to sapphire): the ruby and sapphire material behaved in an identical manner in matrix reactivity studies because the filament surface of the ruby material is depleted in Cr during the flame-polishing process so that the surface is essentially pure Al_2O_3 (Noone and Heuer, 1972).

TABLE IV

Four-Point Bending Strength of Flame-Polished Ruby Filaments (0.5 mm diam) after Receiving Various Coatings and Heat Treatments[a]

Item	Coating[b] and matrix	Heat treatment			Four-point bending strength (MN/m^2)	Number of tests
		Temperature (°C)	Time (hr)	Atmosphere		
1	(Production batch)	—	—	—	5380	16
2	(Control specimens)	—	—	—	4710	2
3	W	—	—	—	5310	2
4	W	1200	20	H_2	2040	6
5	W + Ni plate + NiCr matrix	1200 Hot press	0.5	Vacuum (10^{-5} mm.)	1140	3
6	1.5 μm Y_2O_3 + W	—	—	—	2210	4
7	1.5 μm Y_2O_3	1200	1	H_2	2240	14
8	0.8 μm Y_2O_3	1200	1	Air	1530	5
9	0.8 μm Y_2O_3 + W + Ni	1200 Hot press	0.5 +	Vacuum (10^{-5} mm.)	2010	1

[a] Noone (1970).
[b] All W coatings were 1 μm thick and all electroplated; Ni coatings were 5 μm thick.

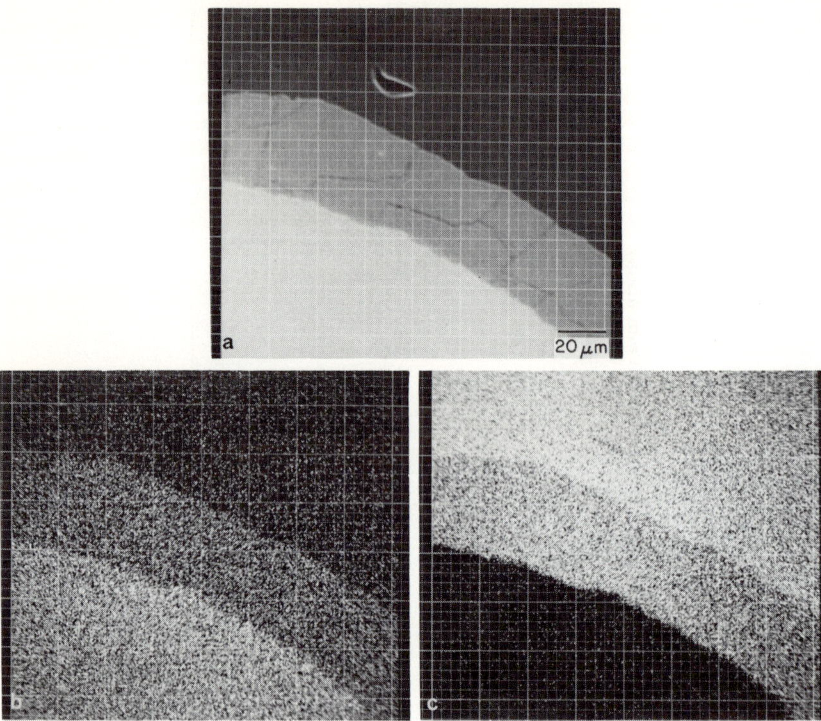

Fig. 20. Electron-probe X-ray images at the interface between an uncoated Mo wire diffusion bonded in 80Ni–20Cr and further heated at 1200° C for 100 hr in hydrogen. An extensive reaction zone is evident. (a) Specimen-current image; (b) Mo X-ray image; (c) Ni X-ray image. (The wire is at the bottom left in each photograph.) (Noone, 1970.)

IV. Fabrication and Properties of Sapphire–Nickel-Alloy Composites

A. Fabrication Considerations

The primary method used to fabricate composites of single-crystal sapphire filaments in a nickel-based matrix is by hot pressing of filaments between sheets of metal, generally referred to as diffusion bonding (Noone, 1970). Other techniques have been attempted, such as electroplating, liquid infiltration (Noone *et al.*, 1969), explosive forming (Snajdr *et al.*, 1968), and the use of powder-metallurgical methods (Kelsey, 1963; Calow and Moore, 1972). To the authors' knowledge, however, the only successful Al_2O_3–Ni-base composites fabricated to date have used some form of hot pressing in an inert atmosphere, although considerable effort has been expended using infiltration techniques.

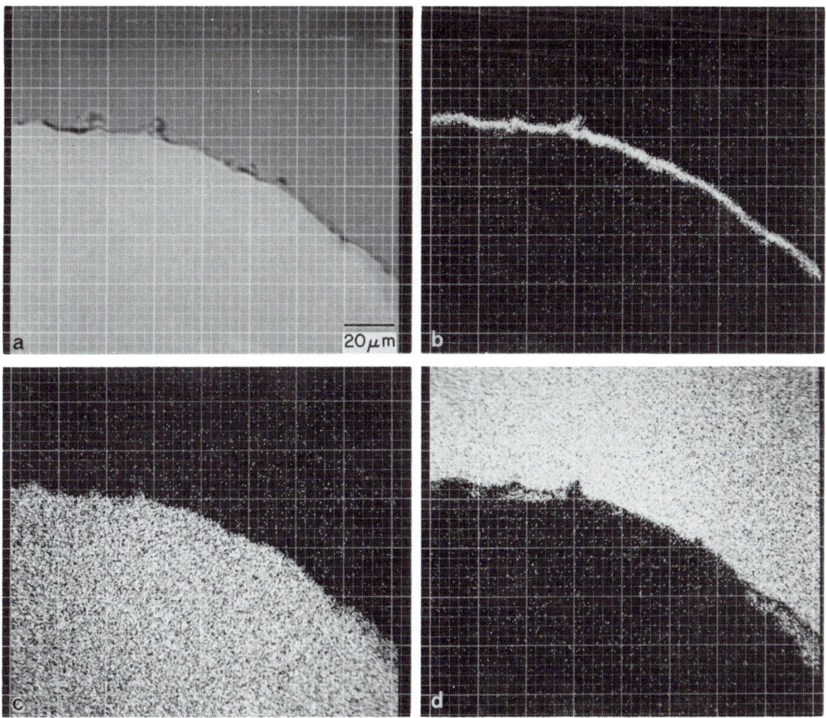

Fig. 21. Electron-probe X-ray images at the interface between Y_2O_3-coated Mo wire diffusion bonded in 80Ni–20Cr and further heated at 1200°C for 100 hr in hydrogen. The Y_2O_3 coating is ~ 1.5 μm thick. (a) Specimen-current image; (b) Y X-ray image; (c) Mo X-ray image; (d) Ni X-ray image. (The wire is at the bottom left in each photograph.) (Noone, 1970.)

As discussed in Section II, many of the early composites using sapphire whiskers as reinforcement were fabricated using a liquid-matrix infiltration process with metal-coated whiskers. When matrices such as aluminum or silver were used, sound composites were readily produced; however, the method was not successful with nickel or nickel-based matrices because of dissolution of the metallic coatings. When larger-diameter single-crystal sapphire filaments became available, thicker coatings could be used, and it was logical to attempt fabrication by the relatively simple and practical process of liquid-matrix infiltration.

At least three methods of infiltration were used, and are illustrated schematically in Fig. 22 (Noone *et al.*, 1969). Of these methods, only rapid vacuum casting was found to be satisfactory. In the other procedures, the time at temperature proved long enough for the coating to be completely dissolved.

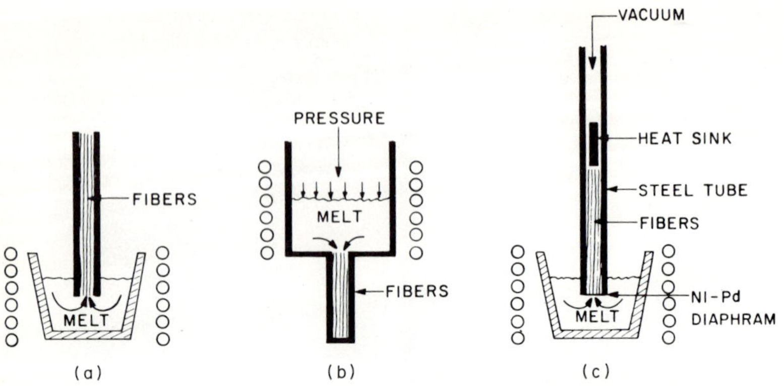

Fig. 22. Schematic representation of three techniques used to prepare composites by liquid matrix infiltration: (a) capillary rise; (b) pressure infiltration; (c) vacuum casting (Noone *et al.*, 1969).

However, the use of high-speed vacuum casting allowed the molten metal to be in contact with the filaments only a matter of seconds before it solidified: this method was used by Dean (1967) for fabricating composites with refractory-metal wires as reinforcement in nickel-based alloys. Experiments using high-speed vacuum casting of nichrome around tungsten wires revealed that at least 6 μm of tungsten would be needed for elemental tungsten to be an effective protective coating on a sapphire surface during composite fabrication, as illustrated in Fig. 23. It will be recalled that in order to prevent reaction due to solid-state diffusion during elevated-temperature exposure of these composites, a thickness of about 100 μm of tungsten was found to be needed: it was concluded that such a thickness led to a severe weight penalty. However, the use of a duplex coating consisting of Y_2O_3 on the sapphire surface (to offer long-term protection) with a 6–10-μm overcoat of tungsten (to facilitate wetting during vacuum casting) would seem to allow the fabrication of a sound composite with little filament degradation and excellent stability at elevated temperatures.

Unfortunately, work to date has not produced sound composites using vacuum casting because of problems associated with thermal shock and the development of slip or twinning in the sapphire by compressive forces induced in the filaments by matrix contraction during cooling from the fabrication temperature (Noone *et al.*, 1969). Typical damage induced by casting molten nichrome around flame-polished sapphire and ruby filaments is shown in Fig. 24. Both cleavage cracks and slip along the basal plane are evident: the basal plane was generally inclined about 30 deg with respect to the filament axis. When experiments were conducted using Tyco filaments, with the basal

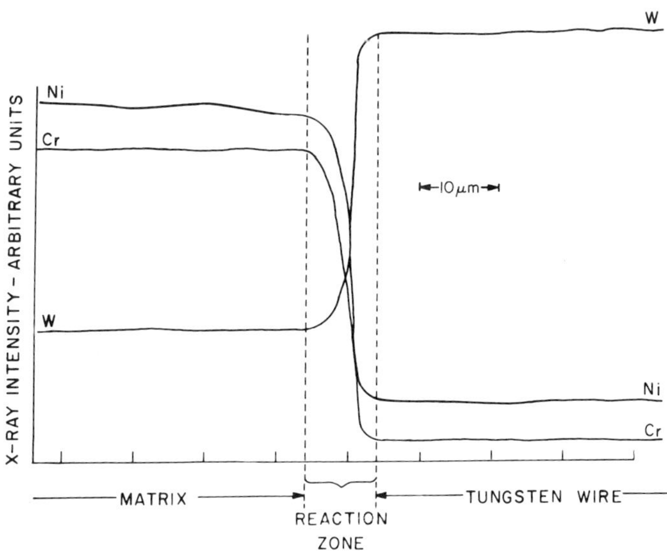

Fig. 23. Electron-probe traces across the interface between tungsten wire and 80Ni–20Cr after vacuum casting (Noone, Feingold, and Sutton, 1969).

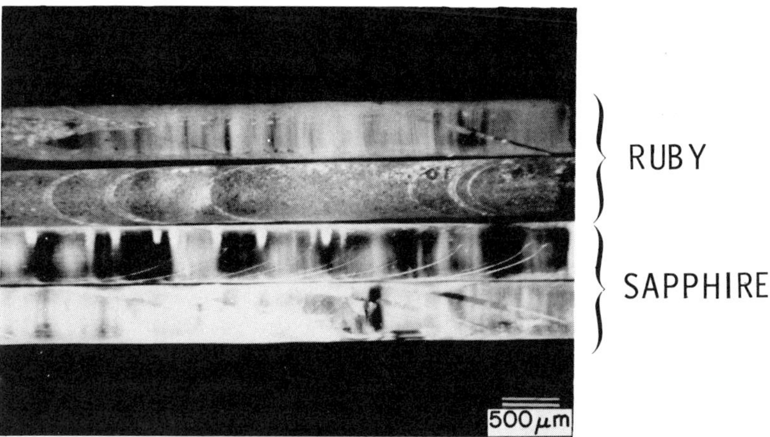

Fig. 24. Flame-polished α-Al_2O_3 filaments (ruby and sapphire) extracted from an 80Ni–20Cr matrix after a vacuum-casting experiment in which the rods were preheated to 1200°C. Note the extensive deformation bands in all filaments. The filaments were also broken into several fragments, but the fractures did not always coincide with the deformation bands (Noone *et al.*, 1969).

plane normal to the filament axis (i.e., not favorably oriented for basal-slip deformation under these circumstances), rhombohedral twinning and cleavage was observed, as shown in Fig. 25. Although a number of modifications were made to the casting procedure, such as preheating the filaments to reduce thermal shock and using thicker coatings to permit somewhat slower cooling, filament damage induced during the process could not be avoided. Vacuum casting remains an extremely important and practical fabrication method, but until the above problems can be circumvented it cannot be used as a method for forming Al_2O_3–Ni-alloy composites.

It seems unlikely that a stress of sufficient magnitude could be induced in sapphire merely by matrix contraction during fabrication to cause the severe filament damage observed. Plastic flow should not occur in sapphire at temperatures less than 900°C, and above 1000°C a material like nichrome is exceedingly weak. However, experiments have clearly shown that slip can be induced in sapphire solely by matrix contraction forces. For example, in one experiment, flame-polished filaments (whose basal plane was inclined about 30 deg to the filament axis) were packed in powdered nichrome alloy, placed in a crucible, and slowly (over 45 min) pushed into the hot zone of a hydrogen furnace until the matrix melted and encapsulated the rods. The crucible was then withdrawn from the furnace slowly (over 45 min). Thermal shock was

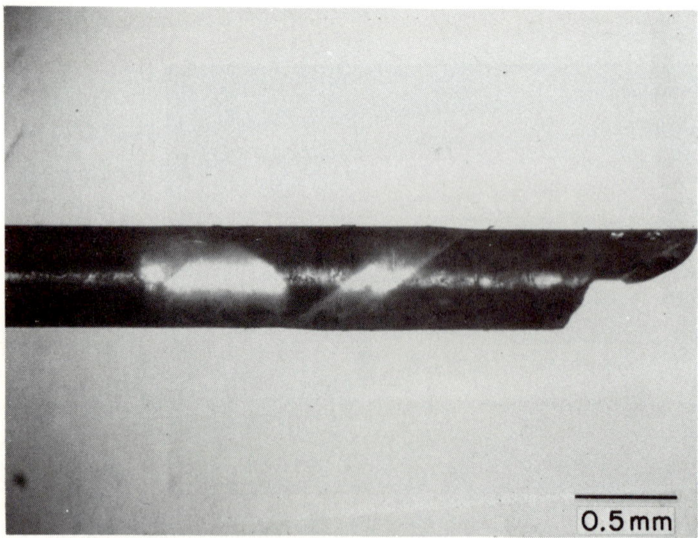

Fig. 25. A large rhombohedral deformation twin induced in a Tyco sapphire filament (*c*-axis parallel to length) as a result of rapid vacuum casting of molten nichrome around the filament. Note the elliptical outline of a second narrower twin (Noone, 1970).

thus avoided, but severe matrix–filament interaction resulted from the extended time (2–5 min) in contact with molten metal, and, as expected, the surfaces of the filaments were severely degraded following this treatment. The significant result was that, despite the absence of severe thermal shock, there was considerable mechanical damage to the rods, as shown in Fig. 26. The presence of slip bands and some cleavage shows that compressive forces generated in the rods by matrix contraction were sufficient to initiate slip at high temperature. Extensive development of "slip lines" on an extracted rod is shown in Fig. 26b. A portion of the end of a rod that was not encapsulated in the metal is shown in Fig. 26c and exhibits no sign of slip or cleavage (Noone *et al.*, 1969).

Similar results were obtained with *c*-axis Tyco filaments during liquid encapsulation into a U-700 alloy matrix, after which rhombohedral slip and deformation twins were observed (Mehan, 1972): Figure 27 shows severe damage induced in a filament following this encapsulation. Some of the surface degradation in this case may be due to attack by interfacially active metals such as chromium and titanium, as discussed by Sutton and Feingold (1966). The surface attack could weaken the sapphire to the extent that subsequent compressive forces could cause fracture: it has been observed that filaments with defect-free surfaces require high stresses (higher than is possible by matrix contraction) to induce plastic deformation at elevated temperatures (Table II).

The result of these experiments indicate that surface-degraded sapphire can be plastically deformed under fairly low loads induced by the matrix constraints at elevated temperature. The compressive strength of as-grown *c*-axis Tyco filaments is low at elevated temperatures, as discussed earlier (Fig. 7), and although *a*-axis rods are much stronger at elevated temperatures (Fig. 7), this orientation favors basal slip. The stress at which basal slip can be initiated in flame-polished Verneuil rods is high if the surface is not degraded (Table II) so that deformation may be avoided in surface-protected rods. It is clear that, even if the thermal shock problem is eliminated, consideration must be given to the expansion coefficient, high-temperature strength, hardness, and stiffness of the matrix alloy, and to the orientation of the sapphire filament in order to avoid the development of slip and/or twinning during cooling from the fabrication temperature. High-strength matrices would be more likely to cause damage of this type than weak, plastic matrices, as was found when U-700 alloy was compared to nichrome.

Hot-pressing processes were successfully used to fabricate Al_2O_3–NiCr composites, and some strength data have been obtained (see Section IV.B). The most successful technique involved the coating of rods with Y_2O_3 (about 1.0 μm thick), followed by a thin (about 0.5 μm) tugnsten coating. The purpose of the tungsten coating, besides providing additional protection,

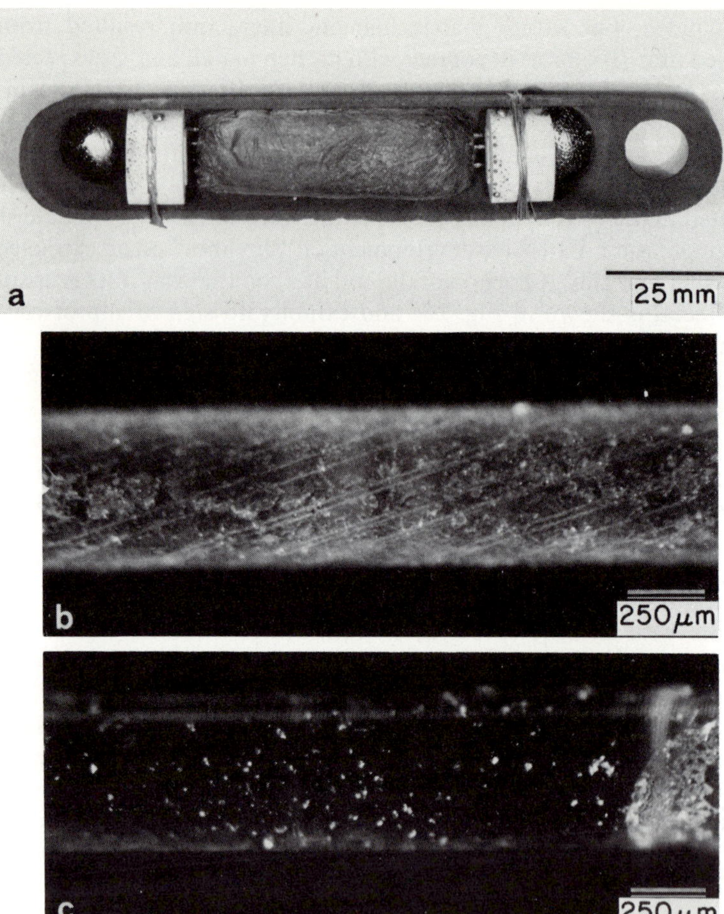

Fig. 26. Results of an experiment designed to evaluate the effects of matrix contraction on a reinforcing α-Al_2O_3 filament after solidification of a molten matrix around the filaments. (a) An alumina boat containing three flame-polished Al_2O_3 filaments supported in BN holders around which 80 : 20 nichrome alloy has been melted; (b) a portion of a filament extracted from the matrix showing extensive development of slip and deformation bands; (c) a portion of the end of a filament that was not encapsulated by the matrix, showing no mechanical damage and no chemical attack (note that the helical polishing marks remain visible) (Noone *et al.*, 1969).

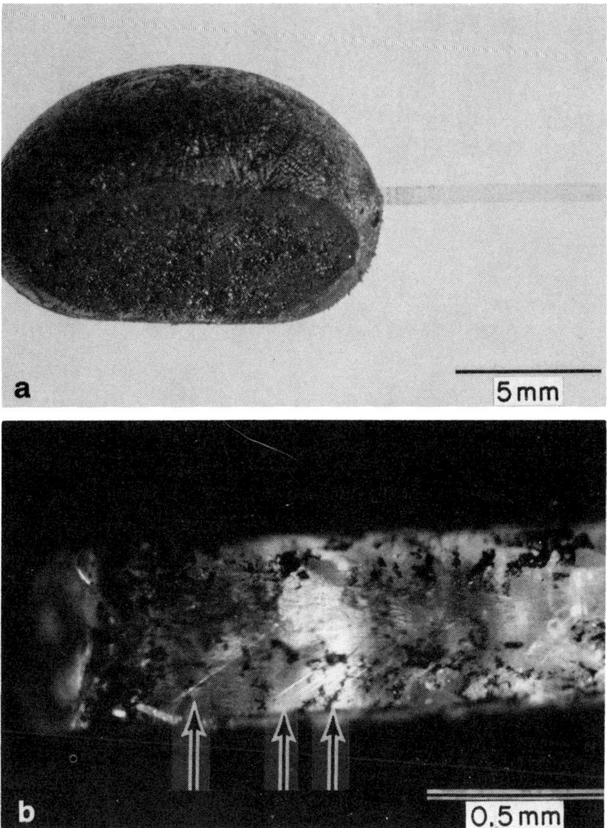

Fig. 27. Appearance of a sapphire filament after contact with molten U-700 for 1 min and cooling to room temperature: (a) melted U-700 and Al_2O_3; (b) extracted filament. Twin bands are indicated by arrows (Mehan, 1972).

was to render the surface electrically conductive so that a relatively thick (50 μm) coating of electroplated nickel could be applied. The purpose of the nickel was to prevent filament–filament contact and minimize possible damage to the Y_2O_3 coating during composite layup and in the pressing operation. The plated rods were placed between sheets of nichrome, which were either grooved or had nichrome wires or strips tack-welded onto them to provide for alignment and spacing of the rods. Pressing was conducted in vacuum and typically at a temperature of 1200°C, using a pressure of 41.4 MN/m^2 (Noone, 1970). A demonstration of the feasibility of the diffusion-bonding process for the production of simple shapes is illustrated in Fig. 28. The small turbine-blade shape (about 5 cm long) was formed by diffusion

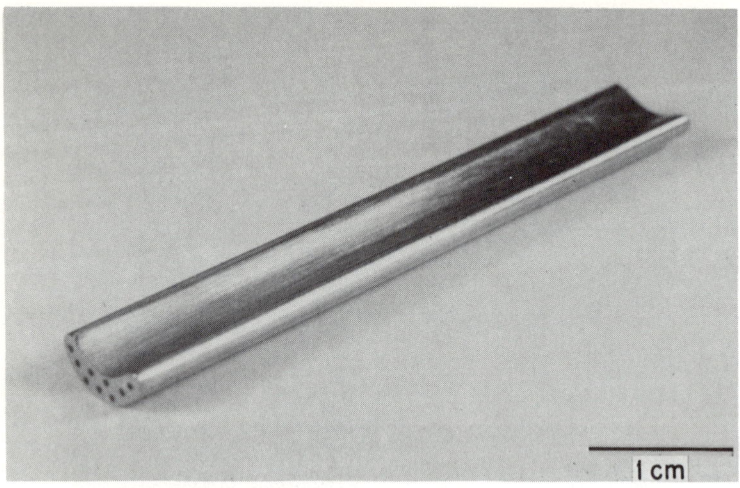

Fig. 28. A three-layer hot-pressed composite of nine α-Al_2O_3filaments in an 80Ni–20Cr matrix ground into the form of a small turbine vane (Noone, 1970).

bonding a three-layer array of nine coated flame-polished ruby filaments suitably positioned between sheets of nichrome matrix, and then machining away the excess matrix to form the shape. The volume fraction of reinforcement in this specimen was 15 v/o; on the basis of degraded filament strength (Table III), it was estimated that the short-term strength-to-density ratio of the composite would be about 2.5×10^5 cm† at 1100°C (Noone, 1970).

Preparation of composites in the above manner using flame-polished rods (basal plane inclined about 30 deg to the filament axis) occasionally resulted in fracture of some filaments in an unpredictable fashion during composite fabrication. However, a sufficient number of the filaments were usually intact to allow subsequent strength tests to be performed. It has not been possible to fabricate composites using uncoated Tyco filaments under similar pressing conditions without excessive filament damage. An overall view of this damage is shown in Fig. 29, and details of one particular fracture mode are shown in Fig. 30. Most of the rods apparently failed by cleavage, although rhombohedral deformation twinning was noted and may have a significant

†Strength-to-density ratios are commonly used by structural engineers, particularly in the aerospace field, to rank various materials. The units of length arise from dividing lb/in.^{2} by lb/in.^{3} to obtain inches, or gm/cm^2 by gm/cm^3 to obtain centimeters.

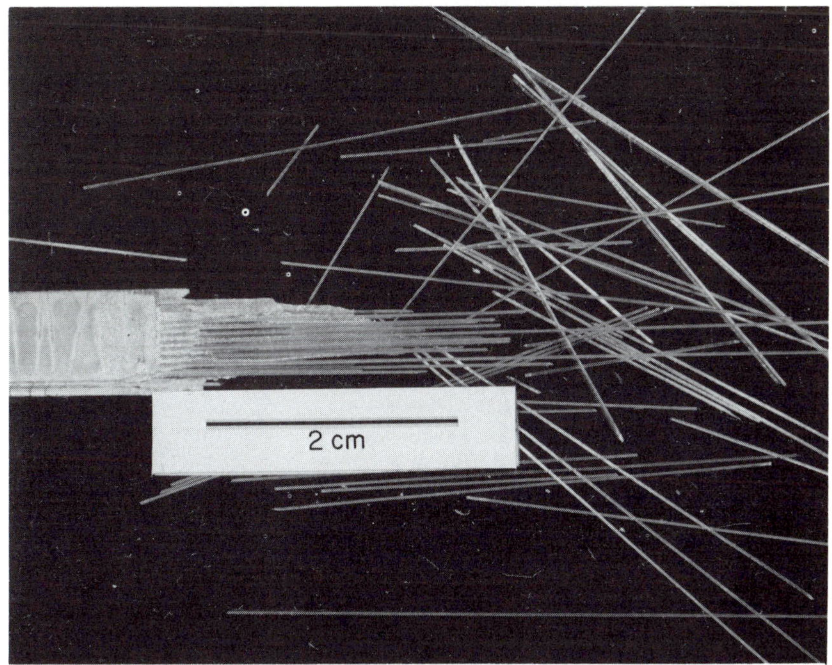

Fig. 29. Typical results obtained from hot-pressed composites of Tyco sapphire filaments in NiCr matrix after partial etching to examine damaged filaments (Mehan and Harris, 1971).

role in the fracture process. Changing the pressing parameters (temperature, pressure, and time) did not lead to any condition that produced sound composites with uncoated *c*-axis Tyco filaments.

An analysis of these failures led to the conclusion that the damage was introduced during the pressing operation and was due to a combination of hydrostatic stresses, local contact stresses, and an induced axial stress (Mehan and Harris, 1971). The weakness of these filaments in compression that results from their orientation and/or surface degradation may well account for their failure during fabrication: the successful fabrication of composites with flame-polished Verneuil filaments is no doubt facilitated by the high strength arising from their more perfect (protected) surfaces.

Another method of fabricating alumina–nichrome composites (using Tyco filaments) has been reported by Brennan (1970). In this method, single-layer tapes were prepared by co-winding (or using hand layup) alternate rows of sapphire filaments and nichrome wire. These arrays were then placed on a nichrome or nickel sheet using an organic bonding agent (polystyrene, for

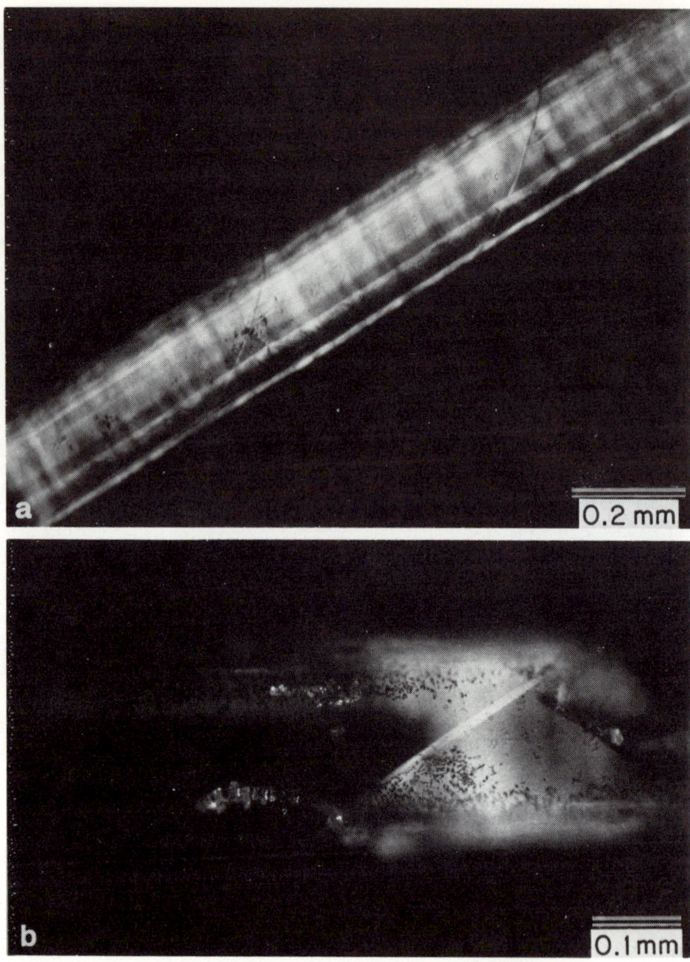

Fig. 30. Examples of rhombohedral twinning in Tyco sapphire filaments extracted from a NiCr matrix after hot pressing (Mehan and Harris, 1971).

example). The tapes were cut to size, stacked in a die, and hot pressed at 1250°C, under 41.4 MN/m^2 pressure for 1 hr. Little or no reaction was observed even after 85 hr at 1200°C: this observation is similar to that noted by Mehan (1970). Although no chemical extractions were conducted to determine whether pressing had damaged the filaments, the composite strength was low (434 MN/m^2 for 30 v/o filament). Fiber breaks were detected acoustically early in tensile tests. This was probably due either to a large scatter in as-grown filament strength or to filament damage introduced during pressing.

Composites of Tyco sapphire filaments in nickel matrices were prepared by Calow and Moore (1972) by hot-pressing procedures using (1) pure nickel powder, or (2) nickel plated onto the fibers by decomposition of nickel carbonyl, or (3) a combination of nickel-coated filaments and nickel powder to form the matrix. With nickel powder as the matrix, pressing was carried out at 1200°C under a pressure of 21 MN/m^2 for 30 min in vacuum to form composites of nominally 20 v/o filaments. Gross filament breakup was a feature of this procedure, although no significant matrix–filament interaction was observed. Using filaments plated with nickel, dense composites could be prepared only if pressing was performed at 1220°C under 31.5 MN/m^2 pressure for 30 min with a coating thickness equal to the fiber diameter (0.25 mm), which limited the volume fraction of filaments to 11–15 v/o. No filament breakup was observed, and no filament–matrix interaction was detected (by strength measurements on extracted filaments), although, as noted above, these conditions inevitably lead to sapphire–nickel interactions. (Composites formed at 1300°C resulted in formation of $NiAl_2O_4$ at the filament–matrix interface.) A matrix formed by the combination of nickel-plated filaments in nickel powder resulted in less fiber breakup for the same pressing conditions than using powder alone; in the best condition, 90% of the fiber fragments were longer than the critical length required for einforcement of nickel at both 20° and 1100°C.

A simple nickel or nichrome matrix was used for almost all composites described in the preceding discussion, but hot-pressing experiments have been conducted using other matrices. Moore (1969), for example, studied compatibility in composites fabricated by incorporating Tyco sapphire filaments into Inco 718 alloy matrix (Ni–0.6Al–0.8Ti–19Cr–5Nb–3Mo–17Fe) and concluded that the filaments reacted to a lesser degree than with either Ni–10Cr or Ni–2Ti matrices. Other studies, however, indicated that it was very difficult to fabricate composites of Tyco sapphire filaments in a U-700 alloy matrix (Ni–0.1C–4.3Al–3.5Ti–14.5Cr–15Co–4.2Mo) (by hot pressing at 1200°C under 41 MN/m^2) without considerable filament breakage (Mehan, 1972). The appearance of a two-filament "composite" after hot pressing between U-700 sheets followed by electrolytic extraction is shown in Fig. 31. No fabrication parameters were found that would permit hot pressing of these composites without severe filament fracture (Mehan, 1972). It will be recalled that merely cooling from the melt was sufficient to induce rhombohedral twinning in Tyco sapphire when in contact with U-700 (see Fig. 27). In view of the difficulty encountered in attempts to fabricate sapphire–nichrome composites, it is not surprising that even more severe problems were encountered with a stronger matrix.

The fact that conventional superalloys like U-700 are not particularly suited as matrices for use with ceramic filaments should not be surprising. The factors that dictate alloying requirements for superalloys are not

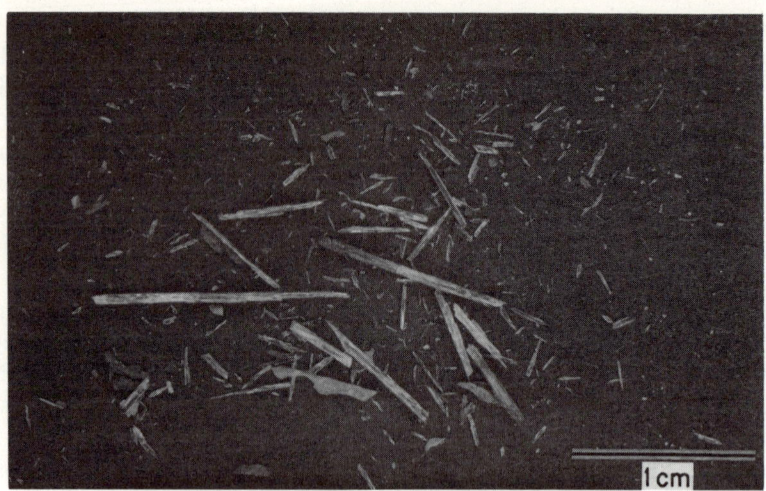

Fig. 31. Severe damage induced in two Tyco sapphire filaments after hot pressing between U-700 sheets at 1200°C under 41 MN/m^2 (Mehan, 1972).

necessarily applicable and are certainly not optimum requirements for composite matrix materials. As Sutton and Feingold (1966) showed, active elements such as titanium tend to migrate to the sapphire–matrix interface and roughen the filament surface (with accompanying filament-strength reduction). Thus titanium, which is commonly added to superalloys to aid in the formation of the hard γ' phase, is not a desirable alloying addition in a matrix for use with sapphire reinforcing filaments. Very strong, hardened alloys are also not desirable for use as matrices because of fabrication difficulties, particularly the increased likelihood of mechanical damage to the filaments during diffusion bonding. The ideal metal matrix for sapphire filaments (or other ceramic filaments) should possess good oxidation resistance (or be readily coated) and good fabricability and should be compatible chemically with the filament, but need only be strong enough at elevated temperature to impart adequate transverse strength (probably supplemented by filaments in the transverse direction) to the composite and to transfer loads to the fibers.

In summary, then, it may be seen that the fabrication of sapphire-filament–nickel-base composites is often a hazardous process. Although composites have been successfully fabricated by hot pressing, it is all too easy to damage the filaments by an incorrect choice of matrix material, filament orientation, or pressing conditions. Considerably more effort will be needed before sound composites can be routinely fabricated by any practical process without inducing filament damage.

B. Properties of Composites of Nickel and Nickel Alloys Reinforced with Al_2O_3 Filaments

There is a paucity of data on the strength of sapphire-reinforced nickel alloys because of the difficulties associated with fabricating composites suitable for use as test specimens. Most of the data that have been obtained consist of tensile tests on small specimens. The property goals for most of these studies were established by calculations using the rule of mixtures, as used in most composite systems. However, for the Ni–Al_2O_3 system, the calculated strengths were rarely achieved. The available data for Ni-alloy composites reinforced with sapphire whiskers, flame-polished sapphire and ruby rods, and Tyco sapphire filament are presented in this section.

1. *Properties of Composites Reinforced with Whiskers*

Sutton and Chorné (1965) assessed the potential of oxide-filament-reinforced metals and presented impressive data obtained from silver reinforced with sapphire whiskers. As discussed in the previous section, fabrication of composites with nickel matrices was thwarted by matrix–fiber interactions in the absence of coatings that could offer protection to the filaments and facilitate fabrication. Whisker-reinforced nickel composites prepared by Chorné *et al.* (1966) by electrodeposition of nickel onto whiskers that were previously sputter-coated with Ni showed some degree of reinforcement at room and elevated-temperatures (for a short time), particularly for continuous-whisker composites.† Table V shows data obtained at room temperature and at 1000°C in air or helium for composites consisting of continuous-whisker composites and discontinuous-whisker composites (these specimens were about 0.5 mm diam and less than 2.5 cm long): the average strength of whiskers used in these composites was about 2760 MN/m^2. High strength-to-density ratios were obtained with continuous-whisker composites—up to 2×10^6 cm at room temperature and 1.1×10^6 cm at elevated temperatures. However, for discontinuous- (overlapping) whisker composites, the elevated-temperature strength was generally poor and often showed little evidence of reinforcement. Fracture behavior in these specimens was characterized by pull-out of whiskers and matrix delamination.

†The term continuous-whisker composite means that carefully selected long whiskers were used and ran the entire length of the composite.

TABLE V

Tensile Properties of Al_2O_3-Whisker–Nickel Composites at 25 and 1000°C[a]

Type of composite[b]	Test temperature (°C)	Whisker volume fraction (v/o)	Composite strength (MN/m^2)	Strength-to-density ratio (10^6 cm)
Continuous	25	22	1230	1.63
	25	51	1050	1.68
	25	39	1350	2.0
	1000	16	282	0.114
	1000	21	495	0.665
	1000	21	495	0.67
	1000	29	759	1.08
Discontinuous	25	28	621	0.845
	25	19	1180	1.52
	25	11	938	1.14
	1000	17	451	0.542
	1000	28	106	0.144
	1000	10	269	0.33
	1000	20	618	0.80

[a] Chorné *et al.* (1966).

[b] Continuous composite: long whiskers running the length of the specimen; discontinuous composite: overlapping whisker ends in specimen.

Composite tensile strength ranged from 106 to 618 MN/m^2 at 1000°C: corresponding strength-to-density ratios were 1.44–8 × 10^5 cm.

In efforts to produce larger composite specimens for further property evaluations, Chorné *et al.* (1968) employed a hot-pressing procedure to consolidate whisker mats that had been partially aligned and then electroplated with nickel. Plated mats consolidated at 1000°C showed evidence of fiber breakage, and reinforcement was poor (room-temperature composite tensile strength was between 62.1 and 552 MN/m^2). In attempts to reduce whisker breakage, consolidation at temperatures up to 1400°C was used, but composite properties remained poor (no doubt as a result of filament surface degradation, as shown by more recent work). This approach to composite fabrication was abandoned because the strength measured at elevated temperature showed no effective reinforcement.

Composites with a nickel matrix plated onto aligned whisker mats by pyrolysis of nickel carbonyl were hot-pressed and evaluated by Calow and Moore (1972). The results of tensile tests on these specimens are summarized in Table VI. It can be seen that reinforcement at room temperature was

TABLE VI

Properties of Composites of Al_2O_3 Whiskers in Nickel[a]

Whisker type	Etched to remove silicon phase	Heat treatment prior to hot pressing	Hot-pressing conditions: Temperature (°C)	Hot-pressing conditions: Pressure (MN/m^2)	Hot-pressing conditions: Time (hr)	Whisker volume fraction (v/o)	Ultimate tensile strength (MN/m^2): At 20°C[b]	Ultimate tensile strength (MN/m^2): At 20°C after heat treatment for 300 hr at 1100°C	Ultimate tensile strength (MN/m^2): At 1100°C
1. FTH[c]	No	2 hr at 1100°C in vacuum (rows 1–5)	1100	15.5	2	20	196	—	—
2. FTH	No		1300	5	1/2	>20	252	—	—
3. TKF[d]	No		1300	7.8	1/2	24	280–306	—	—
4. FTH	No		1300	7.8	3	13	189–210	—	—
5. FTH	No		1175	19.2	1/4	12	359–364	—	—
6. TKF	No	2 hr at 1175°C in vaccum (rows 6–12)	1175	19.2	1/4	15	357–371	—	—
7. TKF	No		1175	19.2	1/4	20	—	—	48
8. FTH	No		1175	19.2	1/4	20	—	—	46
9. FTH	Yes		1175	19.2	1/4	21	—	—	38
10. FTH	Yes		1175	19.2	1/4	20	302	330	—
11. TKF	No		1175	19.2	1/4	13	368–373	—	—
12. Carbonyl nickel matrix	—		1175	19.2	1/4	0	315	—	26

[a]Calow and Moore (1972). [b]Room temperature.
[c]Compagnie Francaise Thompson Houston—Hotchkiss Brandt (up to 10 μm diam).
[d]Thermokinetic Fibers Inc. (USA), (1–3 μm diam).

achieved only after a heat treatment designed to remove carbon monoxide from the matrix; otherwise, the small amount of nickel necessary to form the bond-producing spinel ($NiAl_2O_4$) was reduced by this gas and no reinforcement was observed. Removal of a silicon-rich impurity on the surface of FTH whiskers reduced the reinforcing efficiency: presence of this impurity phase was found to reduce both matrix interactions and whisker-breaking during fabrication, and it also facilitated densification of the matrix. [The whiskers used by Chorné *et al.* (1968) had no surface impurity phase.] There was little change in room-temperature properties after extended heat treatment of these specimens at 1100°C, despite significant changes in microstructure as a result of matrix interaction (specimens 10 and 11, Table VI). However, properties measured at 1100°C (in argon) showed very low reinforcement efficiency (on the basis of the rule of mixtures), as was found by Chorné *et al.* (1968).

Powder-metallurgical processes were studied briefly by Chorné *et al.* (1966) and by Moore and Calow (1969), but the resultant composites suffered from considerable whisker breakage due to imperfect alignment during the powder-compaction operations. Measured properties were poor; room-temperature tensile strength of only 690 MN/m^2 maximum was achieved on composites with whisker volume fractions up to 25 v/o (Chorné *et al.*, 1966).

Liquid-infiltration methods for preparation of test specimens were unsuccessful because of lack of wetting of filaments by the matrix and because of dissolution of filament coatings, as discussed previously. In summary, although some degree of strengthening has been achieved at room temperature by the incorporation of sapphire whiskers into nickel or nickel-based alloys, little retention of this strength has been observed at temperatures of about 1000°C even for short periods of time.

Chorné *et al.* (1966, 1968) believed that mechanical damage to the whisker was the cause of most of the poor properties measured on the composite specimens. Subsequent work by the authors (Noone *et al.*, 1969) attributed the poor properties to filament–matrix interactions, which degraded the surface of the filaments, severely reducing their strength and rendering them ineffective as reinforcement. Calow and Moore (1972) have recently suggested that effective reinforcement is not possible at high temperature because relative movement of matrix and filament (as a result of their difference in expansion coefficient) causes disruption of filament–matrix bonding and therefore prevents transfer of load between matrix and filaments. All of these factors are, of course, important in the achievement of optimum properties in any composite system, but if expansion-coefficient mismatch were most important, then assessment of the future of composites for high-temperature operation would indeed be pessimistic. However, there is room for some optimism, since favorable properties have been measured on material whose components have vastly differing expansion coefficients (e.g., boron filament in aluminum; see Chapter 8) and on Al_2O_3 whiskers in aluminum (Mehan, 1970); these measurements suggest that useful reinforcement may yet be attained in nickel. Many of the problems of achieving useful reinforcement in composites are alleviated, but not eliminated, by using Al_2O_3 in the form of continuous, large-diameter filaments, as discussed in the next section.

2. *Properties of Composites Reinforced with Flame-Polished Sapphire Filaments: Longitudinal Tensile Strength*

Flame-polished sapphire rods were used in studies of interactions between nickel alloys as discussed in Section III (see Tables I–III). The high strength of polished rods made them attractive as reinforcing elements, and their size facilitated the application of coatings and fabrication of composites. Small composite specimens fabricated by an electroplating process (Noone *et al.*, 1969) were prepared for property evaluation. These composites were not considered to be practical materials because of the low bond strength developed in the process and the poor oxidation resistance of the matrix. It was nevertheless a convenient method for demonstrating the reinforcing potential of the relatively large-diameter, continuous sapphire filaments. The filaments were sputtered with tungsten to provide a conductive surface for electroplating of the nickel matrix and to protect the filament from interaction with the matrix during subsequent elevated-temperature tests. The specimens contained either one, three, or six filaments, produced as shown schematically in Fig. 32, and were between 5 and 10 cm long.

Tests at elevated temperatures were conducted in an inert atmosphere with the center of the specimen heated by a small furnace (about a 2-cm hot zone uniform in temperature within 50°C). The specimen ends were soldered to anvils, which in turn were held by chucks in an Instron testing machine (Noone *et al.*, 1969). The tests performed included short-time elevated-temperature tensile tests and elevated-temperature tests following a 30–45-min soak at the testing temperature.

The electroplated matrix generally failed in a brittle intergranular manner within the hot zone at loads much less than the fracture load of the filaments. A feature of the failure of these composites was pull-out of the filaments from the matrix, so that filament fracture was rarely in the same plane as the matrix fracture. There seemed to be no adverse effect of the short soaking of 1000°C before testing at this temperature compared to testing with no preheating. In general, the single-filament composites achieved the highest strength-to-density ratios, but the multifilament composites (three and six filaments had measured strength-to-density ratios within the spread of data from the single-filament composites.

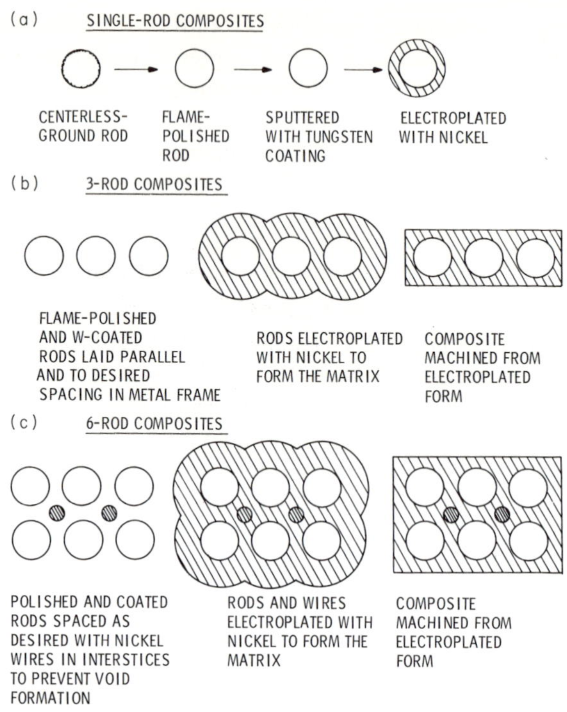

Fig. 32. Schematic representation of the steps involved in the formation of composites consisting of high-strength flame-polished α-Al_2O_3 rods in an electroplated nickel matrix (Noone *et al.*, 1969).

The results are shown in Table VII and in Fig. 33. The dashed curve in Fig. 33 represented the outer envelope of strength-to-density ratios of competitive superalloys (Anon., 1965). Although the strength of these composite specimens was not high, the strength-to-density ratio exceeded that of the superalloys because of the lower density of the composites. A photograph of a six-rod composite prepared for testing is shown in Fig. 34a. Figure 34b is a view of one end of this composite held against a strong light so that the light transmitted down each rod is clearly seen within the outline of the specimen (the rough, grainy appearance of the ends of the rods is because the end of the specimen was ground square but not polished). Figure 34c is a view of this same specimen after tensile testing at 1100°C and clearly shows complete pull-out of two of the rods. The specimen ends are still attached to the steel pull rods in Fig. 34c, and the silver-solder fillets are visible. The remains of the sputtered tungsten coating on the rods after testing can also be seen in Fig. 34c. It seems likely that most of the coating was removed by mechanical

TABLE VII

Elevated-Temperature Tensile Test Results of Al_2O_3 Electroplated-Nickel Composites[a]

Number of filaments	Volume fraction (v/o)	Test temperature (°C)	Tensile strength (MN/m^2)	Strength-to-density ratio (10^6 cm)
1	42	1020	167	0.246
1	37	1000	521	0.75
1	37	1000	842	1.22
1	35	1200	540	0.775
1	46	1000	487	0.735
1	48	1200	391	0.605
1	48	1200	163	0.252
1	45	1200	197	0.605
1	48	1100	376	0.580
1	48	1300	249	0.390
1	49	1400	124	0.194
1	50	1400	153	0.242
1	49	1000[b]	290	0.451
1	48	1000[b]	911	1.41
1	48	1000[b]	322	0.499
1	50	1000[b]	356	0.562
1	51	1000[b]	326	0.514
1	48	1100[b]	486	0.75
1	48	1100[b]	305	0.475
1	50	1100[b]	241	0.382
3	31	1200[b]	275	0.382
3	29	1100	419	0.57
3	27	1100	178	0.248
6	32	1100	217	0.302
6	31	1350	137	0.184

[a] Noone *et al.* (1969).
[b] Held at temperature 30–45 min prior to testing.

abrasion during pull-out and not by dissolution in the matrix in this case. However, some reaction and dissolution between coating and matrix is likely during longer-term heating.

Because an electroplated-nickel matrix was not of practical importance, other fabrication methods were explored, as detailed in Section IV. A. It was found that a hot-pressing process involving diffusion bonding of suitably coated filament (sputtered coatings of Y_2O_3 followed by a tungsten overcoat) between sheets of 80 Ni–20 Cr was the only practical process that produced composite specimens suitable for property evaluation. Several composite specimens were prepared by this method and tensile tested (Noone, 1970). The

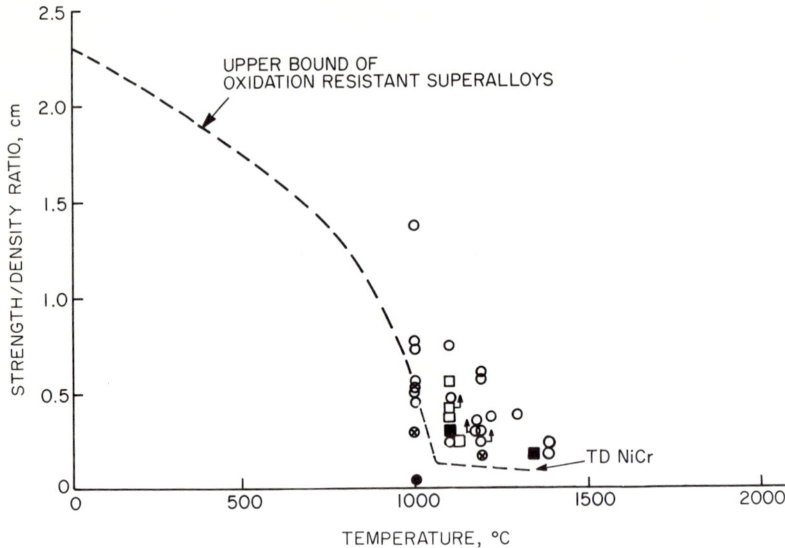

Fig. 33. Strength-to-density ratios versus test temperature for one ○, three □, and six ■ flame–polished α-Al_2O_3 filaments in an electroplated nickel matrix (all of 13-mm hot length). Data for tungsten wire ⊗ and nickel wire ● in nickel electroplate are also shown. The upper bound is for commercial nickel-based alloys (Noone *et al.*, 1969).

method of testing was similar to that used for the electroplated specimens; the results are shown in Table VIII, together with the properties of the unreinforced matrix. The best composite specimen had a strength-to-density ratio of twice that of the unreinforced blank, but there was a scatter in the data, and the other three specimens were only slightly better than the blank. (It must be noted that the blank in this case was inferior to many superalloys available for use at 1100°C.) The strength of the best specimen compared favorably with the results obtained previously with electroplated specimens: the best three-filament composite specimen in that case had a strength-to-density ratio of 5.7×10^5 cm (Table VII).

The difference in properties between hot-pressed and electroplated specimens is primarily caused by the more pronounced filament-surface (strength) degradation in the hot-pressed material. The Y_2O_3 filament coating was obviously not giving sufficient protection to the filaments in the hot-pressing process and may have been mechanically damaged during the process. The improved mechanical bonding in the hot-pressed composites, compared to the electroplated specimens, is evident from the lack of pull-out associated with fracture of filaments in hot-pressed composites. It is interesting

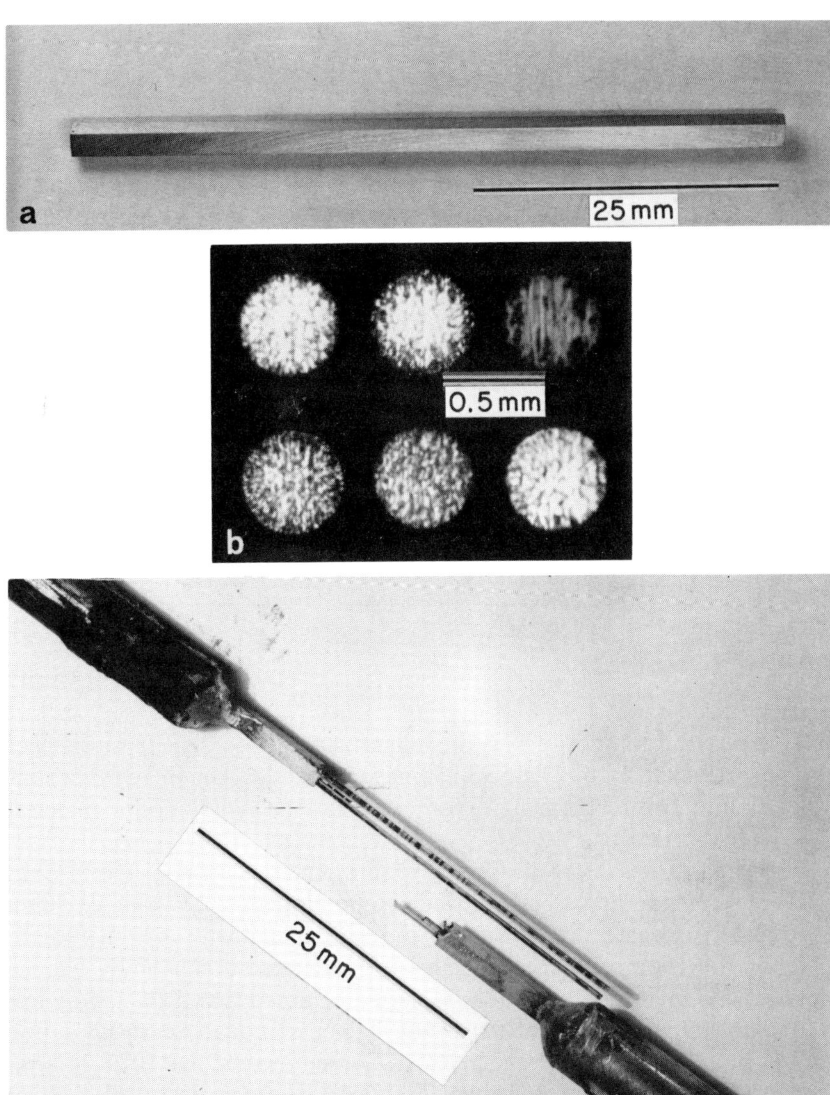

Fig. 34. Photographs of a composite consisting of six flame-polished α-Al_2O_3 rods in an electroplated nickel matrix. (a) Test specimen hand ground from the electroplated form; (b) end view through the single-crystal Al_2O_3 rods and the outline of the composite cross section; (c) view of the same composite after tensile testing at 1100°C showing complete pull-out of two rods (the ends of the steel pull-rods to which the specimen was soldered for the test are also visible) (Noone *et al.*, 1969).

TABLE VIII

Tensile Strength of Composite Specimens of Coated Flame-Polished Ruby Filaments Hot-Pressed in a Nichrome (80:20) Matrix and Tested at 1100°C in Argon[a]

Coating	Number of filaments	Volume fraction of filaments (v/o)	Tensile strength (MN/m^2)	Strength-to-density ratio (10^6 cm)
Hot-pressed NiCr blank	None	—	114	0.140
NiCr wire (1.27 mm diam)	None	—	68	0.083
0.8 μm Y_2O_3 + 1μm W + 50 μm Ni	3	20	105	0.144
1.5 μm Y_2O_3 + 0.25 μm W + 50 μm Ni	3	20	128	0.175
1.5 μm Y_2O_3 + 0.25 μmW + 50 μm Ni	3	21	221	0.305
4 μm Y_2O_3 + 0.25 μm W + 50 μm Ni	2	16	130	0.174

[a]Noone (1970).

to note that the composite with the highest strength-to-density ratio experienced some filament pull-out, which may substantiate the theory that some bond failure before filament failure is desirable in composite materials (Mullin *et al.*, 1968).

Figure 35 shows test data from several of the specimens listed in Table VIII: the stress and crosshead movement are plotted, since strain could not be accurately measured. The fracture of the three filaments is evident on the plot for the composite specimen, as is the extension caused by filament pull-out and matrix deformation. It should be noted that if the failure load was assumed to be carried entirely by the filaments (though clearly it was not), the calculated fracture stress in each filament would be 1070 MN/m^2. Earlier work (Section III.B, Table III) showed that even degraded filaments had higher strengths than this (in bending at room temperature). Fragments of filaments extracted from composites after tensile testing had four-point bending strength of 1680 MN/m^2 (again at room temperature). It seems likely that the severe temperature gradient, misalignment, and perhaps also filament–matrix abrasion during testing contributed to the low fracture stress in the filaments, in addition to the strength degradation associated with chemical interactions at the interface. No plastic deformation was detected in the fractured filaments. Further work is clearly needed to determine the exact failure mechanism and mode for this type of composite.

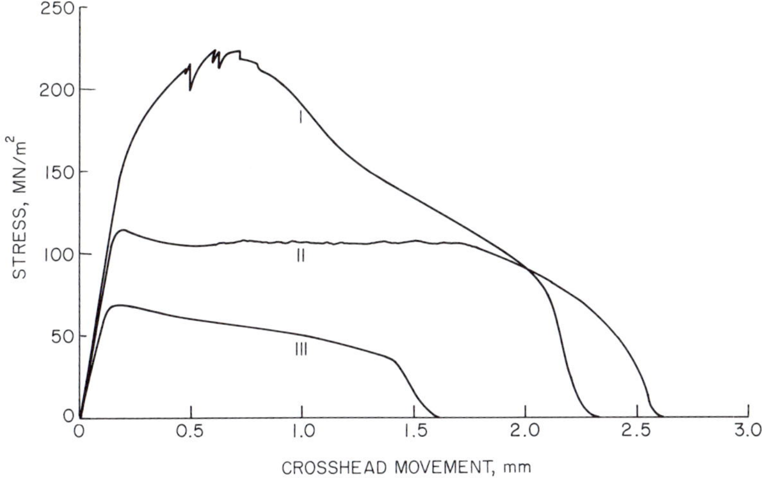

Fig. 35. Stress vs. cross-head movement for tensile tests in argon at 1100°C of (I) a hot-pressed composite of three coated rods in nichrome (20 v/o rods), (II) a hot-pressed 80:20 nichrome-V blank, and (III) 80:20 nichrome-V wire (Noone, 1970).

3. *Properties of Composites Reinforced with Tyco Filaments: Longitudinal Tensile Strength*

As discussed earlier, it was not possible to fabricate Al_2O_3–nichrome specimens by diffusion bonding of more than a single layer of Tyco filaments without excessive filament breakage (Mehan and Harris, 1971). For single-filament composite specimens, it was possible to show that the filament surface was not excessively degraded by the pressing process itself. The *in situ* filament strength was deduced by an indirect method from tensile tests of composites with low filament loading. Strain gauges were mounted on the specimen so that the specimen strain was obtained corresponding to the first filament break: filament fracture was detected both acoustically and by the drop in the load–strain curve. Assuming that the fiber was well bonded to the matrix (which transverse tests showed it was not), this specimen strain would correspond to the filament-fracture strain. By multiplying by the filament modulus, this gives an approximate *in situ* filament strength of $\leq$ 2240 MN/m^2. (The inequality accounts for the fiber debonding, which tends to result in an overestimate of the filament strength).

Composites with about 20–25 v/o of 0.25-mm-diam Tyco filament were prepared of sufficient size for tensile testing. They consisted of three layers of uncoated filaments aligned in pregrooved nichrome sheet 0.38 mm thick, up to 10 cm long, and 6.35 mm wide: a photograph of a typical composite cross section is shown in Fig. 36. The filaments in a similar specimen were partially exposed by etching away the matrix alloy with acid, and good alignment was observed, as shown by Fig. 37. A specimen of this type was tested in tension at room temperature with poor results (i.e., a tensile strength of only 584 MN/m^2): the filament volume fraction was about the critical value (18 v/o) so that appreciable reinforcement was not expected. However, filament fracture was detected (by acoustic-emission monitoring of the specimen during test) early in loading at calculated filament stresses of only 207–345 MN/m^2 (i.e., far below the *in situ* stress of 2240 MN/m^2) estimated from tests on composites with low filament volume fraction. Furthermore, even the elastic modulus was less than that calculated from the rule of mixtures. The low strength was caused by excessive filament damage, as discussed in Section IV.A: for example, out of 36 filaments extracted from one composite, only 10 were unbroken and the remainder were extensively fragmented and showed evidence of rhombohedral slip and deformation twinning.

Calow and Moore (1972) measured the tensile strength of composites prepared by hot-pressing continuous filaments in nickel matrices. Because of poor densification of composites with 20 v/o filaments in carbonyl nickel matrices, only low room-temperature strengths of about 182 MN/m^2 were realized. With reduced filament content (11–15 v/o), densification was facilitated and a tensile strength of 287 MN/m^2 was measured at room temperature. The "soft" carbonyl nickel matrix ensured that filament breakup was avoided during fabrication, but extracted-filament strength was about 1540 MN/m^2, typical of matrix-degraded material (See Table III). Considerable

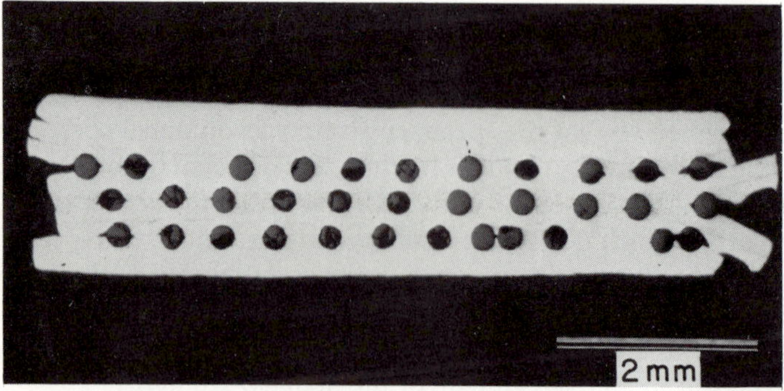

Fig. 36. Cross section of a typical hot-pressed Tyco-sapphire–nichrome composite (Mehan and Harris, 1971).

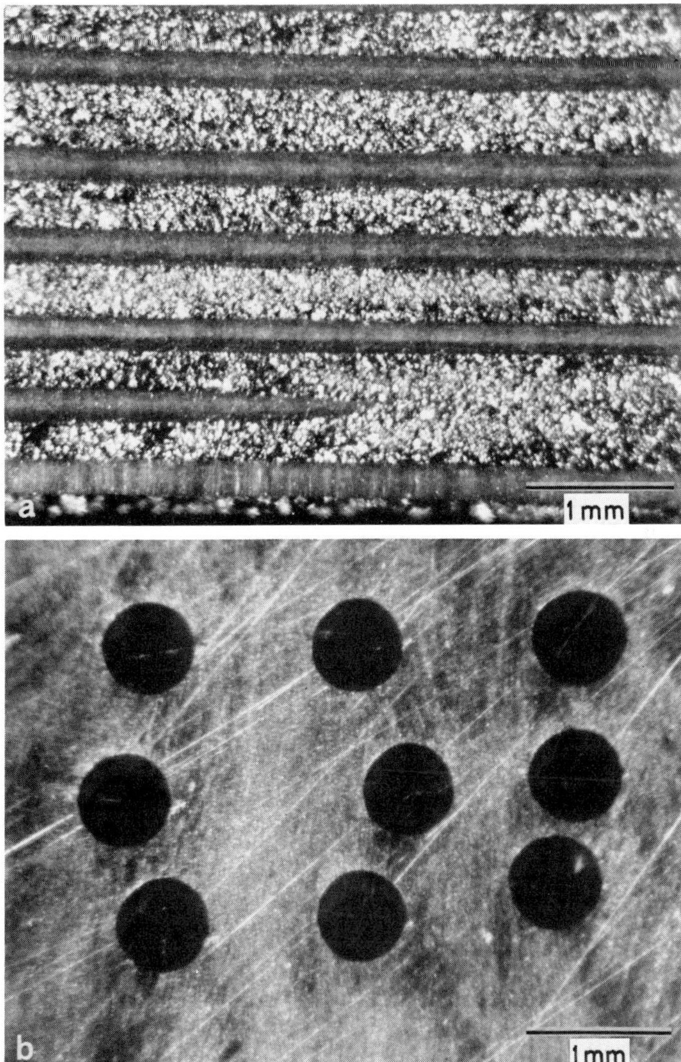

Fig. 37. Tyco sapphire filaments shown partially etched out of a nichrome-matrix hot-pressed composite, and a transverse section of the same composite. Note apparent absence of fractured filaments (Mehan and Harris, 1971).

filament fracture resulted when composites were produced using similar conditions but with nickel powder as the matrix; an ultimate strength of 355 MN/m^2 at room temperature was measured (nominally 20 v/o filament). Calow and Moore used a combination of thinly coated filaments (carbonyl nickel) to prevent filament breakage and a hot-pressed nickel-powder matrix to develop maximum matrix density: the results are summarized in Table IX. Fiber breakage could not be eliminated, but most of the fragments were found to be greater than the critical length required for reinforcement at room temperature. A specimen produced under the best conditions but tested at 1100°C in tension showed little evidence of reinforcement (Table IX). The fracture mode was characterized by high ductility and was shown to be related to the large filament diameter, as predicted by Kelly and Cooper (1967): very low ductility was observed with similarly fabricated specimens but with small diameter whiskers as reinforcements.

The lack of efficient reinforcement in these composites was attributed by Calow and Moore to the disruption of any filament–matrix bonding as a result of relative movement of filament and matrix during cooling from the fabrication temperature because of their different expansion coefficients. Since the matrix has the higher expansion coefficient, a strong mechanical bond is formed as it shrinks around the filaments on cooling after fabrication. Therefore, some strengthening can be observed at room temperature, but the mechanical bond is destroyed on reheating, so that reinforcement (i.e., load transfer across the interface) at high temperatures is not possible. Similar reasoning could be applied to the data in Tables VII and VIII, where the more optimistic properties result from strong mechanical bonding, allowing load to be transferred from the matrix to the fibers in the relatively cold grip section of the test specimens. The poor bonding in the hot zone is evident from the extensive filament pull-out. Also, since the bonding is almost entirely mechanical, the undulating surface of flame-polished fila-

TABLE IX

Composite Properties as a Result of Matrix Constitution Changes[a]

Matrix P/T[b]	Testing temperature (°C)	Density (gm/cm^3)	Ultimate tensile strength (MN/m^2)	Strengthening efficiency (%)
0.975	20	7.48	374	31.4
0.865	20	7.60	335	24.0
0.450	20	7.55	280	13.3
0.995	1100	—	33	5

[a] Calow and Moore (1972).

[b] P is the weight of nickel as powder, T the total weight of nickel.

ments provides more mechanical interlocking with the matrix than as-grown Tyco filament, and this, in addition to the higher inherent strength of the filaments, presumably contributes to the higher strength measured on composites produced with flame-polished filaments as reinforcement.

4. *Properties of Composites Reinforced with Tyco Filaments: Transverse Strength*

Transverse tensile-test specimens of Ni-Cr–Al_2O_3, prepared by diffusion-bonding 0.25-mm Tyco filaments in Ni–18.5Cr matrix, were tested as a function of heat treatment, volume fraction, and presence or absence of a sputtered titanium coating (Mehan and Harris, 1971). The specimens were prepared by pregrooving 6.35-mm-wide nichrome sheets 0.38 mm thick and 50 mm long transversely in the center. Four or five grooves were cut in three sheets, and short lengths of sapphire rods were layed in these grooves and the assembly hot-pressed at 41.5 MN/m^2 and 1100°C for 1 hr. The resulting specimens were about 1.0 mm thick. In several cases, the rods were allowed to extend beyond the specimen edges in order to prevent propagation of preexisting cracks in the ends of the rods along the length of the rods during testing. This arrangement is depicted in Fig. 38a. In all cases the failures occurred at the fiber–matrix interface, as shown in Fig. 38b and c. The result was that for all the transverse tests the composite strength was close to that predicted for the case of a low bond strength between matrix and filament (Cooper and Kelly, 1969; Chen and Lin, 1969). In fact, in no case was the strength different from that which would be predicted if the filament sites were holes. Even a heat treatment at 1200°C, which is known to give rise to a significant amount of interfacial reaction, resulted in composites with essentially zero bond strength when tested at room temperature. Sputtering a thin layer of NiCr or Ti on the filament surface prior to composite fabrication did not improve the bond strength.

The expected behavior in the presence of strong bonding would be a matrix failure with the final failure crack avoiding the filaments (Cooper and Kelly, 1969). The failure mechanism actually observed for the Ni-Cr–Al_2O_3 composite system is thus consistent with the mechanical data presented in Fig. 39 and with the hypothesis that any chemical bonds generated during fabrication are destroyed by relative movement of matrix and filament during cooling from the fabrication temperature.

5. *Thermal-Shock Testing of Flame-Polished Sapphire Rods in a Nichrome Matrix*

Thermal-shock resistance is a critical property of any material developed for gas-turbine-engine applications; therefore, testing of Al_2O_3-filament-reinforced composites in simulated engine thermal shock was attempted.

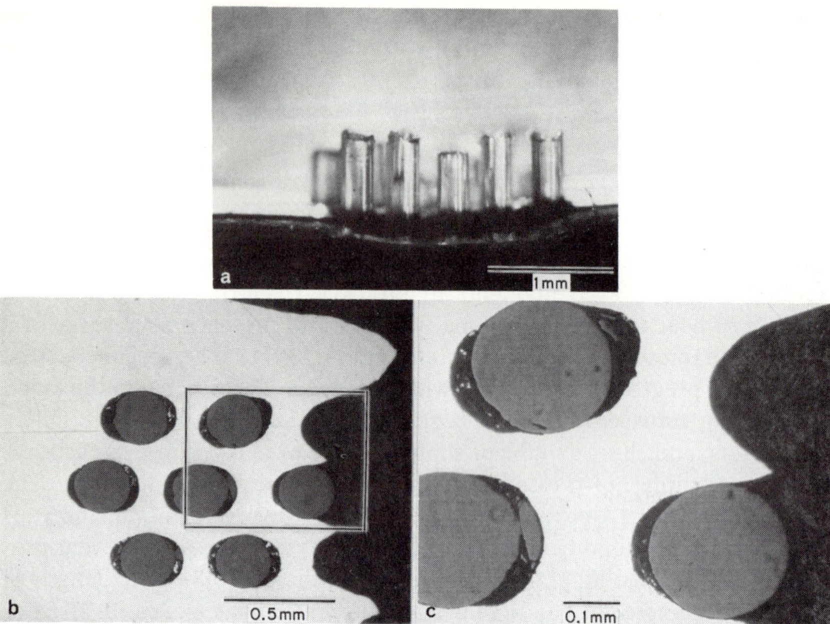

Fig. 38. Longitudinal section after a transverse tensile test of a NiCr–Al_2O_3 composite with filament ends protruding from sides of specimen: (a) side view after pressing; (b) longitudinal section after tensile testing; (c) enlarged detail from (b) (Mehan and Harris, 1971).

Composites fabricated by both electroplating and hot pressing produced specimens amenable to testing under thermal-cycling conditions. The thermal-cycling tests were performed on three-filament electroplated composites, using a heating and cooling schedule that simulated gas-turbine-engine thermal shock. One cycle of this test consisted of inserting the cold specimen into a furnace at 1200°C in 5 sec, holding the specimen in the hot zone for 100 sec, and then withdrawing it in 5 sec into a cold air blast; after 10 sec in the cold air blast, the specimen was immersed in water at 20°C (Noone, 1970).

Several composites were subjected to 20 cycles each and were examined after each cycle by transmitting a bright beam of light through the length of the filaments to determine any mechanical damage (previous experience established that a marked decrease in light transmission resulted when cracks or deformation twins were present in the filaments). One filament exhibited reduced transmission after the first cycle, but no damage was detected in any of the others. Post-test examination of the filaments after dissolution of the matrix showed that a short length had broken off the end of one of them—

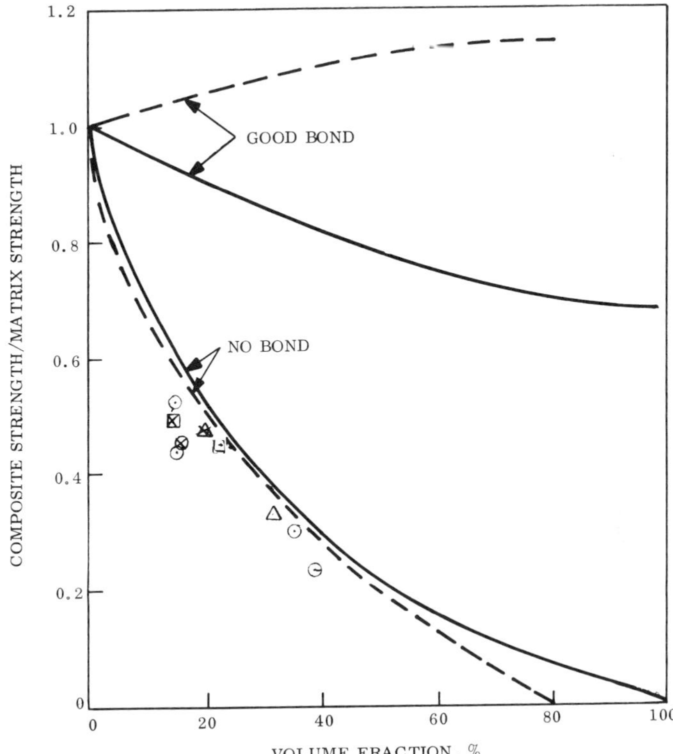

Fig. 39. Transverse strength of sapphire–nichrome composites as a function of filament volume fraction (tests conducted at room temperature): ○ as pressed; □ sputtered (NiCr); △ sputtered (Ti); × heat treated. Solid curves after Chen and Lin (1969), broken curves after Cooper and Kelly (1969) (Mehan and Harris, 1971).

possibly at a flaw in the poorly polished end of that filament. No other damage (slip bands or twins) was observed in any of the filaments, and subsequent four-point bending strengths of over 1380 MN/m^2 were measured.

A progressive increase in the length of the electroplated nickel matrix was observed after each cycle, such that at the completion of the test the ends of the reinforcing filaments were recessed into the end of the composite. One composite 7.1 cm long increased in length at the rate of 0.25 mm per cycle; it was 7.6 cm long after 20 cycles, and the filaments were recessed 2.5 mm at each end. This behavior is attributed to a ratchet mechanism caused by the large difference in expansion coefficient between matrix and filament. The matrix expands more than the filament during the heating cycle, and because bonding in the electroplated system is poor, the matrix slips slightly along the filaments. Upon cooling, the matrix is restrained by the filaments and does not regain its original length; this process is repeated on each cycle.

A different behavior was observed with hot-pressed specimens because the initial interfacial bonding was stronger than in electroplated specimens and, since fabrication was performed at high temperature, the filaments were in compression at room temperature and the matrix could not expand away so readily from the filaments on reheating as it did in the electroplated specimens. Thermal cycling tests were performed on two three-filament hot-pressed composites (flame-polished ruby filaments with W coatings and Ni overcoatings, pressed in nichrome), using the same procedure as described above for electroplated specimens. After one cycle with the first specimen, a reduction in light transparency along all three filaments was observed. This is indicative of filament breakage, and the matrix was dissolved away to reveal that each filament was fractured into three major pieces of varying lengths. The bending strength of one of the fragments was nevertheless 2760 MN/m^2. The second specimen was cycled more slowly (15 sec to insert and withdraw from the furnace) to relieve matrix stresses, and no damage was observed. After 5 cycles, the severity of the shock was increased (10 sec to insert and to withdraw from the furnace) and again no damage was observed. After 10 cycles, the original severe schedule was resumed for 10 further cycles and no damage was observed in the filament.

The matrix–filament bonding (mostly mechanical) appeared to be much improved in the hot-pressed specimens (as expected), since sufficient stresses were transferred to the rods to cause fracture in the first specimen. The behavior of the second specimen suggested that the initial slow cycling annealed out matrix stresses that arose from the pressing operation and no doubt weakened the mechanical bond such that subsequent cycling of greater severity could be performed without damage to the filaments. An encouraging sign from this specimen was the virtual absence of the ratchet mechanism that was observed during cycling of electroformed composites.

6. *Bending Tests at 1100°C of a Composite Reinforced with Flame-Polished Filaments*

High-temperature four-point bending tests were performed on a six-filament hot-pressed composite and on several individual filaments to determine the limitations in both stress and temperature that may be imposed on this composite system by the onset of slip and plastic deformation in the reinforcing filaments. The six-filament composite was fabricated in the form of a rectangular beam $4.5 \times 3.6 \times 3.5$ mm with the filaments (flame-polished ruby with W + Ni coatings) in two layers (each of three filaments) near the top and bottom faces (3.6 mm wide) of the specimen. The filament volume fraction overall was only 12 v/o, but near these faces it was about 20 v/o. This specimen was tested in bending at 1100°C. Considerable plastic deformation was induced in the specimen (5 mm of crosshead movement with an outer test span of 3.8 cm), and when the filaments were subsequently extracted from the matrix, extensive fragmentation of some, and plastic deformation (bending) of others, was observed: the fragments are shown in Fig. 40. A large amount of basal slip and deformation twinning was evident in all filaments, together with surface degradation as a result of interfacial reactions. The stress level during bending at 1100°C was only 48.3 MN/m^2 (outer fiber of the composite beam), which is much less than the strength of similar filaments with as-polished surfaces (Table II).

V. Concluding Remarks

It is evident from Section IV that the quest for useful composites of nickel matrices reinforced with sapphire filaments has not been particularly fruitful. While the authors would not agree that the system is hopeless, the road to a realization of the properties predicted by the rule of mixtures is certainly fraught with difficulties. Many of these difficulties clearly are common to all composite systems of metal matrices reinforced by brittle ceramic filaments, and yet several useful materials have been produced and properties measured that give rise to optimism for the future exploration of other systems, including Ni–Al_2O_3, under appropriate circumstances. For example, boron–aluminum composites are now extensively used in aerospace structure and boron–titanium and carbon–aluminum composites are being studied for similar applications.

By definition, a filament-reinforced composite material with attractive mechanical properties can be achieved only if the filaments have inherently high strength and high elastic modulus. Therefore, a primary objective of the authors' studies with sapphire filaments has been to employ filaments with inherently high strength and then to attempt to preserve that strength while incorporating them into the metal matrix. To this end, avoidance of chemical interactions between matrix and filament by use of appropriate coatings has been emphasized in much of the experimental work described, and development of chemical bonding between filament and matrix was deliberately avoided. It has been shown that preventing matrix interactions in corrosive nickel-based matrices is a difficult technological feat that has not yet been satisfactorily achieved.

In early stages of composite-materials development, it was considered that development of a high-strength chemical bond between matrix and filament was desirable for attainment of maximum composite properties. It is now

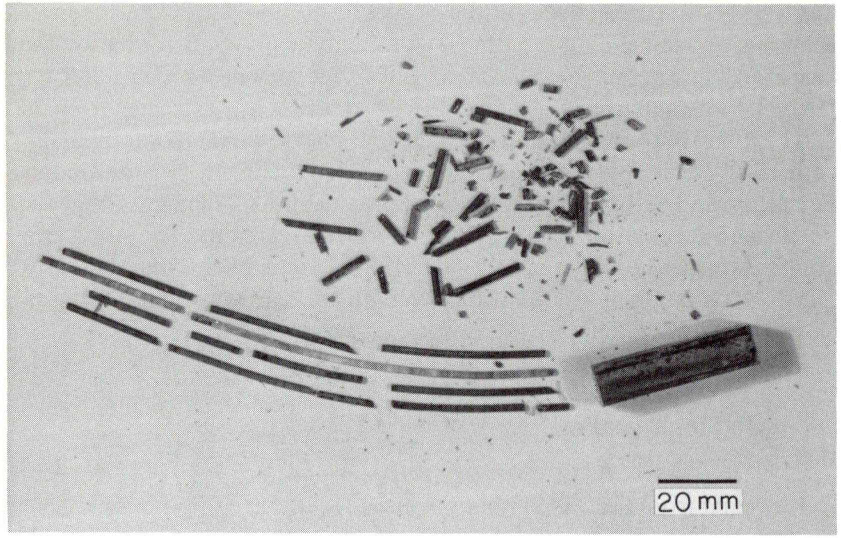

Fig. 40. Fragments extracted after a bending test at 1100°C from a composite specimen of six flame-polished α-Al_2O_3 (ruby) filaments with W coatings and Ni overcoatings hot pressed into a nichrome matrix (a short length of the specimen, shown on the right, was cut off before the remainder was immersed in acid to extract the filaments) (Noone, 1970).

generally accepted that there is an optimum bond strength and that this strength may be much less than the maximum possible chemical bond, and indeed may be satisfied by purely mechanical bonding. Since the results in Section IV show that chemical bonds may in fact be destroyed by relative movement of matrix and filament during thermal cycling, it is also clear that avoidance of interfacial reactions (and preservation of filament strength) justifiably takes priority over approaches that deliberately aim to develop chemical bonds that may subsequently be disrupted. With both approaches the only effective bonding is mechanical, but in the case of the surface-protected filaments, their strength remains high enough to offer effective reinforcement if this mechanical bond can be made adequate to transfer loads to the filaments.

Uncoated sapphire filaments interact with nickel-based matrices at high temperatures and result in filament strength of only about 1500 MN/m^2 at room temperature (i.e., a strength of about 700 MN/m^2 at about 1000°C where reinforcement is desired). It has been suggested that this is adequate for reinforcement purposes and that efforts to preserve pristine filament strength are not warranted. It is clear, however, that to attain a particular composite strength would require a higher volume fraction (about double) of these weaker filaments than of surface-protected filaments. As detailed in Section

III, the maximum practical filament loading for feasible fabrication processes with nickel-based matrices is about 25 v/o (and even at this level the composite may be so highly loaded with the brittle phase as to have poor impact resistance). Calculations based on the rule of mixtures show that with 25 v/o of degraded filaments the predicted tensile properties are not competitive with those of conventional superalloys. As a result of the continued improvement in superalloy development, it is evident that, even with surface-protected, high-strength filaments, it may not be possible to compete (either on strength or even strength-to-density ratios at filament loadings below 40 v/o) with current developmental alloys for gas-turbine service.

Considerable progress has been demonstrated in the ability to protect the surfaces of sapphire filaments from matrix interactions using stable oxide coatings. The filaments therefore have the potential to reinforce the matrix if loads can be transferred to them by adequate bonding. Development of chemical bonds must be avoided, since they are the source of surface (and hence, strength) degradation, and in any event they have been shown to be disrupted readily by subsequent thermal cycling of the resulting composite. Therefore, mechanical bonding must be encouraged, and evidence suggests the possibility of developing efficient mechanical interlocking between filaments and matrix by varying the cross section of the filaments, using exaggerated surface ripples produced during growth or flame polishing (see, for example, the fine-scale surface ripple in Fig. 3d). Better bonding was observed in high-temperature tests using filaments with rippled surfaces than was obtained using filaments with more nearly cylindrical surfaces. The potential and limitations of the bonding that can be attained by matrix shrinkage and therefore mechanical gripping of filaments with deliberately rippled surfaces (yet smooth and defect-free) deserves to be studied further.

It may be noted that the stiffness contributed by the filaments to the metal matrices has almost without exception corresponded to that predicted by the rule of mixtures. There are critical applications in metal systems where high elastic modulus would be valuable even if the increased stiffness provided by reinforcement was not accompanied by a corresponding increase in strength. Situations in which resistance to bending is a critical requirement have not been explored as opportunities for the application of composites: below about 900°C (i.e., before plastic deformation becomes appreciable) sapphire filaments, ribbons, etc. provide a potentially useful stiffness reinforcement.

Plastic deformation in the filaments has been shown to limit the temperature at which sapphire can provide effective reinforcement. Depending on filament orientation and direction of stress, basal slip is activated at about 900°C and rhombohedral slip at 1200°C. There is evidence from short-term bending tests that the stress required to initiate deformation increases for filaments with defect-free surfaces, but whether this effect can be exploited in practice remains to be established.

Acknowledgments

The authors would like to thank their colleagues at the General Electric Company for their support over almost a decade of work relating to sapphire-reinforced metals. The opinions expressed in this chapter are those of the authors and do not necessarily represent the views of the General Electric Company.

References

Anon. (1965). Engineering Alloy Digest, Inc.

Anon. (1970). *Aviat. Week Space Techn.* **92**, 35.

Brennan, J. J. (1970). High-Temperature Metal-Matrix Composites. Rep. Prepared for Refractory Compos. Working Group Meeting, Williamsburg, Virginia, May 1970.

Brattan, R. J. (1969). *J. Amer. Ceram. Soc.* **52**, 417.

Calow, C. A., and Moore, A. (1972). *J. Mater. Sci.* **7**, 543.

Chen, P. E., and Lin, J. M. (1969). *Mater. Res. Std.* **2**, 29.

Chorné. J., Sutton, W. H., Bruch, C., and Feingold, E. (1966). Development of Composite Structural Materials for High Temperature Applications. General Electric Co. Rep. R66SD11.

Chorné. J., Bruch, C. A., Jakas, R., and Sutton, W. H. (1968). Development of Composite Structural Materials for High-Temperature Application. Final Rep., Contract N00019-67-C-243.

Cooper, G. A., and Kelly, A. (1969). Interfaces in Composites, p. 90. ASTM STP 452.

Dean, A. V. (1967). *J. Inst. Metals* **95**, 79.

Glassner, A. (1959). The Thermochemical Properties of Oxides, Fluorides, and Chorides to 2500°K. Argonne Nat. Lab., Rep. ANL-5750.

Golland, D. I., and Beevers, C. J. (1972), *J. Mater. Sci.* **7**, 716.

Heuer, A. H. (1970). *Proc. Brit. Ceram. Soc.* No. 15, 173.

Hurley, G. F., (1972). *Proc. ACS Symp. Reinforcing Fibers, New York* August 1972. To be pub-published as part of the *Proc. Org. Coatings Plast. Div., 1973.*

Kelly, A., and Cooper, G. A. (1967). *J. Mech. Phys. Solids* **15**, 279.

Kelsey, R. H. (1963). Reinforcement of Nickel-Chromium Alloys with Sapphire Whiskers. Final Rep., Contract NOw 53-0138-C.

Kronberg, M. L. (1962). *J. Amer. Ceram. Soc.* **45**, 274.

LaBelle, H. E., Jr., (1971), *Mater. Res. Bull.* **6**, 581.

Mallinder, R. P., and Proctor, B. A. (1966). *Phil. Mag.* **13**, 197.

Mehan, R. L. (1970a). Stability of Oxides in Metal or Metal Alloy Matrices. Tech. Rep. AFML-TR-70-160.

Mehan, R. L. (1970b). *J. Compos. Mater.* **4**, 90.

Mehan, R. L. (1972). *Met. Trans.* **3**, 897.

Mehan, R. L., and Feingold, E. (1967). *J. Mater.* **2**, 239.

Mehan, R. L., and Harris, T. A. (1971). Stability of Oxides in Metal or Metal-Alloy Matrices. Tech. Rep. AFML-TR-71-150.

Mehan, R. L., and Herzog, J. A. (1970). *In* "Whisker Technology" (A. P. Levitt, ed.), pp. 157–195. Wiley, New York.

Moore, A., and Calow, C. A. (1969). *Commonwealth Mining Met. Congr., 9th* **4**.

Moore, T. L. (1969). *In* Metal Matrix Composites. DMIC Memo 243.

Morley, J. G., and Proctor, B. A., (1962). *Nature* (*London*) **196**, 1082.

Mullin, J. V., Berry, J. M., and Gatti, A. (1968). *J. Compos. Mater.* **2**, 82.

Noone, M. J. (1970). Development of Composite Materials for High-Temperature Application. Final Rep., Contract N00019-69-C 0310

Noone, M. J., and Heuer, A. H. (1972). The Science of Ceramic Machining and Finishing, NBS Spec. Publ. No. 348, p. 213.

Noone, M. J., and Sutton, W. H. (1968). Investigation of Bonding in Oxide-Fiber (Whisker) Reinforced Metals. Final Rep., Contract DA-19-066-AMC-330(X).

Noone, M. J., Feingold, E., and Sutton, W. H. (1969). *In* Interfaces in Composites," ASTM STP 452, pp. 59–89. ASTM, Philadelphia, Pennsylvania.

Noone, M. J., Mehan, R. L., and Sutton, W. H. (1969). Development of Composite Structural Materials for High-Temperature Applications. Final Rep. Contract N00019-68-C-0304.

Pettit, T. S., Randklev, E. H., and Felten, E. J. (1966). *J. Amer. Ceram. Soc.* **52**, 199.

Pollock, J. T. A. (1972a). *In* Nat. Bur. Std. Spec. Publ. No. 348, Sci. Ceram. Machining and Finishing, p. 247.

Pollock, J. T. A. (1972b). *J. Mater. Sci.* **7**, 787.

Pollock, J. T. A., (1972c). *J. Mater. Sci.* **7**, 631.

Schuster, D. M., and Scala, E. (1970). *In* "Whisker Technology" (A. P. Levitt, ed.), pp. 403–441. Wiley, New York.

Seybolt, A. U. (1968) *In* "Oxide Dispersion Strengthening" (G. S. Ansell *et al.*, ed.), Vol. **49**, pp. 469–487. Gordon and Beach, New York.

Sicka, R. W., Harkulich, T., Vukasovich, M. S., and Kelsey, R. H. (1964). Reinforcement of Nickel Chromium Alloys with Sapphire Whiskers. Final Rep., Contract NOw 64-0125c.

Snajdr, E. A., and Williford, J. F. (1968). Investigation of Fiber-Reinforced Metal-Matrix Composites Using a High-Energy-Rate Forming Method 14th Refractory Compos. Working Group Meeting, AFML-TR-68-129, pp. 133–162.

Stapley, A. J., and Beevers, C. J., (1969). *J. Mater. Sci.* **4**, 65.

Sutton, W. H. (1970). *In* "Whisker Technology" (A. P. Levitt., ed.), pp. 273–342. Wiley, New York.

Sutton, W. H., and Chorné, J. (1965). *In* "Fiber Composite Materials" (S. H. Bush, ed.), pp. 173–222. ASM, Metals Park, Ohio.

Sutton, W. H., and Feingold, E. (1966). *Mater. Sci. Res.* **3**, 577.

Sutton, W. H., Chorné, J., Gatti, A., and Sauer, W. E. (1965). Development of Composite Structural Materials for High-Temperature Applications. Final Rep., Contract NOw-60-0465 d.

Tyco Laboratories, Inc. (1968). Waltham, Mass. Product Bull. 102, September.

Tressler, R. E., and Moore, T. L. (1970). *Met. Eng. Quart.* **11**, 16.

Wachtman, J. B., Jr., and Maxwell, L. H., (1959). *J. Amer. Ceram. Soc.* **42**, 432.

5

Wire-Reinforced Superalloys

ROBERT A. SIGNORELLI

Lewis Research Center
Cleveland, Ohio

I. Introduction

The need for improved materials at elevated temperatures has stimulated research to develop superalloy-matrix fiber composites. Alternate materials, such as superalloys or refractory-metal alloys, are limited in strength, toughness, oxidation resistance, or ductility. Two of the composites being developed for elevated-temperature use, directionally solidified eutectic superalloys and aluminum-oxide-filament-reinforced nickel alloys, are described in preceding chapters. Refractory-alloy-wire-reinforced superalloys have some advantages over these materials that make them better suited for certain applications. Composites reinforced with high-strength nonmetallic fibers, such as aluminum oxide or carbon, offer the potential for strength-to-density values considerably above those reinforced with refractory-alloy wire. However, serious problems have been encountered in the fabrication of ceramic- and carbon-fiber reinforced superalloy composites. In addition, refractory-alloy wire can be plastically deformed, which can contribute to increased composite ductility and impact resistance and can facilitate fabrication.

Directionally solidified eutectics offer the potential for creep-rupture strength-to-density values greater than those for superalloys, but less than those of refractory-wire–superalloy composites. Also, composites made by combining wire and superalloys have the advantage of controlled, variable, multiaxial strengthening. The variation of amount and angular orientation of cross-ply fibers gives a greater degree of freedom than is possible with as-grown eutectic reinforcement, whether the strengthening phase be fibrous or lamellar. Further, the refractory-wire reinforcement is more easily deformed, which increases ductility and toughness. Refractory-wire-reinforced superalloy composites have been rolled in the axial fiber direction to achieve composite reduction of area values over 70%. Tungsten-wire-reinforced copper composites have been rolled to over 90% reduction in area values, with no apparent fiber damage. This indicates the ability to deform this type of composite to aid in fabrication of components. It should be noted, however, that superalloys or superalloy-matrix composites require an oxidation-resistant outer envelope, which may be a cladding or a coating, for the severe turbine-blade environment.

A significant application for such composites is for turbojet-engine turbine blades for service in the 1000–1200°C temperature range. This application requires a service life of several thousand hours in an environment of high-velocity, hot, corrosive gas with vibratory and thermally induced cyclic stresses and, occasionally, high-velocity impact from debris passing through the engine. Other applications might be for sheet-metal components in hot gas streams. A need for such materials has been encountered in reducing the noise from jet-engine exhausts. Another application might be for thermal protection of reentry bodies such as the space shuttle.

Compatibility of the fiber-reinforcement wire and matrix is a very important consideration in developing metal-wire-reinforced metal composites (Signorelli *et al.*, 1967; Petrasek and Weeton, 1963). The importance of interfacial reaction has been recognized, and mutually insoluble components have been chosen for wire-reinforced model system studies (McDanels *et al.*, 1963; Petrasek, 1965; McDanels *et al.*, 1967; Petrasek *et al.*, 1967; Jech and Signorelli, 1968; Kelly and Tyson, 1965). Excellent stress-rupture properties have been demonstrated to over 0.9 of the matrix homologous temperature with mutually insoluble model system studies of tungsten-wire–copper composites. Unfortunately, the alloying elements of typical superalloys are reactive with refractory-alloy wire. Therefore, reaction must be controlled to minimize loss of strength and ductility. Most refractory-wire–superalloy-matrix combinations easily develop adequate bonding to permit fibers to be stressed to the fracture stress in the composite. While bonding is not a problem, metallurgical reaction can easily degrade properties.

The potential of refractory-wire-reinforced superalloys has been recognized and research conducted to study them, particularly at the Lewis Research Center of NASA and the National Gas Turbine Establishment in England. This chapter will review the problems and progress encountered in developing this material. The problem of matrix–fiber reaction must be overcome to produce composites with reasonable properties. The subject of compatibility is followed by discussion of means of improving composite properties. Fiber development is described, since fibers with improved properties would form the basis for improved composite properties. Matrix composition is discussed with relation to its functions—to improve compatibility and provide strength, oxidation protection, and ductility. Advantages and disadvantages of different fabrication methods are discussed, and the current state of development of the composite material is described by indicating the properties achieved. Areas where little data have been obtained are also indicated. The final section indicates some requirements for further development of the material and for application to engineering components.

II. Demonstration of the Compatibility Problem

A model-system study was conducted by Petrasek and Weeton (1963) to investigate the effect of alloying elements on the tensile and microstructural properties of metal-fiber-reinforced composites. Copper-based binary alloys were used as a matrix for tungsten-fiber composites. Elements were chosen to form binary copper alloys that would isolate the effect of individual alloying elements on matrix-fiber reaction. Data obtained for

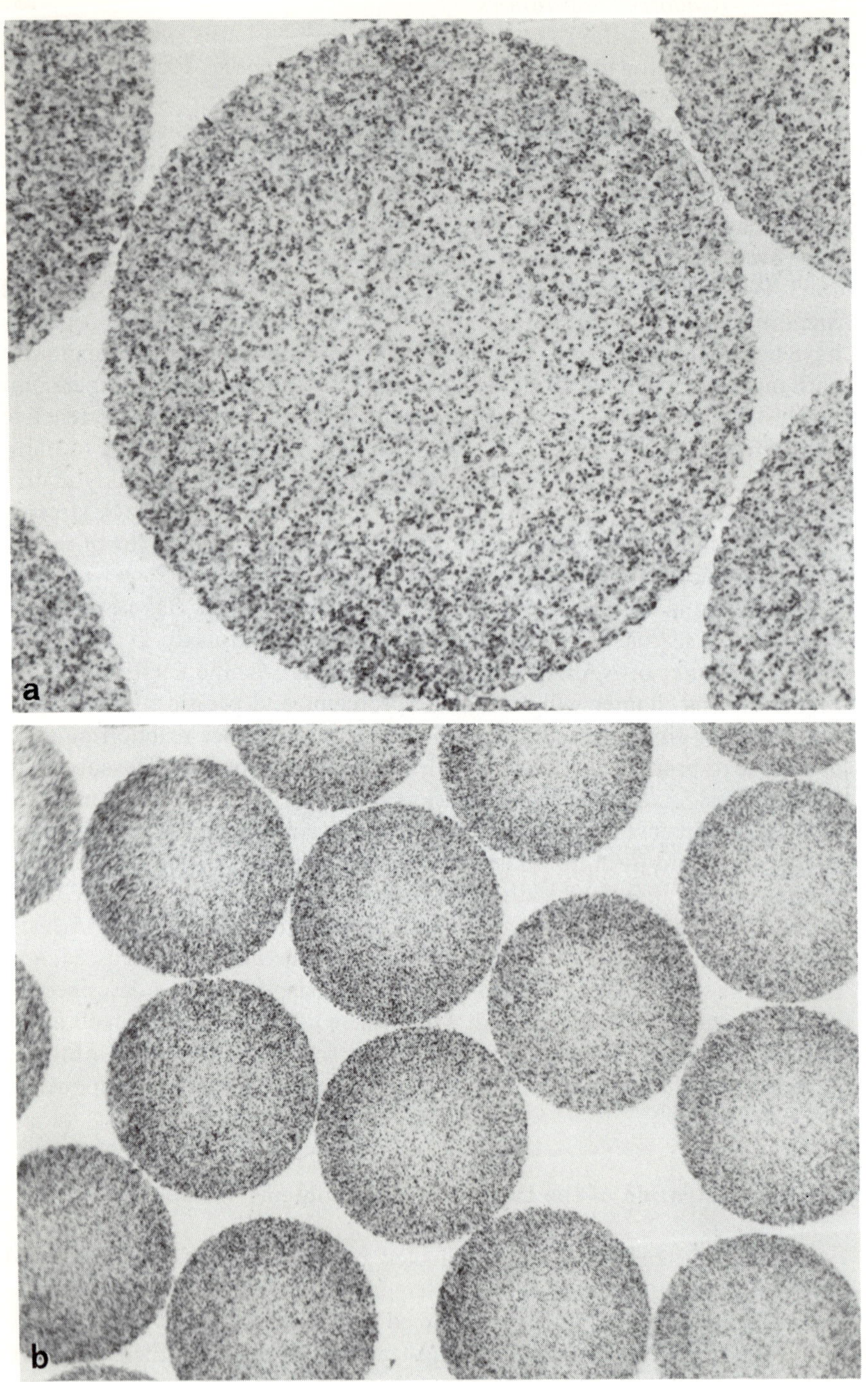

Fig. 1. Tungsten-wire–copper composite: (a) 675 ×, (b) 225 ×.

solute elements in the model system can be related to the behavior of these elements in superalloys. These effects serve as the basis for modifying the superalloy matrix composition to control matrix–fiber reaction.

Copper binary alloys were liquid-phase infiltrated into bundles of tungsten fibers, using infiltration conditions identical to those used previously for pure-copper-matrix composites. The alloying-element content of each copper binary was limited to an amount that would allow a melting temperature of

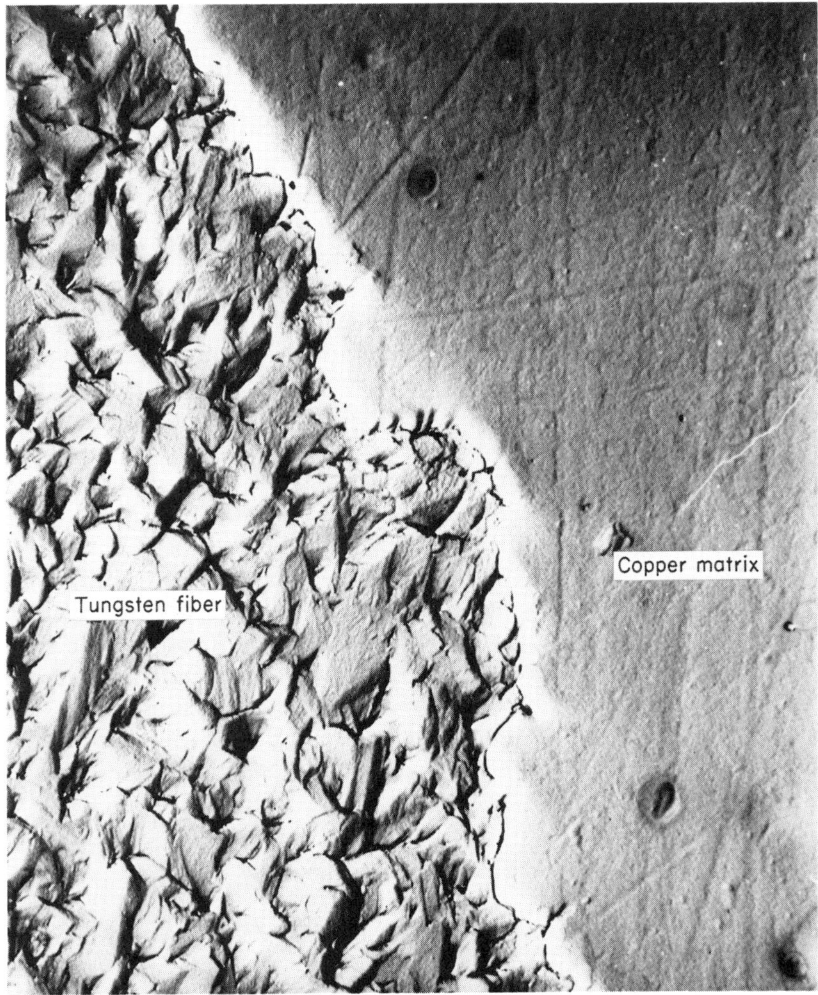

Fig. 2 Interface of as-cast tungsten–copper composite (15,000 ×).

1150°C or lower to permit infiltration at 1200°C. The effects of the alloying elements on the properties of tungsten-fiber composites were compared with the effects of mutually insoluble pure-copper matrix. The alloying elements studied were aluminum, chromium, cobalt, niobium, nickel, and titanium.

A. Microstructure

Composite specimens containing full-length oriented fibers were examined for microstructural variations and tested in tension at room temperature. The microstructural and tensile data obtained for the binary-matrix–tungsten composites were compared with those for the mutually insoluble copper matrix with tungsten. Reactions that occurred at the fiber–matrix interface were related to differences in tensile strength and ductility. Three types of reaction were found to occur: (1) diffusion-penetration reaction accompanied by a recrystallization of a peripheral zone of the tungsten fiber; (2) precipitation of a second phase in the matrix near the periphery of the fiber, with no accompanying recrystallization; (3) a solid-solution reaction with no accompanying recrystallization in the fiber. Photomicrographs illustrating the absence of reaction with a copper matrix and the three types of reaction are shown in Figs. 1–6. No reaction or recrystallization is visible at the copper-matrix–tungsten-fiber interface, Figs. 1 and 2. This would be expected for the mutually insoluble combination. The diffusion of matrix solute into the fiber, which caused recrystallization at the periphery of the fiber, is shown in Fig. 3. The behavior shown for copper–10% nickel matrix was typical for alloy matrix composites containing cobalt, aluminum, or nickel. The greatest depth of penetration occurred with aluminum, followed by cobalt and nickel. Higher concentrations of each additive element caused greater penetration. The second type of reaction, precipitation of a second phase, occurred with composites containing titanium or zirconium additions to the copper matrix. This reaction is illustrated in the photomicrographs in Figs. 4 and 5. Reductions in matrix ductility were attributed to this type of reaction, but no recrystallization of the fiber was noted. The third type of reaction noted, solid solution without recrystallization, is illustrated in Fig. 6. This behavior was observed for composites with chromium and niobium additive elements to the copper.

B. Effect on Room-Temperature Tensile Strength

Tensile properties of the composites having reactive copper-alloy matrices were reduced relative to those having nonreactive mutually insoluble unalloyed copper matrices. The decrease in room-temperature tensile

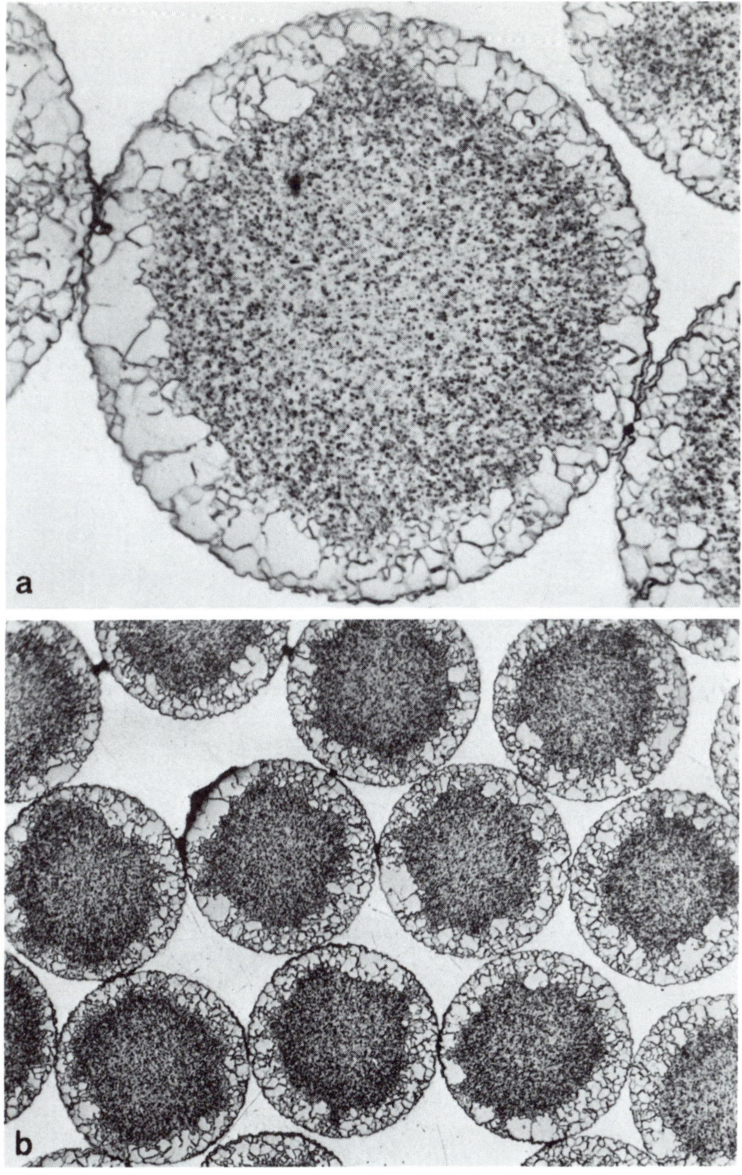

Fig. 3. Recrystallization of tungsten fibers in Cu–10% Ni matrix: (a) 750 ×, (b) 250 ×.

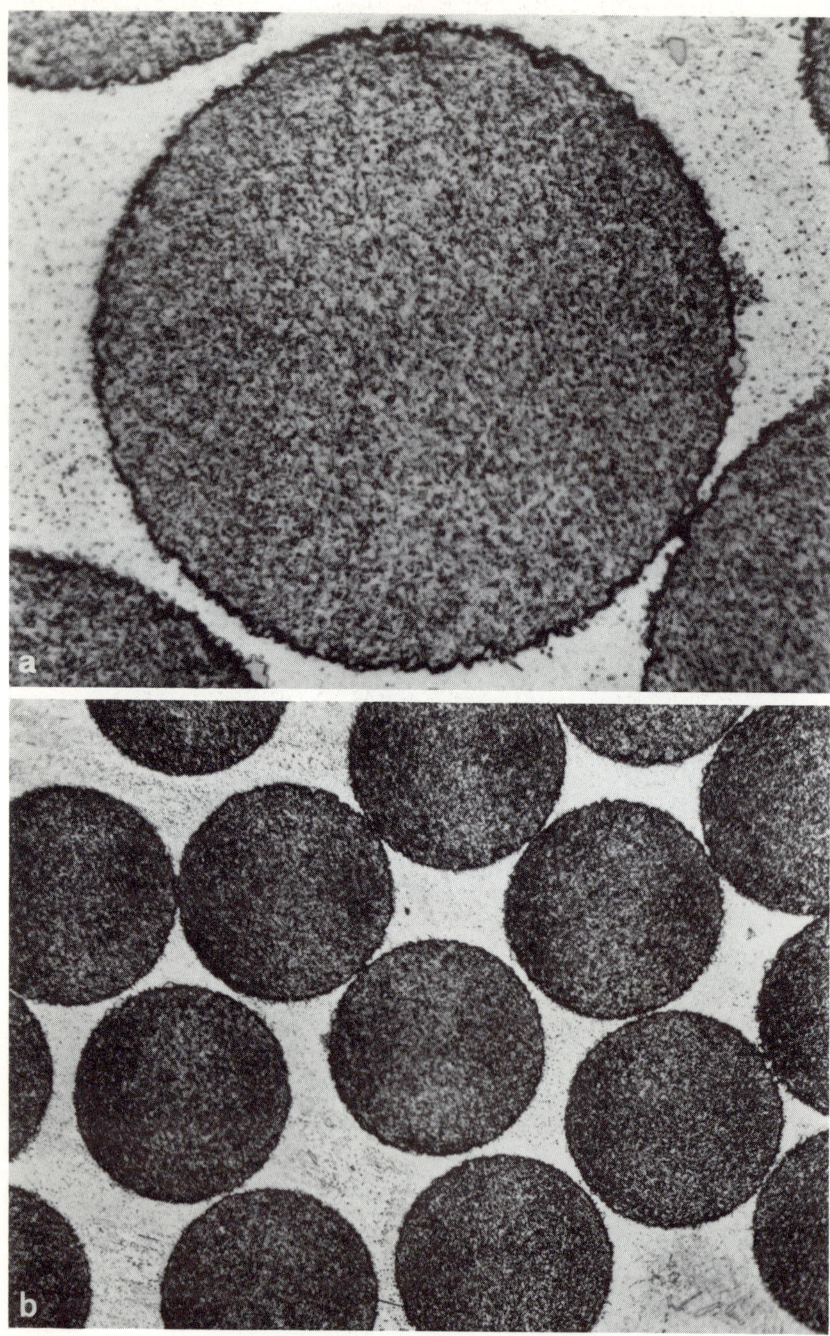

Fig. 4. Tungsten wires in Cu–10 % Ti matrix: (a) 675 ×, (b) 225 ×.

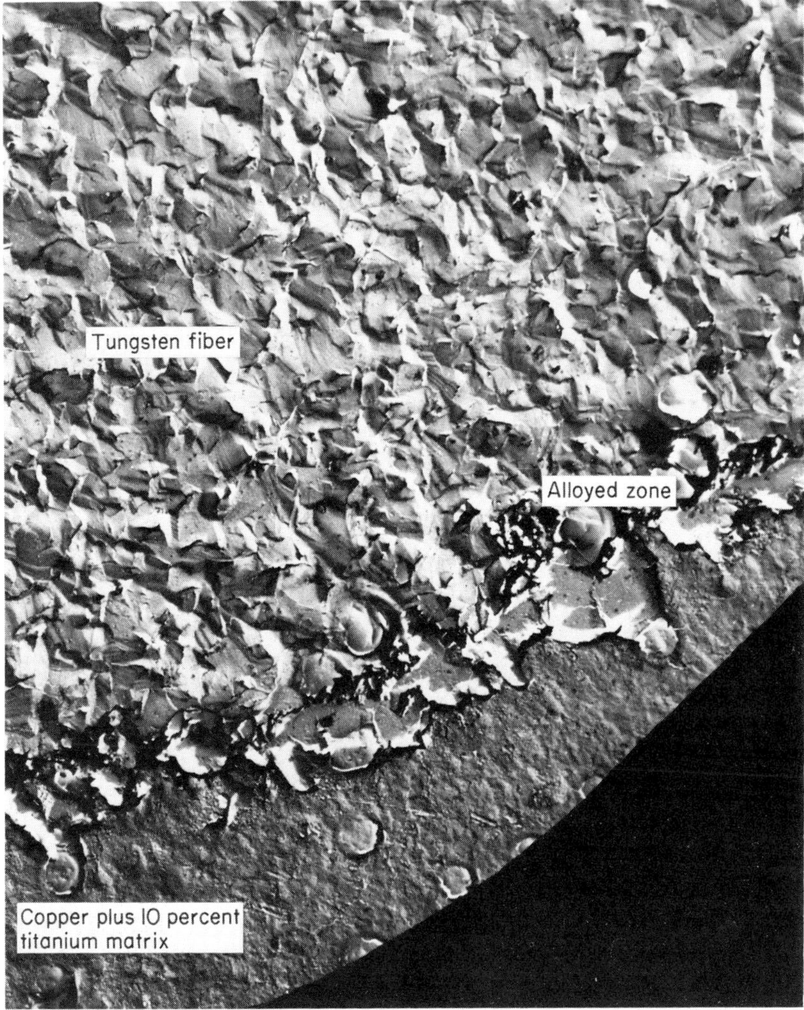

Fig. 5. Electron photomicrograph of tungsten wire in Cu–10 % Ti matrix (9000 ×).

strength varied from 7 to 62 %. The greatest damage occurred with the penetration–recrystallization reaction. The two-phase and solid-solution reactions caused relatively little damage to composite properties. The alloying additions that had low solubility in tungsten at the infiltration temperatures caused the greatest damage. However, all additions caused

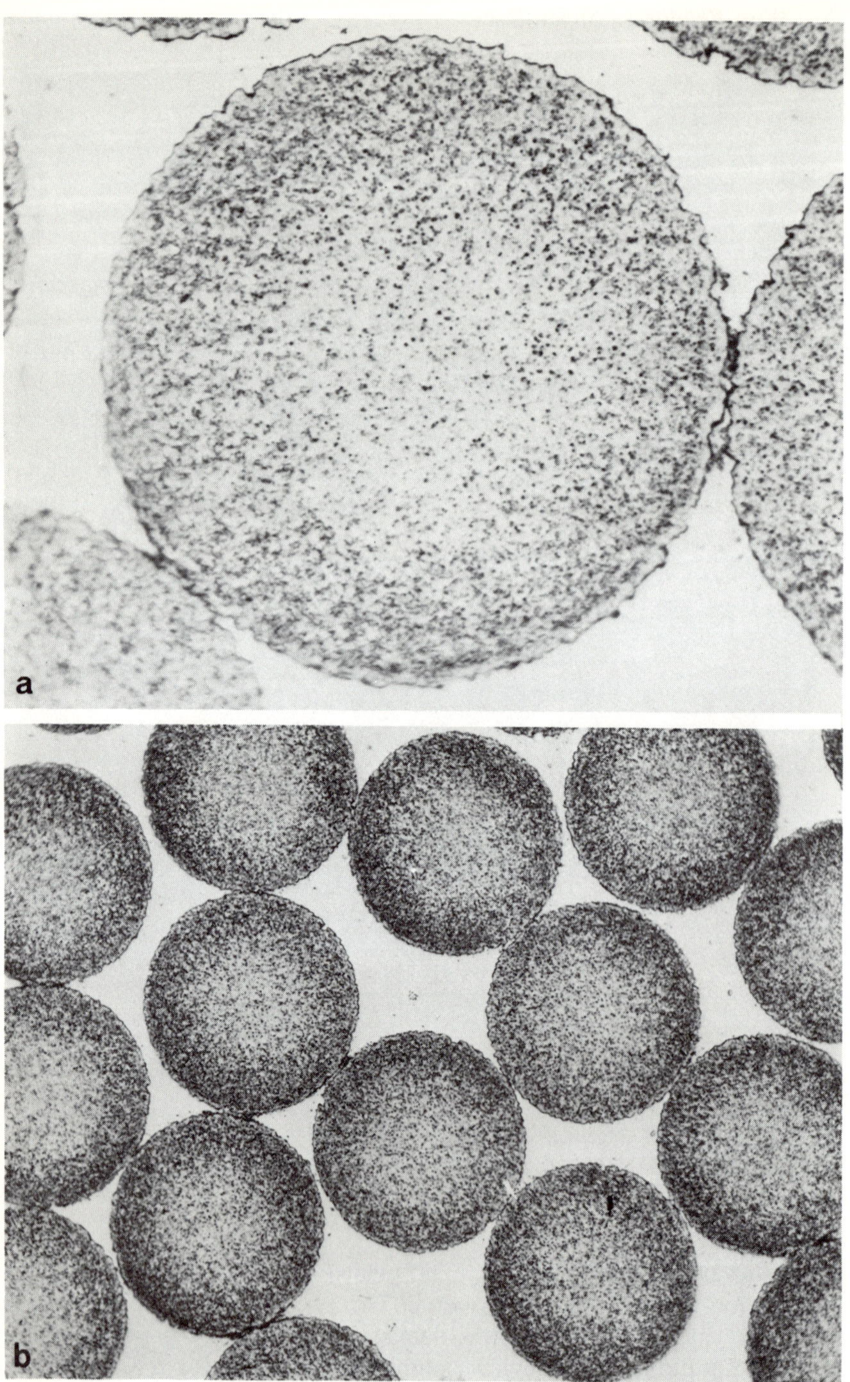

Fig. 6. Tungsten wire in Cu–2% Cr matrix: (a) 675 ×, (b) 225 ×.

damage to tensile properties. The loss was caused by reactions that occurred during the 1-hr period at 1200°C when the fiber bundles were being infiltrated with molten matrix to form composite specimens.

C. *Reaction-Induced Fiber Embrittlement*

Reduction in ductility accompanied reduction in tensile strength, and both were related to the depth of the reaction zone in the fiber. As reaction-zone depth increased, composite properties decreased. Tensile strength is expected to decrease with increasing depth of reaction zone, since the volume of unalloyed and unrecrystallized fiber has been decreased. However, the decrease in strength was larger than that expected from the increased area of recrystallization of the fiber and was associated with brittle failure. The decrease in tensile strength and ductility was associated with a notch effect. Simulation of a notch effect by a brittle surface layer, such as that of fully recrystallized tungsten at room temperature, has previously been shown by Stephens (1961) and Steigerwald and Guarnieri (1962). The fracture of metal specimens with brittle-surface-layer depths was virtually the same as that of the same metal with equivalent notch depths. The brittle surface layer of recrystallized tungsten seems to act as a circumferential notch, reducing the

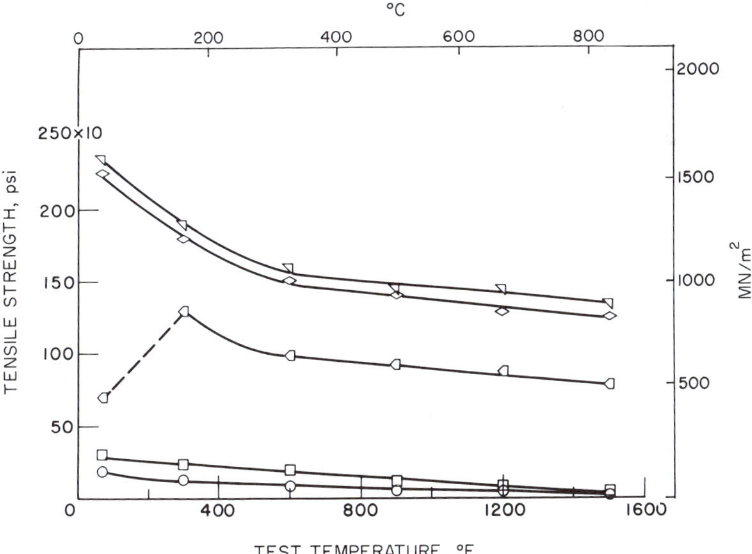

Fig. 7. Tensile strength vs. test temperature for composites containing 70 v/o tungsten fibers and for the individual constituents. ○ copper, □ copper–2% chromium, ◁ copper composite, ◇ copper–2% chromium composite, ◁ copper–10% nickel composite.

strain to failure. The reduction in fiber properties caused by penetration and recrystallization was less severe at test temperatures above the fiber brittle-ductile transition temperature (Petrasek, 1965). Copper-matrix, copper–10% nickel-matrix, and copper–2% chromium-matrix tungsten-fiber composites were tested at temperatures from room temperature to 980°C. The general

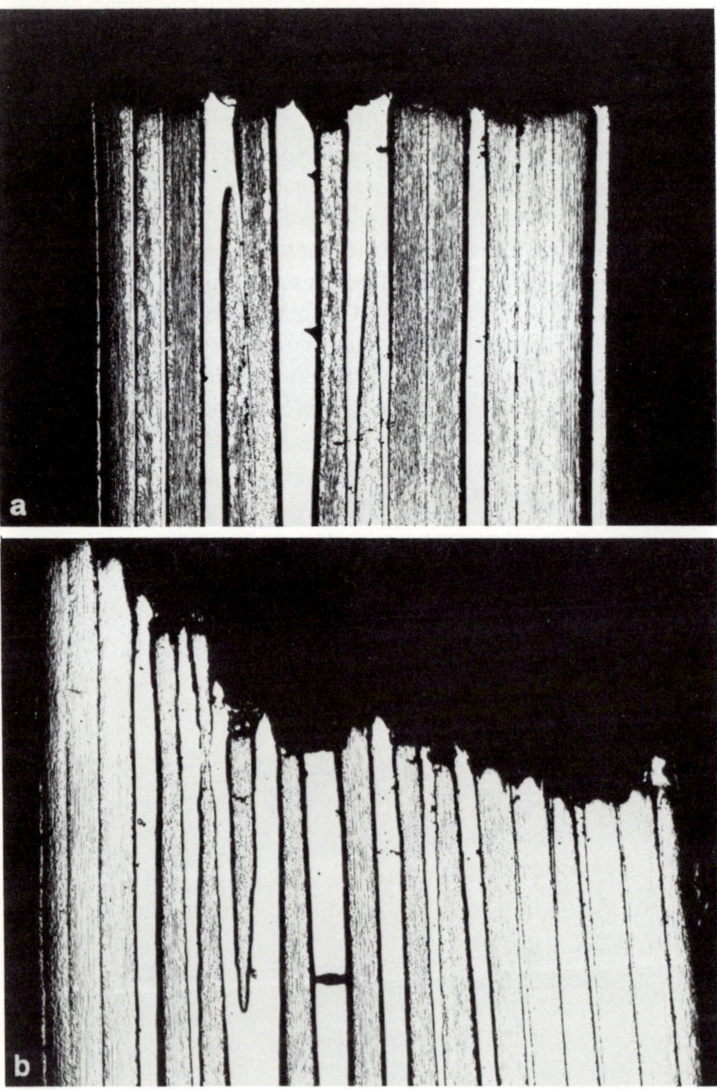

Fig. 8. Fracture edge of tungsten fiber in Cu–10% Ni matrix: (a) room temperature (50 ×); (b) test temperature, 150°C (50 ×).

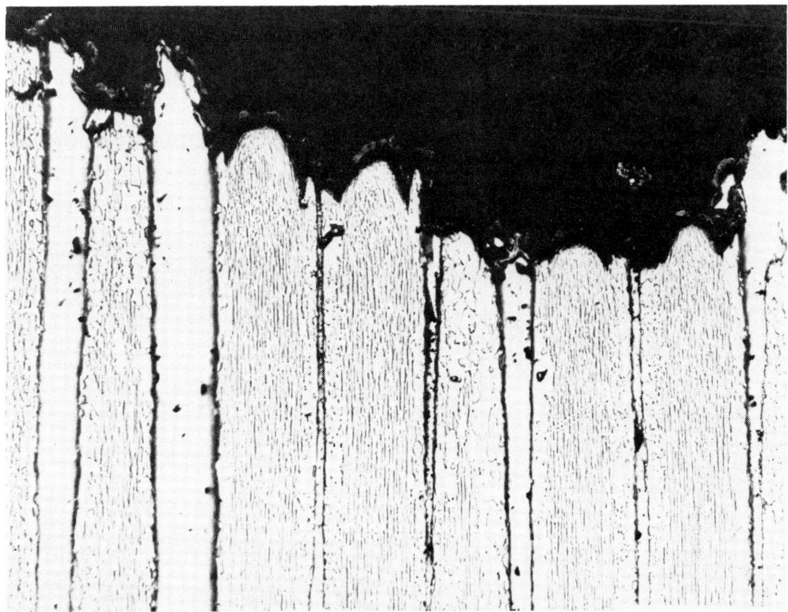

Fig. 9. Fracture edge of tungsten fiber in Cu–10% Ni matrix. Test temperature, 650°C (86.5 ×).

results shown in Fig. 7 were similar to those obtained at room temperature in that the binary-alloy-matrix tensile-strength values were below those of pure-copper-matrix composites. There was one notable difference. The tensile strength of the copper–10% nickel-matrix composite at 150°C was markedly above the room-temperature value. The increase in strength was attributed to a transition in fiber failure from brittle to ductile. As can be seen in comparing the photomicrographs in Figs. 8 and 9, the fracture changed from brittle to ductile.

D. Impact-Strength Variation

Reaction at the matrix–fiber interface also affects the impact strength of composites. The impact strength of tungsten-fiber-reinforced metal-matrix composites was considered by Winsa and Petrasek (1972) with three matrix materials: copper, copper–10% nickel, and a nickel superalloy. The variation of matrix permitted a comparison of the effects of several variables on the pendulum impact strength of composites. The copper represented a ductile nonreactive matrix, and the superalloy a brittle reactive matrix. The pendulum impact strength of reacted fiber composites was lowered relative to that of nonreacted composites. Further, the reduction in impact strength increased with increased reaction depth. The brittle layer of recrystallized tungsten acted to reduce impact strength in a fashion similar to that previously shown for tensile strength.

The results described above are consistent in indicating that reactions between metal wire and metal matrix cause reduction in properties that can severely degrade composite strength, ductility, and impact resistance or toughness. Clearly, the attainment of refractory-wire-reinforced superalloys with superior high-temperature properties is predicated on the control of these reactions.

III. Control of Matrix–Fiber Reaction

A. *Matrix Compositions*

Several approaches can be used to achieve the desired compatibility between matrix and fiber. These include control of matrix composition and impurity content, fiber size, diffusion barriers, and composite fabrication procedures. As noted previously, in relation to the effects of adding alloying elements to copper (Petrasek and Weeton, 1963), some additions were much more damaging than others. Proper selection of the superalloy composition should aid in the control of fiber–matrix reactions. Variations in composite properties with matrix composition has been noted by Baskey (1963) and Dean (1965). In these studies, several standard superalloy compositions were used as the matrix for refractory-wire composites. However, matrix compositions specifically designed for compatibility with refractory wire were found to improve properties. This was demonstrated in a study by Petrasek *et al.* (1968), in which four matrix alloys were fabricated into refractory-wire-reinforced composites. Several approaches were used to control reaction in that study.

Molybdenum-alloy-wire composites were also studied in the initial phase of the program. However, a severe penetration reaction was obtained, which eliminated molybdenum wire from further study. The four matrix alloys had been designed for compatibility with the tungsten fibers, and all were very reactive with the molybdenum. It follows that matrix alloys designed for tungsten would not necessarily be optimal for molybdenum.

Control of matrix impurity content is another aspect of composition that should be mentioned. Since diffusion of impurity elements into the fiber can rapidly degrade properties, removal of them is desirable. A hydrogen reduction step in the sintering of partially densified composite specimens can bc helpful in removing impurities and controlling reaction. Titanium and aluminum oxides are stable at the maximum processing and testing temperatures for superalloy composites and, as such, would be less damaging than other impurities. The degree of matrix–fiber reaction that occurred was reduced by impurity reduction. More rigorous efforts to achieve and maintain very low levels of impurity content may be justified for some special requirements.

B. Consolidation Procedures

The method of fabricating a composite obviously must be planned to control reaction. Liquid infiltration of superalloys has been used by Kotov *et al.* (1971) because of the relative ease of producing complex shapes. Since liquid infiltration of the matrix implies higher processing temperatures than necessary with solid-state processing, increased diffusion rates and more reaction normally result. Chill-casting techniques are often used to minimize the extent of fiber–matrix reactions. Although greater reactions might be expected compared to solid-state fabrication methods, the property loss obtained may be tolerable. With diffusion-barrier-coated fibers or low-activity fiber–matrix combinations, this method can be effectively used. Liquid-phase infiltration has also been used as an expedient way to screen prospective matrix materials for relative compatibility with fibers.

Matrix-alloy homogeneity and lower reactivity have justified the selection of powder-metallurgy techniques for several studies. Powder-metallurgy techniques are well suited to simple specimen shapes and very adequate for the feasibility studies conducted. The temperature schedule selected for consolidation can also serve to control the degree of matrix–fiber reaction. A multistage consolidation process was evolved in the study of Petrasek and Signorelli (1970). Part of the consolidation was performed at a relatively low temperature, 815°C, before heating the body to 1100°C for the final consolidation step. The low-temperature step increased density and reduced porosity to reduce surface-diffusion rates, thereby reducing the rate of reaction. Application of a number of these techniques to control reaction has resulted in composites that have demonstrated excellent high-temperature strength.

C. Diffusion Barriers

Higher temperatures and longer service times than can be obtained with the above techniques can be achieved by coating the fibers with a barrier to minimize or eliminate reaction as a source of fiber-property degradation.

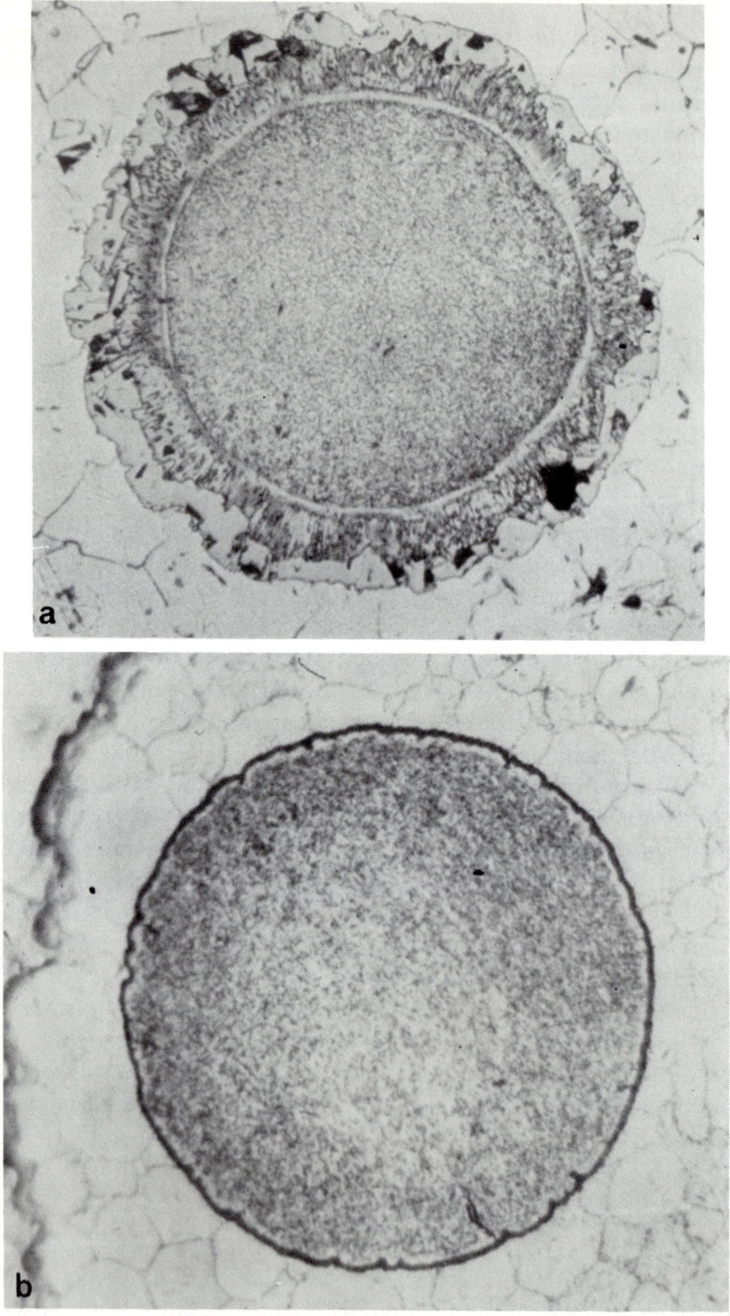

Fig. 10. Coating of wire to minimize reaction. 8-mil tungsten wire heated at 1200°C for 50 hr in nickel (a) uncoated, (b) metal-oxide coated (250 ×).

Silicon-carbide-coated boron is a notable example of success in controlling reaction by a fiber coating. Development of diffusion barriers for refractory-metal-alloy wire has received little attention, and similar examples of successful coatings for refractory wire are not available. The limited efforts undertaken for refractory-wire coatings have demonstrated that they can be very effective in limiting reactions. Figure 10 shows a photomicrograph indicating the effect of exposure for 50 hr at 1200°C on a coated and uncoated tungsten-alloy fiber in a nickel matrix (Signorelli and Weeton, 1969). The coated fiber is unaffected, whereas the uncoated fiber is severely reacted and, judging by previous experience, severely degraded in strength.

Unfortunately, such barriers cannot be reliably and reproducibly coated on fibers. Numerous areas of poor bonding and porosity can lead to general failure of the barrier. The first major obstacle encountered in developing a successful coating is the method of application. The coating must be well bonded, uniform in thickness, and almost free of porosity extending to the substrate surface. The conventional methods of applying coatings to massive substrates are not satisfactory for small-diameter fibers. Chemical-vapor deposition methods are suitable for some coatings but are not broadly applicable to a wide variety of coatings. Sputtering has the flexibility to coat most candidate barrier materials. However, deposition rates are low, and debonding of the coating can be encountered with thicknesses over a few microns. Ion plating has displayed many of the characteristics needed to permit experimental investigations to develop effective diffusion barriers. This method has reasonably high deposition rates, and bonding of coatings can be excellent. A wide range of materials can be coated, and the coating is uniform over all sides of the substrate surface.

IV. Fiber Development

A. Higher-Strength Alloys

Diffusion barriers are designed to protect the fiber from property loss or to minimize the rate of loss. However, the level of strength from which loss occurs is also a vital variable in composite strength. In recognition of this fact, the development and characterization of refractory-alloy fibers for reinforcement of superalloys has been included in the program at the Lewis Research Center for several years. The tungsten and molybdenum fibers used in the early model-system studies were available commercially as lamp-filament or thermocouple wire. These refractory alloys were not designed for use in composites nor for optimum mechanical properties in the 1000–1200°C temperature range. The lamp-filament wire, tungsten alloy 218, was most extensively used in early studies, and it had not been characterized for creep-rupture applications in the desired temperature range. Stress-rupture data for 127-μm-diam 218-alloy tungsten wire were determined (McDanels and Signorelli, 1966) for temperatures from 650 to 1570°C. The rupture properties of 218-alloy tungsten wire were superior to those of rod and bulk forms of tungsten and showed promise for use as reinforcement of superalloys. The need for stronger wire was recognized, and alloys of tungsten, tantalum, molybdenum, and niobium were included in a wire-fabrication program (Petrasek and Signorelli, 1969; Amra *et al.* 1970; King, 1972; and Petrasek, 1972). The chemical compositions of the alloys studied are given in Table I. The alloys selected for the program were limited to those for which rod- and sheet-fabrication procedures had already been developed. This precluded development of new alloys specifically designed for strength at the intended composite-use temperatures. The stress-rupture tensile properties determined for the wires developed are summarized in Table II.

Excellent progress has been made in providing wires with increased strength and strength-to-density ratios compared to the strongest wire available previously. The rupture strength of tungsten-alloy fibers has been increased by a factor of 3 from about 414 MN/m^2 (60,000 psi) to about 1379 MN/m^2 (200,000 psi). Also, lower-density niobium alloys such as FS85, AS30, and B88 have been fabricated into wire. The absolute values of rupture strength for these wires is less than for the best tungsten-alloy wires. However, because of the lower density, the strength-to-density values compare more favorably. The highest-strength niobium wire, B88, has about 1.5 times the 100-hr rupture-strength-to-density value at 1100°C that 218 tungsten lamp filament has. These higher-strength fibers should permit development of composites with improved strength. Still further development of wire strength is expected.

Fiber size is another variable with which to increase composite rupture strength. Since matrix–fiber reaction is the primary cause of reduction of properties and since the degree of property loss for refractory-wire composites has been related to the depth of penetration of reaction into the fiber, composite strength can be increased by increasing the unreacted fiber core area. As shown in Fig. 11, the depth of penetration of reaction was essentially the same for a smaller-diameter fiber as for a larger-diameter fiber. However, the area percent of unreacted core is considerably larger for the larger fiber. The smaller-diameter fiber has a higher rupture stress than the larger, so the two effects must be balanced. For very short-time service where the depth of penetration of reaction is very slight, the smaller fiber results in higher-strength composites; for longer times, the larger is superior. The specific conditions of reaction for 218 tungsten lamp filament dictated that a 381-μm

TABLE I

Chemical Composition of Wire Materials

	Weight percent of component									
Material	W	Ta	Mo	Nb	Re	Ti	Zr	Hf	ThO_2	C
Tungsten alloys										
218CS	99.9	—	—	—	—	—	—	—	—	—
W–1ThO_2	bal	—	—	—	—	—	—	—	0.95	—
W–2ThO_2	bal	—	—	—	—	—	—	—	1.6	—
W–3Re	bal	—	—	—	2.79	—	—	—	—	—
W–5Re–2ThO_2	bal	—	—	—	4.89	—	—	—	1.78	—
W–24Re–2ThO_2	bal	—	—	—	22.54	—	—	—	1.7	—
W–Hf–C	bal	—	—	—	—	—	—	0.37	—	0.030
W–Re–Hf–C	bal	—	—	—	4.1	—	—	0.38	—	0.021
Tantalum alloys										
Astar 811C	8.2	bal	—	—	—	1.13	—	0.91	—	0.027
Molybdenum alloys										
TZM	—	—	bal	—	—	0.45	0.085	—	—	0.031
TZC	—	—	bal	—	—	1.18	0.27	—	—	0.12
Niobium alloys										
FS85	10.44	27.95	—	bal	—	—	0.85	—	—	0.031
AS30	20	—	—	bal	—	—	1	—	—	—
B88	28.3	—	—	bal	—	—	—	1.94	—	0.58

TABLE II

Representative Properties of Refractory-Alloy Wires

Alloy	Density (gm/cm^3)	Wire diameter (mm)	Ultimate tensile strength (ksi)	Ultimate tensile strength (MN/m^2)	Stress for 100-hr rupture (ksi)	Stress for 100-hr rupture (MN/m^2)	Stress/density for 100-hr rupture ($cm \times 10^3$)
A. 1100°C Data							
Tungsten alloys							
218CS	19.1	0.20	126	869	63	434	234
W–1ThO_2	19.1	0.20	142	979	77	531	282
W–2ThO_2	18.9	0.38	173	1193	95	655	356
W–3Re	19.4	0.20	214	1475	69	476	249
W–5Re–2ThO_2	19.1	0.20	176	1213	70	483	254
W–24Re–2ThO_2	19.4	0.20	211	1455	50	345	183
W–Hf–C	19.4	0.38	207	1427	161	1110	584
W–Re–Hf–C	19.4	0.38	314	2165	205	1413	744
Tantalum alloys							
Astar 811C	16.9	0.51	108	745	84	579	351
Molybdenum alloys							
TZM	10.0	0.38	77	531	42	290	295
TZC	10.0	0.13	79	545	38	262	267
Niobium alloys							
FS 85	10.5	0.13	66	455	44	303	295
AS 30	9.7	0.13	61	421	31	214	224
B 88	10.2	0.51	77	531	48	331	328

B. 1200°C Data

Tungsten alloys							
218CS	19.1	0.20	108	745	46	317	170
W–1ThO_2	19.1	0.20	122	841	54	372	198
W–2ThO_2	18.9	0.38	150	1034	70	483	257
W–3Re	19.4	0.20	157	1082	46	317	168
W–5Re–2ThO_2	19.1	0.20	148	1020	44	303	160
W–24Re–2ThO_2	19.4	0.20	147	1014	28	193	102
W–Hf–C	19.4	0.38	201	1386	111	765	404
W–Re–Hf–C	19.4	0.38	281	1937	132	910	480
Tantalum alloys							
Astar 811C	16.9	0.51	71	490	38	262	157
Molybdenum alloys							
TZM	10.0	0.20	77	531	19	131	135
TZC	10.0	0.13	79	545	18	124	127
Niobium alloys							
FS 85	10.5	0.13	40	276	23	159	155
AS 30	9.7	0.13	33	228	—	—	—
B 88	10.2	0.51	50	345	28	193	190

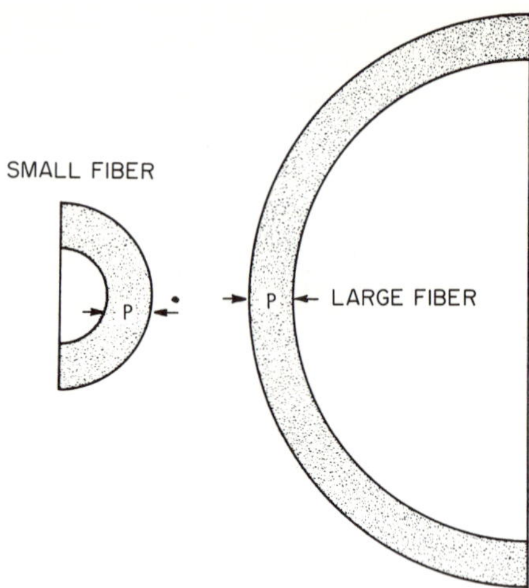

Fig. 11. Effect of increased diameter on retained fiber strength.

(15-mil) diameter fiber was superior for 100- and 1000-hr application at 1090°C. It should be noted that the tradeoff between fiber size, reaction area, and strength is a function of the fiber properties.

B. Thermomechanical Processing of Fibers

The variation in strength with fiber size is related to the heavily worked microstructure resulting from the mechanical processing in drawing the wire. The higher strength of the smaller-diameter 218 tungsten filament discussed above has been attributed to the optimal elongated grain structure or "fibered structure" within the cross section. The processing for the smaller-diameter wire has been evolved as part of the commercial production of lamp-filament wire. The larger-diameter wire is in a less nearly optimal condition. The wire strengthening results from the grain and subgrain structure, not the macroscopic size of the wire. Therefore, it should be possible to develop a thermomechanical deformation schedule to optimize the structure and properties for the larger diameter. The processing schedules for the newer higher-strength alloys shown in Table II are far from optimal.

Much more work is needed to maximize properties. It should be noted that processing can be optimized for a given service temperature and life. As noted previously, the alloys included in the program to develop wire had been developed for other applications, typically for use at higher temperatures. Refractory-alloy compositions and processing schedules should be developed specifically for fiber-composite use. The application of composites developed using less-than-optimal wire will justify the added effort to further improve properties. Similarly, the retention of wire properties with diffusion barriers will increase the incentive to maximize wire properties.

V. Matrix-Alloy Development

The primary function of the matrix is to bind the fibers into a useful body and to protect the fibers. Matrix composition is not restricted by bonding considerations, since excellent bonding of typical superalloy matrix materials to uncoated refractory fibers is readily accomplished using straightforward fabrication techniques. As noted previously, reaction with the fibers is a serious problem and is the primary consideration limiting matrix composition.

A. Compatibility

The model-system work discussed in an earlier section of this chapter demonstrated that matrix composition alone had a marked effect on composite properties. All matrix materials investigated other than the mutually insoluble copper were to some degree damaging to fiber properties. The encouraging result was that several elements were only slightly damaging. Based on these findings, superalloy matrix compositions might be developed that would cause limited reaction and minimal fiber-property loss. The four superalloy compositions given in Table III were selected to reduce reaction with tungsten fibers. The photomicrographs of Fig. 12 indicate the results with the four alloys. Figure 12a, b, d, and e shows the varying degrees of

TABLE III

Nickel-Alloy Matrix Materials

	Nominal Composition of Alloy (w/o)							
Alloy number	Al	Nb	Cr	Mo	Ni	Ti	W	Ta
1	—	—	20	—	55	—	25	—
3	2	—	15	—	56	2	25	—
5	—	1.25	19	4	70.5	—	4	1.25
7	4.2	1.25	15	4	66.8	3.5	4	1.25

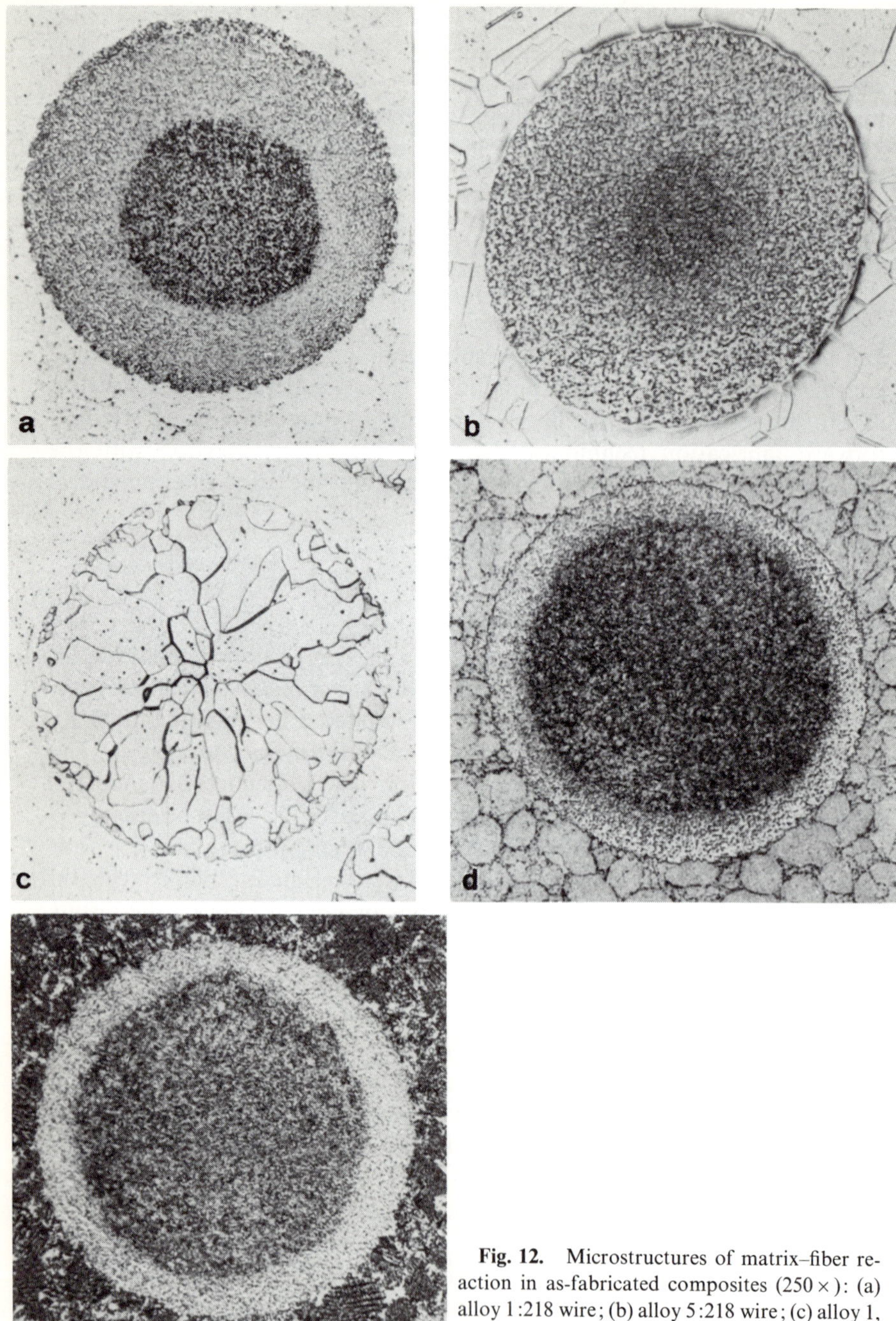

Fig. 12. Microstructures of matrix–fiber reaction in as-fabricated composites (250 ×): (a) alloy 1:218 wire; (b) alloy 5:218 wire; (c) alloy 1, TZM wire; (d) alloy 3:218 wire; (e) alloy 7:218 wire.

reaction of the alloys with tungsten wire. Figure 12c typifies the almost complete reaction and recrystallization of all four alloys with the molybdenum wire. The lack of compatibility with the molybdenum wire was not unexpected, since the matrices had been designed for compatibility with tungsten. No data were available to guide a similar effort for the molybdenum alloy. Alloys 3 and 7, which contained titanium and aluminum, caused less reaction with the tungsten fiber. Evaluation of rupture strength indicated that alloy 3 was the most compatible, and it was selected for further fabrication study. The successful control of reaction by matrix-composition design is inseparable from the other steps taken to control reaction—fiber size and composite fabrication processing. However, it was possible to limit property loss in the fiber in the composite by a combination of such efforts.

The fiber stress to cause rupture in 100 hr at 1090°C was reduced only 10% in the composite compared to the equivalent fiber rupture strength tested in a vacuum outside a composite. Not only had fiber–matrix reaction been controlled during fabrication of the composite, but the damage during a 100-hr exposure to stress at 1090°C had been controlled.

Optimum matrix compositions for use with diffusion-barrier-coated fibers will differ from those for uncoated wire. Diffusion-barrier coatings are selected for thermodynamic stability with superalloy matrix elements. However, the use of such inert coatings may induce wetting or bonding problems with the superalloy matrices. The composite fabrication problem is not unlike that of aluminum-oxide-fiber-reinforced nickel alloys described in Chapter 4 of this volume.

B. Strength

The matrix acts as a binder to permit the geometric array of fibers to perform as a structural unit. The matrix also supports part of the tensile load and resists shear failure at the high service temperature and provides toughness and ductility. The strength requirement of the matrix varies with the service temperature. At temperatures of 1090°C and above, a superalloy matrix contributes a very small portion of the tensile or rupture strength compared to that of the refractory fiber. Matrix materials designed for these high temperatures need sufficient shear strength to permit the fibers to carry the load. The shear strength of the matrix should be high enough to provide for a ratio of 5–10 to 1 between the fiber length used in the composite and the critical fiber length. At temperatures of 1090°C and below, the strength of the matrix can be significant. A stronger matrix can reduce the volume fiber content required to achieve the desired component strength.

Many components, such as turbine blades, operate with a thermal gradient of several hundred degrees. The temperature near the base of a turbine blade may be 100–200°C below the temperature near the tip. A matrix having strength approaching that of a typical superalloy used for blades needs little if any reinforcement in those portions of the blade at lower temperatures. Since the strength augmentation of the matrix varies with temperature along the component, the fiber content along the component can be varied accordingly. For high-density refractory-wire materials, the reduced fiber content can reduce component density. Increased matrix strength can increase the use potential of the composite by reducing component weight, which for rotating components in flight hardware is very important. The increased strength of the matrix should be obtained without sacrifice of ductility or oxidation resistance, which are also needed.

C. Ductility

Plastic deformation of the matrix at room and intermediate temperatures is valuable in providing impact resistance and in relieving thermally induced stresses. Impact strength of tungsten-fiber composites is related to both fiber and matrix ductility (Winsa and Petrasek, 1972). The ability of the matrix to deform plastically was found to be an important factor in achieving adequate impact strength and can improve fatigue resistance. The contribution of the matrix was particularly significant at temperatures where the fibers failed in a brittle manner.

The ability of the matrix to deform plastically is desirable to resist failure from mechanically or thermally induced stresses. Stresses are generated by transient temperature operation of components, which causes time-dependent temperature gradients from one part of the component to another. Thermal-expansion mismatch between tungsten wire and typical superalloys is another source of stresses. Brittle matrix materials can easily crack from these stresses. Superalloys have been developed that can plastically deform sufficiently to demonstrate several thousand thermal cycles without cracking. Superalloy matrix materials with similar resistance to thermal stress failure are necessary for successful composites.

D. Oxidation Resistance

Superalloy compositions normally are designed to provide oxidation and sulfidation resistance for the intended service environment. For some applications, however, such as turbojet-engine turbine blades, coatings or cladding materials are necessary for applications above 980°C. Superalloy

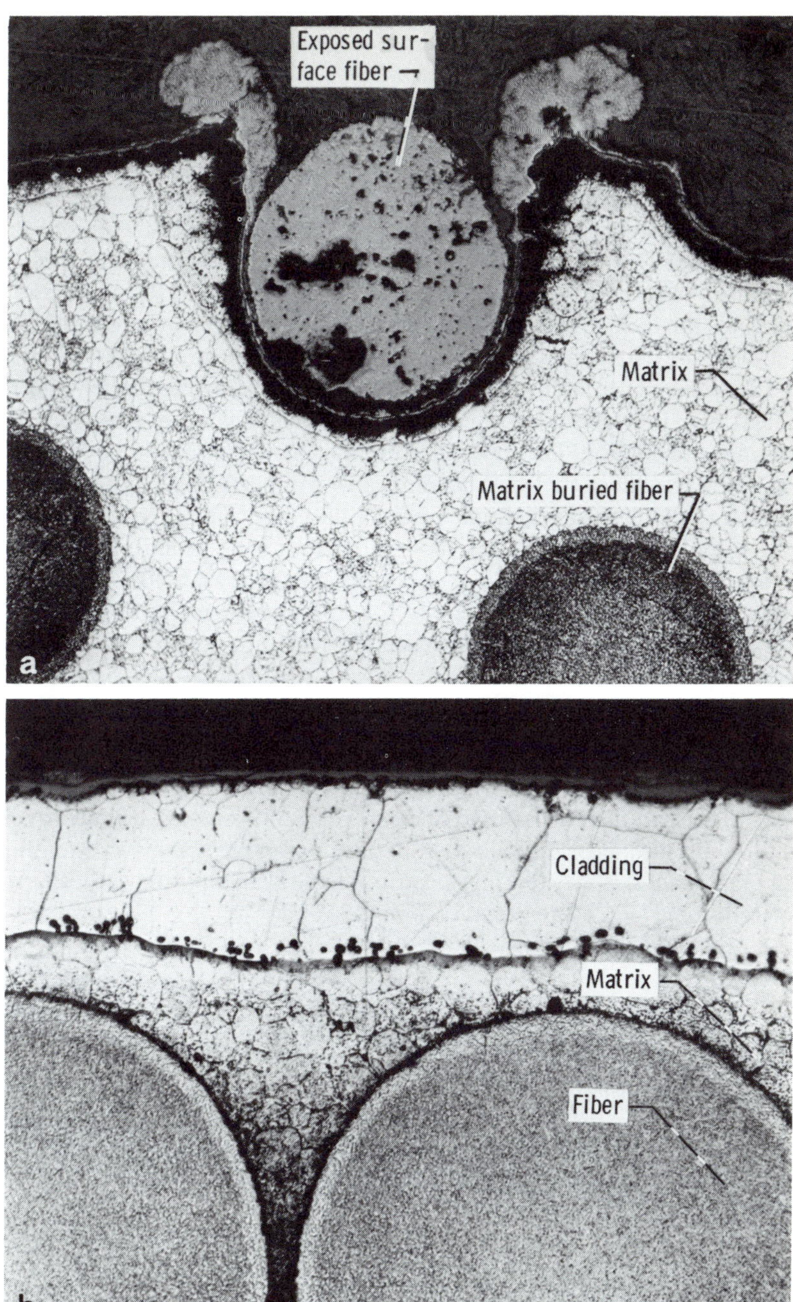

Fig. 13. Transverse section of oxidized W-alloy-fiber–Ni-alloy composite. (a) Fiber, 0.38-mm-diam 218 tungsten. Test condition, 5 hr in air at 1090°C (100 ×). (b) Fiber, 0.38-mm-diam tungsten, 1% thoria. Test condition, 50 hr in air at 1090°C (150 ×).

matrix compositions are designed to provide oxidation resistance consistent with the other requirements. Relatively few data have been obtained to define clearly capabilities and limitations of refractory-wire–superalloy composites; however, the evidence obtained is encouraging. Since the tungsten wires themselves have very poor oxidation resistance, this problem is particularly significant. The photo shown in Fig. 13a indicates the severe oxidation of an exposed tungsten fiber after only 5 hr at 1100°C. Figure 13b indicates the lack of attack on tungsten fibers protected by a few tenths of a milimeter of matrix and cladding after 50 hr of exposure at 1100°C.

The nickel–chromium-alloy (Inconel) cladding resulted when the thin-walled tube in which the composite was fabricated was not removed during specimen machining. The superalloy composite matrix should have adequate oxidation resistance such that small localized imperfections or damage to the cladding will not result in rapid failure of the composite. Cladding or coating alloys are being developed for superalloys that have promise for operation in severe environments for temperatures up to 1200–1300°C. Such alloys may be used as matrix materials for composites with diffusion-barrier coatings and as claddings for uncoated-fiber composites.

VI. Fabrication Methods

The fabrication of matrix and fibers into a composite with useful properties is probably the most difficult task in developing refractory-wire-reinforced superalloys. Relatively few investigations have been reported in the literature for refractory-wire–superalloy composites.

A. Solid State versus Liquid

Both liquid- and solid-state processes have been used effectively to produce composites for these research studies. Fabrication methods for refractory-wire–superalloy composites must be considered to be in the laboratory phase of development. Production techniques for fabrication of large numbers of specimens for extensive property characterizations have not yet been developed as they have for aluminum–boron and resin-matrix composites. Liquid-phase infiltration has been used because of the ease and rapidity of producing uniaxially reinforced test specimens. As discussed in previous sections, fiber–matrix reactions are generally more severe with liquid-phase methods, since diffusion rates are higher than with equivalent solid-state processing. Also, solid-state processing temperatures are lower, further improving compatibility. Compatibility studies to screen a number of candidate matrix compositions for relative reactivity have used liquid-phase consolidation effectively. Casting was used by Dean (1965) to produce refractory-wire–superalloy composites. Tungsten-wire bundles were placed in mold cavities and the casting operation conducted in a vacuum furnace chamber to reduce oxidation of the wires. Reaction between the molten superalloy and wire was reduced by a rapid casting technique approaching chill casting. An alternate means of using liquid-state techniques is to coat the wire with matrix by passing it rapidly through a molten bath of matrix. Coated wires are then diffusion bonded in closed dies to form the composite component. The development of diffusion barriers for refractory-alloy fibers may make liquid-phase techniques attractive for large-scale production of composites.

B. *Powder Matrix Processing*

Powder body fabrication has been used to reduce processing temperatures and reduce fiber–matrix reaction. However, powder processes have several disadvantages. Specific compositions must be melted and cast into ingots and the ingots processed into powder. The additional step causes delay and tends to reduce variation in matrix compositions. The large surface area of fine powders introduces impurities that must be removed. Elaborate equipment is necessary to apply pressure and temperature while maintaining an inert gaseous atmosphere. Most powder-fabrication techniques limit fiber content to 40–50 v/o. Despite these disadvantages, powder processing has been used to achieve control of matrix–fiber reaction and has resulted in excellent composite properties.

Slip casting followed by sintering and hot pressing has been used for several refractory-alloy composite studies at Lewis Research Center. The process used is shown schematically in Fig. 14. Cylindrical composite test specimens have been fabricated with fiber content to over 75 v/o using this method. The process schedule was developed to achieve control of fiber–matrix reaction. Two steps were included in the process to improve reaction control. A preliminary sinter at 815°C with flowing hydrogen gas affected impurity reduction at powder-particle boundaries in an initial isostatic hot pressing at 815°C and reduced the open pore path between matrix and fiber surface before exposure of the composite to high temperature. The combined effect of matrix composition, fiber size, and fabrication process succeeded in controlling matrix–fiber reaction with tungsten lamp-filament composites. The reduction in 100-hr rupture stress at 1090°C was only 10%

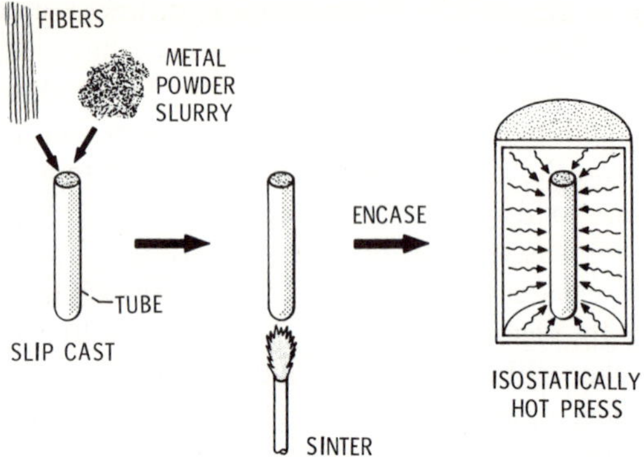

Fig. 14. Composite billet fabrication. Sintering is performed at 815°C for 1 hr in hydrogen. Hot pressing is performed at 138 MN/m^2 at 815°C for 1 hr, followed by 138 MN/m^2 at 1090°C for 1 hr.

for tungsten lamp filament in a nickel superalloy. The highest rupture-strength values reported for metal-matrix fiber composites at 1090°C have been obtained by this fabrication technique.

Although the slip-casting–sintering–hot-pressing technique has demonstrated excellent success for uniaxially reinforced test specimens, it is not an ideal method for component fabrication. Most applications require some cross-ply fiber orientation, which is not easily accomplished with slip casting.

C. *Diffusion Bonding*

Methods for fabricating refractory-alloy-wire-reinforced superalloys basically are similar to those used for other metal-matrix composites, except that the fiber has considerably more ductility, and higher temperatures are required for the superalloy matrix. Monolayer tapes, diffusion bonding, plasma spraying, and roll bonding are all applicable to refractory-wire–superalloy systems. Secondary mechanical working of composites is more easily accomplished with refractory-wire-reinforced systems than with more brittle fibers such as boron, Al_2O_3, SiC, or C. Copper–tungsten composites have been mechanically worked up to 95% by rolling. The reduction in area of the tungsten fibers was over 90% of the value for the composite. This workability can be desirable in producing components. For example, a turbine-blade airfoil may be hot-deformed to final shape, including airfoil twist. Such mechanical deformation can also result in increased strength of the matrix.

VII. Properties of Refractory-Wire–Superalloy Composites

A. Stress-Rupture Strength

The properties achieved with superalloy-matrix–refractory-alloy-wire composites (Petrasek *et al.*, 1968; Petrasek and Signorelli, 1970; Dean, 1965; Baskey, 1967) compare favorably with those of unreinforced superalloys. The most significant property obtained is stress-rupture strength because it is the most critical for most high-temperature applications. The highest stress-rupture properties obtained at 1090°C are shown in Fig. 15. The strength-to-density values of superalloy reinforced with 218 lamp filament, tungsten–2% ThO_2, and tungsten–hafnium–carbon alloy wire are shown along with those of a conventional superalloy and directionally solidified eutectics for comparison. The ordinate units of the figure are the ratio of stress to density to normalize for the high density of tungsten fibers.

After compensating for the density, the composites are up to five times as strong as the superalloy. The stress for 1000-hr rupture value of the W–Hf–C composite is about 414 MN/m^2 (60,000 psi), or almost nine times as strong as the superalloy when not density compensated. The composite

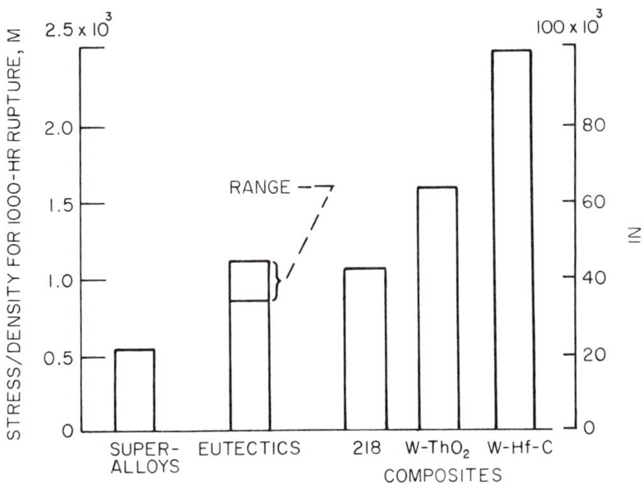

Fig. 15. 1000-hr stress-rupture properties of refractory-wire–superalloy composites. Composites contained 70 v/o of the reinforcement materials shown.

strength also compares very favorably with those of directionally solidified eutectics. The best strength-to-density value at 1090°C for refractory-alloy-wire composites (Signorelli, 1972) is over twice that reported for directionally solidified eutectics.

B. Impact Strength

Although stress-rupture strength is the most important criterion for screening promising high-temperature materials, other properties are necessary for component application. Impact strength is another requirement for use in rotating machinery such as turbines. The impact strength of refractory-wire–superalloy composites has been evaluated in studies by Stetson *et al.* (1966), Petrasek and Signorelli (1970), and Winsa and Petrasek

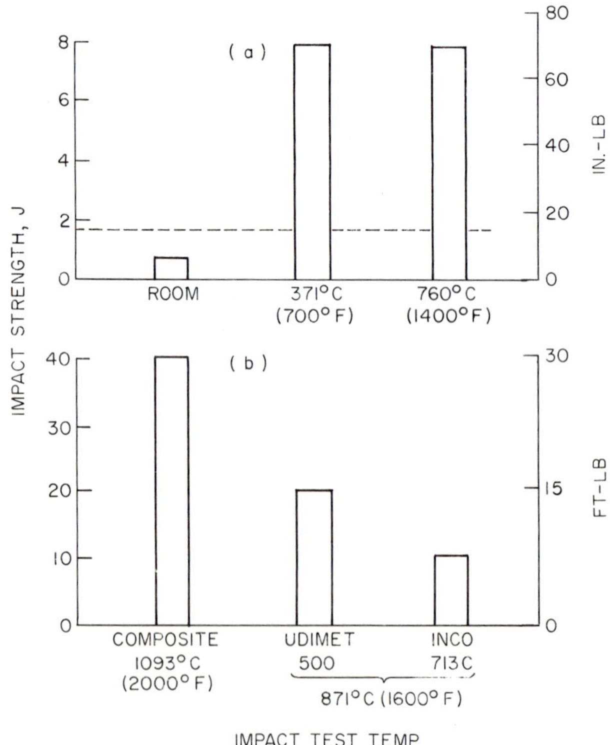

Fig. 16. Impact strength for several test temperatures. (a) Miniature Izod specimen tests of a composite having 60 v/o wire. (b) Charpy specimen tests.

(1972). The results of pendulum impact evaluation have indicated that refractory-wire–superalloy composites can be made with adequate impact strength for turbine-blade use. Turbine blades represent a component with one of the more stringent requirements for impact resistance. The impact strength of refractory-wire–superalloy composites above the ductile–brittle transition temperature was greater than that of superalloys presently used as turbine blades (Fig. 16). The composite impact strength was lower at test temperatures below the ductile–brittle transition temperature of the tungsten fiber. The brittle reaction zone that formed by matrix–fiber interaction also embrittled the fiber. However, as shown in Fig. 17, a 30-in.-lb room-temperature impact-strength value was obtained for tungsten-fiber–superalloy composite. That test value was obtained after mechanically working the composite to improve the powder-metallurgy matrix bond. The notched- and smooth-bar values in the worked condition were both about 30 in.-lb and attributed to improved matrix impact strength. This point was demonstrated by the increase in impact strength of a heat-treated bar over the as-hot-pressed bar. The thermal treatment of 250 hr at 1090°C would be expected to decrease the tungsten-fiber contribution to impact strength by increasing the depth of matrix–fiber reaction. However, the bonding of matrix powder particles would be expected to improve, thereby increasing matrix impact strength. Scanning electron micrographs of the respective specimens confirmed the improvement in matrix deformation at the fracture surface. At temperatures below the ductile–brittle transition

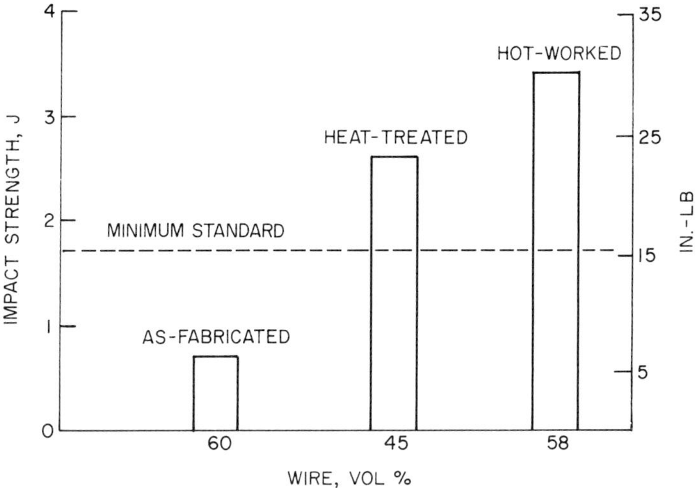

Fig. 17. Improved room-temperature impact strength (miniature Izod tests).

temperature of the fiber, the contribution to impact strength of the matrix was significant if the composite matrix was properly selected and fabricated to deform plastically. It should be noted that the contribution of the matrix can be reduced by constraining matrix plastic deformation. High volume fiber content, which leads to a thin film of matrix between a relatively stiff high-modulus brittle reinforcing phase, can cause such constraint.

A decrease in matrix thickness or interfiber distance occurs with a decrease in diameter of the reinforcing fiber or phase for a given volume fiber content. The effect can be very startling, as illustrated by the following example. Compare two 40 v/o fiber content composites, the first with 20-mil- or 500-μm-diam fibers and the other with 5-μm-diam fibers. An accurate calculation of interfiber distance would require use of a hexagonal or square array and an average matrix thickness obtained. A rough approximation based on comparison of a cylindrical fiber and a uniform matrix coating will suffice here. The large-diameter (500-μm) fiber would have about a 150-μm- or 6-mil-thick matrix coating, whereas the 5-μm fiber would have about a 1.5-μm-thick matrix coating. Matrix constraint in a tensile pull-out test was indicated by increased matrix shear strength when the interfiber distance was 25 μm, and the effect was more pronounced with values of less than 10 μm (Jech and Signorelli, 1968). From this behavior, it might be expected that the matrix would be constrained in a higher-strain-rate impact test. If the effect of decreased reinforcement-phase thickness dimensions acts to constrain matrix deformation, the toughness or impact strength can be reduced. Composites wherein the fiber failure is brittle with low impact-strength values may achieve satisfactory composite impact by promoting a large matrix impact-strength contribution. Matrix deformation may not be possible with very thin phase matrix composites, such as directionally solidified eutectics. For those materials, efforts must be made to achieve impact strength by means other than matrix or fiber deformation. Delamination and crack blunting and arresting are alternate means that might be used to increase impact strength of these composites.

C. Oxidation

Relatively few oxidation data are found in the literature for refractory-wire–superalloy composites. Token testing has been conducted and reported in studies by Dean (1969) and Morris and Burwood-Smith (1971). The basic design of the composite material assumes that the superalloy will provide oxidation resistance, including protection of the tungsten fibers. Superalloy compositions that can be used for a matrix with uncoated fibers can be reasonably oxidation resistant for operating temperatures up to about 980°C. However, above that temperature, it may be necessary to provide improved oxidation resistance. Composites containing diffusion-barrier-coated fibers may use matrix materials that are weak and ductile but oxidation resistant to temperatures above 1090°C.

Coatings or cladding materials invariably develop localized points of failure. Where the superalloy matrix is reasonably oxidation resistant, this can probably be tolerated. A more serious problem is where the oxidation-prone refractory-alloy wire is exposed. More data are necessary for such exposure and for all oxidation conditions. However, the preliminary results obtained for refractory-alloy-wire exposure are encouraging. Tungsten-wire–superalloy specimens were exposed to slow-moving furnace air and to 1.85-m/sec air at about 1090°C. The wire was exposed at both cylindrical end surfaces of the composite specimens. The penetration after 100-hr exposure was less than 1.2 mm for slow-moving air and about 2.5 mm for 1.85-m/sec air. These penetration depths are very small compared to the fiber length for most practical components. Such a reduction in fiber length would probably not be detected as a reduction in strength in a tensile specimen. Exposure of tungsten fibers would not occur deliberately in a composite, but could occur to a few fibers near the surface of a component because of local damage to the matrix or cladding. The results obtained, while inconclusive because of limited data, are encouraging.

D. Mechanical and Thermal Fatigue

A limited number of high-cycle fatigue studies of refractory-wire–superalloy composites have been conducted. Additional laboratory data are needed to evaluate the quantitative behavior of the materials and to estimate their potential for high-temperature component service. High-cycle fatigue resistance of Hastelloy X and Nimocast 258 reinforced with tungsten wire was improved relative to unreinforced Hastelloy X and Nimocast 258 in studies by Baskey (1967) and Dean (1965). However, tungsten-wire reinforcement did not increase the fatigue resistance of Nimocast 713C (Morris and Burwood-Smith, 1971). The lack of improvement in fatigue strength found by Morris and Burwood-Smith in contrast to the results of Baskey and of Dean was attributed to low resistance to crack propagation in the relatively brittle 713C alloy matrix and the poor crack-stopping ability of the large-diameter wires used by Morris and Burwood-Smith. The small number of large-diameter (0.13-cm) wires used would be expected to be less effective as a crack stopper than the smaller-diameter (0.025-cm) wires used in the investigations of Baskey and Dean. It should be emphasized that increased matrix ductility, which would probably increase resistance to crack initiation, also would be advantageous in enhancing fatigue resistance.

Thermally induced low-cycle fatigue requirements may be critical for turbine-blade applications of composites. Stresses may be generated by the thermal gradient between different portions of a component during transient-temperature operation as well as by the thermal-expansion difference between the refractory wire and the superalloy matrix. It will be necessary to use ductile matrix materials, which hopefully can relieve such induced stresses by plastic deformation.

Some variation in results was obtained for the very few data points published for thermally cycled refractory-wire–superalloy composite specimens by Glenny (1970) and Morris and Burwood-Smith (1971). Matrix cracking has been observed with cylindrical specimens of tungsten-wire–Nimocast 713C composites heated in fluidized beds. Similar specimens containing 20 v/o wire of 1- or 1.3-mm diam showed no cracking after several hundred fluidized-bed heating cycles between 550 and 1050°C. Although matrix cracks were observed after as few as two thermal cycles, tensile tests conducted on specimens thermally cycled and cracked showed no strength loss. The relatively brittle 713C alloy, which cracked in thermal fatigue, also demonstrated poor resistance to mechanical fatigue (Morris and Burwood-Smith, 1971). More ductile matrix alloys may be more resistant to crack propagation and provide better thermal and mechanical fatigue resistance.

Further testing is necessary to indicate the seriousness of this failure mode and to evaluate possible corrective measures. Ability to resist thermal fatigue may be critical for aircraft-engine service but may be of little concern for blades for use in gas-turbine-powered electric-power-generation systems. Continuous or very long time periods of operation are the norm for base-load power-generating plants. These advanced power systems also require very high turbine-inlet temperatures (Robson *et al.*, 1970), and refractory-wire–superalloy composites may be well suited for such service.

VIII. Summary

This review suggests that refractory-fiber–superalloy composites have demonstrated sufficiently high strength and impact values to have considerable potential for application to advanced turbine-engine blades. The data obtained thus far indicate a potential for increasing operating temperatures of turbine blade materials to 1200°C (2200°F) and above. However, few data have been obtained thus far as to their oxidation, erosion, and thermal and mechanical fatigue resistance. Additional testing is necessary to demonstrate the performance of refractory-wire–superalloy composites under all of these imposed environmental and loading conditions to indicate where improvements may be needed. The excellent mechanical properties already obtained and the promising potential for increased turbine life and operating temperatures justify the effort to complete the evaluation of these materials. A number of generalizations can be made:

1. Successful development and application of refractory-wire–superalloy composites to aircraft gas-turbine blades could permit blade-use temperatures as high as 1150°C (2100°F) without diffusion-barrier-coated fibers and as high as 1260°C (2300°F) with diffusion barriers. Oxidation protection for the blade airfoil is a critical requirement for such an increase in operating temperatures, particularly for aircraft use where temperature cycling causes spalling of protective oxide films. The composites have equal or better potential for application to land-based power-generating gas turbines operated at increased use temperatures because limited thermal cycling of the system eases oxidation and thermal-fatigue conditions and lessens the need for low density.

2. Density-normalized 1000-hr stress-rupture values at 1090°C (2000°F) for tungsten-fiber–superalloy composite specimens were over four times those for conventional superalloys and over twice those for the best published values for directionally solidified eutectics. Further increases in the strength of refractory-wire–superalloys are possible to increase this advantage. The potential strength of refractory-wire–superalloy composites using diffusion-barrier-coated wire could be from four to six times the density-normalized values for directionally solidified eutectics at 1090°C (2000°F).

3. Very few data have been obtained for failure mechanisms such as oxidation, fatigue, and erosion. The limited data obtained indicate sufficient promise to justify further research to develop the composite system. The need for oxidation protection at 1090°C (2000°F) and above has been indicated.

4. Matrix cracking has been observed for some brittle cast-superalloy-matrix composites thermally cycled to simulate aircraft-engine operation. Further testing is necessary to investigate the seriousness of this problem. The matrix must exhibit sufficient ductility to resist low-cycle fatigue failure generated by thermal-expansion mismatch between matrix and fiber in order to be suitable for aircraft-engine service. This failure mode would be of little concern for refractory-wire–superalloy composite blades used in advanced electric-power-generation systems where continuous or long-period operation is the norm. These advanced power systems also require very high turbine-inlet temperatures for efficient operation—precisely the advantage afforded by refractory-wire–superalloy composites.

5. Refractory-wire–superalloy composites can be fabricated to display pendulum impact-test values well above minimum requirements for turbine blades. Charpy and miniature Izod impact-test values for hot-pressed composites compare favorably with those for superalloys at temperatures above the ductile–brittle transition temperature of the refractory fiber, about 260–370°C (500–700°F) for the tungsten wire used in the composites. Impact-strength values of composites adequate for turbine-blade appli-

cations below the fiber ductile–brittle transition temperature have been obtained for tungsten-wire–superalloy composite specimens fabricated by using techniques suitable for turbine blades.

References

Amra, L. H., Chamberlin, L. F., Adams, F. R., Tavernelli, J. G., and Polanka, G. J. (1970). Development of Fabrication Process for Metallic Fibers of Refractory Metal Alloys. General Electric Co. (NASA CR-72654).

Baskey, R. H. (1963). Fiber Reinforcement of Metallic and Nonmetallic Composites. Clevite Corp. (ASD-TDR-63-619).

Baskey, R. H. (1967). Fiber-Reinforced Metallic Composite Materials. Clevite Corp. (AFML-TR-67-196, AD-825364).

Dean, A. V. (1965). The Reinforcement of Nickel-Base Alloys with High-Strength Tungsten Wires. Rep. R-266, Nat. Gas Turbine Establishment, England.

Glenny, R. J. E. (1970). *Proc. Roy. Soc. (London), Ser. A* **319** (1536), 33–44.

Jech, R. W., and Signorelli, R. A. (1968). The Effect of Interfiber Distance and Temperature on the Critical Aspect Ratio in Composites. NASA TN D-4548.

Kelly, A., and Tyson, W. R. (1965). *Int. Mater. Conf., 2nd, Berkeley, California* (V. F. Zackay, ed.), pp. 578–602. Wiley, New York.

King, G. W. (1972). Development of Wire-Drawing Processes for Refractory-Metal Fibers. Westinghouse Electric Corp. (NASA CR-120925).

Kotov, V. F., Fonshtein, N. M., and Shvarts, V. I. (1971). Heat-resisting composite material: nichrome–tungsten-filament. *Metalloved. Term. Obrab. Metal.* (8), 20–22.

McDanels, D. L., and Signorelli, R. A. (1966). Stress-Rupture Properties of Tungsten Wire from 1200° to 2500° F. NASA TN D-3467.

McDanels, D. L., Jech, R. W., and Weeton, J. W. (1963). Stress–Strain Behavior of Tungsten-Fiber-Reinforced Copper Composites. NASA TN D-1881.

McDanels, D. L., Signorelli, R. A., and Weeton, J. W. (1967). Analysis of Stress-Rupture and Creep Properties of Tungsten-Fiber-Reinforced Copper Composites. NASA TN D-4173.

Morris, A. W. H., and Burwood-Smith, A. (1971). Fibre-Strengthened Nickel-Base Alloy. High-Temperature Turbines. AGARD-CP-73-71.

Petrasek, D. W. (1965). Elevated-Temperature Tensile Properties of Alloyed Tungsten-Fiber Composites. NASA TN D-3073.

Petrasek, D. W. (1972). High-Temperature Strength of Refractory-Metal Wires and Consideration for Composite Applications. NASA TN D-6881.

Petrasek, D. W., and Signorelli, R. A. (1969). Stress-Rupture and Tensile Properties of Refractory-Metal Wires at 2000° and 2200° F (1093° and 1204° C). NASA TN D-5139.

Petrasek, D. W., and Signorelli, R. A. (1970). Preliminary Evaluation of Tungsten-Alloy-Fiber–Nickel-Base Alloy Composites for Turbojet Engine Applications. NASA TN D-5575.

Petrasek, D. W., and Weeton, J. W. (1963). Alloying Effects on Tungsten-Fiber-Reinforced Copper-Alloy or High-Temperature-Alloy Matrix Composites. NASA TN D-1568.

Petrasek, D. W., Signorelli, R. A., and Weeton, J. W. (1967). Metallurgical and Geometrical Factors Affecting Elevated-Temperature Tensile Properties of Discontinuous-Fiber Composites. NASA TN D-3886.

Petrasek, D. W., Signorelli, R. A., and Weeton, J. W. (1968). Refractory-Metal-Fiber–Nickel-Base-Alloy Composites for Use at High Temperatures. NASA TN D-4787.

Robson, F. L., Giramonti, A. J., Lewis, G. P., and Gruber, G. (1970). Technological and Economic Feasibility of Advanced Power Cycles and Methods of Producing Nonpolluting Fuels for Utility Power Stations. Rep. UARL-J970855-13, United Aircraft Corp. (PB-198392).

Signorelli, R. A. (1972). Review of Status and Potential of Tungsten-Wire–Superalloy Composites for Advanced Gas-Turbine Engine Blades. NASA TM X-2599.

Signorelli, R. A., and Weeton, J. W. (1970). Metal-Matrix Fiber Composites for High Temperatures, Aerospace Structural Materials. NASA SP-227, pp 287–206.

Signorelli, R. A., Petrasek, D. W., and Weeton, J. W. (1967). "Modern Composite Materials" (R. Krock and L. Broutman, eds.), pp. 146–171. Addison-Wesley, Reading, Massachusetts.

Steigerwald, E. A., and Guarnieri, G. J. (1962). *Trans. ASM* **55**(2), 307–318.

Stephans, J. R. (1961). Effect of Surface Condition on Ductile-to-Brittle Transition Temperature of Tungsten. NASA TN D-676.

Stetson, A. R., Ohynsty, B., Akins, R. J., and Compton, W. A. (1966). Evaluation of Composite Materials for Gas-Turbine Engines. AFML TR-66-156 pt. 1, AD-487290, Solar.

Winsa, E. A., and Petrasek, D. W. (1972). Pendulum Impact Resistance of Tungsten-Fiber–Metal-Matrix Composites. Compos. Mater., Testing and Design. Spec. Tech. Publ. No. 497, ASTM, pp. 350–362.

6

Fiber-Reinforced Titanium Alloys

ARTHUR G. METCALFE

Solar Division of International Harvester Company
San Diego, California

I. Introduction

One of the first applications of boron filaments was to reinforce titanium alloys. Two important reasons for interest in titanium were that this matrix would allow composites to be used above the temperature limit set by plastic matrices and also that titanium alloys have the highest strength-to-density ratio of the common structural materials. Unfortunately, early work to reinforce titanium was unsuccessful because it was found that the reactivity of titanium caused excessive reaction with boron (as well as with other filaments). Attention turned to aluminum as a matrix for boron with encouraging results, and interest in titanium waned. However, slowly there has been resolution of the problems of compatibility, and interest in titanium-matrix composites has been renewed. The causes of degradation in properties resulting from this incompatibility have been identified, so that the science of titanium matrix composites is, perhaps, more firmly established than that for most other matrices and systems. Solution of the compatibility problem by various approaches has delayed the technical development of titanium-matrix composites, and it has only been at the beginning of this decade that significant progress could begin to be made in the area of technology. At the present time, several different systems are under development and offer potential for various applications. The presentation in this chapter will follow the historical development to some extent because this approach leads to better appreciation of some of the problems and provides the background necessary to understand the course of the technical development.

Titanium has a higher strength-to-density ratio than any of the other common structural materials. At the easily attainable strength of 150,000 psi, titanium is equivalent to steel, which has a strength of 300,000 psi, or to aluminum, with 95,000 psi, but the latter are difficult or impossible to attain. In addition, titanium retains its strength to intermediate temperatures much better than aluminum alloys, so that it offers increasing advantage over aluminum alloys for aircraft structures as the speed is increased from subsonic to supersonic. An increase in skin temperature with increase in speed is not the only effect. The increase in speed also requires a design change toward more slender wings and other airfoils, for which higher-stiffness materials are essential. In this regard, titanium offers little advantage over the other structural materials, all of which have stiffness-to-density ratios of

TABLE I

Expansivities of Matrices and Reinforcements

Metal	Expansion $(10^{-6}/°F)$	Expansion $(10^{-6}/°C)$	Filament	Expansion $(10^{-6}/°F)$	Expansion $(10^{-6}/°C)$
Aluminum	13.3	23.9	Boron	3.5	6.3
Titanium	4.7	8.4	SiC-coated boron	3.5	6.3
Iron	6.5	11.7	Silicon carbide	2.2	4.0
Nickel	7.4	13.3	Alumina	4.6	8.3

100×10^6 in. However, reinforcement of titanium by high-stiffness filaments would produce a material with the necessary stiffness. In addition, titanium was recognized to have two other advantages over other potential matrices for boron filaments. The first was that titanium alloys have thermal expansivities lower than those of most other structural materials and approaching that of boron (see Table I). The second advantage was that the high strength of titanium would allow composites to be fabricated with less reinforcement in off-axis directions than would be required for weaker matrices.

Although this introduction to the subject of titanium matrix composites has been based largely on the titanium–boron system, the remarks are of more general applicability. For example, reference to Table I shows that titanium alloys have less incompatibility in expansion with other commonly available reinforcements than most of the other potential matrices.

II. Development of Approaches

The chemical compatibility problem delayed development of titanium-matrix composites because chemical degradation leads to low-strength composites.

A. *The Compatibility Problem*

Hot pressing of Ti–6Al–4V foils with boron filaments was investigated in 1965 as a method to produce titanium–boron composites by Sinizer *et al.* (1966). Pressures of 3600 psi and temperatures between 1500 and 1650°F were used for times of 1.75–27 hr. Significant strengthening was obtained in compression with as little as 12 v/o of boron filaments with an increase in the compression ultimate from 125 ksi for the matrix to 170 ksi for the composite. But weakening was obtained in tensile testing. In a parallel study of titanium–boron composites by Kreider *et al.* (1966), attempts were made to plasma-spray the titanium matrix onto an assemblage of boron filaments, but again, tensile tests showed that no strengthening was obtained.

Metallography presented by Sinizer *et al.* (1966) showed that the boron filaments were surrounded by a uniform band of a white reaction compound of thickness equal to 1–2 μm, with a dark, rapidly etching zone of a two-phase structure beyond the compound.

Figure 1 shows a typical stress–strain curve obtained for a titanium-alloy–12-v/o boron composite with a reaction zone of thickness equal to 1.3 μm, or 13,000 Å. Filament failure does not lead to immediate composite failure with this low volume percentage of filaments, and this permits the stress–

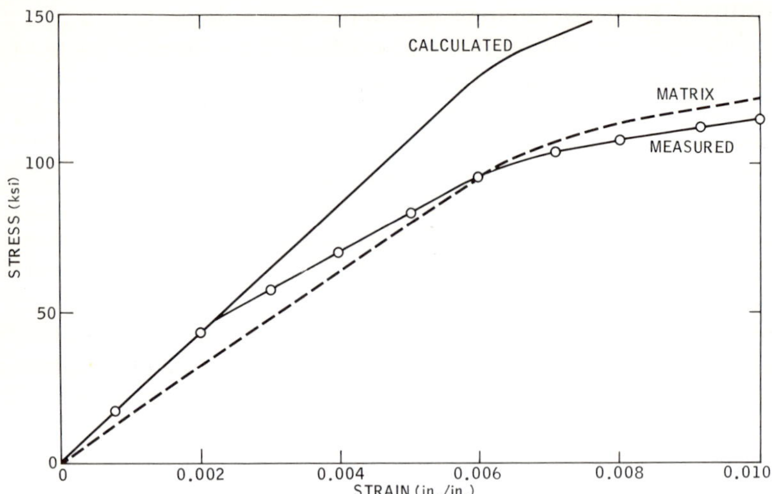

Fig. 1. Typical stress–strain curve of Ti–B composite with boride thickness of 13,000 Å (1.3 μm). (After Metcalfe, 1967.)

strain curves to be analyzed to follow the steps in failure. The calculated stress–strain curve is based on the rule of mixtures. For a volume fraction V_F of continuous filaments,

$$S_C = S_F V_F + S_M(1 - V_F) \tag{1}$$

where S_C is the stress in the composite, S_F the stress in the filament, and S_M the stress in the matrix, all measured at the same strain. It will be noted that the observed curve follows the calculated curve up to a strain of 2500 μin./in. Beyond this strain, the apparently elastic behavior is characterized by a reduced elastic modulus equal to 88% of that of the matrix, i.e., the composite acts as if the 12 v/o of filaments contribute nothing to the stiffness. The reason for this lack of stiffening is immediately apparent from Fig. 2, where

Fig. 2. Failure of overheated Ti–5Al–2.5 Sn–B composite. (After Metcalfe, 1967.)

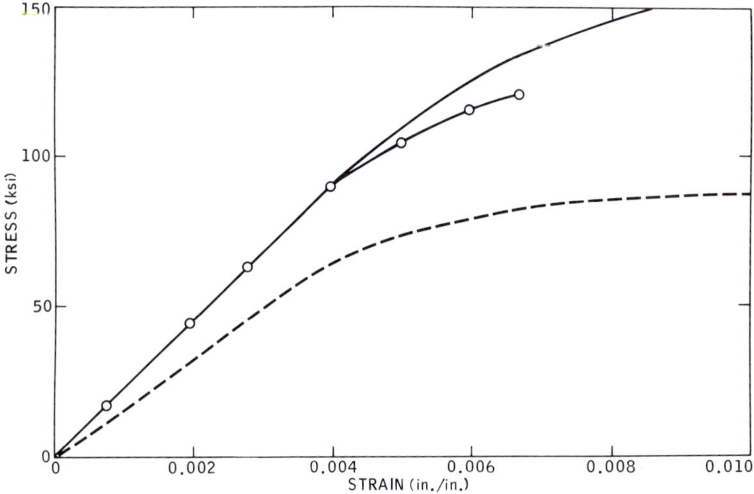

Fig. 3. Typical stress–strain curve of Ti–B composite with boride thickness of 250 Å. —— calculated, –O– measured, – – – matrix. (After Metcalfe, 1967.)

the taper section shows an example of excessive reaction between Ti–5 Al–2.5 Sn and boron as a white zone of titanium diboride around the filaments surrounded by a dark etching zone. This specimen was taken from a tensile specimen near the fracture point, and shows how the filament has broken up into short, ineffective lengths that are incapable of stiffening the titanium alloy matrix.

Figure 3 shows the stress–strain curve for a titanium–22-v/o boron composite made about the same time (1966) as the specimen used for Fig. 1 but for which the processing parameters were given as 10 sec at 1800°F by Metcalfe (1967). Under these conditions, the thickness of the reaction zone was estimated to be approximately 250 Å. This composite was made with unalloyed titanium matrix, and the lower strength of this matrix can be seen by comparing the plots of matrix strength in Figs. 1 and 3. The stress–strain curve follows the calculated values up to 4000 μin./in. with progressive departures beyond this point that could be related to residual stresses and to premature rupture of some of the filaments. The demonstration was very significant in another respect; it marked the first time that rule-of-mixtures behavior had been observed in a composite with appreciable reaction at the interface. Previously, the view had been held generally that composite behavior would be possible only in nonreactive systems. For example, in summarizing the state of the art, Hibbard (1964) concluded: "There should be little or no solubility or other reaction between the matrix and the fiber, which should wet each other."

B. Analysis of Interaction Effects

In addition to demonstrating that rule-of-mixtures behavior was possible when the amount of reaction was controlled, Metcalfe (1967) analyzed the mechanical properties of titanium-matrix composites in terms of the amount and properties of the interaction layer. The approach taken in this analysis was that the reaction zone introduced new sites for crack initiation within the composite. These new sites would operate in addition to the already existing sites present in the original filaments. It was also assumed that no crack-initiation sites occur in the matrix and that the distribution of flaws within the filament is unchanged in the fabrication. Under these conditions, the strength of the composite will be unchanged so long as the existing population of flaws in the filament control failure. This requires that the flaws introduced by the reaction layer remain smaller than the existing population. When the reaction layer remains thin, this requirement can be met, but a transition in the origin of fracture will occur above a critical amount of reaction.

If the theoretical strength of boron filaments is equal to $E_F/10$, where E_F is the elastic modulus, then the observed strengths of boron S_F will require defects with stress concentration factors K_F given by

$$K_F = E_F/10S_F \tag{2}$$

This relationship gives stress-concentration factors between 10 and 20 for the typical defects controlling the failure of boron filaments with strengths of 300–600 ksi. If located on the filament surface, such defects would approximate sharp-ended cracks, and reaction of filament and matrix may not affect this defect intensity. In addition, evidence has been presented by Wawner (1967) that the predominant source of failure of boron filaments is

TABLE II

Strain at Fracture of Components of Composites[a]

Material	Modulus, E (10^6 psi)	Strength, S (psi)	Strain at fracture ($\times 10^6$)
Boron	60	360,000	6000
Silicon carbide	70	350,000	5000
TiB_2	77	193,000	2500
$TiSi_2$	37.7	168,000	4460
TiC	66	197,000	3000

[a]Based on data given by Samsonov (1964).

frequently internal. Therefore, in either case, the condition will be satisfied such that the prior defects in the filament are unaffected by the surface reaction.

The white zone shown in Fig. 2 has been identified as titanium diboride, TiB_2, by Blackburn *et al.* (1966). Metcalfe (1967) assumed that this phase contained growth defects that would limit its strength to values characteristic of more massive material. This assumption allows the strain to fracture of this compound to be determined from published strength data. Table II summarizes values for several compounds that may be formed by reaction between titanium matrices and various filaments. It will be noted that the phase titanium diboride fails at a strain of 2500 μin./in., or equal to the strain at which one composite with heavy reaction lost its effective reinforcement, as illustrated by Fig. 1. (The strain in the reaction zone will equal the strain in the composite if plane sections remain planar under load.) Composites with larger amounts of boron filaments fail completely at this strain.

Figure 4 shows a schematic cross section of a titanium–boron composite after straining to greater than 2500 μin./in. The titanium diboride has cracked and has not caused the boron to fail. At the instance of rupture of the boride to form cracks, the latter will exert a stress concentration given by a factor K_I of the type

$$K_I = B \left(\frac{\text{depth of crack, } x}{\text{root radius, } r}\right)^{1/2} \tag{3}$$

The exact value of the constant B will depend on the stress distribution around the crack and particularly on the degree of support provided by the titanium matrix. However, this stress concentration can be less than that due to the intrinsic defects in the filament ($K_F > K_I$). In other words, there will be no decrease of filament strength until $K_I > K_F$. Solution of the appro-

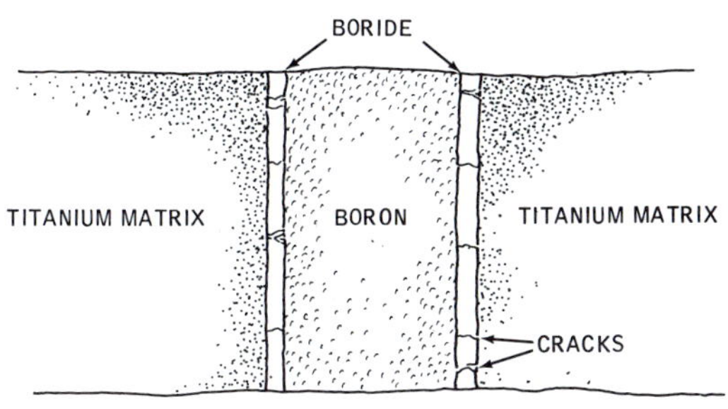

Fig. 4. Cracks in boride layer on boron. (After Metcalfe, 1967.)

priate equations leads to the critical value of interaction thickness $(x_{\text{crit}})_{\text{I}}$ when interaction begins to control failure as

$$(x_{\text{crit}})_{\text{I}} = (E_{\text{F}}/10BS_{\text{F}})^2 r \tag{4}$$

When the thickness of the interaction layer exceeds this critical amount, failure of the filaments occurs at a stress that decreases as the thickness of the layer increases. With increasing reaction, a second critical thickness is reached where the filaments fail at the instant the boride layer cracks. This simple theory gives this second critical thickness to be

$$(x_{\text{crit}})_{\text{II}} = r/(0.025B)^2 \tag{5}$$

Assuming that $B = 1$ and that $r = 3$ Å (the cell size of TiB_2), these equations give the two critical thicknesses equal to 1000 and 5000 Å. Figure 5 shows the relation between strain-to-fracture and reaction thickness. This figure shows that any thickness of boride cracks at a strain of 2500 μin./in. Filaments that fail at a strain of 5000 μin./in. (300,000 psi) are unaffected by the cracks in the diboride if the thickness of the latter does not exceed 1000 Å, and the boron continues to fail at its intrinsic defects. Higher-strength boron with less-severe intrinsic defects will tolerate less reaction, as shown for filaments that fail at a strain of 6000 μin./in. Between 1000 and 5000 Å of diboride, the boron fails on continued loading at the cracks in the boride that formed earlier in the loading cycle. Above 5000 Å of interaction, simultaneous failure of boron and diboride takes place.

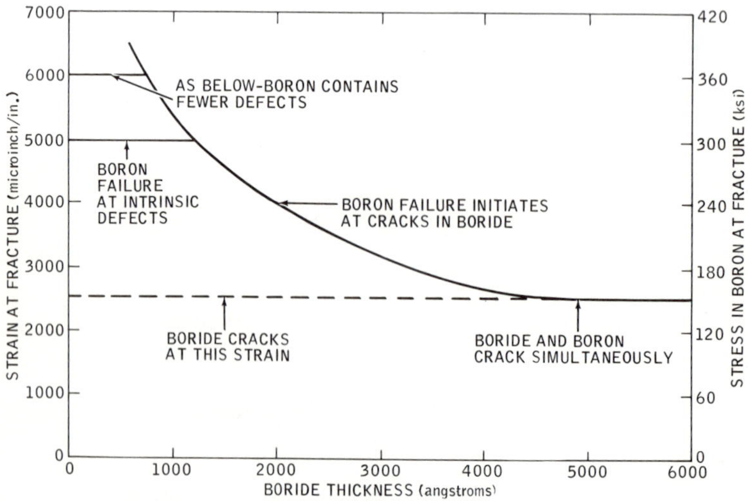

Fig. 5. Failure mechanisms in Ti–B composites. (After Metcalfe, 1967.)

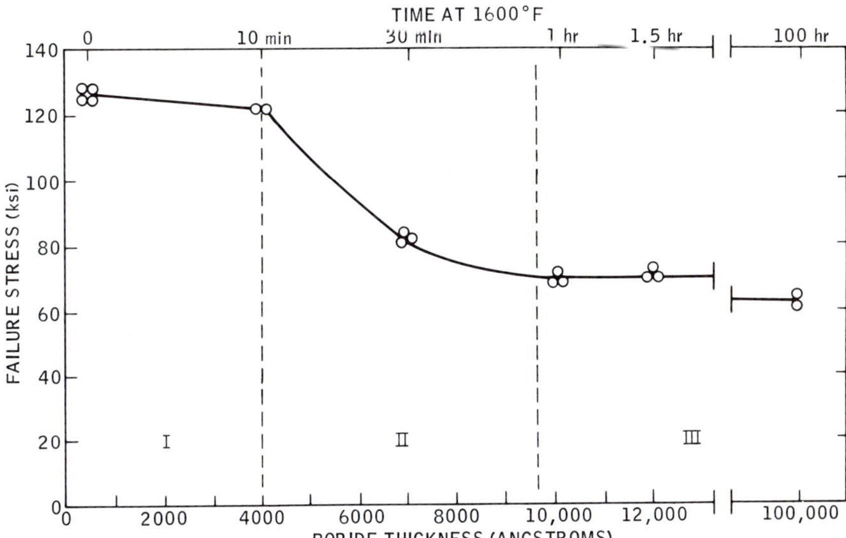

Fig. 6. Strength vs. diboride thickness for Ti(40A–B). In region I, filament flaws control failure; in region II, boride flaws cause filament failure on continued loading; in region III, filament strength is limited to the strength of the boride.

The prediction presented in Fig. 5 was evaluated in subsequent work with two unalloyed titanium matrices of different strengths by Klein *et al.* (1969). Figure 6 shows results for Ti(40A) matrix with 25 v/o of boron filaments. The general form of the plot in Fig. 5 is followed, but the thicknesses at each of the critical values are not in exact agreement. The individual stress–strain curves were analyzed for loss of stiffening on the assumption that this point marked the failure of the filaments. The strain-to-fracture values of the filaments within the composite were determined and plotted against amount of reaction in Fig. 7. Both boron and BORSIC† filaments in unalloyed titanium were analyzed in this way. The plot shows that the first critical thickness is at 4000 Å and the second critical thickness occurs at approximately 7000 Å for this matrix. The excellent agreement of the strain-to-fracture of both filaments with excess reaction with the predicted values of 2500 and 4500 μin./in. for boron and BORSIC respectively leads substantial support to the theory. The results for titanium–boron are also in agreement with the tensile test shown earlier in Fig. 1, where loss of stiffening occurred at the same strain.

The second matrix evaluated was Ti(75A) by Schmitz *et al.* (1970) and had somewhat higher strength. Specimens contained 25 v/o boron. Analysis

† BORSIC is a registered trade name of United Aircraft Company for its silicon carbide-coated boron.

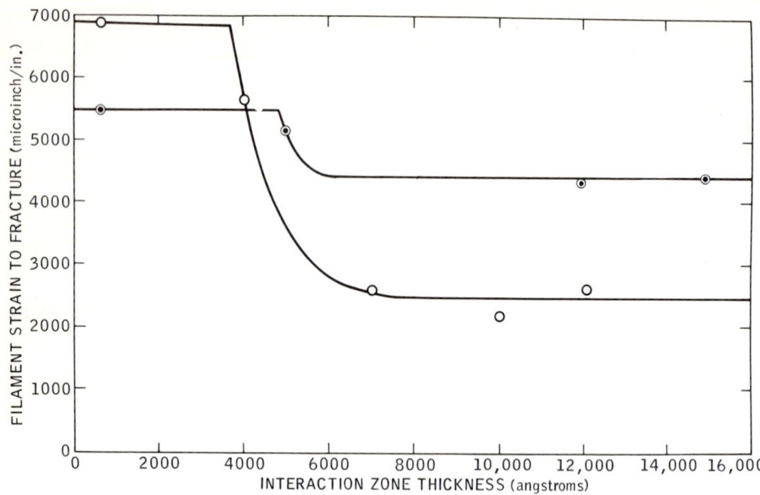

Fig. 7. Filament strain to fracture for Ti–B and Ti–BORSIC composite tapes as a function of interaction-zone thickness. O Ti–B composites; ⊙ Ti–BORSIC composites. Curves show predicted values.

of 34 tensile tests of specimens with 500–12,000 Å of diboride interaction led to the conclusion that this higher-strength matrix could withstand as much as 5500 Å of interaction before suffering degradation. Similarly, the second thickness was in excess of 10,000 Å.

It was concluded from these two studies that the general form of the reaction and fracture relationship of Metcalfe (1967) was correct. Further, the theory appeared to apply equally well to another reactive system, titanium–silicon-carbide-coated-boron. The prediction for the strain-to-fracture of the titanium-silicide reaction product was 4500 μin./in. (see Table II), and excellent agreement with this value is shown in Fig. 7.

Prior to the demonstration that titanium-matrix composites could have rule-of-mixtures properties if the reaction was kept below the first critical amount, it had been found that filaments extracted from these composites were markedly weakened. Indeed, this observation strengthened the belief that useful composites could not be made with reactive systems, that is, with systems where compounds were formed at the interface. Investigations by Klein *et al.* (1969) confirmed this strength loss and found that a boron filament with an initial strength of 466 ksi was reduced to a strength slightly under 150 ksi after extraction from composites having a Ti(40A) matrix. These filaments retained their surface films of approximately 500 Å of titanium boride after extraction, so that it was not surprising that the strain at failure was 2500 μin./in. (equal to a strength of 150,000 psi for a filament with an elastic modulus of 60×10^6 psi). Hence, it can be concluded that without the support of a surrounding titanium matrix, the first critical thickness of titanium diboride is less than 500 Å. This increases to 4000 Å for a matrix of Ti(40A) and to 5500 Å for the higher-strength matrix, Ti(75A). Figure 8 shows a plot of these data against the proportional limits of these matrices and the strains at the proportional limits. It has been assumed that no matrix corresponds to a proportional limit of zero. The results suggest that the support of the matrix has an important influence on the stress concentration exerted by the cracks in the diboride layer. This seems reasonable because without support, a crack will behave as if it were open ended, whereas with full elastic support the crack will approach a closed-end condition. This will cause a change in the constant B in Eq. (3).

In summary, the theory of interface interaction predicts that all brittle interaction layers will crack at a strain determined by their strength and elastic properties. The severity of these cracks is determined by their length, which in turn is determined by the thickness of the reaction layer. When the severity of the stress concentration exerted by the crack is less than the stress concentrations caused by the already existing defects in the filament, the composite strength is unchanged. Progressive loss of strength occurs as the length increases above a critical length given by the equality of these two types of stress concentrator. Above a second critical length, failure of

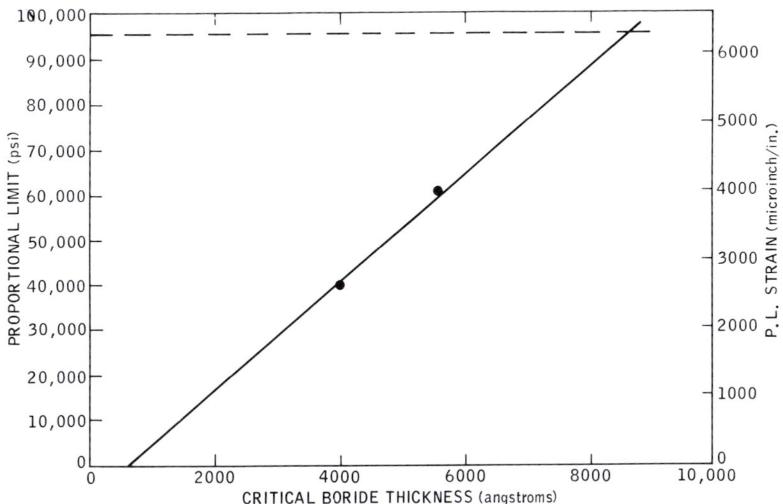

Fig. 8. Effect of elastic properties of matrix on critical boride thickness. The dashed line indicates typical filament-fracture strain.

the reaction zone results in immediate failure of the filament. Filaments with this excessive amount of reaction fail at 2500 μin./in. (150,000 psi) for boron and at 4500 μin./in. (270,000 psi) for silicon-carbide-coated boron. It has been shown for titanium–boron that the elastic support provided by the matrix affects the amount of reaction that can be tolerated. A plot of this relationship indicates that as much as 8000 Å of interaction should be tolerable for a matrix that remains elastic up to the strain limit of the filaments. Many titanium alloys will remain elastic to this point, which corresponds to an elastic limit of 96,000 psi, assuming an elastic modulus of 16×10^6 psi.

No effect of interaction-zone thickness was found on the transverse strength by Klein *et al.* (1969) or by Schmitz *et al.* (1970).

C. Requirements for Fan Blades

A considerable stimulus was given to the titanium-matrix composite development by work in England on carbon-reinforced plastic fan blades at the end of the decade of the sixties. Typical analyses show considerable weight saving is possible with stiff reinforced fan blades because vibration-damping shrouds can often be eliminated. This leads to a cascading effect because the disk to which the blades are attached can be reduced in weight. These weight savings can reach as much as 45% of the weight of the stage in the most advantageous circumstances.

The carbon-reinforced plastic blades were developed and tested in flight; however, these blades were found deficient in one significant area. The key problem that defied solution was protection against impact from foreign objects, sand, and rain. Although no comparative data are available on the erosion resistance of the English material, Stusrud *et al.* (1971) have presented comparative data on a series of composites. These authors simulated the stress conditions in fan and compressor blades by deadweight loading before conducting the ballistic impact test. The static stress was sufficient to create rapid propagation of the fine cracks produced by the impact. Repeated impacts at increasing energy levels were applied until the sample was penetrated or fractured. Panel thicknesses up to 0.125 in. were investigated to simulate the various thicknesses along the section of the blade. Figure 9 compares the different materials. It will be noted that the graphite–epoxy material was stressed much less than the other materials (20 ksi versus 35–40 ksi for the other materials). Table III gives estimated properties for the composite materials used in this evaluation.

The requirements for fan blades can be reviewed briefly with advantage at this point because titanium-matrix composites are being considered for this

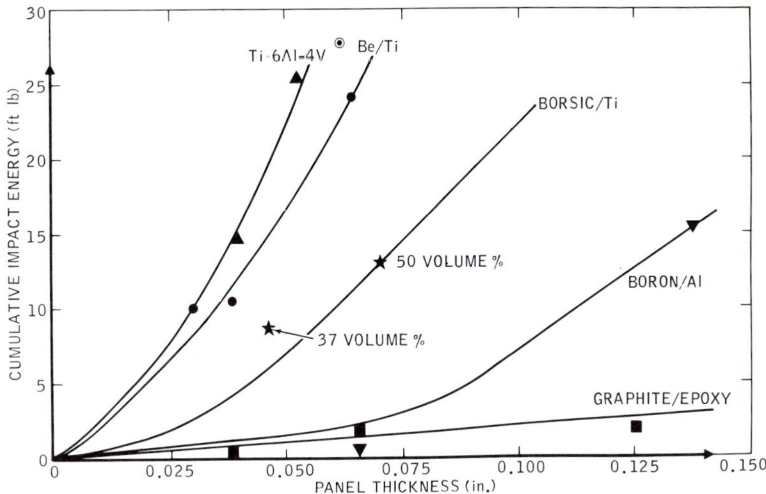

Fig. 9. Ballistic impact strength of composites for $\frac{1}{2}$-in. wide panels under static stress (Stusrud *et al.*, 1971).

● 50 v/o B–Al (6061) at 40 ksi
◆ 37–50 v/o BORSIC–Ti at 50 ksi
⊙ 53 v/o Be–Ti at 35 ksi (58-mil wires)
● 50 v/o Be–Ti at 50 ksi
■ 60 v/o graphite–epoxy at 20 ksi
▲ monolithic Ti–6Al–4V at 50 ksi

application. The important property of erosion and foreign-object damage resistance has already been mentioned. In addition, the airfoil sections of fan and compressor blades must be designed to withstand a combination of centrifugal, bending, untwist, and random vibratory stresses. But even more important in the design is tuning the blade airfoil frequencies to avoid coincidence with the frequencies imposed by the engine throughout its operating range, and to avoid self-induced vibration or flutter. In general, fan blades are critical in bending and torsional stiffness, and their associated

TABLE III

Comparative Properties of Materials Evaluated by Ballistic Impact

Material	Density (lb/in.^{3})	Elastic modulus (10^6 psi)	Strength (ksi)
Ti–6Al–4V	0.16	16	145
Ti–6Al–4V–50 v/o Be	0.11	27–32	100–120
Ti–6Al–4V–50 v/o BORSIC	0.13	38–40	140–160
6061Al–50 v/o B	0.095	35	180
Epoxy–60 v/o Morganite I graphite	0.061	33	120

frequencies, rather than stress. This point is very important because analyses show that certain titanium-matrix composites can be used effectively for this application even if their strength does not reach that predicted by the rule of mixtures so long as their stiffness attains the full predicted value.

D. Approaches to the Compatibility Problem

The renewed interest in titanium-matrix composites coinciding with possible application to fan blades has led to six approaches to solve or avoid the compatibility problem:

1. high-speed processing to minimize reaction;
2. low-temperature processing to minimize reaction;
3. development of low-reactivity matrices;
4. development of coatings to minimize reaction;
5. selection of systems with increased reaction tolerance;
6. design to minimize effect of reduced strength.

1. High-Speed Processing

High-speed processing was the first method to be demonstrated in work by Schmitz and Metcalfe (1968), and was the method of manufacture employed in the demonstration that rule-of-mixtures properties were possible in a system that underwent reaction. The procedures used will be discussed further under the system titanium–boron, but briefly, the composites were fabricated in tape form by electrical heating as the foils and filaments were passed between heated wheels. Typical temperatures were estimated to be 1800°F, and the time at temperature was estimated to be 1–2 sec. The amount of titanium diboride was measured to be less than 500 Å and was in reasonable agreement with the amount calculated for this exposure from the reaction data determined by Schmitz and Metcalfe (1968). Tapes made by this process contained 30 boron filaments and had predictable properties. The strength of such tapes equalled 140–145 ksi with approximately 25 v/o boron in a matrix of Ti(75A). The latter has a strength of 75 ksi at the strain where these composites fail, so that significant strengthening was demonstrated. The elastic modulus was 27×10^6 psi. This work will be discussed further in the section on titanium–boron composites.

2. Low-Temperature Processing

The earliest work on the low-temperature approach appears to be that of Hamilton (1969). The approach differs from the previous method in fabrication technique. Hot pressing was used so that very short times were not possible and lower temperatures were necessary to restrict the amount of reaction. The author selected Ti-6Al-4V–BORSIC for this development which was based on a three-step approach. First, the limit of reaction for BORSIC was established; then, the times and temperatures for hot pressing to remain within this limit were calculated; and, finally, optimum hot-pressing parameters were developed to remain within the calculated envelope. Tests were made of BORSIC filaments exposed in vacuum with and without contact with titanium. The strengths of these filaments were measured and converted to strain-to-fracture values from their elastic modulus. These strains were fitted to the theory of Metcalfe (1967) using the type of curve given in Fig. 5. Good agreement was obtained with the general shape of the curve except that the strain-to-fracture of heavily reacted filaments was estimated to be 4000 μin./in. rather than the higher value of 4500 μin./in. established for composites (Fig. 7). In the second step, Hamilton analyzed the flow-stress versus strain-rate curves for a Ti–6Al–4V matrix to determine the hot-pressing boundary that would not cause excessive reaction (500 Å). Hot-pressing parameters were optimized within this boundary, but no details are given of these parameters. Good quality composites were made with strain-to-fracture values exceeding 6000 μin./in. using BORSIC filaments with an average strength of 435 ksi and a standard deviation of 55 ksi. The elastic moduli also met expectations and were between 23 and 28×10^6 psi for composites with 21–27 v/o filaments.

The permissible times and temperatures for hot pressing to limit reaction to less than 500 Å were calculated by Hamilton from published reaction data. These data indicate that reasonable times for hot pressing, for example, 15 min, would require a limit on the maximum temperature of approximately 1400°F.

3. *Development of Low-Reactivity Matrices*

The work of Blackburn *et al.* (1966), Snide (1968), Bonnano and Withers (1966), and Schmitz and Metcalfe (1968) showed that the reactivity between boron or silicon carbide and titanium matrices was influenced to an appreciable degree by the composition of the matrix. Increasing alloy content, typical of beta alloys, led to reduced rates of reaction with the filaments. Selection of lower-reactivity matrices from available alloy compositions was the first step in use of this approach, but a longer-range goal to develop a matrix designed specifically for composites was the next step. The immediate application of this approach was limited by the availability of titanium alloys in foil form suitable for hot pressing, but two compositions have received much attention. These are the alpha–beta alloy, Ti–6Al–4V, and the beta alloy, Beta III, with a composition Ti–11Mo–5Zr–5Sn. An important advantage of beta alloys is that they can be rolled to foil with few intermediate anneals, so that they are more economical to produce in foil form than alpha–beta alloys.

Development of a compatible matrix for boron was undertaken by a group at the Solar Division of International Harvester Company under U.S. Air Force sponsorship (Klein *et al.*, 1969; Schmitz *et al.*, 1970). The steps in this development were: (1) to establish the reaction kinetics between boron and unalloyed titanium; (2) to establish specific rate constants for dilute solutions of candidate alloying elements in titanium; (3) to establish rules for additive behavior of more than one alloying element; and (4) to combine alloying elements to achieve a rate of reaction less than 1% that of unalloyed titanium. As in all other alloy development, attention had to be directed to other aspects of viable alloys, such as producibility, stability of the alloy, mechanical properties, and suitability for manufacture of composites.

Reaction kinetics with unalloyed titanium were measured over a wide range of conditions. Ti(40A) matrix tape was made by the high-speed process with both boron and BORSIC filaments. Heat treatments between 1000 and 1900°F and for times up to 6820 hr were used, but no data were included where the amount of reaction exceeded 100,000 Å. A limit to the amount of reaction is essential for BORSIC where the silicon-carbide layer is penetrated before this amount of reaction has occurred, but, more important, the limitation should be imposed to restrict observations to the range of reaction-layer thicknesses that include the thicknesses of interest. Reference to Figs. 6 and 7 shows that this limit should be closer to 20,000 Å, but such a limitation introduces acute experimental difficulties. In support of this viewpoint may be cited the work of Ratliff and Powell (1970), where an abrupt change in mechanism of diffusion was observed for the system titanium–silicon-carbide at a reaction-zone thickness of 100,000 Å. Reaction-zone thicknesses were measured by optical metallography with taper and vertical sections, supplemented by electron microscopy in some cases. Figure 10 shows the results of this study for unalloyed titanium matrices. The reaction rate constant is defined from thickness x and time t by the relationship $x = k\sqrt{t}$.

The single temperature of 1400°F was selected for much of the work on alloy development, which was limited to development of a compatible alloy for boron. The results indicated the elements in solution can be separated into three groups. In the first, the alloying element had no effect on the reaction-rate constant so that the alloy behaved as if the titanium remained at unit activity. Tin and silicon fall into this group of alloying elements. The second type causes a moderate reduction of the reaction rate proportional to the amount of element added, and may be regarded as a true diluent. Germanium and copper fell into this category. The third type caused more reduction of the reaction constant than found for the diluent

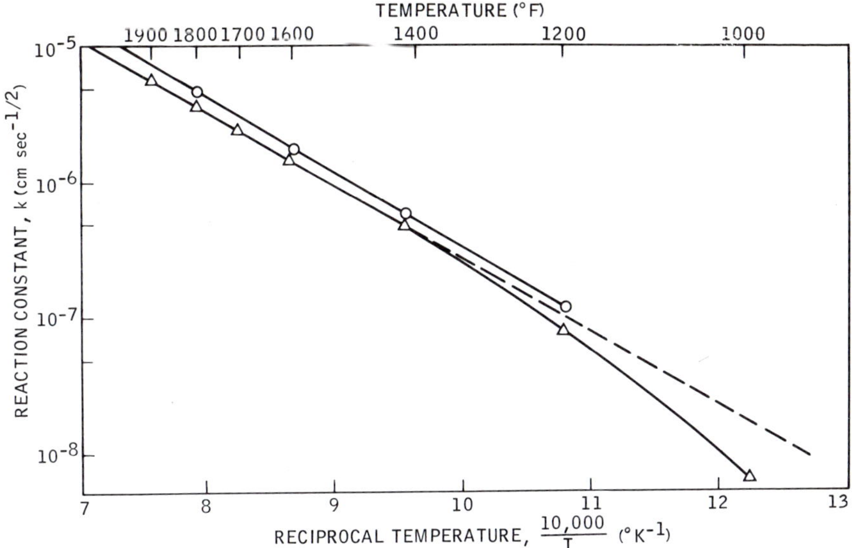

Fig. 10. Variation of rate constant k with temperature, where k = (zone width in cm)/(time in sec)$^{1/2}$. O Ti–BORSIC; △ Ti–B.

or second type. Measurements on binary alloys indicated that aluminum, molybdenum, and vanadium were of type 3, and subsequent analysis of more complex alloys indicated that zirconium fell into this type.

The reaction constant for unalloyed titanium and boron at 1400° F was found to be 5.0×10^{-7} cm/sec$^{1/2}$ so that a diluent of type 2 would have a specific rate constant of -0.05×10^{-7} cm/sec$^{1/2}$ per atomic percent of the

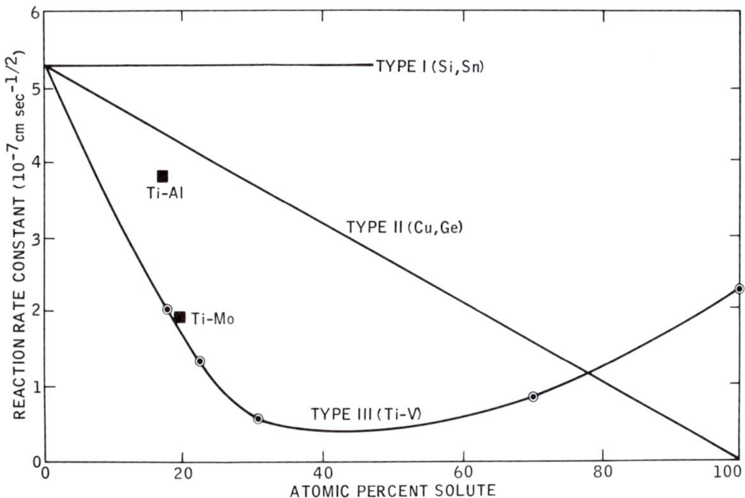

Fig. 11. Effect of alloying on rate constant for Ti–B reaction.

element. (The specific rate constant was defined as the decrease in rate constant per percent of alloying element.) Figure 11 shows the effect of each type of alloying element on the reaction rate constant for titanium. A type-3 alloying element is illustrated by vanadium. The initial decrease in reaction-rate constant exceeds that of the diluent type 2, but the rate gradually decreases until it begins to increase above 50 at. % and rise to the rate constant for vanadium.

Some confirmation of these results for individual specific rate constants was obtained by comparing observed values for alloys with those calculated from the appropriate values for dilute alloys. Simple additivity of individual effects was assumed. Table IV presents results for two commercial alloys. Agreement is good and suggests that the results are additive at least for dilute alloys.

Earlier work by Blackburn *et al.* (1966) had shown that aluminum was rejected ahead of the growing diboride in the reaction between boron and the alloy Ti–8Al–1Mo–1V. This result was confirmed in the work at Solar. Similar indications were obtained that molybdenum was being rejected ahead of the diboride by the presence of a new phase that appeared in alloys such as Ti–30Mo. Rejection of aluminum and molybdenum from the diboride would be expected from considerations of factors such as free energy of formation, melting point, and phase stability. On the other hand, similar considerations suggest that the more stable diboride formers that are isomorphous with TiB_2 should be largely present in the diboride phase, and should influence the reaction rate by alloying with the diboride. Vanadium and zirconium were believed to act in this manner. Judicious combinations of the two types of reaction-controlling elements were expected to offer the most effective means to reduce the rate of reaction, because each would operate in a more dilute alloy range. This viewpoint was borne out by results for alloys containing several elements. An alloy with a total of 30.5 % of the four alloying elements, aluminum, molybdenum, vanadium, and zirconium, had a reaction-rate constant of 0.2×10^{-7} cm/sec$^{1/2}$, whereas the binary alloy containing 30 % of the most effective individual alloying element, vanadium, had a rate constant of 0.6×10^{-7} cm/sec$^{1/2}$. Similarly,

TABLE IV

Comparison of Calculated and Observed Rate Constants

Alloy	Rate constant (10^{-7} cm/sec$^{1/2}$) Calculated	Observed
Ti–6Al–4V	2.96	3.6
Ti–8Al–1Mo–1V	3.56	3.4

the alloy Ti–30Mo had a rate constant of 1.9×10^{-7} cm/sec$^{1/2}$. The steps in the alloy development are complex, and the reader is referred to the original reports for details of the work. It should be noted that a reduction in the rate constant from 5 to 0.2×10^{-7} cm/sec$^{1/2}$ represents an increase in the time for a given amount of reaction by a factor of 625. Figure 12

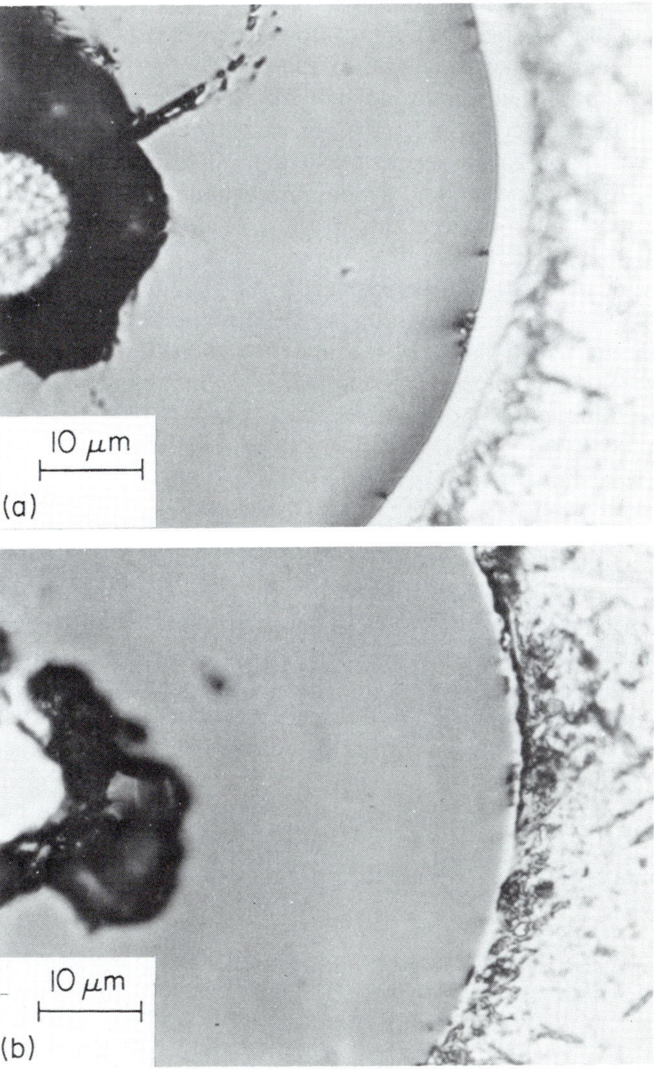

Fig. 12. Comparison of reaction with boron in 100 hr at 1400°F. (a) Unalloyed titanium ($k = 5 \times 10^{-7}$ cm/sec$^{1/2}$); (b) compatible alloy ($k = 0.2 \times 10^{-7}$ cm/sec$^{1/2}$).

compares the reaction of boron in 100 hr at 1400°F with unalloyed titanium and with one of these compatible alloys.

Metcalfe (1970) proposed that the effect of alloying elements in the titanium-diboride phase on the rate of reaction is through a change in the stoichiometry. Available evidence from the work of Rudy (1969) suggests that TiB_{2-x} is a better representation of titanium diboride than the stoichiometric compound, where x varies from 0.05 to 0.02 at 2000°C (no data are available for lower temperatures). On the other hand, vanadium and zirconium diborides include compositions where x is negative, that is, the phases extend to the boron-rich side of the ideal composition. If it is assumed that the transport of boron through the diboride occurs on the boron lattice sites, then the reduction in number of vacant boron lattice sites as stoichiometry is approached through alloying will explain the reduction in reaction rate by vanadium and zirconium. A minimum in Fig. 11 for alloys containing vanadium corresponds to those compositions that include the stoichiometric.

Development of an alloy specifically for titanium-matrix composites appears to offer the best solution to the compatibility problem. A wide variety of requirements must be satisfied for a successful alloy, but the start that has been made on this problem marks a coming of age in the field of filament-reinforced composites.†

4. Development of Coatings to Minimize Reaction

Klein *et al.* (1969) investigated several coatings on boron to reduce the rate of reaction. The comparison was performed at 1400°F and extended up to reaction-zone thicknesses of 40,000 Å, but no departures from parabolic growth rates were found, as shown in Fig. 13 for the reaction-zone thickness versus square root of the time. All reaction studies were with the unalloyed titanium matrix Ti(40A). No significant improvement was found for any of the coatings. An important difference exists in the reaction product. This product is principally titanium diboride in each case except for reaction with BORSIC, where the products are largely titanium silicides but may also include titanium carbide. Although no mechanical tests were made after reaction, it is not expected that the tolerance for the reaction in any of these cases would be any greater than for reaction with boron. The view has been

† Footnote added in proof: Recently, Schmitz and Metcalfe (1973) published results on composites with the compatible matrix, Ti–13V–10Mo–5Zr–2.5Al, containing 27v/o of 5.6-mil-diam. boron. The longitudinal strengths equalled the predicted values up to 1100°F, confirming the advantage of controlling reaction. The ultimate strength of 129 ksi and the 100-hr stress-rupture strength of 72 ksi for the temperature of 1000°F were equal to the corresponding values for the Ti–6Al–4V–45v/o BORSIC composites.

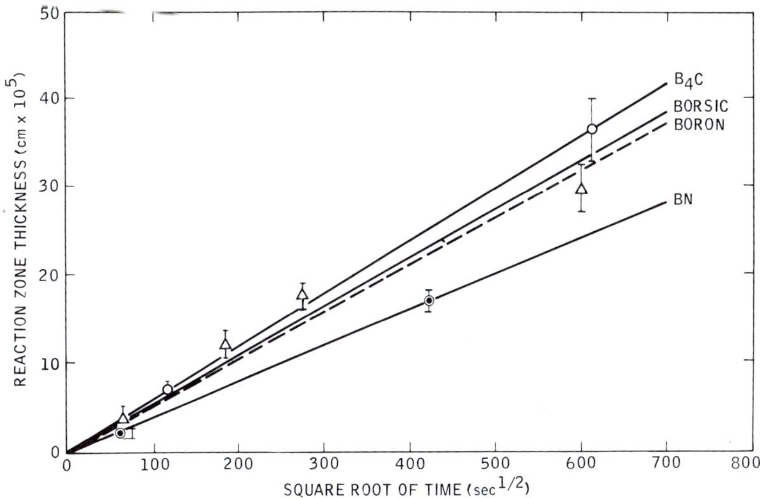

Fig. 13. Rate of reaction of coated filaments in Ti(40A) matrix at 1400°F. B_4C coated; ⊙ BN coated; △ BORSIC.

expressed that the silicon-carbide layer on boron acts as a sacrificial layer that delays reaction between the titanium and boron until the silicon carbide has been consumed. However, this viewpoint ignores the possible harmful effect of the reaction products that form in the sacrificial reaction. As shown earlier, these products cannot be tolerated to a greater extent than the borides, although once degradation sets in, the loss of strength and of strain-to-fracture are less for the Ti–SiC reaction than for the Ti–B reaction (see Fig. 7).

Review of the thermodynamics of reactions involving titanium shows that the metal forms compounds with most elements and that these compounds are characterized by very high free energies of formation. Also, titanium dissolves many elements and these solid solutions are at least as stable as the compounds. Hence, the likelihood of finding a coating for boron that is thermodynamically stable in the presence of titanium seems remote. In addition, the cost of application of this coating and the reliability of the coating do not favor this approach.

5. *Selection of Systems with Increased Reaction Tolerance*

Table II presented strain-to-fracture data for some interaction compounds that can be formed between titanium and some common filaments. The results suggest another solution to the compatibility problem. If the strain-to-fracture value of the interaction compound were to exceed that of the filament, then filament failure would occur before cracks could form at the interface. For a compound that fails without any plastic flow, and with a typical strength of 150,000 psi, the elastic modulus must be less than 25×10^6 psi to match a filament strain of 6000 μin./in. However, compounds with low elastic moduli tend to have low strengths, so that a maximum modulus of 20×10^6 psi would appear to be a better value. This restriction places a considerable penalty on the chances of success.

The situation would be much improved if the compound were to possess a very small amount of ductility. Certain intermetallic compounds are known to have some ductility. Noteworthy are the intermetallic phases in the systems Ni–Al and Ti–Al. Unfortunately, inadequate data are available for complete analysis of the system, but it is possible that the tolerance of the titanium–alumina composites for reaction shown by Tressler and Moore (1971) derives from these types of consideration. This system will be discussed later.

Another class of system with much increased tolerance for reaction is represented by the ductile filaments. Jones (1968) has shown that an iron aluminide forms in the system aluminum–stainless-steel when the processing conditions are not optimum. This iron aluminide cracks during tensile loading in the same way as the titanium boride was shown to crack in Fig. 4. But there is sufficient ductility in the stainless steel wire so that the stress concentration at the root of the crack is attenuated by plastic flow, revealed by slip lines in the steel. Continued plastic flow in the steel causes the wire to contract away from the interaction zone until it fails by necking. Although the exact mechanism has not been studied in detail, it is believed that titanium reinforced with beryllium wire may act in the same way.

Titanium reinforced with molybdenum wire differs in some respects from titanium–beryllium because reaction leads to a series of solid solutions rather than a brittle interaction zone. This system is believed to have tolerance for extensive interaction.

6. *Design to Minimize the Effect of Reduced Strength*

This approach has been discussed in Section II.C. Fan blades are stiffness-critical and can be designed to use composites with less than rule-of-mixtures strengths.

III. Composite Development

Development of titanium-matrix composites has been in progress for several years, but the six lines of approach detailed earlier were not clearly formulated throughout this period. The first approach—high-speed processing was proposed and demonstrated in 1967. Low-temperature processing was introduced by Hamilton (1969), and recent results show that strengths of 200,000 psi have been obtained by this approach in the Ti-6Al-4V–BORSIC system. Low-reactivity matrices began to be developed in 1969, but this line of research is slow moving because of the very many steps required in alloy identification and alloy development prior to the development of composites. The fourth approach through coatings started earlier than any of the others, with the initial work in 1966, but has not been successful because of the extreme reactivity of titanium. Preliminary work on the fifth approach through systems with increased reaction tolerance has been rapid and shows some promising results. Finally, the approach through design to make use of composites with full stiffness although degraded strength, has been advanced the most of all systems discussed.

The approach through design has permitted the most rapid exploitation of the advantages of titanium-matrix composites. The reduced strength of degraded composites often corresponded to the ultimate tensile strength of the matrix alloy, so that the view has been expressed that strengthening of titanium alloys by the principal filamentary reinforcements was not possible. Because this erroneous viewpoint has gained some acceptance, it will be worthwhile to examine the basis for this viewpoint in terms of the current theories.

The strength of a composite S_C with volume fraction V_F of filaments of elastic modulus E_F is given by the rule of mixtures as

$$S_C = V_F S_F + V_M S_M$$

or

$$S_C = V_F E_F \varepsilon_F + (1 - V_F) E_M \varepsilon_F \tag{6}$$

for the high-strength titanium alloy matrix of elastic modulus E_M that remains elastic to the failure of the composite at strain ε_F. But the strain ε_F is determined by the reaction layer for fully degraded filaments and equals 2500×10^{-6} for boron and 4500×10^{-6} for silicon carbide in contact with titanium (assuming formation of titanium disilicide). Figure 14 shows the variation in composite strength for the two cases calculated from Eq. (6), assuming a typical titanium-alloy matrix with ultimate tensile strength of 150,000 psi and elastic modulus of 16×10^6 psi. The strain-to-fracture for boron with a titanium-diboride reaction layer is 2500×10^{-6} or equal to a strength of 150,000 psi (the elastic modulus of boron is assumed to be 60×10^6 psi). Hence, degraded titanium–boron composites never exceed 150,000 psi. The reinforcement capability is generally better for silicon carbide (or BORSIC), where the value of 150,000 psi may be exceeded when more than 40 v/o of filaments are added, even if the filaments are degraded.

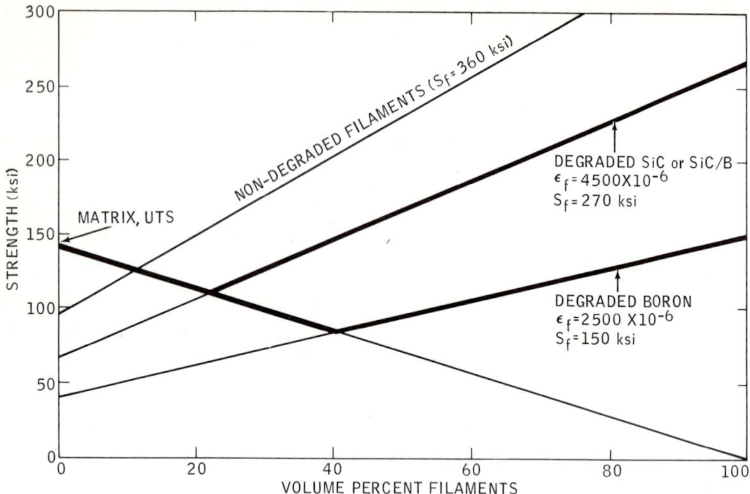

Fig. 14. Effect of filament degradation on strength of composites with titanium-alloy matrices.

However, in both cases, much higher strengths can be obtained for non-degraded filaments.

A. Titanium–Boron Composites

Preliminary work on composites with unalloyed titanium revealed the compatibility problem as discussed in Section II.A. One solution to this problem was obtained by use of the high-speed processing approach. Figure 15 shows a schematic of the process. Roll-grooved foils of unalloyed titanium were assembled with 30 filaments of boron in two layers on 0.006-in. centers. This achieved precision placement of filaments. A stack of three foils and 30 filaments was fed into a continuous diffusion-bonding machine using hot refractory-metal wheels. The bonding speed was 6 in./min and exposed the composite to the equivalent of a static heating at 1800°F for a little over 1 sec. The tape was trimmed to a size of 0.1×0.013 in. with approximately 25 v/o of boron.

Taper sections of these tapes showed that the amount of diboride formed in manufacture was 250–500 Å in thickness. This amount is well below the amounts shown to be critical in Figs. 5 and 6. However, the residual stress was expected to be high in view of the rapid processing, and in agreement with this suggestion, it was found that a stress-relief anneal increased the

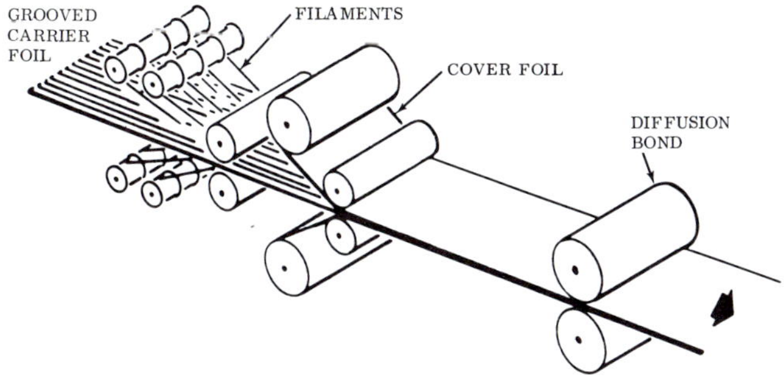

Fig. 15. Schematic of tape manufacture. (Metcalfe and Schmitz, 1967; reproduced by permission of SAE.)

strength and the strain to fracture, and decreased the scatter of results. Subsequently, it was demonstrated that an anneal of 30 min at 1200°F decreased the coefficient of variation from 8% to approximately 1%, while the average strength was increased from 119 to 125 ksi for composites with a Ti(40A) matrix and 25–28 v/o of boron. Even more important was the fact that the strain-to-fracture increased from 5200 to between 6000 and 7000 μin./in. The strength of this matrix is 72,000 psi, so that very significant strengthening was obtained. The elastic modulus was slightly higher than the rule-of-mixtures value for the appropriate volume percentage of filaments.

Properties of the annealed tape were consistent from lot to lot. Typical for mechanical properties for Ti(75A) with 25 v/o boron were: longitudinal strength, 140 ksi; coefficient of variation of strength, 2%; longitudinal modulus, 25.7×10^6 psi; transverse strength, 60 ksi; and transverse modulus, 22×10^6 psi. Off-axis strengths at 30 and 60 deg were 55–60 ksi with elastic moduli of $22–23 \times 10^6$ psi. Filament splitting appeared to be the principal cause of the limit on transverse strength. Preliminary tension–tension fatigue strengths and compression strengths were reported. The compression strength of a composite fabricated from several pieces of the tape was in excess of 300,000 psi for a 22 v/o composite, corresponding to filament stresses in excess of 800,000 psi.

Based on the reaction kinetics presented for unalloyed titanium in Fig. 10, the life of the as-fabricated tape was calculated to be nearly 10,000 hr at 1000°F (or 3.5 hr at 1400°F) before the critical amount of diboride was formed. This amount was shown to be 5500 Å for the Ti(75A) matrix. In agreement with such calculations, tape composites were annealed for over

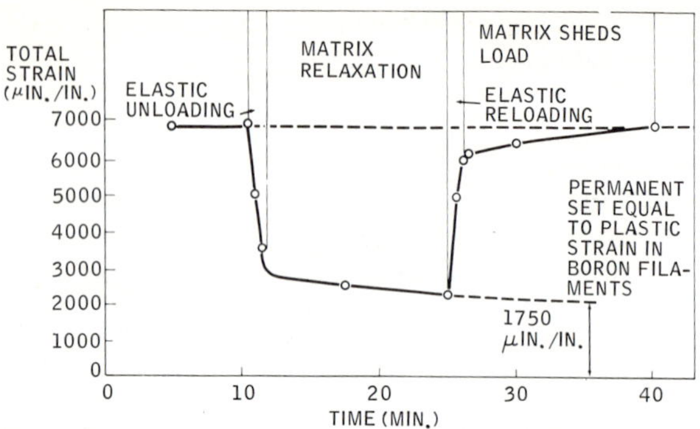

Fig. 16. Unloading and reloading at 1000° F of 17-v/o Ti–B composite after 81 hr exposure. Composite stress, 34.6 ksi; maximum filament stress, 206 ksi. (Reproduced by permission of ASME.)

4 hr at 1400° F without change of strength. Also, creep tests of tape specimens at 1000° F were performed for 116 hr without loss of strength.

Relaxation studies made by Metcalfe and Schmitz (1969) showed that the unalloyed titanium matrix relieved its load in creep in 15–30 min during the course of creep tests at 1000° F. Figure 16 shows a typical relaxation curve. After loading, the matrix begins to relax and the filaments must take up the load that was formerly supported by the matrix. This caused a slow additional extension of the filaments that is detected as an additional creep strain of the composite specimen. The situation is reversed when the specimen is unloaded, so that the elastic contraction of the filaments drives the matrix into compression. Complete unloading of the filaments is prevented until the compression strain in the matrix has relaxed. These results indicate that the unalloyed titanium matrix would be inadequate for use at 1000° F under conditions involving creep and relaxation because internal and external load transfer would be compromised over extended periods of time.

The results for unalloyed-titanium-matrix composites with boron lead to several important conclusions:

1. Very consistent properties can be obtained by the continuous-tape method of manufacture. It is not known if this results partly from the small size of the tape, but it is certain that the very uniform spacing and the uniform residual-stress pattern contribute markedly to the consistency. The increase in properties after annealing has already been cited to support the latter.

2. Control of the interaction between the filament and matrix allows composites with rule-of-mixtures properties to be made.

3. Reaction between filament and matrix can be controlled to leave several thousand hours of useful life at 1000° F.

4. Rule-of-mixtures properties can be obtained in titanium–boron composites, although the extracted filaments have markedly reduced strengths.

The creep and relaxation studies cited above are important in another aspect. The results show that boron-filament-reinforced titanium possesses good creep strength at 1000°F. Early work had shown loss of strength of boron at this temperature, but generally this work was performed in atmospheres containing residual air. Under such conditions, the strength appears to decrease rapidly with temperature, particularly above the melting point of boric oxide.

Tensile and stress-rupture tests on boron filaments by Ellison and Boone (1967) performed in vacuum show little strength loss up to 1000°F and support the viewpoint that surface interaction with the atmosphere affects the strength of boron filaments. These results and similar results by Metcalfe and Schmitz (1969) for filaments tested in vacuum are in general agreement with the high tensile and creep strengths exhibited by boron filaments when they are completely surrounded by a titanium matrix. Results from both sources indicate that the stress for rupture in 100 hr is between 265 and 315 ksi at 1000°F and between 250 and 275 ksi at 1500°F.

Recently, similar results have been demonstrated for a titanium-alloy matrix and the large boron filaments of 5.6-mil diam.

B. Systems with Silicon-Carbide–Titanium Interface

Full degradation of BORSIC-filament-reinforced titanium composites was shown to reduce the strain-to-fracture value to 4500 μin./in. according to Fig. 7. Table II showed that this strain equals that at which titanium silicide is expected to fracture, and supports the concept that the intermediate compound controls the fracture of such composites. The stress in the filament at this strain is 270,000 psi for BORSIC filaments, or 315,000 psi if the titanium silicide is formed from reaction with silicon-carbide filaments (because of the higher elastic modulus). These strengths are close to the strengths of early batches of filaments so that the conclusion was drawn from some of the early work that either silicon carbide formed less deleterious products or the rate of formation was less than for the reaction with boron. Another view that has been expressed is that the silicon-carbide coating on boron (BORSIC) is sacrificial and prevents any degradation until reaction is complete when reaction can begin between the titanium and the underlying boron.

Reaction kinetics for the silicon-carbide–titanium system have been investigated in some depth, although many questions remain unanswered.

This subject will be the first to be reviewed here. Preparation and properties of both BORSIC- and silicon–carbide-filament-reinforced composites are other areas of significant study and will also be discussed at length.

1. Reaction Kinetics

Full discussion of reaction kinetics is given in Chapter 3, Vol. 1 of this treatise. The reader is referred to this volume for a fuller review; but whereas the emphasis in the referenced chapter is to the physical chemistry, the emphasis in the account that follows will be on the implications of these results to the general field of titanium-matrix composites.

There are certain limitations on the conditions under which useful kinetic data can be generated. Three important conditions are (1) similarity of geometrical conditions to those in composites, (2) similarity of thicknesses of reaction product to those encountered in composites, and (3) similarity of temperatures to those of interest to composite fabrication and use. It will be recalled that interaction zones of up to 20,000 Å are of major interest in the case of titanium–boron composites according to Fig. 6. Ratliff and Powell (1970) studied the reaction of cylinders of titanium and silicon carbide. They found that the reaction path changed markedly once the titanium had become saturated with carbon, but this saturation took considerable time with massive pieces of titanium. Ratliff and Powell made no observations at thicknesses less than 44,000 Å and found a change of reaction-rate constant at approximately 100,000 Å of reaction product. Also, these authors studied reaction at 1000°C (1832°F) to 1200°C (2192°F), although these temperatures are considerably above those that are expected to be encountered in composite work with this system. In fairness to these authors, it must be added that the object of their study was wider than to investigate these reactions from the point of view of composites. An important contribution by Ratliff and Powell (1970) was to show that the reaction product was complex and changed with conditions (e.g., time, composition of matrix, temperature, volume percentage of matrix, and composition of silicon carbide). Hence, a single strain-to-fracture value may not be appropriate, as suggested earlier. Table II shows that this could vary from 3000 μin./in. for TiC to 4500 μin./in. for titanium disilicide. The greater complexity of the silicon-carbide–titanium interface compared with the boron–titanium interface requires more critical evaluation of reaction-kinetics data for this system.

The major study using silicon-carbide filaments and unalloyed titanium was made by Ashdown (1968) before the results of Ratliff and Powell (1970) were available on the saturation of the titanium with carbon that affects the early stages of the reaction. Ashdown used 0.030-in. titanium sheet with only two filaments (0.08 v/o of filaments), so that it is certain that carbon

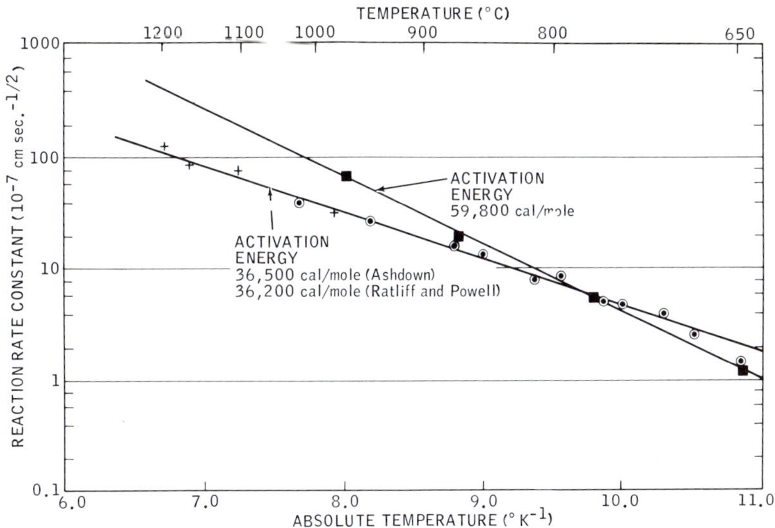

Fig. 17. Arrhenius plot for Ti–SiC reaction. + Ratliff and Powell (1970), stage 1; ⊙Ashdown (1968); ■ Klein *et al.* (1969).

saturation would be delayed. Figure 17 shows that Ashdown (1968) was in excellent agreement with Ratliff and Powell (1970) at the higher temperatures where the results overlap. At lower temperatures, Ashdown (1968) is in good agreement with Klein *et al.* (1969). The latter obtained data with composites containing 25 v/o of BORSIC filaments in an unalloyed titanium matrix. Saturation of the matrix occurred readily and was indicated by a general hardening and some loss of ductility. The higher values of the reaction constant obtained by Klein *et al.* (1969) at higher temperatures are now readily explained. Early saturation of the matrix with carbon will cause the rate of growth of compound to increase because both the silicon and carbon from the reacted filament will form compounds instead of the carbon dissolving in the matrix.

Table V presents collected data for reaction between silicon carbide and titanium alloys. Again, only the results of Schmitz and Metcalfe (1968) and Klein *et al.* (1969) refer to actual composites, but examination shows that there is clearly a reduction in the reaction rate constant for titanium alloys compared with unalloyed titanium. This reduction is most marked for the heavily alloyed beta alloy B120VCA (Ti–13V–11Cr–3Al). There are not enough data to assess the effectiveness of each alloying element in reducing the rate of reaction, as could be done in the case of reaction with boron, but the heavily alloyed beta compositions appear to be more effective.

TABLE V

Constants for Titanium-Alloy—Silicon-Carbide Reaction[a]

Alloy	650°C 1200°F	677°C 1230°F	760°C 1400°F	800°C 1472°F	850°C 1562°F	871°C 1600°F	900°C 1652°F	927°C 1700°F	982°C 1800°F	1000°C 1832°F	1038°C 1900°F	Reference
Ti–6Al–4V	—	—	—	—	—	—	7.6	—	18.7[b]	—	—	Blackburn *et al.* (1966)
	—	—	—	3.8	9.5	—	8.3	—	13.5[b]	—	—	Snide (1968)
	—	—	—	—	—	—	—	10	15	—	20	Schmitz and Metcalfe (1968)
	—	—	—	—	—	—	—	—	—	22[c]	—	Ratliff and Powell (1970)
Ti–13V–11Cr–3Al	—	—	—	—	—	—	—	—	9	—	8	Schmitz and Metcalfe (1968)
Titanium	1.1	—	6	—	—	21	—	—	70	—	—	Klein *et al.* (1969)
	1.3	2.5	6	—	—	16	—	—	32	—	—	Ashdown (1968)

[a]Units are 10^{-7} cm/sec$^{1/2}$. [b]991°C. [c]Stage-I kinetics.

2. *Silicon-Carbide-Reinforced Titanium*

Silicon-carbide continuous filament grown on a tungsten-wire substrate was evaluated for reinforcement of titanium alloys in several studies. One of the first filaments available was made by General Technologies Corporation, Reston, Virginia, and had the following typical properties:

strength: 350 ksi,	standard deviation of strength: 60 ksi
diameter: 0.0038 in.	elastic modulus: 70×10^6 psi

Composites with 22 v/o of this filament in Ti–6Al–4V matrix were made with a typical tensile strength of 130,000 psi and an elastic modulus of 30×10^6 psi. The strength-to-density ratio was 930,000 in. and the elastic modulus-to-density ratio was 214×10^6 in. The stiffness exceeds the rule-of-mixtures value for this composition (28.3×10^6 psi); it is not known whether this results from synergistic effects due to triaxial stresses or to solution of carbon in the matrix, or to some other cause. However, the strength was well below predictions.

Tsareff *et al.* (1969) compared the General Technologies Corporation (GTC) filament with silicon carbide produced by United Aircraft Research Laboratories (UARL). Composites were made by diffusion bonding 0.004-in. Ti–6Al–4V foils with silicon-carbide-filament mats under nominal conditions of 1600°F, 1 hr, at 6 ksi pressure. Powder additions were used in most cases to aid in control of spacing. The mats were held together with polystyrene that was baked off during the heating cycle. Excellent bonding was obtained with all filaments, but the reaction zones were approximately 1 μm with boron and 0.5 μm with silicon carbide. The tensile strength measured in the filament direction was typically within the range 120–140 ksi for 18–38 v/o of filaments, or below the ultimate tensile strength of the matrix. On the other hand, the elastic modulus followed the rule-of-mixtures law or exceeded it by as much as 20% in some cases. Dynamic-flexure moduli were found to remain high at elevated temperatures. For example, a 30-v/o composite with an elastic modulus of 32×10^6 psi at room temperature retained a value of 29×10^6 psi at 1200°F. Fatigue strength was measured with a resonant-cantilever-beam test. The composites tended to have slightly lower fatigue strengths than the matrix under the conditions of test, and several factors were believed to contribute to this. The notch effect of free filaments at the surface was thought to be the most important factor, but a difference in matrix due to oxygen content and preferred orientation, as well as residual stress, were included as factors. The strain in the matrix appeared to be a controlling factor.

The off-axis strengths of composites with 28 v/o reinforcement were high. Average tensile properties are summarized in Table VI. These results gave

TABLE VI

Off-Axis Tensile Properties of Ti-6Al-4V–28 v/o SiC

Filament orientation (degrees)	Average Strength (ksi)		Elastic modulus (10^6 psi)	Poisson's ratio
	Ultimate tensile strength	Proportional limit		
0	142	117	36	0.275
15	135	117	35	0.277
30	113	104	32	0.346
45	107	75	31	0.346
90	95	53	28	0.250

an excellent fit to the equation developed by Tsai (1966). The high off-axis strength and stiffness provide substantial support for the viewpoint that unidirectional titanium composites may be used in applications that would require multidirectionally reinforced aluminum- or plastic-matrix composites.

The creep properties of 28-v/o SiC–Ti-6Al-4V composites were determined for a strain of 0.2% at temperatures between 780 and 1000° F. Significant improvement in strength was found; for example, the stress to cause 0.2% creep strain in 100 hr at 900° F was increased from 15,000 psi for the matrix to approximately 80,000 psi for the composite. This strength was superior to that of the best available commercial alloy.

Williford and Snajdr (1969) applied the high-energy-rate forming process to fabrication of Ti-6Al-4V–SiC composites. Metal powders and filaments were contained in an expendable mold and sealed under vacuum in a billet container. After heating to the desired temperature, the billet assembly was subjected to a suitable pulse of energy in the impaction machine. The pulse duration was short and varied from 3 to 15 msec. There was little opportunity for interaction of powder and filament and little change in the grain size of structure of the metal matrix during processing. Complete consolidation of the composites could be achieved at 1000°C. Typical specimens were found to have reacted to the extent of 6500 Å.

In spite of the small amount of reaction, the tensile strength of a single specimen with 21 v/o of silicon-carbide filaments was 132,000 psi, compared to a rule-of-mixtures strength of 175,000 psi for matrix prepared from the same powder. However, two properties showed marked improvement. The elastic modulus of 25.7×10^6 psi was over 90% of the rule-of-mixtures prediction. A creep test of a composite with 16 v/o filaments at 1022° F (550° C) was made under 55 ksi and resulted in failure in 197 hr. This performance represents a 200° F improvement over that of unreinforced Ti–6Al–4V. The authors report absence of an observable reaction zone following this creep exposure at 1022° F (550° C).

Other results for the system were not as encouraging, particularly for composites with high filament loadings. Composites with 28 v/o reinforcement had less than 50 % of the anticipated strength, and this decreased to 10 % for composites with 60 v/o of filaments. A low-strength transverse to the filaments was interpreted to mean no useful amount of fiber-to-matrix bonding. Although filament breakage was acute in some cases, extracted filaments from composites containing 60 v/o of reinforcement were free of breakage in other cases. Hence, the cause of poor results with high-volume-percentage composites is not clear, but comparison with the results of Tsareff *et al.* (1969) suggests that the high-energy method of preparing composites is not as satisfactory as hot pressing.

The investigations by Tsareff *et al.* (1969) and by Williford and Snajdr (1969) are in good agreement on the improvement in tensile elastic modulus and creep strength, and on the poor tensile strength at room temperature. Although application to stiffness-critical components is possible, full use of titanium-matrix composites requires that the filaments contribute their full strength potential. Figure 18 shows several stress–strain curves due to Tsareff *et al.* (1969), from which it can be seen that the composites fail at strains-to-fracture between 3000 and 5100 μin./in. Extracted filaments from specimen

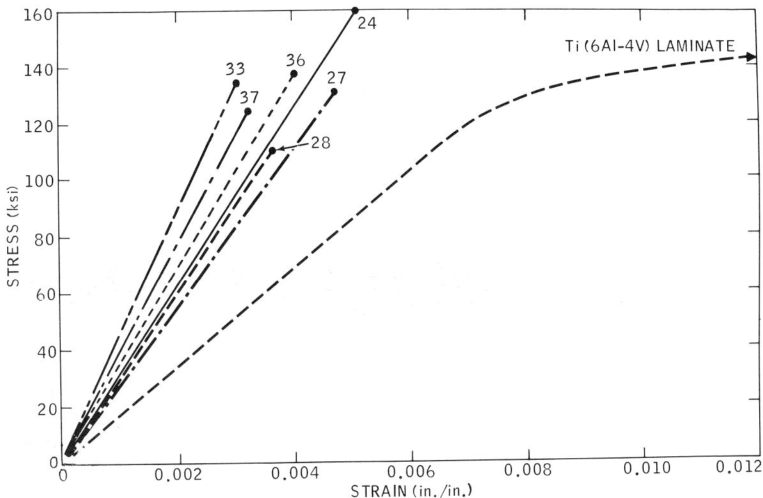

Fig. 18. Stress–strain curves for Ti–6Al–4V composites containing GTC SiC filaments (Tsareff *et al.*, 1969).

number 24 retained an average strength of 300,000 psi or a strain-to-fracture value of 4500 μin./in., so that the performance of the composite specimen was excellent. On the other hand, filaments from another composite were weakened to 200,000 psi. Such variation in amount of degradation makes analysis of the results difficult. To this variation must be added the complexing factor of residual stress that places the filaments under an initial compression, thereby favoring an increased contribution from the filaments. Tsareff *et al.* (1969) suggest that the oxygen composition of the matrix may be important, because this affects the local flow and work hardening at filament breaks. On the other hand, Williford and Snajdr (1969) believe that filament spacing and the triaxial stress state resulting from close filaments is responsible for poor tensile strength, particularly in specimens with 60 v/o of reinforcement.

Before leaving the subject of reinforcement of titanium by silicon carbide, it is worth noting that some efforts have been made to incorporate whiskers into titanium matrices. Spark sintering with powder matrices of Ti–6Al–4V and Ti–13V–11Cr–3Al led to excessive reaction, including explosion in one case, according to workers at Melpar, Incorporated, of Falls Church, Virginia. Similar problems with whiskers were found by Williford and Snajdr (1969) in attempting to use high-energy-rate forming methods, even at temperatures as low as 1110°F (600°C).

3. *Titanium Reinforced with Silicon-Carbide-Coated Boron*

Silicon-carbide-coated boron was developed by United Aircraft Research Laboratories and given the trademark BORSIC. It is now marketed by the Hamilton Standard Division of United Aircraft. It was introduced to solve the problem of reaction between boron and aluminum in hot pressing, but has found increasing use with titanium instead of the more expensive, weaker, and denser silicon carbide.

Tsareff *et al.* (1969) had investigated Ti-6Al-4V–BORSIC composites at the same time as their work on silicon-carbide filaments. These composites were fabricated by hot pressing at 1600°F for 1 hr as described earlier, and showed the same amount of reaction. The composites contained between 26 and 34 v/o of reinforcement. Tensile tests gave very consistent results, with strengths between 106 and 138 ksi. The strain-to-fracture value was also quite consistent and averaged 4000 μin./in. The latter was only 54% of the strain-to-fracture value estimated from the average strength of the as-received filaments (475 ksi). The elastic modulus closely approached the rule-of-mixtures values.

Similar results were obtained by others in the field. For example, Davis (1969) reported strengths of 125–142 ksi for Ti-6Al-4V–BORSIC containing 25 and 48 v/o of the reinforcement.

At this point, work ceased on attempts to improve the properties of titanium-matrix composites in favor of application of the existing material. This approach followed the directions described in Section II.D.6, where the fan blade for a bypass jet engine was selected as a structure that is stiffness critical rather than strength critical. The higher stiffness of a composite blade has several beneficial effects, but the principal effect is to raise the critical frequencies. This permits removal of vibration-damping devices such as midspan bumpers, thereby reducing the mass of the blade. The lower density of the composite compared with the monolithic titanium alloys lowers the centrifugal stress at all points along the blade, and removal of the bumper causes another large reduction in centrifugal stress in the section of the blade between the root and the original radial position of the bumper. Hence, the strength requirement for a composite blade is lower than that of the monolithic blade. On the other hand, removal of the bumper reduces the stiffening against untwisting under the aerodynamic loading, but the increased stiffness of the composite is more than adequate to compensate for this and may even permit the blade to be made thinner.

Stevens and Hanink (1970) selected Ti-6Al-4V–50 v/o BORSIC for fabrication development of fan blades. The composite was fabricated from prewound mats of 4.2-mil BORSIC filaments coated with a mixture of polystyrene and Ti–6Al–4V powder. Titanium-alloy foils of 2.5-mil thickness were creep-formed to shape before laying up with the mats. The blade layup assembly was encapsulated in a thin stainless-steel jacket designed to allow a dynamic vacuum during the hot-press diffusion bonding. Typical hot-pressing procedures were 1600° F for 30 min under 12,000 psi. Specimens for material characterization were prepared under the same processing conditions used for fan blades. Properties of these composites are given in Table VII.

The average room-temperature longitudinal tensile strength of 140 ksi is less than predicted by the rule of mixtures. The off-axis strengths are also lower than expected for a titanium-matrix composite, but may reflect the transverse weakness of the underlying boron filaments. On the other hand, the small decrease of strengths with temperature up to 850° F is encouraging. The failure strains are low, and do not indicate as efficient use of the filaments as found by Tsareff *et al.* (1969) in composites with fewer filaments. This point is particularly important in connection with the development of high-strength composites and is discussed in the next section. The tensile and flexure moduli are high and remain so up to at least 850° F. The tensile elastic modulus exceeds that predicted from the simple rule of mixtures.

Creep-rupture tests showed that the material preserved its strength up to 850° F, but the scatter of data was somewhat high. Contraction occurred in some specimens at 500° F but was eliminated by annealing prior to test, suggesting that high residual stresses are present in as-fabricated specimens.

TABLE VII

Properties of Ti-6Al-4V—50 v/o BORSIC Composites

Temperature (°F)	Orientation (degrees)	Tensile strength (ksi)	Failure strain (μin./in.)	Elastic modulus (10^6 psi)		Coefficient of expansion (10^{-6}/°F)
				Tensile	Flexure	
70	0	140	3340	41.5	34.4	2.50
70	15	100	3220	36.8	33.3	—
70	45	66	4220	31.2	31.8	—
70	90	42	3130	29.8	31.2	3.17
500	0	119	—	—	33.2	2.80
700	0	107	—	—	32.4	—
850	0	109	—	—	31.5	3.17
850	15	86	—	—	29.9	—
850	45	53	—	—	27.6	—
850	90	35	—	—	24.4	3.64

Also, the amount of creep rarely exceeded 0.1 %, and the scatter of data for times up to 1000 hr matched the scatter of tensile-strength data. It was concluded that there is no creep at 500°F and designing to the minimum tensile strength would suffice, at least for times up to 1000 hr. Some creep was found at 850°F, with a decrease in load-carrying ability with time. Although some specimens did not fail in 1000 hr under 95 ksi, the data indicate that 70–80 ksi may be a safer stress for this duration.

The fatigue tests were performed on specimens clad on all surfaces with Ti–6Al–4V to simulate the condition at the surface of the fan blades. The results are representative of the special conditions employed, but the latter does avoid the surface notch effects described by Tsareff *et al.* (1969) when filaments are exposed. In summary, the fatigue tests at room temperature, 500 and 700°F, showed good agreement with predictions based on the Goodman diagram. The straight-line Goodman diagram joining the ultimate tensile strength to the alternating stress endurance limit provided a good fit to the data.

Recently, some detailed studies by Toth at TRW of the processing of Ti–6Al–4V containing 45–50 v/o of the 5.7-mil-diam B–SiC filaments have been disclosed (Collins, 1972). Hot pressing for a time of 30 min was used at a series of increasing temperatures. The pressure was varied to obtain well-bonded composites at each temperature. Figure 19 shows the results for an average of four specimens plus some repeat tests for the transverse strength. The strength measured at the pressing temperature of 1600°F is in good agreement with that found by Stevens and Hanink (1970) who used the same temperature, but the failure strain is higher (see Table VII). It is noteworthy

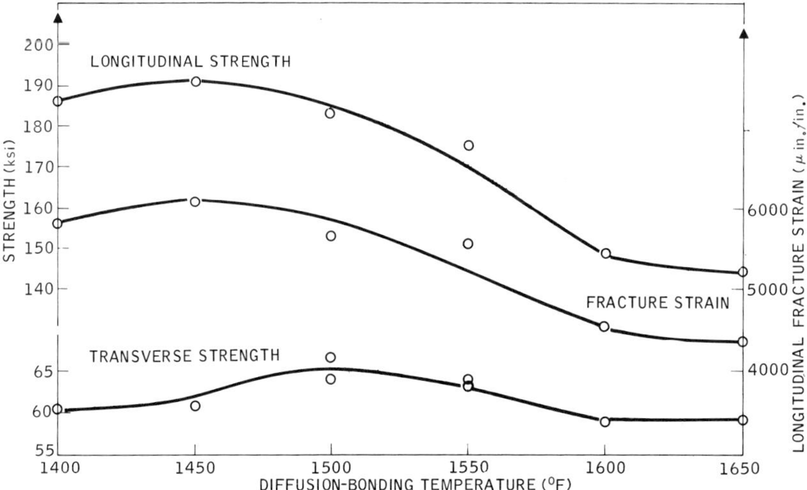

Fig. 19. Effect of diffusion-bonding temperature on properties of Ti-6Al-4V–BORSIC composites. (After Collins, 1972.)

that the fracture strain of the composites processed at the highest temperatures (1600 and 1650°F) are equal to the values predicted for fully degraded composites. Reduction of the processing temperature to below 1500°F appears to reduce the amount of reaction to the point where effective use of the filaments can be achieved.

These results represent a considerable improvement over previous data and show the importance of processing in the development of composites. The combination of longitudinal and transverse strengths obtained in this study far exceeds that obtainable with plastic or aluminum matrices, where a proportion of the filaments must be placed in off-axis positions to realize adequate off-axis strengths, but at the expense of the longitudinal strength. This advantage of a titanium matrix was pointed out in the introduction to this chapter in reviewing the position of titanium-matrix composites. A further advantage of this matrix was shown in high-temperature tests of these composites (Collins, 1972). A room-temperature strength of 180,000 psi decreased to 122,000 psi at 1000°F, while the transverse strength only decreased from 64,000 to 40,000 psi over the same temperature interval.

Similar results for longitudinal and transverse tensile strengths have been presented by Prewo and Kreider (1972) for this system. In addition, it was shown that the notch tensile strength was high. The center-notched specimens gave notched-to-unnotched tensile ratios in the range 0.7–0.9 for composites with 28–48 v/o of BORSIC, and for both longitudinal and transverse testing.

The authors found that BORSIC-reinforced composites with a Ti–6Al–4V matrix were less sensitive to the presence of notches than those with aluminum-alloy matrices. On the other hand, the impact energy equalled the impact energy of aluminum-matrix composites when the crack ran through the filaments but was below that of aluminum-matrix composites when the crack ran between the filaments. However, the magnitude of the values was thought to be adequate for reliable high-performance, lightweight structures.

4. *Discussion of Strength in Composites with Silicon-Carbide–Titanium Interface*

Four factors had been identified in prior work as possible contributors to degradation of these composites so that less than rule-of-mixtures properties were obtained. These were interaction, triaxiality of stress, residual stress, and filament damage. The importance of each of these was assessed differently by prior workers, depending, in part, on the method of processing used. The recent demonstration that fracture strains exceeding 6000 μin./in. could be obtained in this system (Collins, 1972) has shown that interaction is the most important of these factors. However, the average composite strength of 190,000 psi corresponds to an average filament stress at fracture of 305,000 psi. This is below that obtained in aluminum-matrix composites with the same filament, although many years of development effort were required before aluminum-matrix composites could be made with failure strains corresponding to filament stresses of 400,000 psi. Continuation of the improvement in titanium-matrix composites will depend on understanding of all of the contributors to composite degradation. These factors will be reviewed here.

The interaction-product factor appears to be very well established. However, the agreement with predictions may be partly fortuitious because the reaction products are more complex with silicon carbide than assumed in the prediction of the fracture strain according to the theory of Metcalfe (1967). Based on the results of Ratliff and Powell (1970), the reaction product(s) seems likely to depend on the following factors: temperature of reaction, amount of reaction products, solubility of carbon in matrix, composition of matrix, silicon-to-carbon ratio in filament, and volume fraction of matrix. Reference to Table II shows that conditions leading to the predominance of TiC as the reaction product would decrease the strain-to-fracture of degraded composites to 3000 μin./in. But another effect of reaction is to add carbon to the matrix and reduce its ductility.

Williford and Snadjr (1969) pointed to triaxial stress between filaments as a major contributing factor to premature failure of composites with 60 v/o of silicon-carbide filaments. This effect is likely to be enhanced if the ductility of the matrix is low; for example, as a result of solution of oxygen or carbon. (One advantage of pressing at a low temperature is that the amount of interstitial elements in a titanium matrix is reduced.) If this effect is important, it would imply that the optimum percentage of filaments is less than 60 v/o. On the other hand, the same limitation on volume percentage was thought to exist in the case of aluminum-matrix composites, but was found to move to higher percentages as the precision of manufacture was improved.

Residual stress is the third factor cited as limiting the strength of these composites. The results of Klein *et al.* (1969) in demonstrating the increase in strain-to-fracture of unalloyed titanium-matrix composites with boron as a result of annealing 30 min at 1200°F provide support for this suggestion. The increase in strain-to-fracture was from 5200 to nearly 7000 μin./in. as a result of this anneal. There will be considerably greater difficulty in reducing the residual stress in composites made with high-strength titanium-alloy matrices. Such reduction may be a key to further improvements in the performance of these composites.

The final factor is filament damage. This can be assessed on a gross scale by solution of the matrix to determine broken filaments, but more subtle modes of damage may occur that do not result in broken filaments. Measurement of the strength of extracted filaments is not a dependable technique because of the possible damage on extraction and because the state of stress on the fiber is not the same as in the composite. Also, Crane and Tressler (1971) have shown that alumina filaments may be degraded from nearly 400 to 200 ksi by self-abrasion but that the strength of these filaments at 1000°C is unaffected by this prior treatment. Similarly, the high stress-rupture strength found for boron by Ellison and Boone (1966) in spite of the low room-temperature strength of these early filaments may also reflect an increase in tolerance for damage with temperature. In both cases, there is believed to be sufficient ductility at the elevated temperature to offset deleterious effects due to surface notches. Another factor to be considered in the assessment of filament performance is the presence of the matrix; Fig. 8 showed how the support provided by the matrix could influence the behavior of the filaments.

C. *Beryllium Reinforcement of Titanium*

Beryllium wire became available in the mid 1960's with strength and ductility greatly in excess of that obtainable in more massive material. In spite of its high cost, great interest was expressed in many laboratories, and attempts were made to use this material to reinforce titanium and aluminum. It was found that the properties of this wire could be changed by heat treatment and by surface etching. For example, 5-mil-diam wire with a strength of 150,000 psi and over 10% elongation increased in strength by 13–15% after surface etching, with little change in ductility. Heat-treatment response appeared to depend on the characteristics of the batch of wire, and appeared to be an aging phenomenon. In general, heat treatments at low temperatures (e.g., 600°F) tended to increase the strength of the wire, particularly the yield strength, while high-temperature heat treatment (e.g., 800°F) tended to decrease the strength but increased the stability of the wire at high temperatures. The latter is important in connection with processing because the strength of highly worked beryllium wire begins to be lost above 900°F.

Attempts to make composites with unalloyed titanium at temperatures near 900°F were unsuccessful for several reasons. Recovery of the worked structure in both the titanium and beryllium reduced the composite strength. Also, bonding was incomplete at this temperature, even when the pressure was raised to 100,000 psi. In addition, the beryllium wire was severely deformed in the processing. In view of these problems, work effort was redirected to a matrix of Ti–6Al–4V alloy, because this alloy could provide strength rather than the beryllium, while the latter could provide stiffness.

It has been pointed out earlier that strength requirements are not high in certain stiffness-critical applications such as fan blades. Moreover, the ductile-filament–ductile-matrix composite provides impact resistance nearly equal to that of monolithic titanium fan blades, as shown by Stusrud *et al.* (1971) in Fig. 9.

Two major problems were found with Ti-6Al-4V–Be composites: (1) reaction at the higher bonding temperatures, and (2) flattening of the beryllium wire so that it combined to form a monolithic sheet of beryllium rather than discrete wires. Goodwin and Herman (1969) show that cowound titanium alloy and beryllium wires could be hot pressed between separator foils of titanium alloy to eliminate the spreading and coalescence of the individual beryllium wires. The hot-pressing temperature chosen was the lowest possible to achieve bonding, but was in the range where the beryllium lost strength rapidly. For example, beryllium wire with a room-temperature strength of 153 ksi is weakened to 121 ksi at 1250°F and to 98 ksi at 1325°F. Composites with 33 v/o of beryllium had a longitudinal strength of 147 ksi after pressing at 1350°F. The transverse strength was 84 ksi with elastic moduli of 24×10^6 psi in both directions. These results were in excellent agreement with theoretical predictions. Subsequently, improved cleaning procedures allowed hot-press bonding to be performed at 1275–1325°F with further improvement in properties. Fatigue tests showed that the endurance limit was controlled by the matrix strain at the surface and was the same for all orientations.

Higher-volume-fraction composites were made by the same process to attain higher torsional stiffness and lower density. Somewhat higher press-bonding temperatures (1450° F) were required to improve flow and bonding, and this decreased the strength of the beryllium wire. Typical properties of 50-v/o composites were as follows:

tensile strength: 100 ksi	failure strain: 3%
tensile modulus: 38.6×10^6 psi	torsion modulus: 10.6×10^6 psi

Other composites were made at the lower press-bonding temperatures used previously and allowed the strength to be increased to 147 ksi. These results illustrate the principal compromise necessary with these composites. The higher-volume-percent beryllium composites desired for high stiffness depend to an increasing extent on the beryllium for their strength, so that processing temperatures must be kept low to preserve the strength of the beryllium wire. Fabrication represents the major problem with these promising composites. In addition to the fabrication problems discussed above, cost is extremely important. The original cost of $5000/lb for 0.005-in. beryllium wire in the mid 1960's had decreased to approximately half this amount by 1971, with projections for volume quantities still remaining as high as $800/lb. On the other hand, the cost of 0.1-in. diam beryllium rod is $250/lb, so that this material was examined as a starting point for manufacture. Because the fabrication conditions generally led to loss of much of the strength of the fine beryllium wires, new methods of manufacture using the lower-cost rods would give equivalent properties. Schmidt (1971) presented some approaches to using such materials in a patent assigned to the U.S. Navy. Others contributing to this field include Dynamet Corporation (1970), Bufford and Pickett (1970), and principally the laboratories of Detroit Diesel Allison Division and the Naval Air Development Center. The methods suggested include:

1. Extrusion of beryllium rods or powder inside titanium tubes.
2. Extrusion of beryllium rods contained in drilled holes in a titanium billet.
3. Fabrication from large-diameter wires, followed by forging to shape.
4. Development of lower-cost methods to make beryllium wires by hydrostatic drawing.
5. Hot rolling of wire–foil assemblies.
6. Extrusions of large-diameter beryllium powder in a matrix of titanium powder to produce elongated beryllium.

The forging and extrusion processes introduce a new freedom in these composites. The beryllium rods (or compacted powder) may be reduced more in one direction than another, so that the cross section of the filaments may have an aspect ratio as high as 10 : 1. This geometry will benefit the transverse and torsion moduli favorably.

The powder extrusion (method 6) performed by Dynamet (1970) for Aerospace Corporation elongates the large beryllium grains to give reinforcements. Typical properties for a 40-v/o composite were: ultimate strength, 85 ksi; elongation, 4%; tensile elastic modulus, 25×10^6 psi; and density, 0.124 lb/in.^{3}. These properties are encouraging for an unalloyed titanium matrix.

It is believed that the principal advantage of these new fabrication methods will lie in the reduction of cost rather than in improved properties. In comparison with composites fabricated from the nonductile filaments, the tolerance of titanium–beryllium for these new and lower-cost fabrication methods is a major factor in their favor. For example, if titanium–beryllium can be forged to shape using conventional equipment, it seems very likely that this factor alone will suffice to ensure that this titanium-matrix composite will be the first to find application.

D. Alumina Reinforcement of Titanium Alloys

The problems of compatibility experienced with other filaments in titanium matrices led Tressler and Moore (1971) to investigate single-crystal alumina filament. It was recognized from thermodynamic data that some problems with compatibility would occur, so that one purpose of the study was to determine the minimum diffusion-bonding conditions for successful fabrication of composites. Another purpose of the work was to establish the mechanical properties attainable with this system.

Tressler and Moore (1971) used continuous sapphire filaments 10–11 mils in diameter in matrices of Ti–6Al–4V and Ti(40A) (unalloyed). Specimens were diffusion bonded in vacuum. A bonding time of 15 min was selected, and good bonds were obtained at 1500°F for a pressure of 14 ksi. Lower temperatures required much higher pressures, so that the conditions selected represent an application of the low-temperature processing approach (Section II.D.2) to the compatibility problem. However, the reaction zone measured 1–1.5 μm after this bonding cycle.

The principal reaction products were identified to be the compound Ti_3Al and oxygen-stabilized alpha phase extending into the matrix. The authors present evidence that the reaction is more complex than suggested by these products, but were unable to identify the minor phases present. The compound zone grew more rapidly in a matrix of Ti–6Al–4V than in unalloyed titanium, suggesting that aluminum is dissolving to form the alpha-stabilized zone in the case of unalloyed titanium but cannot dissolve as rapidly when the matrix contains aluminum. Hardening of the matrix was detected, and extended for distances equal to interfilamentary spacings for a 20 v/o composite after an anneal of 69 hr at 1500°F. Activation energies for a diffusional controlled reaction were 50 and 51 kcal/mole for the reaction with the alloy and the unalloyed titanium matrices. The reaction rates were somewhat higher than for boron with titanium but similar to the rates for silicon carbide with titanium.

Mechanical properties of these composites were not entirely consistent. This was attributed to lack of complete foil–filament and foil–foil bonding, filament misalignment, and filament degradation. The average filament strength was 300 ksi, so that the strain-to-fracture for these filaments was expected to be 5000 μin./in. (the elastic modulus of alumina is 60×10^6 psi). The maximum value for the elastic modulus of the composites was 27×10^6 psi for 22 v/o of filaments, or slightly above the rule-of-mixtures value. This specimen had a strength of 125 ksi and a strain-to-fracture value of 5600 μin./in. so that it exceeded the expected values for effective utilization of the filaments. Filament misalignment was identified in this specimen, but the bonding was good, and filament degradation, if occurring, was believed to be small. The high strain-to-fracture value means that the reaction product, unlike that in the titanium–boron and titanium–silicon-carbide systems, is not limiting the composite strength. It has been suggested earlier that this may result from the small amount of ductility believed to be characteristic of the intermetallic phase Ti_3Al.

Transverse strength of these composites was approximately 50 ksi (maximum 57 ksi) for 22 v/o of filaments. Both the strain-to-fracture value and the strength decreased appreciably on annealing, and, the strength was reduced to 25 ksi after 69 hr at 1500°F. Almost complete longitudinal fracture of the filaments occurred under transverse loading. The fracture went through the reaction zone and into the region of the embrittled matrix, but became a ductile fracture as the crack emerged into the unaffected matrix.

E. Other Systems

Some preliminary work has been performed on directionally solidified alloys involving a titanium matrix, and on molybdenum-reinforced titanium.

1. Directionally Solidified Alloys

Yue *et al.* (1967) reported on in-house work at Lockheed (Palo Alto) to develop directionally solidified titanium alloys. The problem of reaction with crucibles was avoided by a floating-zone technique using electron-beam heating. The technique was difficult because of the low thermal conductivity of titanium. A directionally solidified Ti–8.5Si alloy had a strength of 147,000 psi with compressive strengths twice this value. No other details have been given of these materials, and it seems reasonable to conclude that the experimental difficulties will make such development very slow.

2. Molybdenum-Reinforced Titanium

Molybdenum-reinforced titanium has been examined at intervals over the last 15 years, but it is believed that it should be treated as a model system rather than as a practical material. Baskey (1967) showed that composites could be made with either continuous or discontinuous filaments using powder metallurgy, but that significant improvement in properties was not obtained. Tumey and Goodwin (1969) prepared composites from TZM† wire with a strength of 370,000 psi by hot pressing with Ti–6Al–4V foils. The tensile strength of composites with 30 v/o of TZM was 200,000 psi in the longitudinal direction, decreasing to 90,000 psi in the transverse direction, with elastic moduli of 26.8 and 22.1×10^6 psi for the two principal directions. Torsional rigidity and creep strength were also improved. However, when the density of 0.21 lb/in.^{3} is considered, the advantages are seen to be much smaller. The strength-to-density ratio is slightly below that of the matrix alloy, although the elastic modulus-to-density ratio is somewhat above.

IV. Status and Future of Titanium-Matrix Composites

The directions of work on titanium-matrix composites have already begun to change markedly. Work prior to 1971 was directed largely to preliminary studies of a large number of systems from which clear indications have emerged of the viable composites for continued work. At the same time, an adequate working knowledge of the factors governing performance has been established, and this will provide the background required to assist in fabrication development and application. Fabrication methods will receive emphasis in the next few years, and will be directed to economic realization of the properties demonstrated on laboratory-size samples. The work is expected to lead to significant applications within the next five years. Beyond this point, it is likely that attention will revert to more basic studies of the material as application problems appear that require more basic knowledge.

This brief synopsis of the present status and future of titanium-matrix composites will be discussed and developed further in this section.

† TZM is a molybdenum alloy containing 0.5w/o Ti, 0.1w/o Zr, and 0.03w/o C.

A. *Review of Systems*

A minimum of eight reinforcements for titanium have been examined in work to date. Three of these are ductile filaments: molybdenum, tungsten, and beryllium. Four potential filaments remain elastic up to fracture at room temperature, although there is good evidence that there is some ductility at higher temperatures: boron, silicon-carbide-coated boron, silicon carbide, and alumina. In addition, reinforcement by eutectic constituents has been examined. The reinforcements have included whiskers, discontinuous filaments, and filaments of different diameters.

Table VIII summarizes comparative data for composites fabricated from the principal reinforcements. The entries for strength and stiffness represent a comparison with the properties of a typical titanium alloy such as Ti–6Al–4V. In some cases, the comparison is with projected properties expected to be attained when fabrication problems are overcome. The high-temperature specific strength refers to 600–1200°F, and stability is also referred to this temperature range. The last four reinforcements—boron, silicon-carbide-coated boron, silicon carbide, and alumina—are in order of increasing densities and decreasing strengths, but the potential room-temperature strengths of composites made from the first three are similar and have been given equal rating. The significantly higher density of alumina (4 g/cm^3) adversely affects the potential strengths and stiffness of composites made from this reinforcement.

TABLE VIII

Summary of Properties of Titanium-Matrix Composites

Reinforcement	Specific strength: Room temp.	Specific strength: High temp.	Specific stiffness	Ductility	Stability	Fabricability
Beryllium	Same as matrix	Same to low	High	Good	Good	Good, may be shaped by forging
Molybdenum	Same as matrix	High	Same as matrix	Moderate	Good	Good
Boron	High	High	High	Low	Good	Difficult
Silicon carbide–boron	High	High	High	Low	Good	Difficult
Silicon carbide	High	High	High	Low	Good	Difficult
Alumina	Same as matrix	Moderate	Moderate	Low	Fair	Very difficult

Two systems offer good combinations of fabricability and properties. These are titanium–beryllium and titanium–boron, including both uncoated and silicon-carbide-coated boron. Titanium–beryllium composites have good ductility and are readily fabricated. The ductility of the composite is responsible for the good fabricability because the composite may be consolidated and shaped by hot-working methods such as extrusion and forging. This ease of fabrication is expected to lead both to economical methods of manufacture of hardware and to lower-cost material through use of larger beryllium sections that are reduced in size during hot working. The ductility is also responsible for the excellent toughness of this composite.

The titanium–silicon-carbide-coated-boron systems offer the greatest potentials in both stiffness and strength. The principal problem remains control of fabrication. The reaction and permissible limits of reaction to avoid strength loss are similar in the two systems. Fabrication methods have been developed for Ti-6Al-4V–BORSIC that have given longitudinal strengths of over 200,000 psi with 45–50 v/o reinforcement and transverse strengths of 65,000 psi. Work with uncoated boron has followed a different route involving development of a compatible matrix alloy. The composition of this alloy falls within the beta field. Production of foil is more readily performed with such an alloy, but the high alloy content makes the alloy very strong at the usual hot-pressing temperature.

B. Fabrication of Composites

Table VIII showed that major differences between the various composite systems occur in the area of fabricability. However, fabricability studies have not been published to the same extent as results on other aspects of composite technology, so that meaningful discussion of details of the various processes is difficult. However, all evidence points to the importance of controlling every aspect of the processing to achieve the best results. An apparently small change in the source or size of the filaments will have a marked effect on the properties obtained. For example, introduction of the 5.6-mil-diam filament was a significant step in reducing fabrication problems with both boron and BORSIC compared to the previously available 4-mil filaments. Filament breakage was reduced because the larger filaments had higher strength in diametral loading, while pressures required in hot pressing were reduced because the matrix flowed more readily into the larger gaps between the larger filaments.

The problems in the case of both boron- and BORSIC-reinforced composites can be summarized by reference to Fig. 19. As the hot-pressing temperature is increased from 1400° F, the strength of the composite increases as a result of improved consolidation. But the strength begins to decrease once the limit of permissible reaction is exceeded. This strength decrease continues until the lower limit is reached at the predicted strain-to-fracture value for fully degraded composites. The "window" of temperature between these competing effects is quite narrow in this system and makes it difficult to reach the strength predicted from the rule of mixtures. The latter is 200,000 psi for 50 v/o BORSIC filaments with a strength of 350,000 psi and a fracture strain of 6000 μin./in. Development of a more compatible matrix will raise the upper temperature bound for the "window." However, the matrix alloy must not be more difficult to hot press, or the lower temperature bound of the window may be raised. The latter can be decreased, in theory, by higher pressures, but the already high pressures used (typically 20 ksi) limit the increase possible before filament damage begins to occur and before the process becomes impractical. The combination of these restraints explains the difficulty in hot-pressing titanium matrix composites.

Fabrication of titanium–beryllium composites is not as difficult as with boron or BORSIC because the beryllium deforms rather than ruptures. Advantage can be taken of this feature to produce composites with lath-shaped reinforcement instead of rods.

C. Future Development of Titanium-Matrix Composites

The principal advantages of titanium-matrix composites over composites using a plastic or aluminum matrix have been discussed earlier in the chapter. Briefly, these advantages are higher-temperature of service, higher off-axis strengths without cross plies, high erosion and damage resistance, more efficient use of reinforcement by reduced need for cross plies, reduced manufacturing costs by use of unidirectional composites, reduced residual stresses because of closer expansion match, and lower anisotropy of strength and modulus, particularly in unidirectional composites. On the other hand, there are a number of disadvantages, including higher density, fabrication difficulties, and cost. The fabrication problems and cost are interrelated.

The most important of the factors that may affect choice of a titanium-matrix composite over one with a lower-strength matrix are the off-axis properties and the cost-related fabrication difficulties. The advantage of the greater isotropy attained with the titanium matrix can be illustrated with titanium–beryllium. Cowound and hot-pressed Ti-6Al-4V–35 v/o Be has been made with an elastic modulus of 24×10^6 psi in both principal directions and strengths of 147 and 84 ksi in the longitudinal and transverse directions. Unidirectional composites with boron and coated boron also show similar values in stiffness in the two principal directions but greater disparity in strength that results from filament splitting. It seems certain, therefore, that one direction of future work will be an effort on the part of filament manufacturers to raise the diametral strength of these types of filament. As pointed out earlier, a significant start has been made through introduction of the 5.6-mil filaments.

Cost and fabrication difficulties represent the major obstacles to application of the titanium-matrix composites. Cost arises from the high cost of the filaments or reinforcement; the matrix, because this is usually in the form of foils; and the fabrication process. The fabrication process represents the area where cost saving can be made most readily, because these costs are many times the cost of the raw materials in the case of most metal-matrix composites. Some of the directions under study for titanium–beryllium composites have been detailed in Section III.C, and continued work along these lines can be predicted. The problems are more acute in the composites with boron or coated filaments, and new approaches are needed for their solutions. One approach under investigation is the manufacture of a monolayer tape that can be used as an intermediate in the manufacture of hardware. However, such a tape requires thinner titanium foils than are needed for multilayer tapes of the same volume percentage, yet the cost of titanium foils rises rapidly with reduction of gage and reaches several hundred dollars per pound for an alloy such as Ti–6Al–4V when the thickness approaches 0.001 in. Beta-alloy foils represent an alternate lower-cost route, but the fabrication costs remain high for both types of foil.

One answer to the problem of fabrication costs might be development of a method to coat filaments with a uniform layer of the titanium-alloy matrix desired. This would provide a means to replace the expensive titanium foils and, at the same time, provide a positive method of spacing the filaments. (The latter appears to be more important in titanium-matrix composites than it is with lower-strength matrices.) Attempts to control filament spacing by plasma-arc spraying of the matrix have not been successful up to now, because the increase in oxygen content has been excessive and reaction between filaments and the molten spray has been excessive. High-speed coating methods for single filaments are required that can be part of the filament-manufacturing cycle. Electro-beam evaporation from multiple sources might be one such method.

The fabrication costs are affected by many variables. One of the most important is the narrow "window" of temperature available for hot pressing. A more compatible system offers one approach to raise the upper temperature bounding this window. Lower pressures could be used if composites could be pressed at higher temperatures. The pressure may be 20 ksi or more in present practice. It seems certain that these high pressures must be reduced because production of the largest fan blades for jet engines will require a minimum press size of 5000 tons, and structural panels of 2–3 ft^2 in area will be the limit for most available presses. More compatible matrices are only one approach to this problem. Another is to develop continuous methods. These have two advantages: one is that the time at temperature may be reduced drastically so that higher temperatures can be used; the second is that the composite is processed locally so that the consolidation force is reduced. Examples of such methods are the continuous method demonstrated for titanium–boron and extrusion or hot rolling of titanium–beryllium composites.

In summary, it is believed that the titanium-matrix composites have excellent potential to solve many problems in advanced systems using high-performance materials, but that cost is the major obstacle standing in the way of their application. This cost factor arises more from fabrication than from the raw materials, although an ideal fabrication method will use lower-cost precursors. It is believed that the major effort in the future will be in the area of fabrication and that continuous methods of fabrication will begin to be developed. These will include methods to coat reinforcements with the appropriate titanium-alloy matrix as well as continuous methods of consolidation.

References

Ashdown, F. A. (1968). "Compatibility Study of SiC Filaments in Commercial Purity Titanium." GSF/MC/68–1. Thesis, Air Force Inst. of Technology, Dayton, Ohio.

Blackburn, L. D., Herzog, J. A., Meyerer, W. J., Snide, J. A., Stuhrke, W. F., and Brisbane, A. W. (1966). MAMS Internal Research on Metal-Matrix Composites. MAM-TR-66-3.

Bonnano, F., and Withers, J. C. (1966). *Advan. Fibrous Reinforced Compos., SAMPE, Nat. Symp., 10th, San Diego, California* **10**, F105–127.

Bufford, A. S., and Pickett, J. J. (1970). Beryllium–Titanium Composites, presented at 17th Refractory Composites Working Group Meeting, Williamsburg, Virginia.

Collins, B. R. (1972). Private communication of preliminary results of Dr. I. Toth on Contract F33615-71-C-1044.

Crane, R. L., and Tressler, R. E. (1971). *J. Compos. Mater.* **5**, 537–541.

Davis, L. W. (1969). Titanium-Matrix Composite Development, Amer. Ceram. Soc. Meeting, Hudson, Ohio.

Dynamet Corporation (1970). Commercial Data from Dynamet Corporation, Burlington, Massachusetts.

Ellison, E. G., and Boone, D. H. (1967). *J. Less Common Metals* **13**, 103.

Goodwin, V. L., and Herman, M. (1969). Beryllium-Wire–Metal-Matrix Composites Program. Final Rep. EDR 6518 on Contract N00019-69-C-0234.

Hamilton, C. H. (1969). *AIME Symp., Pittsburgh, Pennsylvania.*

Hibbard, W. R. Jr. (1964). "Fiber Composite Materials," pp. 1–10. ASM, Metals Park, Ohio.

Jones, R. C. (1968). "Metal Matrix Composites," STP 438, pp. 183–317. ASTM, Philadelphia, Pennsylvania.

Klein, M. J., Reid, M. L., and Metcalfe, A. G. (1969). Compatibility Studies for Viable Titanium-Matrix Composites. AFML-TR-69-242.

Kreider, K., G., Salkind, M. J., and Cheatham, R. (1966). Services and Materials Necessary to Develop a Process to Produce Fibrous Reinforced Metal Composite Materials. United Aircraft Res. Lab., East Hartford, Connecticut, AF33(615)-3209, Summary Tech. Rep.

Metcalfe, A. G. (1967). *J. Compos. Mater.* **1**, 356–365.

Metcalfe, A. G. (1970). Interaction and Fracture in Metal Matrix Composites. Lecture given at Univ. of Utah, June 12.

Metcalfe, A. G., and Schmitz, G. K. (1967). Development of Filament Reinforced Titanium Alloys. SAE Paper 670862, presented at Aeronaut. and Space Eng. and Manufact. Meeting, Los Angeles, California.

Metcalfe, A. G., and Schmitz, G. K. (1969). Current Status of Titanium–Boron Composites for Gas Turbines. Presented at the Gas Turbine Conf. and Products Show, Cleveland, Ohio, Trans. ASME Paper No. 69-GT-1.

Prewo, K. M., and Kreider, K. G. (1972). Rep. L110833-1, United Aircraft Res. Lab. East Hartford, Connecticut.

Ratliff, J. L., and Powell, G. W. (1970). Research on Diffusion in Multiphase Systems: Reaction Diffusion in the Ti–SiC and Ti-6Al-4V–SiC Systems. AFML-TR-70-42.

Rudy, E. (1969). Compendium of Phase Diagrams. AFML-TR-65-2, Part V.

Samsonov, G. V. (1964), "Handbook of High-Temperature Materials, No. 2, Properties Index." Plenum Press, New York.

Schmidt, R. (1971), U.S. Patent 3,609,855.

Schmitz, G. K., and Metcalfe, A. G. (1968). Development of Continuous-Filament-Reinforced Metal Tape. AFML-TR-68-41.

Schmitz, G. K., and Metcalfe, A. G. (1973). Evaluation of Compatible Titanium Alloys in Boron Filament Composites. AFML-TR-73-1.

Schmitz, G. K., Klein, M. J., Reid, M. L., and Metcalfe, A. G. (1970). Compatibility Studies for Viable Titanium-Matrix Composites. AFML-TR-70-237.

Sinizer, D. I., Toy, A., and Attridge, D. A. (1966). Development and Evaluation of the Diffusion-Bonding Process as a Method to Produce Fibrous Reinforced Metal-Matrix Composite Materials. North Amer. Aviat. Inc., Los Angeles, California, AF33(615)-3210, Tech. Rep. AFML-TR-66-350 (AD804268).

Snide, J. A. (1968). Compatibility of Vapour Deposited B, SiC, and TiB_2 Filaments with Several Titanium Matrices. AFML-TR-67-354.

Stevens, E. C., and Hanink, D. K. (1970). Titanium Composite Fan Blades. Final Rep. AFML-TR-70-180.

Stusrud, R. W., Tumey, M. J., Sippel, G. R., and Herman, M. (1971). Titanium-Matrix Composites. Presented at 18th Refractory Compos. Working Group Meeting, Huntsville, Alabama.

Tressler, R. E., and Moore, T. L. (1971). Mechanical Properties and Interface Reaction Studies of Titanium–Alumina Composites. Presented at *Westec Conf. ASM, Los Angeles, Calif., Metals Eng. Quart.* **11** (1), 16–22.

Tsai, S. W. (1966). Mechanics of Composite Materials. AFML-TR-66-149, Parts I and II.

Tsareff, T. C., Jr., Sippel, G. R., and Herman, M. (1969), *AIMME Symp. Metal-Matrix Compos., Pittsburgh, Pennsylvania.*

Tumey, M. J., and Goodwin, V. L. (1969). *AIMME Symp. Metal-Matrix Compos. Pittsburgh, Pennsylvania.*

Wawner, Jr., F. E. (1967). "Modern Composite Materials," Chapter 10, pp. 244–269. Addison-Wesley, Reading, Massachusetts.

Williford, J. F., Jr., and Snajdr, E. A. (1969). *AIMME Symp. Metal-Matrix Compos., Pittsburgh, Pennsylvania.*

Yue, A. S., Crossman, F. W., Davinroy, A. T., and Jacobson, M. I. (1967). Advanced Metallic Composites. Lockheed Aircraft Corp., Palo Alto, California.

7

Development of Metal-Matrix Composites Reinforced with High-Modulus Graphite Fibers

E. G. KENDALL

The Aerospace Corporation
El Segundo, California

I. Introduction

Carbon and graphite fibers are of great interest as reinforcements in metal-matrix composites because of their high strength and stiffness, low density, and potential for large-scale production at low cost. Many structural applications exist for such high-strength lightweight composites. Graphite fibers are in an intermediate stage of development: fibers with a strength of approximately 2070 MN/m^2 (300,000 psi) and modulus of elasticity of

380 GN/m^2 (55 × 10^6 psi) are commercially available at the present time. Considerable improvements in the properties and general quality of both carbon and graphite fibers are expected in the next few years. By comparison with the fibers, the technology of graphite-reinforced metals is in an early stage of development. All of the work conducted to date has been on a laboratory scale.

Theoretical calculations by Eatherly, as cited in Bacon (1960), have predicted that the true modulus of elasticity of graphite may be as high as 1000 GN/m^2 (145 × 10^6 psi). Experimental work by Bacon (1960) has confirmed the potentially high modulus of graphite. Graphite whiskers were found to have tensile strengths as high as 20.7 GN/m^2 (3 × 10^6 psi) and moduli in excess of 690 GN/m^2 (100 × 10^6 psi). Although graphite fibers fabricated by graphitizing suitable polymeric precursor materials do not have such high strength and modulus as these whiskers, significant progress has been made toward predicting the theoretical strength and modulus of graphite. For example, Bacon and Schalamon (1968) have obtained laboratory samples of graphite fiber with strengths greater than 3450 MN/m^2 (500,000 psi) and moduli up to 760 GN/m^2 (110 × 10^6 psi).

Graphite fibers are attractive as reinforcements for metals. However, a lack of chemical compatibility between graphite and many metal systems, as determined by Morse (1966), and problems encountered in the fabrication of composites have hampered the development of graphite-reinforced metals. One of the major problems stems from difficulties in handling the fine multifilament (tow) form in which the fibers are currently available. The high-modulus graphite filaments are typically 7 μm in diam in strands of 1000–100,000 fibers. Large monofilament graphite fibers (approximately 0.004-in. diam) that could be fabricated into composites by well-established diffusion-bonding techniques are under development. However, these are likely to be significantly more expensive than the multifiber strands in large-volume production.

Aluminum, magnesium, nickel, and cobalt have been found to be fairly compatible with graphite fibers. However, metals such as titanium, which is a strong carbide former, cannot be considered unless a stable diffusion barrier is placed at the matrix–fiber interface of the composite to separate the titanium and the graphite.

Predictions by McDanels *et al.* (1965) based on the rule of mixtures have shown that graphite-reinforced metals have good potential properties for structural applications. Aluminum–graphite composites have been prepared with a uniaxial tensile strength at 20–500°C of approximately 690 MN/m^2 (100,000 psi). The specific uniaxial strength (strength-to-density ratio) of this composite was approximately 3.3 × 10^4 m (1.3 × 10^6 in.) at 500°C, which is

1.5 times that of a typical high-temperature titanium alloy at 500°C. A considerable amount of work has been carried out on the reinforcement of nickel with graphite fibers. However, the application of these composites has been unsuccessful, primarily because of the oxidation of the fibers at high temperatures. The use of oxidation-resistant nickel- or cobalt-based alloys instead of the pure metals may solve this oxidation problem, but no work has been reported as yet on the reinforcement of such alloys with graphite fibers.

In addition to structural applications, graphite-reinforced metals such as copper, aluminum, and lead have other properties of interest, e.g., high strength combined with good electrical conductivity, low coefficient of friction and high wear resistance, and high dimensional stability over a range of temperature. Copper–graphite and aluminum–graphite composites are of interest for high-strength electrical conductors, and aluminum–graphite, lead–graphite, and zinc–graphite composites have potential for use as bearing materials.

Graphite-reinforced plastics will offer strong competition to graphite-reinforced metals for many structural applications. Since the density of the polymeric matrix is less than that of relatively lightweight metals such as aluminum (1.5 gm/cm^3 for the plastics and 2.7 gm/cm^3 for aluminum), the plastic composites have higher specific properties than the metal composites at room temperature. However, the higher temperature resistance and strength of the metal matrix will probably make the metal composites more desirable for many applications.

Aluminum alloys appear to offer the most potential as matrix materials for graphite-reinforced metals and consequently have commanded most of the research effort. The use of aluminum–graphite composites will be largely controlled by their cost, which in the next decade could be near $20–50/lb. This would make the composite competitive for many applications in aircraft, missiles, electrical machinery, rocket propulsion systems, launch-vehicle structures, and spacecraft. Aircraft applications are in skins, struts, spars, wing boxes, and helicopter blades. Important applications that require the 500°C temperature capability of the aluminum–graphite composites include reentry shielding for missiles and compressor blades for gas-turbine engines. Launch-vehicle structures require lightweight materials for stiffening of large-diameter cylindrical sections, interstages, adapters, and tank- and equipment-support structures. Spacecraft applications are in shells and trusses in the primary structures, and booms, solar-cell panels, equipment mounts, and antennas in auxiliary structures. When lower-cost composites become available, widespread nonaerospace applications may develop for aluminum–graphite composites in structures in rapid-transit and deep-submergence vehicles and rotating parts in electrical generators.

II. Graphite Fibers

Most carbon fibers or filaments are produced from carbonaceous precursors (polymers, natural bitumens, or modified cellulose) in the form of yarn or monofilaments, which are thermally pyrolized to produce the residue of carbon in a similar form. Because the volatiles evolved from this carbonizing process need traverse only the radial distance of the fiber, higher densities, elastic moduli, and strengths are possible in these than in the bulkier forms of graphite obtained from the same process. Because carbon and graphite forms can range from very hard, brittle, and rigid diamondlike structures to softer, less rigid, but strong and refractory graphites, some of these filaments can, by modification of initial composition and thermal treatment, be tailored to specific properties. Descriptions of these process methods and fiber properties have been reported by Bacon (1966, etc.), Ezekiel (1971), Brown (1968), and Watt (1972).

The strength of commercially available graphite fibers varies from 1380 to 3450 MN/m^2 (200,000 to 500,000 psi), although laboratory materials with strengths up to 6.9 GN/m^2 (10^6 psi) have been produced. The commercial strengths are accompanied by moduli ranging between 166 and 518 GN/m^2 (25 and 75×10^6 psi), although considerably higher values have been obtained. These values, coupled with a density of 1.8 g/cm^3, give highly desirable specific properties. Reviews of the mechanical properties of commercially available fibers have been presented by Ross (1971), Rauch (1971), and Galasso (1969, 1971). Although graphite fibers are polycrystalline, they exhibit a high degree of preferred orientation. This preferential arrangement of hexagonal crystallites provides both high strength and modulus, and affects the thermal and electrical properties and the density.

At the present time, rayon and polyacrylonitrile (PAN) are the precursor materials generally used, although others are being investigated. The exact techniques used to convert these precursor fibers into high-strength, high-modulus graphite are proprietary and probably differ somewhat with each manufacturer. In all cases, however, thermal decomposition of the organic fiber occurs under closely controlled conditions. Also, at some point in the process, stress is applied in order to obtain the highly preferred orientation previously mentioned. Rayon can be stretched only in the latter stages of pyrolysis, but PAN can be stretched at any stage in the process.

The typical structure and shape of fibers from these two processes have been described by Diefendorf (1969) and are shown in Fig. 1. At Rennssalaer Polytechnic Institute, graphite fibers of 8-μm diam have been studied by optical-, electron-optical-, X-ray-, and electron-diffraction techniques. Since electrons penetrate the fibers to a depth of only 2000 Å, electron diffraction is useful for studying the structure near the fiber surface. X-ray diffraction, on

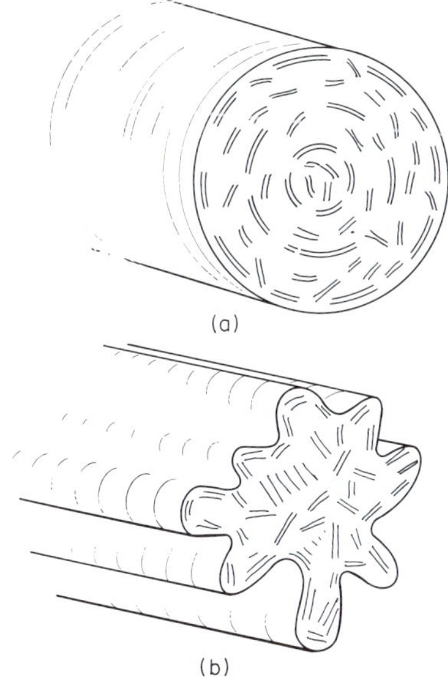

Fig. 1. *c*-axis orientation in (a) PAN and (*b*) rayon precursor graphite fibers.

the other hand, averages the orientations present throughout the fiber. Hence, by using both electron- and X-ray-diffraction techniques, the surface and interior structures can be studied and analyzed separately. These studies have shown that the graphite basal planes are more parallel with the fiber longitudinal axis near the surface than in the interior of the fibers. Preferred orientation in the fiber radial direction was examined optically with polarized light. The *c*-axis orientation in PAN- and rayon-based fibers is shown in Fig. 1. The PAN-based fibers are symmetrical, with a low preferred orientation of the *c*-axis perpendicular to the surface. The preferred orientation of some fibers is high near the surface when graphitization occurs at local points on the surface. Rayon-based fibers have a lower degree of preferred orientation within the fiber. Some preferred orientation exists following the crenulated surface. Rayon-based fibers show areas of high preferred orientation of the *c*-axis perpendicular to the surface. These areas provide an easy shear path, with resultant fiber splitting. Determination of moduli in conjunction with structural studies indicates that an increase in preferred orientation of the *c*-axis perpendicular to the surface is associated with increased modulus.

The temperature used for graphitization has a marked effect on the tensile strength and modulus that can be attained in the fiber. Unfortunately, the optimum temperature is not the same for both properties, and therefore compromises must be made. In order to demonstrate this point, a specific graphitization cycle may produce a combination of properties as indicated in Fig. 2, as given by Moreton (1967).

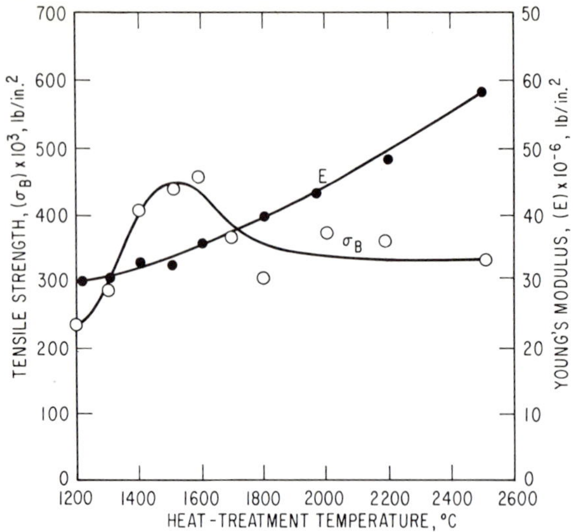

Fig. 2. Graphitization cycle, tensile strength, and modulus of PAN fiber.

The strength of carbon fiber decreases during oxidation or heating in an inert atmosphere with increasing temperature to 400°C. Above 400°C, the strength increases with rising temperature. During graphitization, both the modulus and strength of carbon fibers produced from rayon increase with increasing temperature. However, when PAN is used as the precursor fiber, the strength of the resulting carbon fiber increases to 1500°C, decreases to 1900°C, and then levels off, while the modulus increases continuously with increasing temperature over the entire graphitization temperature range (Fig. 2).

Processes for the production of graphite fibers from coal tar, pitch, or asphalt are undergoing development. These fibers are first extruded by a process similar to the spinning of nylon and are then carbonized and graphitized much the same as with the other precursors. They differ in that they have no noticeable orientation and appear to be glassy. In research to date, their properties have not been consistent, but this is potentially a low-cost process.

The form in which the fibers are offered varies from small yarn bundles of 1000 filaments with some twist to several 100,000-filament tows spread into wide tapes, depending on the manufacturer. All of these fibers differ considerably in mechanical properties and are known by a variety of trade names (Table I). However, testing techniques and procedures for these fragile fibers are far from being standardized, and the resultant data are quite sensitive to specimen preparation and handling methods. In addition,

TABLE I
Manufacturer's Data

	FRACTURE STRESS, 10^3 lb/in.^{2}	MODULUS, 10^6 lb/in.^{2}	% STRAIN TO FAILURE	BULK DENSITY, lb/in.^{3}	TOW LINEAR DENSITY, 10^{-5} lb/ft	FIBERS/ TOW	TOW AREA, in.^{2}
MODMOR I[a] UNSURFACE TREATED	340	59		0.0668	59	10,000	7.36×10^{-4}
CELION 70, 1 in.[b]	300	88		0.0708	277	40,000	3.25×10^{-3}
CELION 70, 3 in.[b]	300	88		0.0708	830	120,000	9.74×10^{-3}
THORNEL 50[c]	320 ±20	57 ±3	0.5	0.0603	4.65 ±0.14	1,440	6.44×10^{-5}
COURTAULDS HMS[d]	250-350	50-60		0.070	62.5	10,000	7.44×10^{-4}
FORTAFIL 6T[e]	420 ±20	59 ±3	0.70	0.069	250	38,000	3.02×10^{-3}
THORNEL 75[c]	385 ±17	76 ±3	0.5	0.0656	3.88 ±0.1	1,440	4.92×10^{-5}
MODMOR I[a] SURFACE TREATED	308	54		0.0668	59	10,000	
FORTAFIL 5T[e]	400 ±20	48 ±3	0.83	0.065	250	38,000	3.2×10^{-3}
THORNEL 400[c]	425	30	1.3	0.0643	5.65×10^{-5}	1,000	7.31×10^{-5}
COURTAULDS HT[d]	325-375	35-40		0.067	62.5	10,000	7.78×10^{-4}
MODMOR II[a]	380	40		0.0614	59	10,000	8.0×10^{-4}
HERCULES[f]	317	55		0.0688	53.2	10,000	6.43×10^{-4}
HOUGH[g]	300	27				1.0	1.25×10^{-5}
PANEX 30A[h]	400-470	30		0.0643	910-1000	160,000	1.13×10^{-2}
THORNEL 300[c]	325	34	1	0.0612	12.3	3,000	
FORTAFIL 4R[e]	450	38	1.2	0.065	38.4	6,000	
HERCULES[f] TYPE A	400	28-32		0.064	564-624		
PRD-49[i]	400	19		0.0522			

[a]Whittaker-Morgan Inc, Costa Mesa, California

[b]Celanese Corporation, Summit, New Jersey

[c]Union Carbide Corporation, New York, New York

[d]Courtaulds Ltd, Coventry, England

[e]Great Lakes Carbon Corporation, New York, New York

[f]Hercules Inc, Wilmington, Delaware

[g]Hough Laboratories, Springfield, Ohio

[h]Stackpole Fibers Company, Lowell, Massachusetts

[i]DuPont Company, Wilmington, Delaware

TABLE II
Aerospace Corporation Test Results

	FRACTURE STRESS, 10^3 lb/in.^{2}	MODULUS, 10^6 lb/in.^{2}	% STRAIN TO FAILURE	FIBER AREA, 10^{-8} in.^{2}	BULK DENSITY, lb/in.^{3}	TOW LINEAR DENSITY, 10^{-5} lb/ft	FIBERS/ TOW
MODMOR TYPE 1							
(untreated)	322 ±117	38.3 ±10.91	0.863 ±0.179	8.75 ±3.36	0.0553	60.1	8919
(surface treated)	380 ±104	53.3 ±12.11	0.74 ±0.12	6.34 ±1.72			
CELION 70	222 ±91	54.15 ±15.31	0.42 ±0.14	9.63 ±3.71	0.0705	833	37,600 ±1100
THORNEL 50	315 ±92	63.53 ±10.01	0.5 ±0.12	3.90 ±0.65	0.0604	4.98	1370 ±11
COURTAULDS HMS	249 ±108	69.14 ±13.41	0.36 ±0.12	5.31 ±1.13		59.2	9600 ±91
HERCULES HM	475 ±109	57.53 ±7.59	0.82 ±0.12	5.87 ±1.23	0.064	56.7	
HOUGH	244			980			1.0
FORTAFIL - 6T	461 ±134	70.5 ±15.7	0.65 ±0.1	6.83 ±1.85			
THORNEL 75	358 ±101	83.8 ±20.0	0.43 ±0.14	3.41 ±0.65			
PRD-49	555 ±91	18.9 ±2.0	2.9 ±0.4				
PANEX 30A	405 ±130	33.5 ±6.2	1.2 ±0.2	7.1 ±1.2			
THORNEL-300	452 ±119	32.1 ±5.2	1.43 ±0.26	5.43 ±1.04			
THORNEL-400	375 ±76	29.0 ±4.3	1.3 ±0.3				
FORTAFIL-4R	349 ±93	31.6 ±4.3	1.1 ±0.3				
HERCULES TYPE A	507 ±111	35.3 ±4.27	1.4 ±0.2	6.42			

manufacturing processes require close control, and variations from batch to batch can be noted in some fibers. Consequently, independent laboratory tests as given in Table II are recommended by the author to characterize fully the specific fiber to be used in fabricating a metal–graphite composite. The wide selection available in the overall size, shape, and geometrical cross section of some of the fibers listed in Table II is demonstrated in Figs. 3–5.

The PAN-based straight-tow Panex fiber (Stackpole Fiber Co.) is shown in Fig. 3a. This fiber is 3 in. wide and contains 160,000 filaments. The hairlike nature and accompanying fragility of this fiber are readily apparent. Such a fiber presents handling problems when fabricating a metal composite,

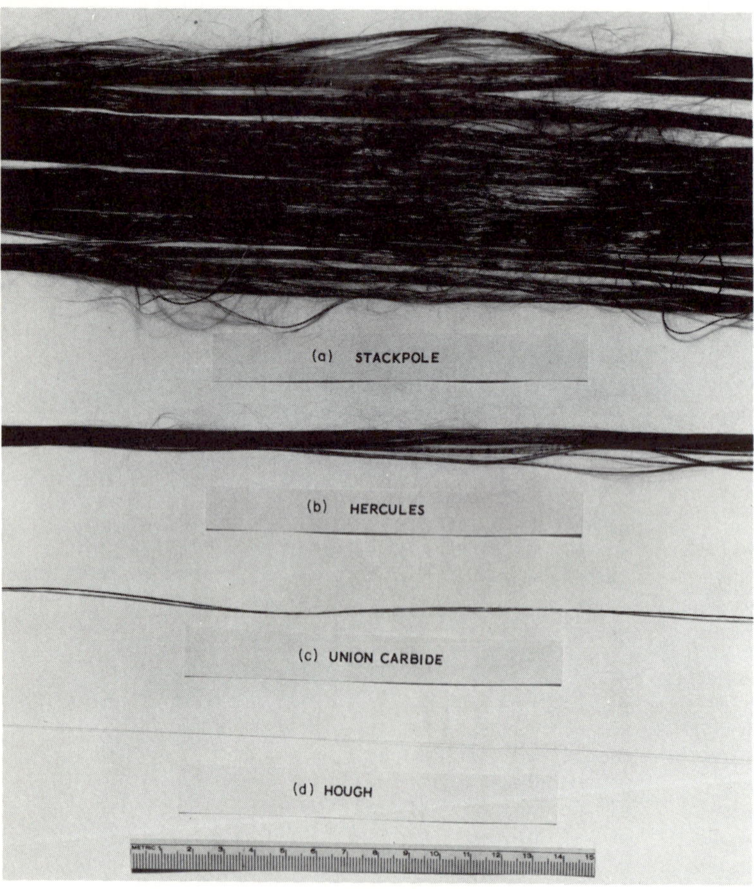

Fig. 3. Macrophotograph of (a) Stackpole, (b) Hercules, (c) Union Carbide, and (d) Hough graphite fibers.

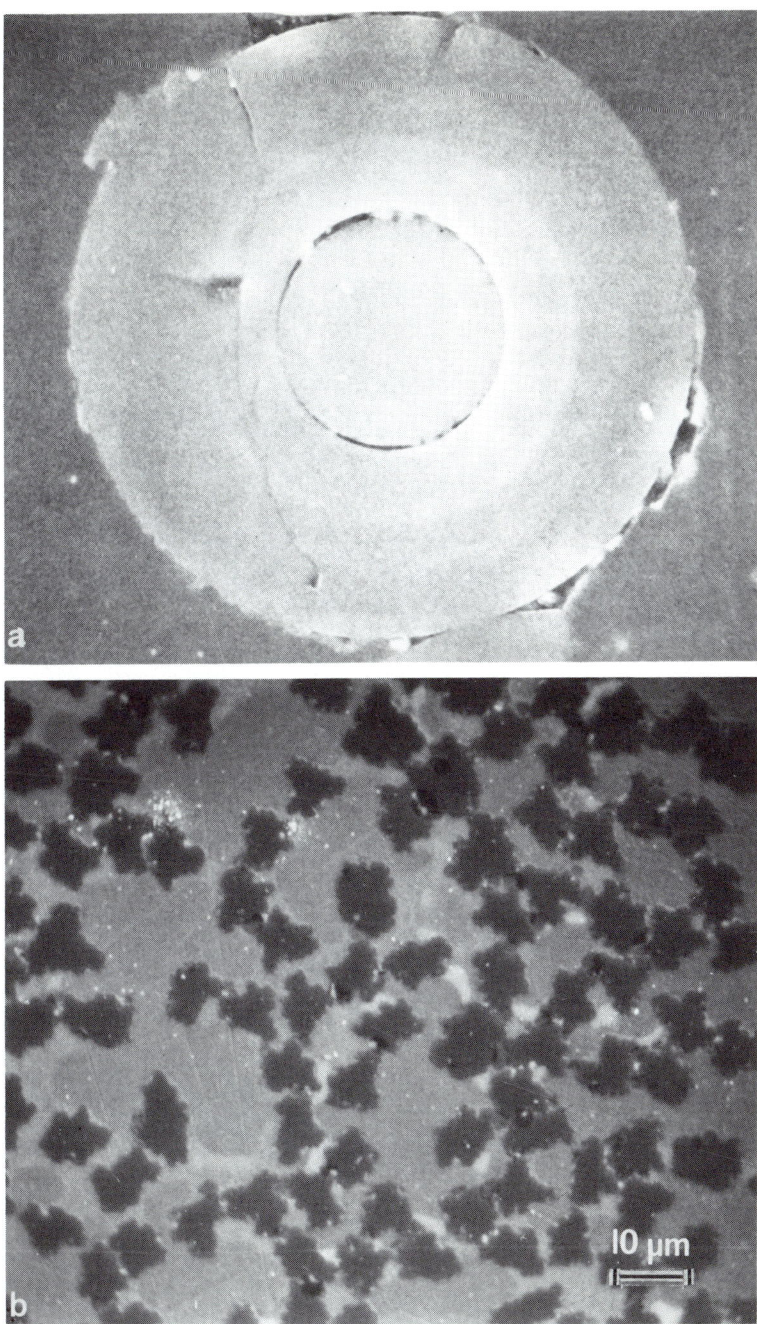

Fig. 4. Photomicrographs of cross section of (a) Hough and (b) Thornel-50 fibers.

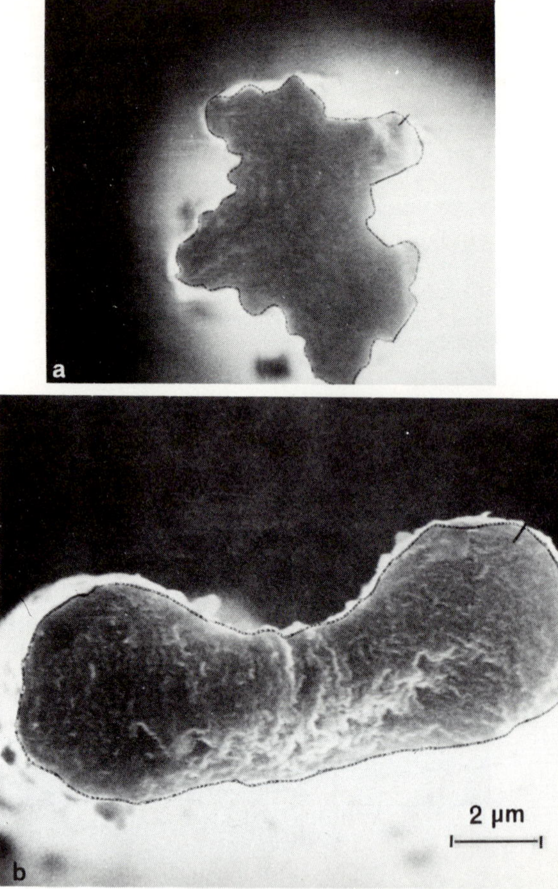

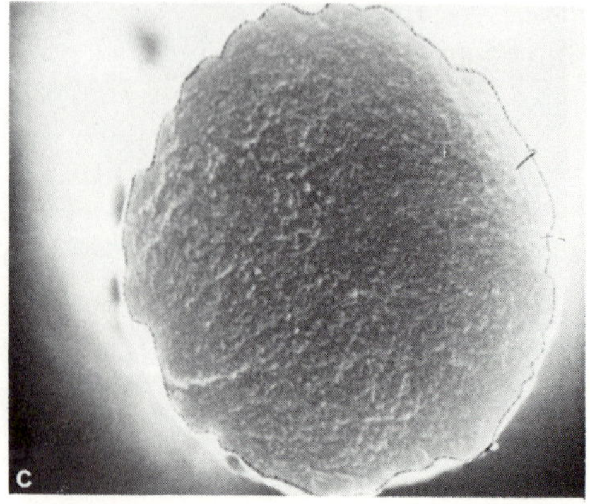

Fig. 5. Photomicrographs of (a) Thornel-50, (b) Fortafil, and (c) Grafil fibers.

particularly by liquid-infiltration methods. The Hercules fiber (Hercules, Inc.) (Fig. 3b), also from a PAN precursor, is much thinner, equally as fragile, and contains 10,000 filaments. It can be obtained in three grades: high modulus (HM), high tensile (HT), or with a still lower modulus (Type A). The Thornel-50 fiber (Union Carbide), from a rayon precursor, contains 1440 filaments, and is double stranded with a slight twist (Fig. 3c). The individual filament size in each of the above three types of fibers is 7–8 μm. The Hough monofilament (Hough, 1970, 1971) (Fig. 3d) is fabricated by chemical-vapor deposition and contains 25–30 mol% boron. This filament has been developed as a potential reinforcement for metals by fabrication techniques similar to those being used for boron–metal composites, such as diffusion bonding.

Scanning electron microscopy (SEM) disclosed the wide variation in the fiber cross sections due partly to the type of precursor and partly to the graphitization process. In Fig. 4, the Thornel-50 fiber is shown in relation to the Hough fiber to indicate the size difference. Three distinct zones are apparent in the Hough fiber. The core consists of a 1-mil-diam carbon fiber. A layer of pyrolytic graphite (PG) was deposited on this core by chemical-vapor deposition methods. At this point, the third layer was then built up by simultaneously codepositing 25–30 mol% boron from an atmosphere of triethylborane with the PG. A marked increase in strength level was noted after the addition of boron as compared with aluminum and silicon. Optimum properties for this fiber are noted in Table II. Strength levels of 2070–2760 MN/m^2 (300–400 ksi with a modulus of 186 GN/m^2 (27 $\times$ 10^6 psi) have been attained in laboratory quantities. The Thornel-50 fiber is also shown at higher magnification in Fig. 5a. This is the most widely used fiber of a family, first introduced in 1965, that includes Thornel-25, -40, -50, and -75, all made from a rayon precursor. They are easily identified by the crenulated-appearing cross section caused by stretching during the high-temperature graphitization. Fortafil (Fig. 5b) and Grafil (Fig. 5c) are two commercially available fibers made from a PAN precursor. Their dog-bone and circular cross sections, respectively, are typical of this type of fiber.

In addition to the work of Hough, other work is under way to develop a large-diameter (about 5-mil) graphite monofilament in order to preclude the difficulties in fabrication of metal–graphite composites as reported by Meiser and Davison (1969). Quackenbush (1970) impregnated bundles of graphite fibers with a high-char polymer and then converted the polymer to carbon by a high-temperature treatment. His best results were obtained using a PAN precursor with a furfuryl alcohol binder. The fiber diameter was 8 mil with a tensile strength of 1210 MN/m^2 (175,000 psi) and modulus of 345 GN/m^2 (50 $\times$ 10^6 psi). Flom *et al.* (1972) have investigated the use of polyacetylene as a precursor. Their best resulting fibers, 1-mil diam, had an

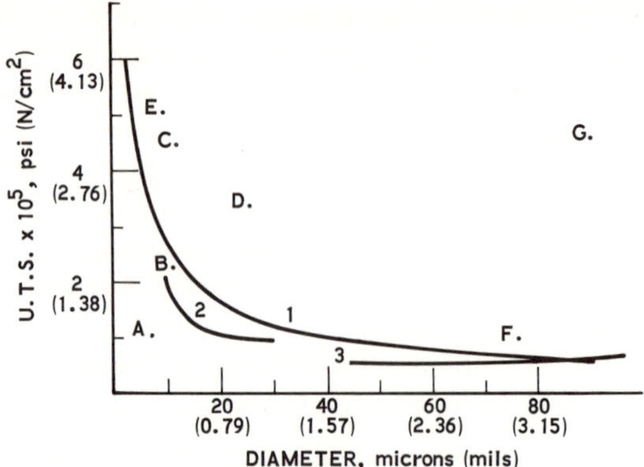

Fig. 6. Uniaxial tensile strength vs. diameter for selected carbon monofilaments. A, sublimed whisker; B, polycrystalline whisker; C, pyrolytic whisker ($E = 92.4 \times 10^6$psi,63.7×10^6 N/cm^2); D, pyrolytic whisker ($E = 8.7 \times 10^6$psi, 6.0×10^6N/cm^2); E, stress-graphitized carbon fiber; F, pyrolytic carbon on 12-μm (0.5-mil) tungsten; G, carbon alloy fiber. 1, pyrolytic graphite whiskers; 2, carbon monofilament from PVC pitch; 3, pyrolytic carbon on 25-μm (1-mil) tungsten.

average strength of 1900 MN/m^2 (275,000 psi) and a modulus of 324 GN/m^2 (47×10^6 psi). In his review of the strength of carbon filaments and whiskers, Hough (1970) concluded that when the diameters exceed 10 μm ($\sim$0.5 mil), the strength of the filament degrades rapidly. This is substantiated by Fig. 6, which shows tensile strength versus diameter for a variety of monofilaments. A similar effect is shown in Fig. 7 for the tensile modulus. This effect appears to be independent of the source of the filament, i.e., PAN, pitch, polyvinyl-alcohol (PVA), or any other polymer. Hough attributed this loss in strength to the presence of gas-borne dirt particles, soot, and internal porosity.

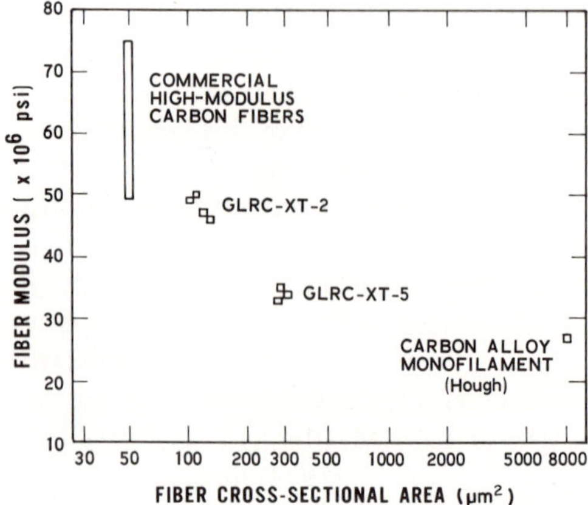

Fig. 7. Tensile modulus vs. cross-sectional area for various carbon fiber types.

The multifibril tow fibers are not without defects of their own. Investigators have shown that both the PAN- and rayon-derived fibers contain internal and surface defects. Figure 8, which is a SEM micrograph of a tensile fracture in an aluminum–graphite composite, shows internal defects within the Thornel-50 fiber. Bacon and Silvaggi (1968, 1971) have shown that the ultrafine microstructure of graphitized rayon fibers consists of a honeycomb of micropores, 60–90 Å apart, separated by long ribbonlike layers of turbostratic graphite. The structure is highly aligned in the stress-graphitized, high-modulus fibers, where pores in the carbon or rayon precursor are

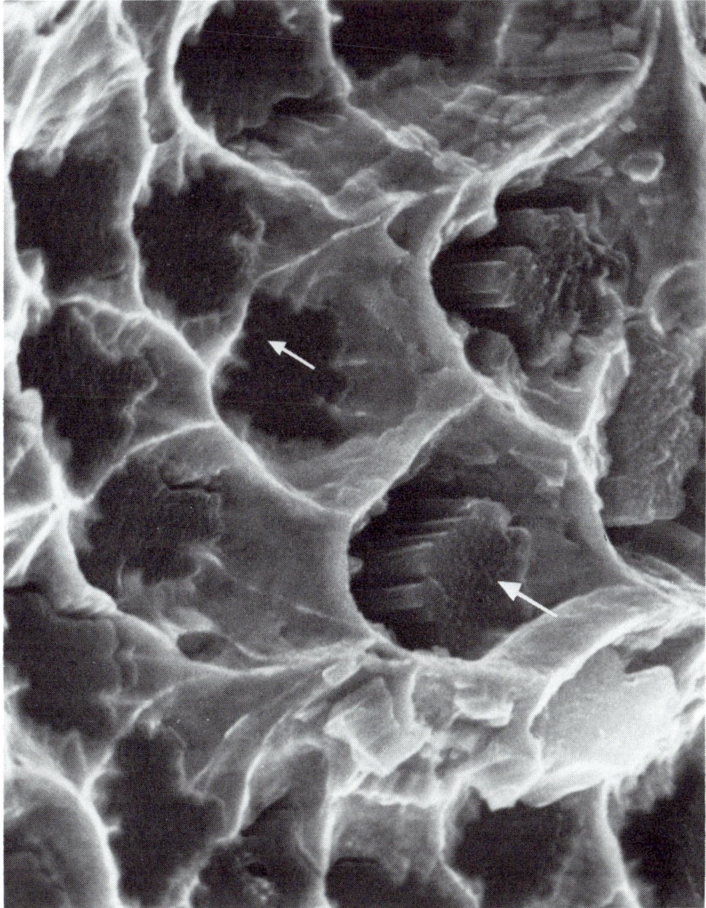

Fig. 8. SEM micrograph of tensile fracture in Thornel-50 yarn composite showing internal defects.

elongated to lengths of 1000 Å during the stretching operation. On the other hand, Fourdeux *et al.* (1970) and Perret and Ruland (1969) have indicated that the pores do not change in diameter and, rather than all elongating, many collapse, thereby contributing to an increase in density and a concurrent increase in modulus. The graphite fibers consist of bundles of fibrils 500 Å in diameter. Failure of the bonds between these fibrils is responsible for the grainy appearance of tensile fractures shown in Fig. 8.

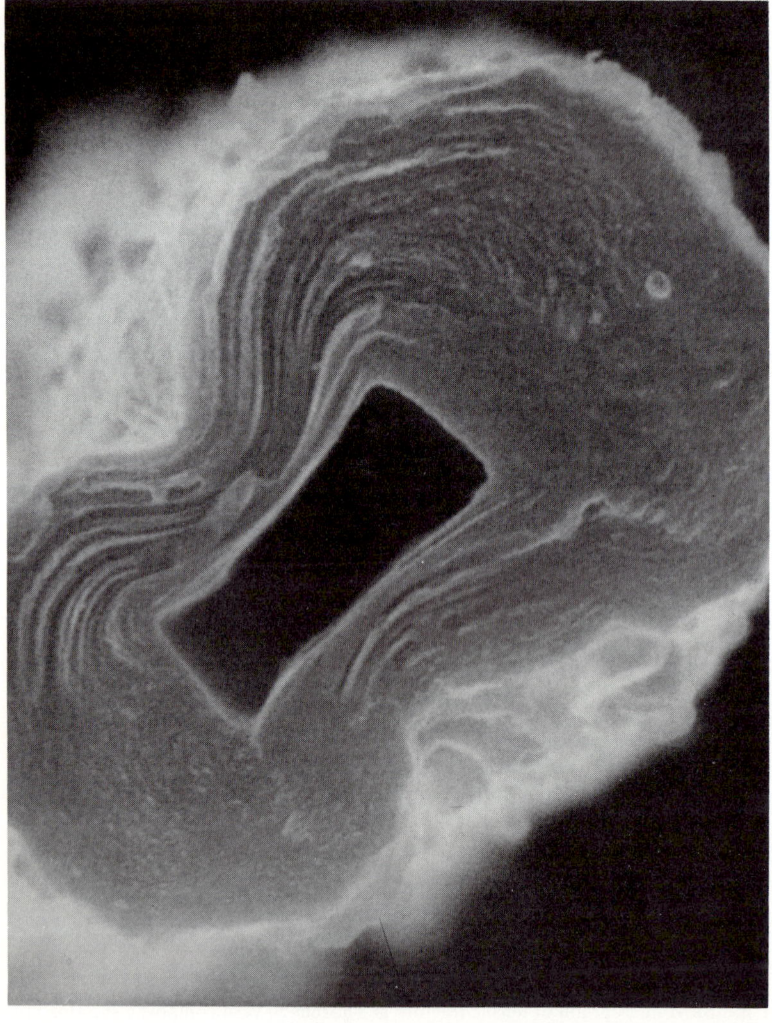

Fig. 9. SEM micrograph of internal defect in Fortafil dog-bone fiber.

Similar investigations into PAN-derived fibers by Johnson and Watt (1967), Watt and Johnson (1969), Badami *et al.* (1967), and Fourdeux *et al.* (1970) now confirm the basic similarity between the ultrafine microstructure of both the rayon and PAN precursor fibers. A fibrillar structural unit, 250–1000 Å in diameter, is shown to be present in both fibers. A very unusual internal defect was found in a PAN precursor dog-bone-shaped Fortafil fiber (Fig. 9). With such defects in mind, continuous monitoring of the fiber quality is recommended for metal–graphite composite researchers. Aside from internal defects, mechanical properties of the fibers and ultimately the composites themselves are most significantly affected by the fiber surface characteristics. The morphology of the surface (surface area, roughness, and pore-size distribution), nature of surface compounds (nitric oxide, carbon dioxide, and atomic oxygen), wetting properties (thermodynamics and contact angles), and adsorption properties are some of the basic considerations included in defining the interfacial phenomena. These subjects have been extensively reviewed by Didchenko (1968).

All of the commercial high-modulus fibers presently available have to be incorporated within a resin matrix. Research has been directed toward improving the condition of extremely low interlaminar shear properties (5000 psi). Thus, all fibers now receive a surface-oxidation treatment (gaseous or liquid) that affects surface morphology. A carbonlike surface is established for low-modulus fibers and a graphitelike surface for high-modulus fibers. Definite improvements in shear strength coupled with a slight decrease in tensile strength for resin-matrix composites have been indicated by Larsen *et al.* (1971). An interesting observation was of the differences in the nature of the PAN- and rayon-based fiber surfaces after this treatment (Fig. 10). The PAN fiber showed considerable pitting, whereas the rayon fiber was relatively smooth.

In summary, it must be stressed that most investigations of the surface properties of graphite have been performed on bulk graphite with a randomly oriented atomic surface. On the other hand, the high-modulus fibers are highly textured, with the basal planes aligned parallel to the longitudinal axis of the fiber and, needless to say, possess quite different surface characteristics. One should not attempt to draw conclusions from one type of fiber and translate to the other. Much has been accomplished in studying the microstructure and surface of the fibers. However, much more remains to be accomplished, particularly with the metal–graphite composite system as an objective. The ability of the graphite fibers to withstand extremely high temperatures with no loss of strength or modulus qualifies them as desirable reinforcements in metals. Certain difficulties are presented to fabricators by the highly reactive nature of graphite in contact with many metals and the very small diameter of the present commercial fibers. On the other hand, the potential low cost of the composites is a great incentive for their use.

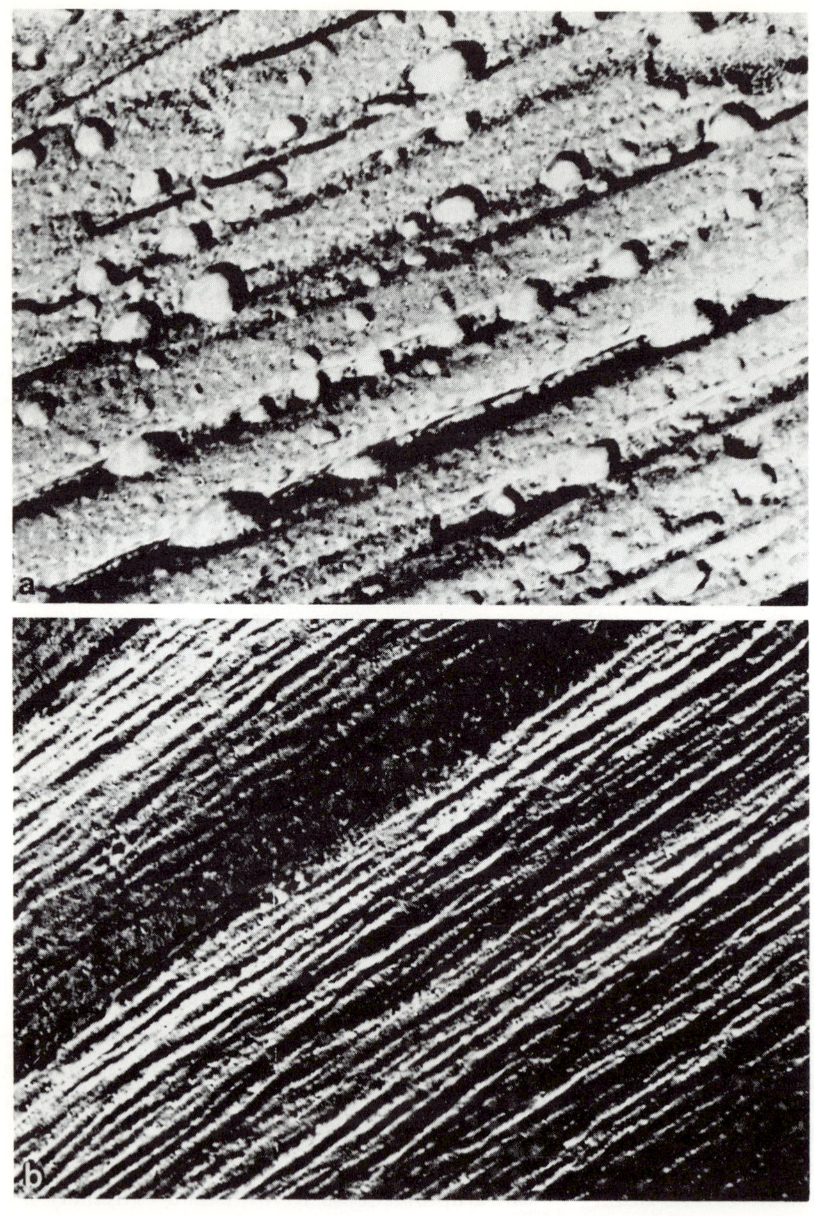

Fig. 10. Micrographs of fiber surfaces showing effects of oxidation treatment by the manufacturer: (a) intermediate-modulus PAN-based fiber; (b) Thornel (rayon) fiber. (40,000 ×)

III. Metal–Graphite Composite Systems

A. General Considerations

In the case of whisker-type reinforcements, the synthesis of metal–graphite composites has been of great interest to materials scientists for many years. The optimum properties (as measured by rule-of-mixtures strengthening) that the graphite fibers can impart to a metal composite have been realized in only a few instances, primarily because of the very fine size of the tow fibers (7 μm) as well as the high degree of reactivity between carbon and most metals at elevated temperatures.

The preparation of metal–graphite composites has been attempted by methods quite familiar to materials-fabrication specialists, and descriptions of these methods have been published in several works elsewhere. It will suffice here to emphasize how they have been used in preparing the individual graphite–metal composite systems. These methods are summarized in Table III. The application of these techniques to other composite systems, e.g., boron–aluminum and titanium-matrix composites, has been described extensively elsewhere by Lynch and Kershaw (1972), Kelly and Davies (1965), and Davis and Bradstreet (1970), and in this volume. Boron–aluminum is the metal-matrix composite in the most advanced state of development. The diffusion-bonding process by which it is made has been effectively used in incorporating the 4-mil-diam boron within the aluminum matrix. Plasma spraying provides a means of making an initial bond between aluminum and boron for the production of a laminate tape. The inclusion of brazing foil enables the use of much lower temperatures and pressures with cost reduction as an objective.

The processes are not readily adaptable to the fabrication of aluminum–graphite composites simply because they do not permit the aluminum to penetrate all of the interstices of the individual small (7-μm) fibrils that make up the commercial tow fibers. Of the processes listed in Table III, only the

TABLE III

Methods for Fabrication of Metal–Graphite Composites

Metal powder	Metal melt
Molten press	Pressure feed
Diffusion bond	Draw through
Hot extrusion	Electroplate
Metal foil	Vacuum evaporation
Molten press	Chemical-vapor deposition
Diffusion bond	

metal-melt (liquid-infiltration) and electroplate methods have shown the capability to penetrate the interstices of the fiber. Chemical-vapor deposition has a strong potential, and research is in progress. The following sections present details, with respect to individual composite systems, of the fabrication methods and the resulting properties.

In a study of the basic problem of graphite compatibility with metals, Morse (1966) has given a theoretical approach and definition of the model composite by the application of thermodynamic principles to chemical inertness and thermal stability. A concept of "mutual insolubility" of the two elements at the melting point provides a criterion for both chemical inertness and the range of thermal stability. This concept was applied to predict possible metal matrices and coatings for graphite fibers. The results of these predictions are given in Table IV. The first seven metals listed, especially nickel, cobalt, and rhenium, offer the greatest potential for a thermally reliable system. This conclusion is based partly on limited solid solubility of carbon in the metal without carbide formation coupled with high-melting-point eutectic formation, and partly on "wetting" experiments of Naidich and Kolesnichenko (1961, 1963, 1964). Wetting is the ability of a liquid metal to flow over the graphite substrate, i.e., that condition when the

TABLE IV

Metals Chemically Inert toward Graphite at Their Melting Points

Metal	Melting point (°C)	Graphite–metal eutectic	Maximum carbon solid solubility
Rhenium	3180	16.9 at. % (2486°C)	11.7 at % (2486°C)
Iridium	2440	Undetermined (2296°C)	—
Rhodium	1966	Undetermined (1694°C)	—
Platinum	1773.5	Undetermined (1736°C)	4.0 at % (solid)
Palladium	1550	Undetermined (1504°C)	—
Cobalt	1495	12.75 at. % (1309°C)	5.0 at. % (1309°C)
Nickel	1452	10.0 at. % (1318°C)	8.9 at. % (1318°C)
Copper	1083	None	0
Gold	1063	None	0
Silver	960.5	None	0
Germanium	940	None	0
Arsenic	817	None	Trace
Antimony	629.4	None	0.033 wt % (1055°C)
Zinc	419.4	None	0
Lead	327	None	0
Cadmium	320.9	None	0
Bismuth	271	None	0
Tin	231.9	None	0
Mercury	−38.9	None	0

surface energy of the graphite exceeds that of the liquid metal. The surface free energy γ of the molten metal can be expressed as a function of the contact angle θ of the liquid with the graphite with respect to a given vapor phase:

$$\gamma_{SV} = \gamma_{SL} + \gamma_{LV} \cos \theta$$

where S, V, and L stand for solid, liquid, and vapor. The relation is illustrated in Fig. 11.

The contact angle not only varies with temperature but also is extremely sensitive to the adsorption of impurities on the graphite or the liquid metal or both, particularly the dissolution of carbon into the metal. Wetting will occur when the contact angle is less than 90°. Naidich reported that, of the metals listed in Table IV, only palladium, nickel, and cobalt had contact angles of 50–70°, but these increased 50% when saturated with carbon. Although conflicts can be found in the literature, as a general rule, the transition metals wet graphite well. Of the nontransition metals, only aluminum and silicon appear to wet graphite, because they form stable carbides and most metals will wet the carbides. Most of the remaining metals listed in Table IV are nontransition metals, and, although they possess no solid solubility in carbon, their contact angle is greater than 90° and thus they exhibit no wettability. It should be noted, however, that the foregoing is based entirely on data derived from bulk graphite and, as previously pointed out, should be accepted with this in mind. Interpretation with respect to high-modulus fibers should be made with caution.

Morse has pointed out that, in addition to the potential of the seven metal–graphite binary composite systems, these same seven metals might be adopted as diffusion-barrier coatings for other matrix metals such as niobium, titanium, tungsten, iron, aluminum, and magnesium, which all form carbides

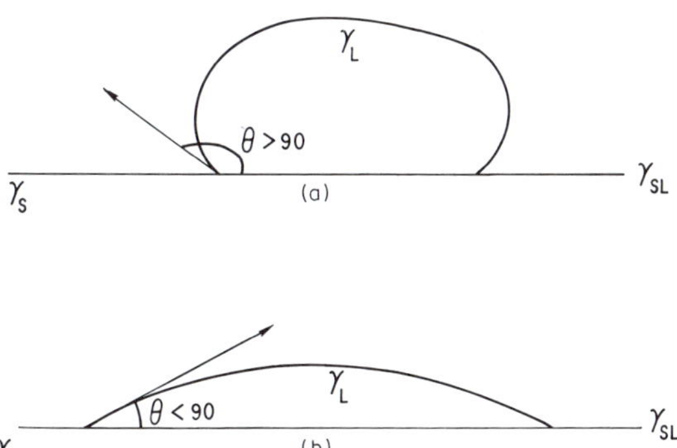

Fig. 11. Wetting of graphite by metals: (a) no wetting; (b) wetting. The balance of surface energies is given by the Dupré equation $\gamma_S = \gamma_{SL} + \gamma_L \cos \theta$.

easily. This approach has been taken, as described later, for the preparation of Al–Ag–G, Mg–Ti–G, and Al–Ni–G composites. Rashid and Wirkus (1972) have described a unique method in which a liquid-metal transfer agent (tin) is used to form carbide coatings of tantalum and niobium (Ta_2C and Nb_2C) on the surface of commercial PAN and rayon graphite filaments of 7-μm diam.

Whatever the coating material may be, it must possess, among many individual requirements, the overall capability to establish a compatible interface. The interface is described as the region between the filaments and the matrix where the bond is developed, and its thickness may vary from a few to many angstroms to a few mils. The interface (bond) has two primary functions. In a mechanical sense, it must be capable of transferring stresses from the matrix to the fibers. Thus, it must have good shear properties. In a chemical sense, it must possess satisfactory adhesive properties, which are dependent on the degree of wettability ($\theta < 90°$). The adhesive strength of the joint may be defined by the work of adhesion W_{SL}, which is the work per square centimeter required to pull apart a solid–liquid joint:

$$W_{SL} = \gamma_{LV} + \gamma_{SV} - \gamma_{LS}$$

Using the data of Naidich, Munson (1967) has calculated this quantity for many liquid metals in contact with diamond and graphite. The results are given in Table V. The Group-VIII metals, i.e., iron, nickel, and cobalt, all wet diamond and graphite well and have very high works of adhesion compared to copper, aluminum, and lead. In most all cases, the metals wet diamond better than graphite.

Although the major effort in reinforcing metals with graphite has been with an aluminum matrix, considerable effort has been directed toward other systems, primarily nickel and, to a lesser degree, copper, magnesium, lead, zinc, tin, and beryllium as matrices.

In more recent studies, various investigators, including Widdows and Nicholas (1967), Manning and Gurganus (1969), and Rhee (1970), have studied the wetting of graphite and ceramics by liquid aluminum and aluminum alloys very precisely. The results of their work are quite uniform and can be summarized in seven major points:

1. Liquid aluminum does not wet graphite, as measured by contact angle θ to about 1000°C. Above 1000°C, θ decreases to less than 90° (Fig. 12).
2. Below 1000°C, aluminum forms stable carbide (Al_4C_3) and appears to be relatively inert; θ may decrease to less than 90° as a function of time.
3. Aluminum alloyed with 1 w/o vanadium, uranium, niobium, titanium, or zirconium does not produce wettability up to 960°C but does at 1120°C.
4. Aluminum with 1 w/o scandium, titanium, vanadium, manganese, iron, or cobalt does not produce wettability at 800 and 925°C. Aluminum with

TABLE V

Surface Energies and Works of Adhesion of Liquid Metals on Carbons

	Diamond				Graphite			
Metal	Temperature (°C)	$\gamma_{LV}{}^{a}$ (erg/cm^2)	$\gamma_{SL}{}^{a}$ (erg/cm^2)	W_{SL} (erg/cm^2)	Temperature (°C)	$\gamma_{LV}{}^{b}$ (erg/cm^2)	$\gamma_{SL}{}^{b}$ (erg/cm^2)	W_{SL} (erg/cm^2)
Copper	1150	1150	4700	210	1100	1300	3400	300
Silver	1000	900	4210	450	980	920	3060	260
Gold	1150	1120	4730	150				
Germanium	1150	650	4080	340	1100	650	2950	100
Tin	1150	500	4050	210	900	500	2860	40
Indium	800	500	4130	130	800	500	2800	100
Antimony	900	360	3940	180	900	360	2680	80
Lead	1000	400	3900	260	800	400	2700	100
Aluminum	800	850	4500	110	800	850	3180	70
Iron	1550	1860	2400[c]	1860	1550	1860	1200	3060
				3200[c]				
Nickel	1550	1700	2700[c]	1700	1550	1700	1470	2630
				2800[c]				
Cobalt	1550	1840	2900[c]	1840	1550	1840	1710	25[illegible]0
				2700[c]				
Palladium	1550	1500	2600[c]	1500	1560	1500	1400	2500
				2700[c]				

[a] $\gamma_{SV} = 3760$ erg/cm^2.
[b] $\gamma_{SV} = 240$ erg/cm^2.
[c] Probable values.

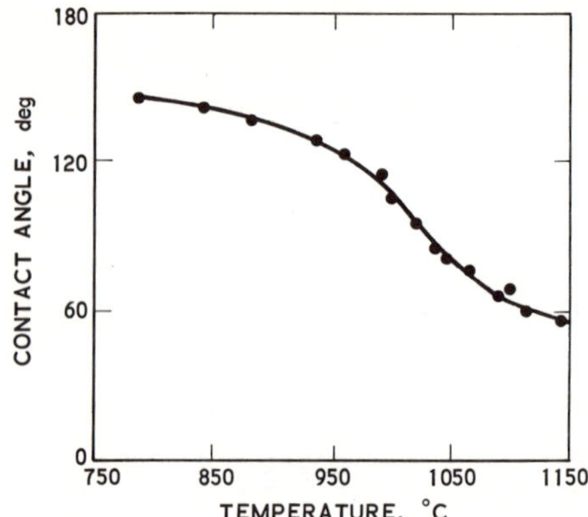

Fig. 12. Contact angle of aluminum on carbon as a function of temperature.

1 w/o calcium, chromium, or nickel does produce wetting at 925°C but does not at 800°C.

5. Aluminum alloys with magnesium, copper, and silicon do not decrease θ up to 840°C (all were less than 90°).

6. Titanium-coated graphite (TiC) with liquid aluminum produces a sharp decrease in θ (improved wettability) (Fig. 13).

7. Wettability of four compounds by liquid aluminum increases with temperature in order of TiC > TiN > TiB_2 > AlN.

Considerable discussion has been presented of the wetting of carbon by liquid aluminum, since it is believed to be quite relevant to the successful fabrication of aluminum–graphite composites.

B. Aluminum–Graphite

Aluminum-graphite composites were first attempted in 1961 when Koppenaal and Parikh (1961) produced a composite of graphite fibers in an aluminum-4% copper alloy. Chopped 0.5-μm-diam graphite fibers were

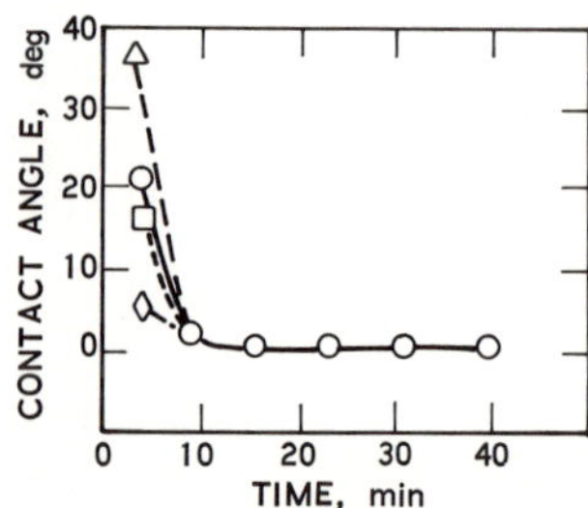

Fig. 13. Contact angle vs. time for binary aluminum alloys on titanium-coated graphite ○ Al-1.62Cu, □ Al-9.82Cu, ◇ Al-10.38Cu, △ Al-2.58Si.

mixed with the alloy powder in a ball mill, and this mixture was hot-extruded. Extrusion temperatures varied from 690 to 1100°F. The composite extrusion containing 20–40 w/o of graphite was considerably stronger than extruded alloy powder alone, but no values exceeded 242 MN/m^2 (35,000 psi). This research was discontinued because of the low strength of the graphite fibers available at that time.

In a series of investigations of aluminum–graphite composites, Sara (1968, 1969, 1971), Sara and Winter (1967), and Sara *et al.* (1966) investigated the use of external pressure to force aluminum into graphite-fiber bundles as depicted in Fig. 14. The mold was preheated to 750°C and then force was applied to the graphite piston. Two experiments were conducted at pressures of 2.48 and 7.46 MN/m^2 (360 and 1080 psi). The transverse microstructure of the resulting composite is shown in Fig. 15. Aluminum metal was forced into spaces 20–30 μm wide formed at the ply junctures, but the plies (each containing 720 individual fibers) were not infiltrated. It was concluded from this work that infiltration could not be accomplished by using pressure alone, and that improvement of the wetting of graphite by aluminum was necessary in order to achieve infiltration.

Sara also investigated several techniques for cladding graphite fibers to prevent chemical interaction with aluminum and promote wetting and infiltration of fiber bundles. He coated fibers with TiC by passing them through a reducing atmosphere of $TiCl_4$ and hydrogen at 1000°C. Satisfactory coatings were obtained (Fig. 16), but tensile tests of the fibers showed them to be

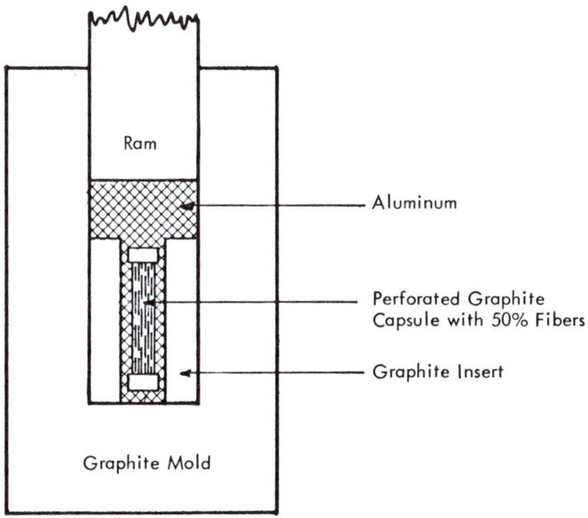

Fig. 14. Mold configuration for pressure infiltration of aluminum around graphite filaments.

severely degraded by the coating process. Bundles of the TiC-coated fibers were then infiltrated with aluminum. A transverse microstructure of the resulting composite is shown in Fig. 17. A large reaction zone, containing Al_4C_3, was observed between the fibers and the aluminum matrix. Monocarbides, such as TiC and CbC, reacted readily with aluminum to form Al_4C_3. It was concluded that they did not have potential as diffusion barriers.

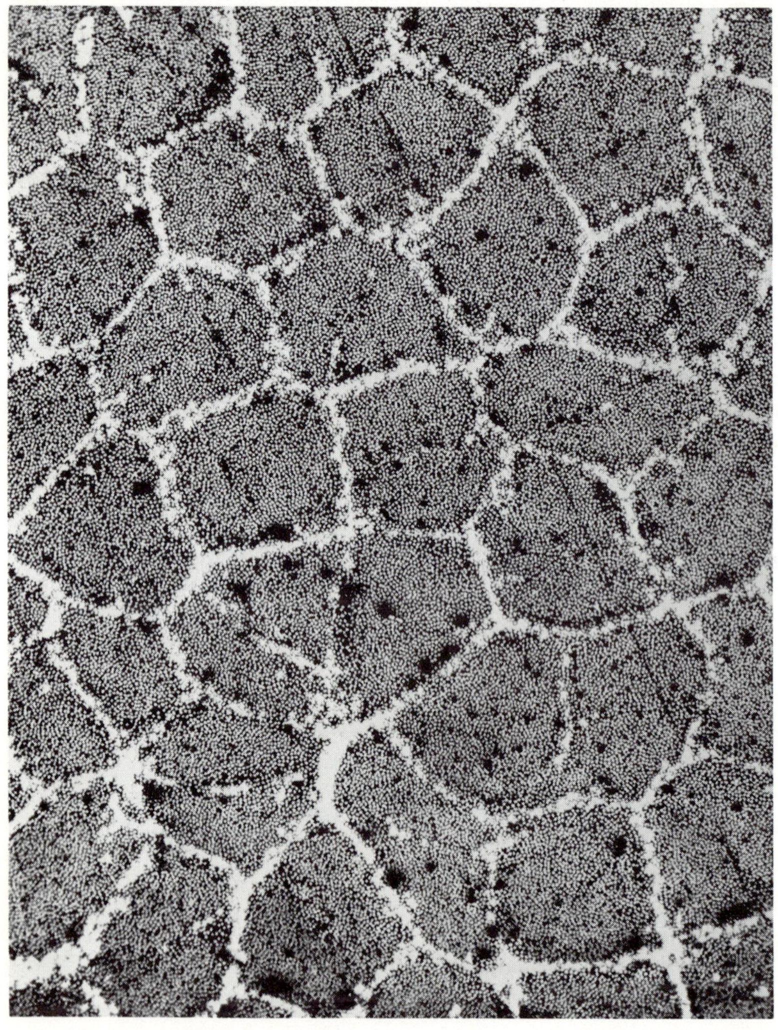

Fig. 15. Pressure-infiltrated aluminum around plies of graphite yarn.

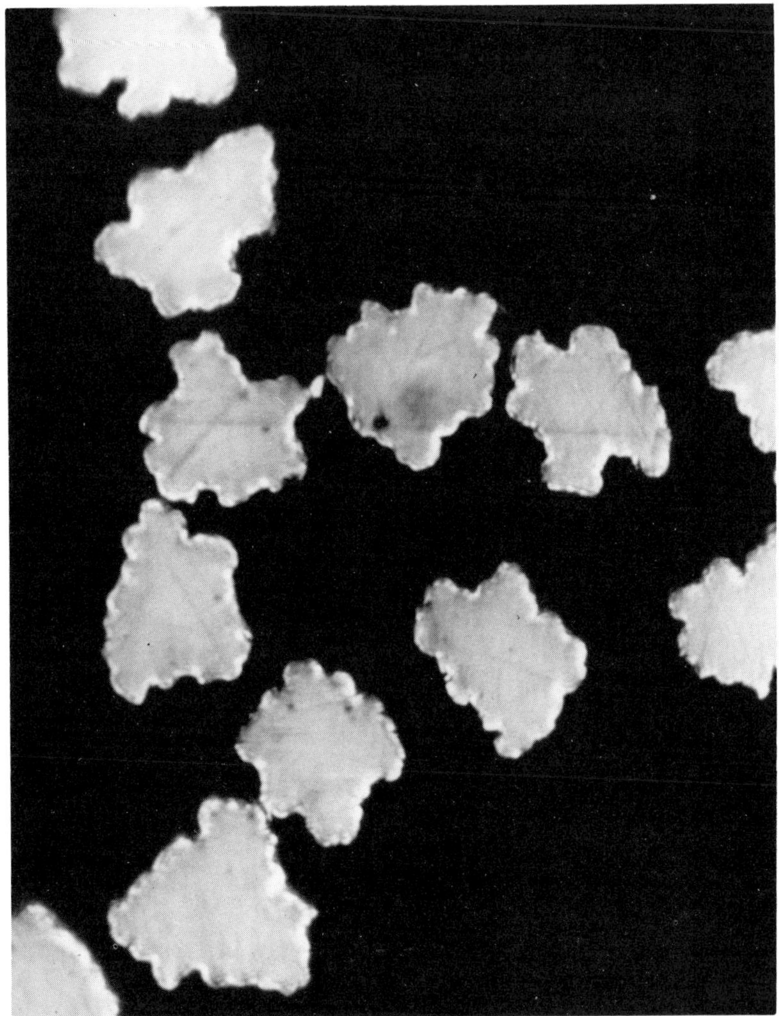

Fig. 16. TiC-coated fibers.

Sara also studied the use of tantalum and nickel as coupling agents. Tantalum- and nickel-coated fibers in an aluminum matrix are shown in Figs. 18 and 19, respectively. The tantalum coating was found to have potential as a coupling agent; no reaction was observed with either the graphite or the aluminum, and the coating was readily wetted by aluminum. The nickel coating reacted with the aluminum to form significant quantities of Al_3Ni

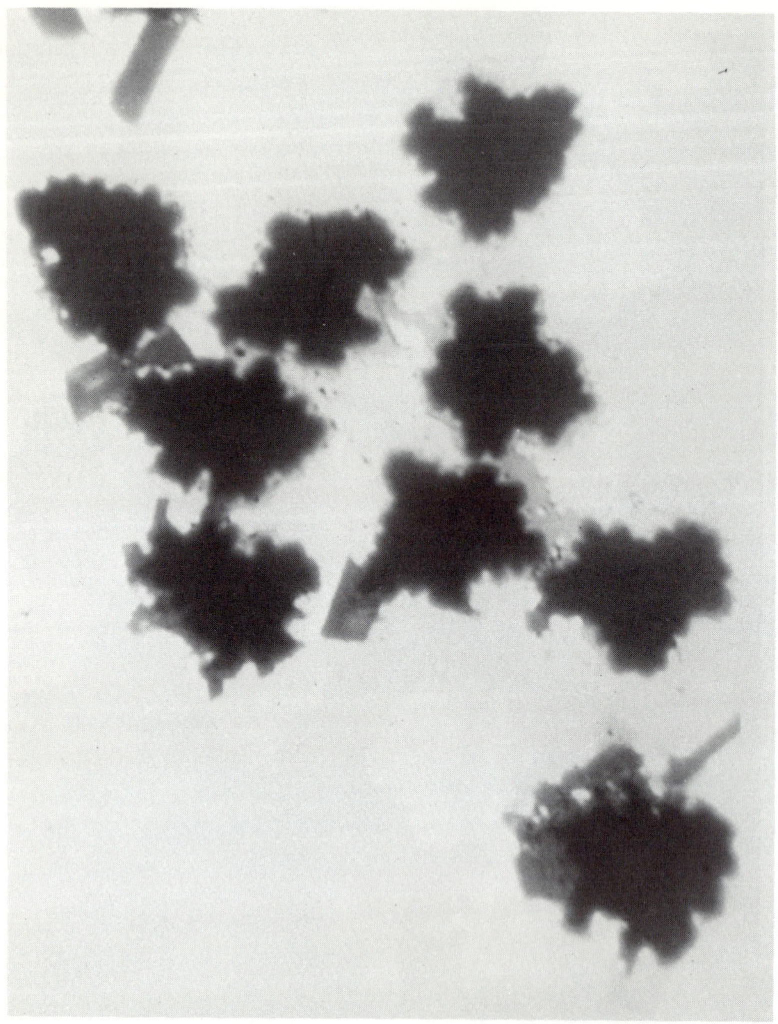

Fig. 17. TiC-coated fibers infiltrated with aluminum.

and was therefore considered unsuitable as a coupling agent. Silver coatings were also investigated, but solution of the silver in the aluminum during infiltration caused dewetting of the fibers.

Mehan and Gebhardt (1969, 1970) have investigated the use of chemical-vapor deposition methods in order to coat Morganite II fibers and improve their wettability by liquid aluminum. The uniaxial tensile strength obtained,

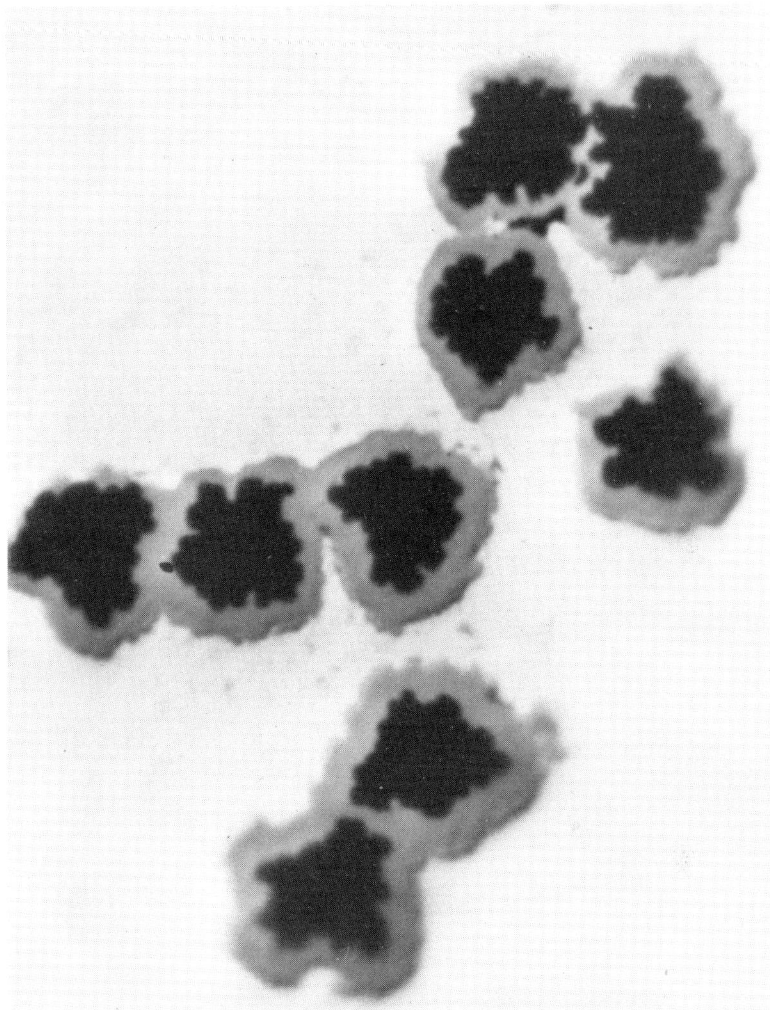

Fig. 18. Tantalum-coated fibers infiltrated with aluminum.

504 MN/m^2 (73,000 psi), was 70% that of the rule-of-mixture values. Blankenburgs (1969) has investigated the preparation of composites by mixing 5–8-μm aluminum powder with 7-μm fiber chopped to approximately 0.1-in. lengths. The mixture was extruded at 600°C and contained 8–15 v/o fiber. Severe fragmentation of the fibers occurred during extrusion, and the fragments, apparently well aligned, were approximately 30–50 μm in length.

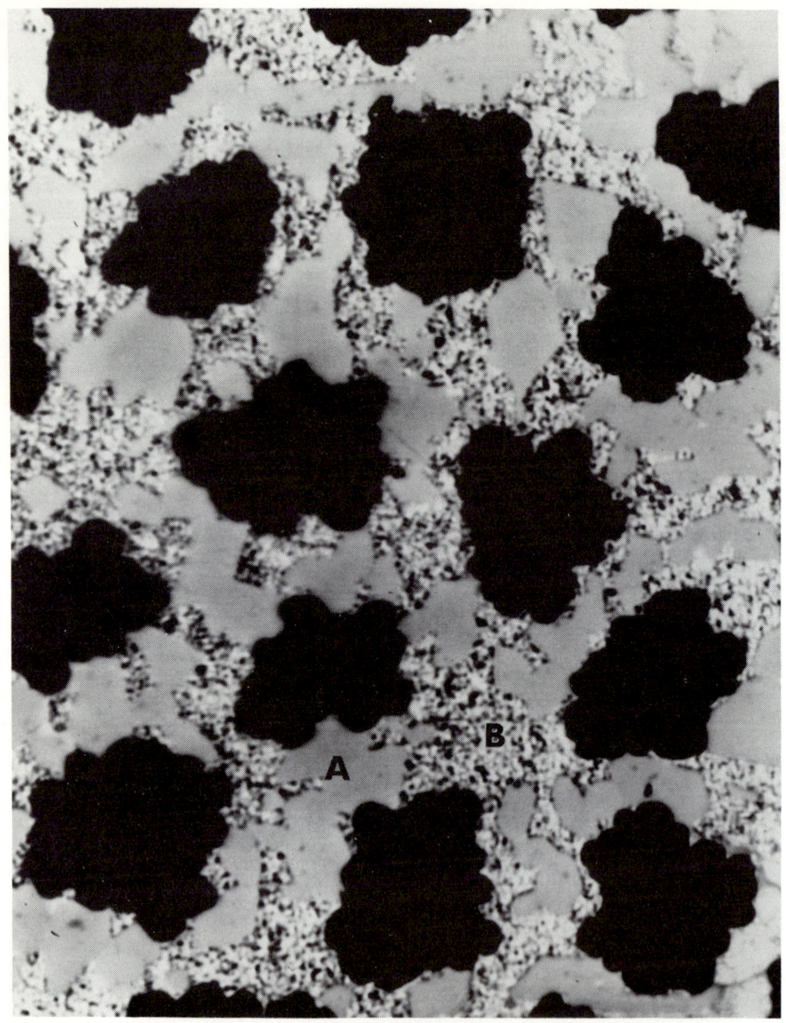

Fig. 19. Nickel-coated fibers infiltrated with aluminum showing (A) Al_3Ni and (B) Al.

The fragmentation increased with increasing volume fraction. Properties reported were about 90 MN/m^2 (13,000 psi) for extruded powder with no graphite. This was increased to about 120 MN/m^2 (17,000 psi) for the composite as extruded and 170 MN/m^2 (25,000 psi) after heat treatment. Evidence was presented on the formation of Al_4C_3 as low as 550°C.

Buschow and Esola (1969) have investigated a process for electroplating aluminum from a nonaqueous electrolyte. They modified this process by codepositing fiber and aluminum simultaneously to produce aluminum–graphite composites. Successful codeposits were made that contained short filaments with l/d ratios up to 100:1 and in small volume fractions, never exceeding 0.11. These composites had a strength of 184 MN/m^2 (26,700 psi) and a modulus of 65 GN/m^2 (9.4 $\times$ 10^6 psi) as compared with 97 MN/m^2 (14,100 psi) and 48 GN/m^2 (7 $\times$ 10^6 psi) for plain electroformed aluminum. This was believed to be a significant increase, and it was specially noted that the fibers were not surface pretreated.

Schmidt and Hellman (1969) have developed a method for plating aluminum on substrates by using aluminum hydride as a source material and contacting it with a decomposition catalyst. Several hundred feet of graphite yarn were coated, and it was found that the maximum coating thickness with this process was 0.3 μm of aluminum. Although the yarn appeared to be uniformly covered with aluminum, the bond was very poor and the coating could be wiped off.

LaIacona (1971) has developed a process for combining coated fibers and aluminum foil to produce a composite. Thornel-50 composites were coated with electroless nickel, and these coated fibers were then spread on 1100 aluminum foil and sprayed with an acrylic solution to hold them in position. Layers were consolidated by hot-pressing at 1000°F. The composites produced had about 8 v/o graphite (approximately equal to the critical fiber volume fraction, V_F) and developed strengths of about 160 MN/m^2 (23,000 psi), which was a significant increase over pure aluminum foil. There is some doubt that this process is amenable to the production of a composite with a high volume fraction of graphite and thus commercial application.

Howlett *et al.* (1971) have studied various methods for the fabrication of graphite–metal composites, including coating and hot pressing, electroforming, hot extrusion of powder–fiber mixtures, and liquid-metal infiltration. In the case of electroplating, a reasonable degree of success was achieved with copper, lead, nickel, and tin. However, some difficulty was encountered with aluminum. Attempts were made to coplate aluminum and chopped fibers from an ethereal plating solution in an inert atmosphere. The carbon fiber was chopped to lengths of about 0.1 cm (0.04 in.) and was used in both the pretreated and untreated condition as well as with a 2-μm copper plating. The experimental results were disappointing. The main problem was the flocking of the fiber, which shorted out the electrodes. This necessitated reducing the fiber content of the bath to very low levels. Only very low volume fractions, approximately 1 v/o, were incorporated in the plate. No solution was found to these experimental difficulties, and the method was considered to have no potential.

Vacuum-evaporation techniques were successful in depositing bulk aluminum onto fibers. However, the problem of shadowing was formidable, and methods for opening the tow mechanically should precede any successful deposition. Coated fibers were consolidated into test specimens. Fiber distribution was good, but the strength of the composite was quite low. Peeling of the matrix away from the fiber was evident. Because of the difficulty of penetration, this method was discontinued. Coating by thermal decomposition of triisobutyl aluminum proved effective. Penetration and uniformity of the coat were very good after the shadowing problems were overcome. Subsequent cold- and hot-pressing resulted in bonding, but there was not enough material available to determine any properties.

Several techniques were attempted to hot-work mixed aluminum powder and chopped carbon fiber. Barrel- and dry-mixing in a fast-moving air stream were quite unsuccessful, and it was decided that wet-mixing should be used. An organic binder in a highly volatile carrier liquid was used and yielded mixtures containing 12–25 v/o fibers. This mixture was packed in an aluminum can and the binder evaporated. The cans were evacuated, sealed, and extruded at 400–600°C with extrusion ratios from 12 to 24. A combination of high extrusion temperature, low extrusion ratio, and low fiber volume was necessary in order to avoid fiber damage. A mixture containing 11 v/o fiber attained properties as high as 25,000 psi. Although this method of fabrication has some promise, other methods appear to have more potential.

DeLamotte and coworkers (1972) have reported the fabrication of continuously cast aluminum–carbon composites with an apparatus illustrated in Fig. 20. In this method, a 1-μm carbon fiber tow previously coated with nickel is drawn through a die located at the bottom of a crucible that contains

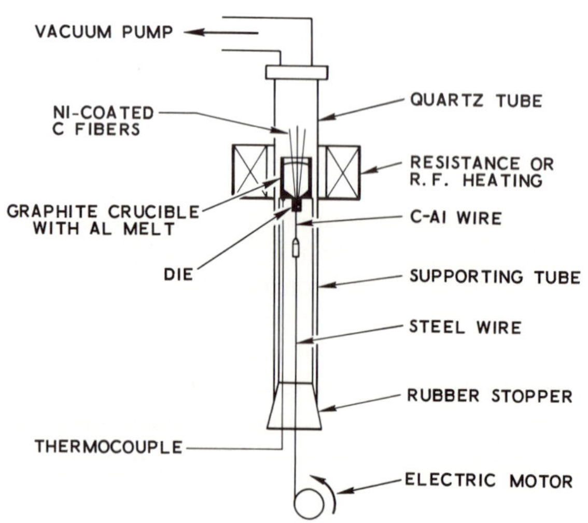

Fig. 20. Wire-drawing apparatus.

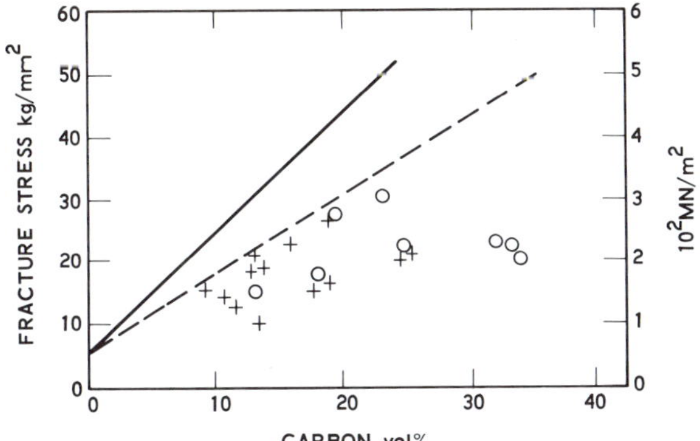

Fig. 21. Ultimate tensile strength of carbon–aluminum composites fabricated with nickel-coated fibers as a function of fiber volume fraction.

the aluminum melt. The liquid metal wets and permeates the tow and solidifies in the die, producing continuous wire. Fracture stress as a function of V_F for these composites is plotted in Fig. 21. The strength of the composites falls off above 30 v/o fiber. As the volume fraction of fiber increases, the proportion of the Al_3Ni phase in the matrix of the composite increases. Under these conditions, as soon as weak fibers break, cracks may spread through the brittle Al_3Ni phase and lead to premature fracture at low stresses.

Davis and Bradstreet (1970) obtained aluminum-electroplated Thornel fibers and consolidated them to produce composites by diffusion bonding. In the initial experiments, the coated fibers were placed between layers of aluminum foil and then hot-pressed. Solid, nonporous composites were produced, but the content of fibers was only about 9 v/o. In other tests, the foil was eliminated and coated fibers were consolidated in a cavity die. This produced a sound composite with about 25 v/o fibers. Microexamination showed a few groups of uncoated fibers and an occasional fiber with an unusually thick plating. In general, distribution was reasonably uniform with little fiber-to-fiber contact. Tensile strengths of 40–50 ksi were obtained, but fiber pull-out during testing indicated unsatisfactory bond between the fiber and plating.

In research sponsored by Rolls Royce, Ltd., for applications to aircraft, Jackson and coworkers (Jackson, 1969; Jackson and Marjoram, 1968, 1970; Jackson *et al.*, 1972) have investigated the preparation and properties of aluminum–graphite composites in detail. Fabrication techniques included chemical-vapor deposition by thermally decomposing triisobutyl aluminum, and electroforming from ethereal baths. Baker *et al.* (1971) formed sheets of composite by winding a layer of carbon fibers on a stainless-steel drum and then electroplating the drum with aluminum. For good penetration of the metal into the tow, very careful prespreading of the fibers during winding was essential. They found that fibers should not be more than five layers deep on the mandrel surface. Plating was accomplished in an organic bath based on diethyl ether. In order to exclude water and because of the danger of explosions, it was necessary to work under a dry inert atmosphere. After plating, the sheet was slit and removed from the mandrel, frequently mounted in a frame, and replated to remove irregularities on the underside. Composites were built up from the plated sheets by either diffusion bonding or liquid-phase bonding. With fiber contents of 30–40 v/o, tensile strengths on the order of 400–500 MN/m^2 have been obtained, which are considerably lower than rule-of-mixtures values.

In his initial work, Jackson (1969) reported on compatibility studies conducted on composites prepared by vacuum evaporation of the aluminum onto the graphite fibers. Heat-treatment tests indicated that the fibers did not lose strength after exposure at 500°C for one week. Data presented in Fig. 22 indicate a sharp loss in strength after one day of exposure at 600°C. Jackson postulated that the composite would have long life at 400°C. Follow-up work by Baker *et al.* (1972) investigated the compatibility of two types of carbon fibers (high and low modulus) with aluminum deposited as described previously. Microstructures of the fiber surface after exposures revealed that the Al_4C_3 initiated on the surface as fine platelets with a random orientation. The most significant result from the study was that the (low-modulus) type-C fiber was more resistant to degradation by aluminum (Al_4C_3 formation) than the higher-modulus fiber at 730°C for 10 min.

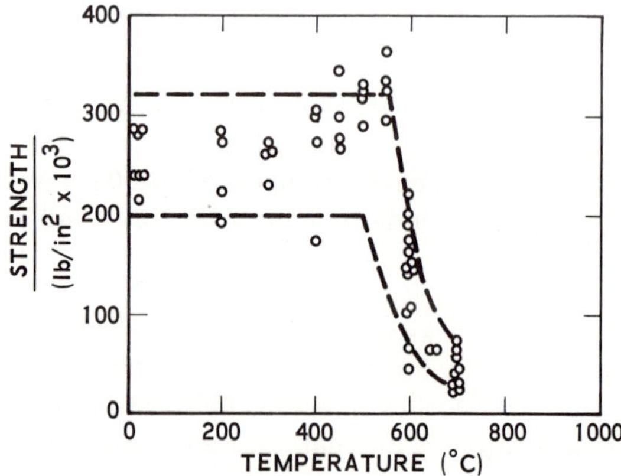

Fig. 22. Strength of aluminum-coated graphite fibers after high-temperature exposure for 24 hr.

In a study of the tensile properties of aluminum–graphite composites prepared by thermal decomposition of triisobutyl aluminum and hot-pressing methods, Jackson *et al.* (1972) reported disappointment in that composites of 30–40 v/o fibers fell far short of rule-of-mixtures strengthening. The average value was about 410 MN/m^2 (60,000 psi) with a single maximum value of 550 MN/m^2 (80 ksi) (80% of the rule-of-mixtures value) occurring at 30 v/o. In an effort to determine the reason for these results, the variables in fabrication were studied. Porosity, fiber breakage, and Al_4C_3 formation were optimized, and although higher values were attained, they fell far below expectations, and it was concluded that some other mechanism was operative.

In summarizing the work described in the literature cited herein, the following conclusions can be drawn:

1. No single method or combination of methods of fabrication has consistently attained rule-of-mixtures strengthening in aluminum–graphite composites.
2. Insufficient attention has been given to characterization of the specific graphite fiber as to surface phenomena.
3. Many preparation methods that have been attempted have ignored obvious end effects of porosity, fiber damage, and carbide (Al_4C_3) formation.
4. Chemical-vapor deposition and electroplating are candidates for coating technique because they avoid mechanical damage of the fiber.
5. Hot-working of the composite in the solid state defeats the principle of fiber strengthening by destroying the integrity of the composite.
6. Insufficient attention has been given to the thermodynamics and kinetics of the solid, liquid, and vapor species reactions involved during the individual fabrication phases.
7. Metallic and nonmetallic fiber coatings have promise as diffusion barriers for the reaction $Al + C \longrightarrow Al_4C_3$.
8. Liquid infiltration shows the most promise as the method technologically suited to accomplish the desired objectives.

The most significant work in the field to date has been the development of an aluminum–graphite composite-fabrication process based on liquid-infiltration methods. Composites that meet rule-of-mixtures strength levels over a wide range of V_F have been produced by Kendall *et al.* (1971), Pepper and coworkers (1970a,b; 1971a,b), and Kendall and Pepper (1970a,b). In 1966, powder-metallurgy methods were first used to evaluate the problem. It quickly became apparent that liquid-metal methods held the most promise for infiltrating the very fine interstices of the graphite fiber. Even though it was well understood at the time that molten aluminum did not wet graphite, the effort proceeded. Initial work with Thornel-25 fibers indicated that high-temperature annealing caused degradation of strength properties above 650°C. Consequently, the binary eutectic alloy (Al–12Si) that depressed the melting point of aluminum (660°C) to 580°C was selected as the matrix material. Early experiments indicated that special techniques had to be developed that involved a simultaneous, coordinated treatment of the matrix alloy and the fiber to promote wettability. The result was a batch process of distinct individual operations performed in a step function as shown schematically in Fig. 23. Fiber was carefully, rigidly spaced on a hanger device (Fig. 24), which was used to expose the fibers to each unit operation. It was imperative to avoid mechanical damage of the fiber. A vacuum glove box and an inert-gas atmosphere were necessary to avoid contamination. Infiltrated fiber plaquettes of Thornel-50 were sectioned for mechanical and microstructural evaluation. Fiber volume fractions near 28 v/o were obtained. Uniaxial tensile strengths ranged from a low of 380 MN/m^2 (55,000 psi) to a high of 1070 MN/m^2 (155,000 psi), with an average value of 730 MN/m^2 (106,000 psi) at 28 v/o fiber. A relative-frequency chart summarizing these test results is shown in Fig. 25. A rule-of-mixtures chart showing ultimate tensile strength (σ_C) versus V_F calculated from strength properties of the A13 aluminum alloy and Thornel-50 is shown in Fig. 26. At 28 v/o fiber, the experimental test value of 730 MN/m^2 is shown to be 100% of the calculated σ_C.

An optical micrograph showing the transverse microstructure of the aluminum–silicon-alloy-infiltrated yarn is shown in Fig. 27. The completeness of infiltration is readily apparent. The gray rodlike second phase is hypereutectic silicon. An SEM micrograph of a polished section is shown in Fig. 28. There is no interaction zone between the fiber and the matrix. The interface is very clear and is indicative of a strong bond between the matrix and the fiber. This is also confirmed by the appearance of a scanning electron fractograph, Fig. 29, showing the absence of fiber pull-out accompanied with ductile yielding of the matrix. Care in handling of the fiber is evidenced in Fig. 30, which shows separation and alignment of the fibers through the cross section of the infiltrated strand.

In the highly anisotropic composite, there is considerable mismatch between the coefficient of thermal expansion of the matrix and the graphite fibers. At room temperature, the linear expansion of the graphite filaments in the longitudinal direction is approximately $-0.9/°C$, whereas the linear expansion of the Al–Si alloy matrix is $+22/°C$. Although deformation in the longitudinal direction is minimal, a degree of thermal stability of the composite is indicated by the results of thermal-cycling tests shown in Table VI. Infiltrated-strand specimens were thermally cycled 20 times between -193 and $+500°C$ and then tested in uniaxial tension. No strength degradation, matrix–fiber interface failure, sliding at the interface, or plastic deformation

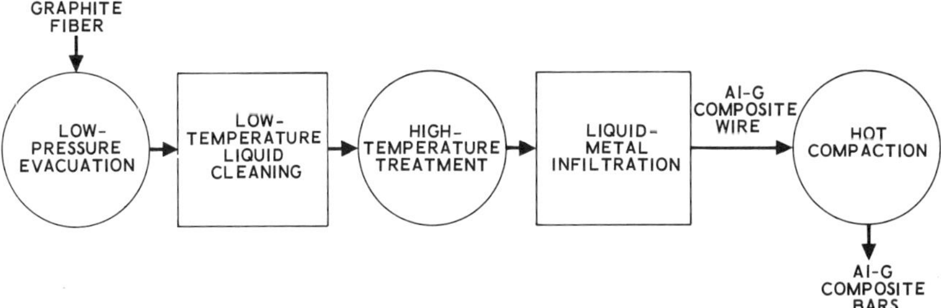

Fig. 23. Schematic representation of aluminum–graphite liquid-infiltration process.

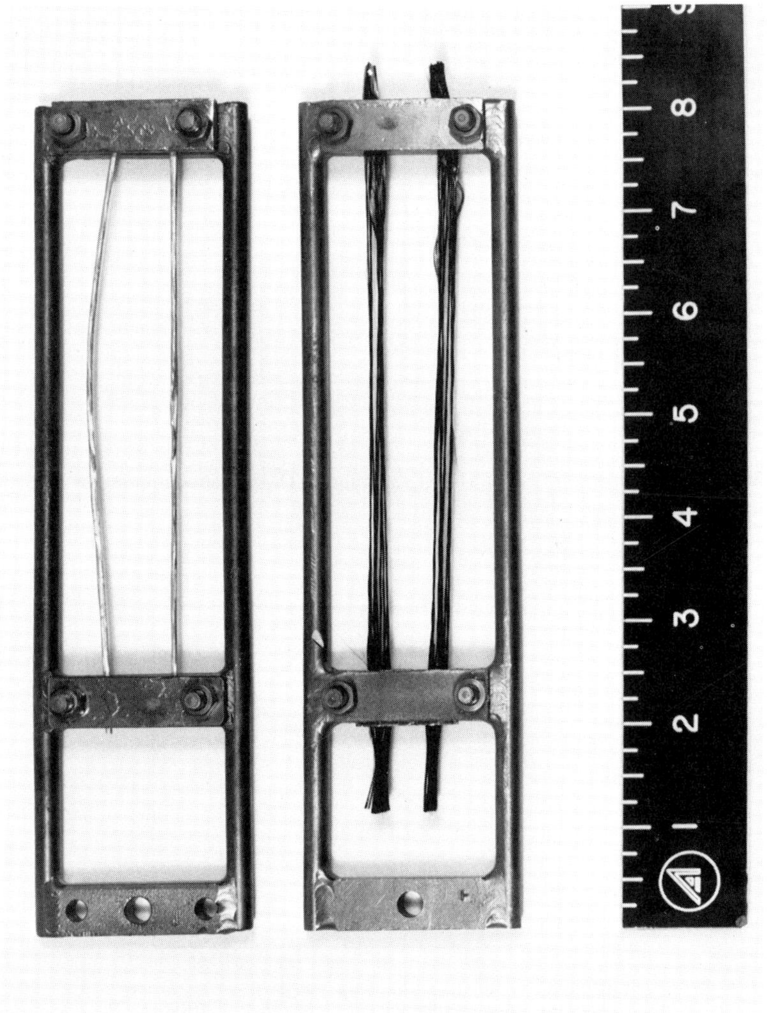

Fig. 24. Batch-process hanger device.

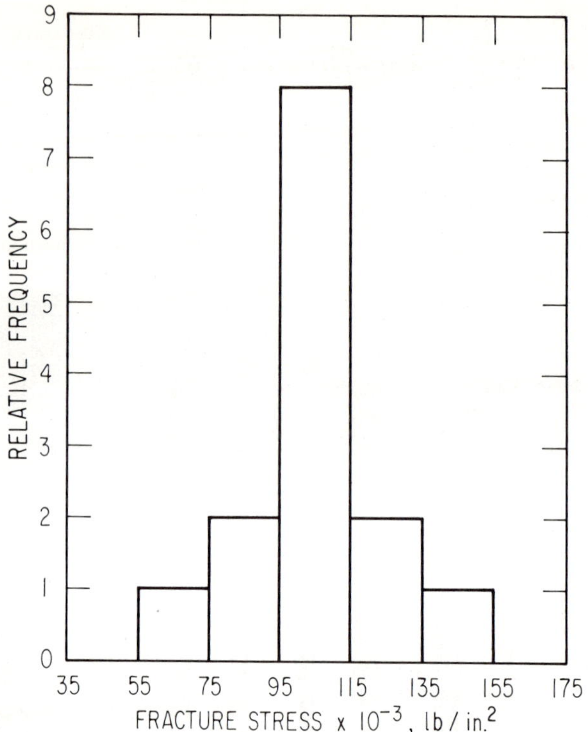

Fig. 25. Room-temperature fracture of A13 aluminum–Thornel-50 composite (28 v/o fiber).

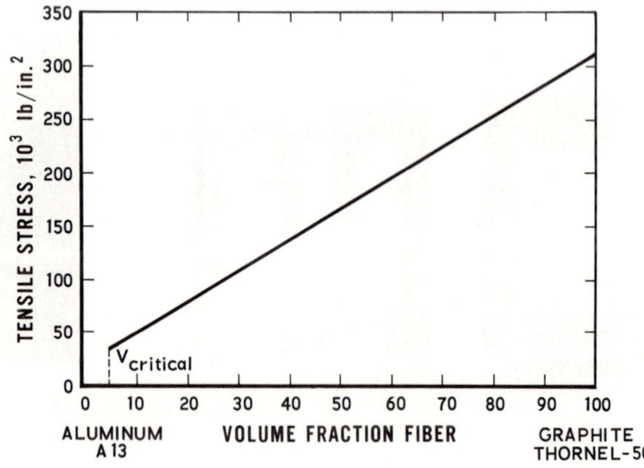

Fig. 26. Rule-of-mixtures strengths for A13 aluminum–Thornel-50 graphite-fiber composite.

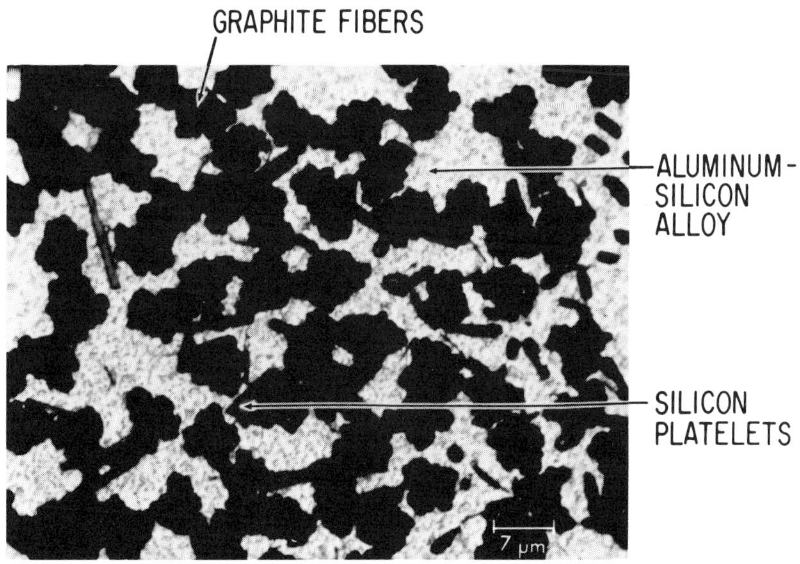

Fig. 27. Transverse microstructure of aluminum–graphite composite strand.

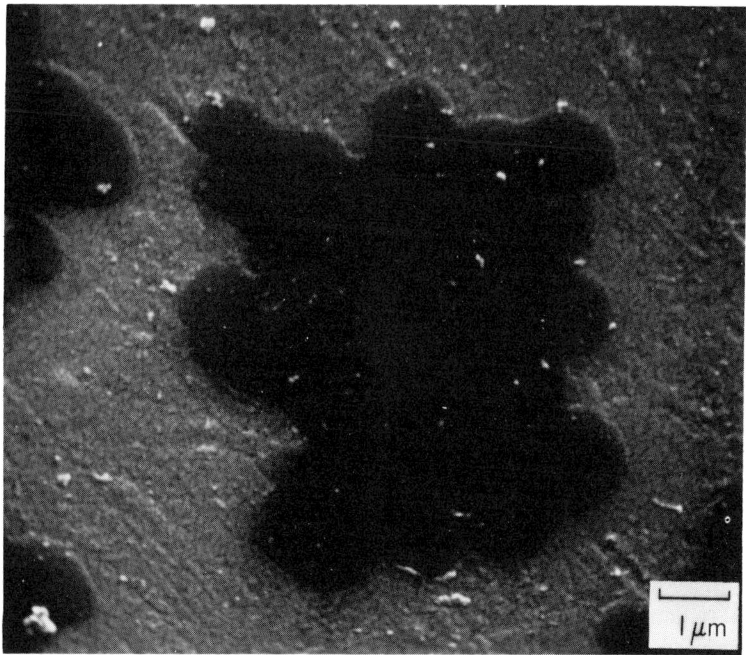

Fig. 28. Transverse section of a fiber and surrounding matrix in aluminum–silicon alloy–Thornel-50 fiber composite.

Fig. 29. Scanning electron micrograph of aluminum–graphite composite tensile fracture.

in the matrix was noticed. Evaluation of the behavior in the transverse direction remains to be done in order to determine whether a "thermal rachetting" effect truly does exist as has been demonstrated in other composite systems when a poor interfacial bond exists.

The integrity of the bond is further demonstrated by the lack of matrix yielding prior to fracture. Typical stress–strain behavior for all composites tests showed a linear relationship over the range of 0.1–0.5 % strain (point of fracture), which yielded a tensile modulus of 145 GN/m^2 (21×10^6 psi) for a 28-v/o Thornel-50 composite and 164 GN/m^2 (23.8×10^6 psi) for a 22-v/o Thornel-75 composite. These moduli corresponded to rule-of-mixtures values of 148 and 171 GN/m^2, respectively.

As an additional indication of thermal stability, high-temperature tests to 540°C indicated no degradation of strength properties. Figure 31 shows a flat temperature dependence of the fracture stress for the composite with a 13 w/o silicon–aluminum alloy, which has a melting point of 580°C. The matrix would be expected to have a negligible tensile strength at 540°C, but

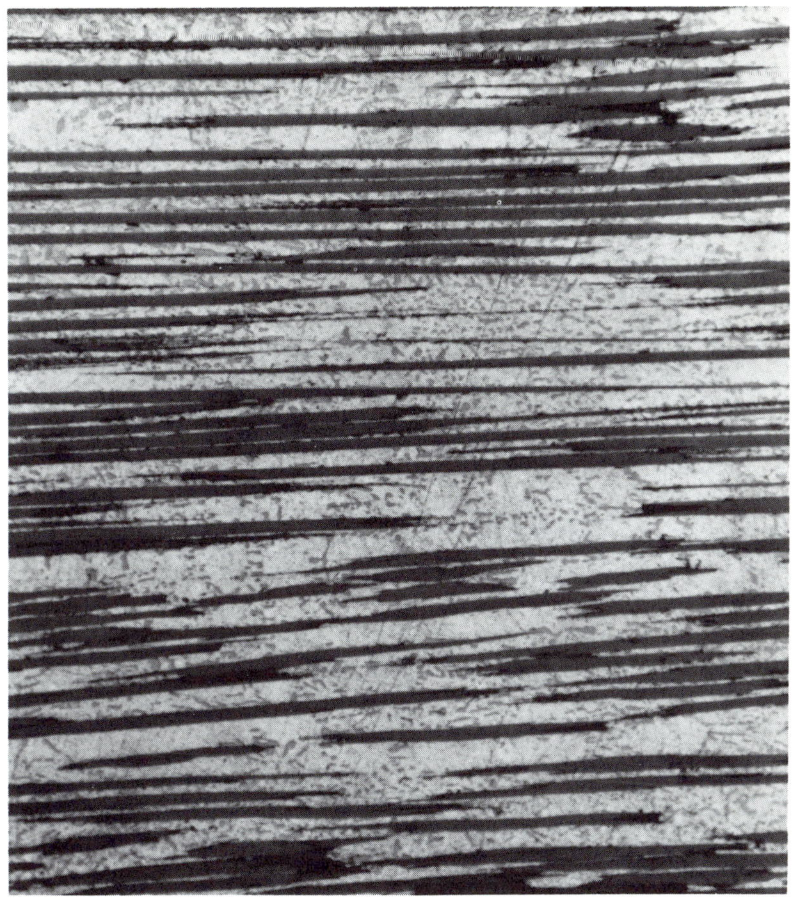

Fig. 30. Longitudinal section of a fiber in aluminum–silicon–Thornel-50 fiber composite.

the tensile strength of the fibers would be expected to increase slightly. At 540°C, then, the strength of the composite can be expressed as

$$\sigma_C = \sigma_F V_F$$

For the Thornel-50 fiber at 28 v/o, σ_C will equal 550 MN/m^2 (80,000 psi). The test data for the composite at 540°C indicated a σ_C of 680 MN/m^2 (99,000 psi). Examination of fractures indicated considerable ductile necking of the matrix at the point of fracture from room temperature to 425°C. However, at 540°C, fracture surfaces revealed anomalously brittle matrix failure with very little interface separation.

TABLE VI

Uniaxial Tensile Date for Aluminum–Silicon Alloy–Thornel-50 Composite Thermally Cycled 20 Times from −193 to +500°C

Sample number	Ultimate tensile strength (psi)	Rule-of-mixtures strength (%)
C7	103,000	103
C8	100,000	99
C9	100,000	99
C10	99,000	99
Average	101,000 (696 MN/m^2)	100

Experiments have been conducted to determine the important transverse tensile-strength properties. Aluminum–graphite composite tapes 6 in. long and 1 in. wide were fabricated for evaluation. Considerable scatter was obtained within each individual composite composition as well as from other matrices and other fiber systems (Table VII). Impurities that were interjected into the aluminum matrix during the fiber-cleaning portion of the liquid-metal infiltration process were found to be the cause of poor transverse strengths. Scanning electron microscopic examination of transverse tensile fractures of the impure composites revealed that failure was initiated in the aluminum matrix. Tensile tests of aluminum alloys containing impurities intentionally added to simulate the matrices of the composites revealed that

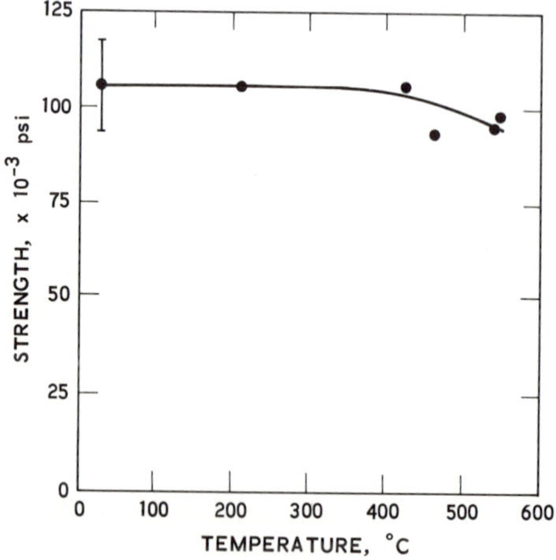

Fig. 31. Tensile strength of A13 aluminum–Thornel-50 composite (28 v/o fiber) as a function of temperature.

TABLE VII

Summary of Transverse Tensile Strengths of Various Aluminum–Graphite Composite Systems

Composite		Average		High	Low	Number of
Fiber	Matrix	(MN/m^2)	(psi)	(psi)	(psi)	tests
Thornel-50	Al–12Si	26	3777	6500	433	9
Courtaulds	220 Al	42	6117	8690	3760	20
Courtaulds	356 Al	70	10,008	14,600	5500	26
Courtaulds HM	Al–10Mg	29.5	4280	4500	3600	5
Whittaker–Morgan	356 Al	50	7300	11,300	4100	5
Whittaker–Morgan	7075 Al	21	3040	5100	400	5

the presence of the impurities caused a severe loss of strength and ductility. For example, the addition of the impurities to an Al–7Mg alloy reduced its strength from 276 to 46 MN/m^2 (40,000 to 6700 psi). Improvements in processing techniques have allowed the preparation of composites of low impurity content, and transverse strengths comparable to those observed in other metal composites such as aluminum–boron have been obtained (Table VIII). Metallographic examination of composites with various aluminum-alloy matrices has revealed that certain matrix alloys form an excessive amount of shrinkage voids. These shrinkage voids decrease the transverse strength of the composite. The 356 aluminum alloy Al–7Si–3Mg

TABLE VIII

Transverse Tensile Strengths for 356 Aluminum–Courtaulds HM Graphite Composite

Sample number	Transverse strength (MN/m^2)	Transverse strength (psi)
808 A	91	13,100
808 A	88	12,700
808 A	74	10,700
808 A	68	9900
808 B	79	11,500
837 A	67	9700
837 A	67.5	9800
837 A	76	11,100
837 B	67	9700
837 B	72	10,400

TABLE IX

Summary of Mechanical Properties of A13 Aluminum—28 v/o Thornel-50 Composite

Property	Value
Ultimate tensile strength	730 MN/m^2 (106,000 psi) at 20°C
	660 MN/m^2 (95,000 psi) at 500°C
Rule-of-mixtures strength	700 MN/m^2 (101,000 psi) at 20°C
	550 MN/m^2 (80,000 psi) at 500°C
Transverse tensile strength	~83 MN/m^2 (~12,000 psi)
Tensile elastic modulus, E	145 GN/m^2 (21.0×10^6 psi) at 20°C
Shear modulus, G (calculated)	55 GN/m^2 (7.9×10^6 psi)
Density	2.4 g/cm^3 (0.0805 $lb/in.^3$)
Strength-to-density ratio	2.4×10^6 cm (1.25×10^6 in.)
Modulus-to-density ratio	620×10^6 cm (248×10^6 in.)
Poisson's ratio, μ (calculated)	0.306

has been found to have the least tendency toward shrinkage-void formation. Recent tests for 356-aluminum-alloy–Thornel-50 composite have yielded transverse strength values of approximately 83 MN/m^2 (12,000 psi).

The mechanical properties of A13 aluminum–28 v/o Thornel-50 composite are summarized in Table IX. The Poisson ratio was calculated from the expression

$$\nu_{12} = \nu_F V_F + \nu_M(1 - V_F)$$

where

ν_{12} is the composite Poisson ratio,
ν_F is the fiber Poisson ratio $= 0.25$ (after Sara *et al.*, 1966),
ν_M is the matrix Poisson ratio $= 0.33$,
V_F is the volume fraction fibers $= 0.3$,
$\nu_{12} = 0.25 \times 0.3 + 0.33 \times 0.7$, and
$\nu_{12} = 0.306$.

This value compares favorably with the value of 0.29 obtained by Sara for a composite of a nickel matrix ($\nu = 0.301$) with Thornel-50 fibers. A more accurate predication of the Poisson ratio would involve finding mathematical solutions to the stress fields in the fibers themselves and also in the interaction between the fiber and matrix. The shear modulus G was also calculated from the expression

$$G = E/2(1 + \nu)$$

where ν is the Poisson ratio $= 0.31$.

The liquid-metal-infiltration process lends itself to a wide variation in the composition of the matrix alloy. As previously discussed, the eutectic

Al–12Si alloy was selected for early work because of its low melting point of 580°C. With the improvements in fiber manufacture and heat treatment, it was found that higher-melting-point alloys could be used without degradation of the fiber strength. Consequently, research was directed toward determination of the effect of aluminum alloys of various compositions on the end properties of the composites. Simultaneously, Thornel-75 fiber became available commercially and was used as the reinforcement. The matrices were composed of commercially pure aluminum, Al–7Mg, Al–7Zn, and Al–13Si alloys. Tensile strengths as high as 1070 MN/m^2 (155,000 psi) and a modulus of elasticity of 190 GN/m^2 (27.5×10^6 psi) were obtained on hot-pressed samples of the Al-7Zn–Thornel-75 composite. The results of this work are shown in Table X. Note the increase in V_F in the pressed condition over the as-infiltrated condition. Excess aluminum is squeezed out during the liquid-phase hot-pressing cycle, and with the increase in V_F, an increase in the ultimate tensile strength can be expected—provided of course that no fiber damage is experienced during the operation. Specimens $15 \times 2.5 \times 0.6$ cm (Fig. 32) have been consolidated from the infiltrated strand by a special procedure as follows (Fig. 33):

1. Sandblast surface of wire before pressing;
2. Prepare 0.05-cm (0.020-in.) 356 aluminum foil and use as a filler metal; sandblast surface of foil;
3. Use graphite die (Fig. 33) lined with aluminum-foil wrap;
4. Degrease composite wire and filler metal foil;
5. Lay up die as shown in Fig. 33;
6. Use 280 N/m^2 (40 psi) pressure at 650°C for 10 min; evacuate die before pressing, backfill with high-purity argon, and maintain flowing argon atmosphere during pressing;
7. Apply pressure when molten (650°C) and keep pressure on while cooling;
8. Approximately 60 pieces of wire (containing eight strands of Thornel, 1400 fibers to each strand) will make a 1.2×0.3 cm section bar at 35 v/o fiber.

If the compacting temperature becomes much higher than the liquidus temperature, the kinetics of the reaction between aluminum and graphite will be accelerated. The strength of the composite will degrade because of the formation of excessive Al_4C_3. It is believed that when the total amount of carbide exceeds 1000 ppm, a decrease in composite strength can be expected. In Fig. 34, a longitudinal section of a hot-pressed composite shows a lathlike growth of the carbide phase.

In contrast, at lower temperatures, the extent of interaction has been controlled such that a definitive reaction layer is not visually discernible. The

TABLE X

Tensile Properties of Various Aluminum-Alloy–Thornel-75 Composites

Matrix composition	Specimen condition	Volume percent fiber	Strength: Average MN/m^2	Strength: Average psi	Strength: Number of samples	Strength: Low value (psi)	Strength: High value (psi)	Average modulus (GN/m^2)	Average modulus (psi)
Commercially pure aluminum	As-infiltrated	32	68	99,000	8	65,000	116,000	178	25.7
	Pressed	35	65	95,000	7	85,000	104,000	147	21.3
Aluminum–7 w/o zinc	As-infiltrated	32	71	103,000	7	59,000	132,000	166	24.1
	Pressed	38	87	126,000	10	102,000	155,000	190	27.5
Aluminum–7 w/o magnesium	As-infiltrated	31	68	98,000	4	87,000	124,000	195	28.1
Aluminum–13 w/o silicon	As-infiltrated	22	55	80,000	7	73,000	88,000	165	23.8

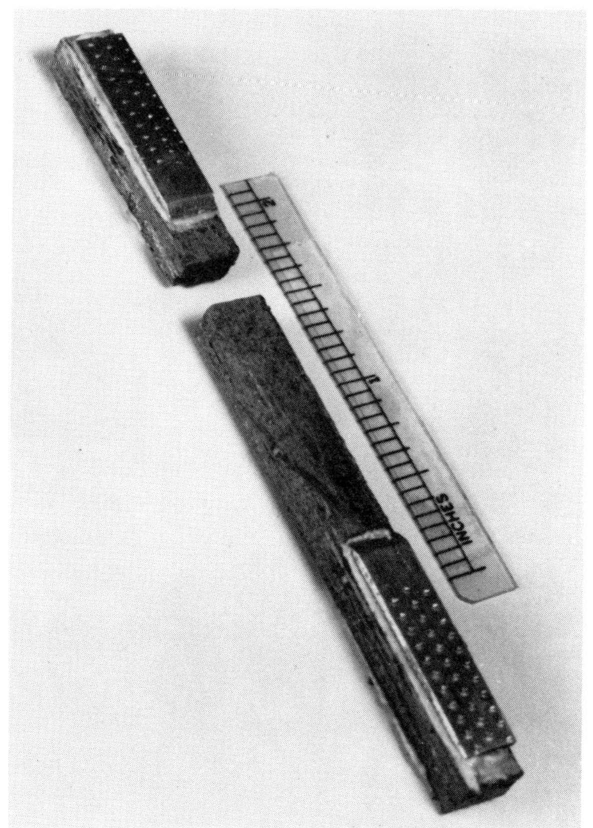

Fig. 32. Hot-pressed mechanical-property composite specimen.

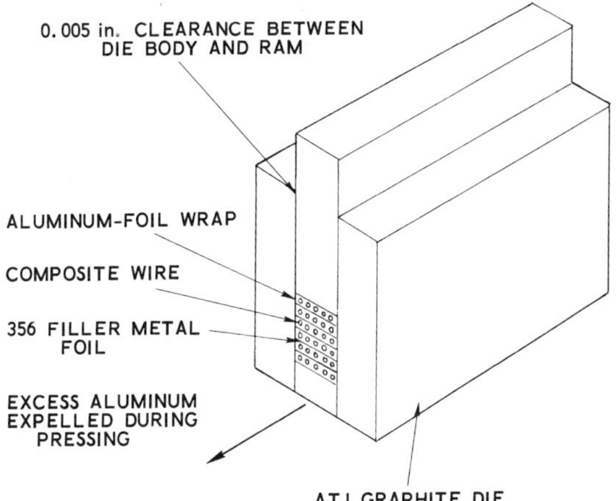

Fig. 33. Hot-pressing die configuration.

Fig. 34. Massive carbide (Al_4C_3) formation at fiber–matrix interface.

solubility of carbon in molten aluminum approaches 0.1 at. %; at saturation limits of carbon solubility, the aluminum-carbide phase, Al_4C_3, is stable. The lathlike carbide phase apparently forms in the molten aluminum and, by so doing, enriches the molten metal with silicon. The gray phase in the center of the micrograph is the eutectic alloy that was not resolved in this micrograph. It is imperative then that close control of carbide formation be maintained. Chemical analysis procedures for the determination of Al_4C_3 and total carbon in the composites were especially developed for this purpose.

In other work on matrix-composition variations, a ternary alloy of Al–9Si–4Cu was studied. During solidification, the primary phase to solidify

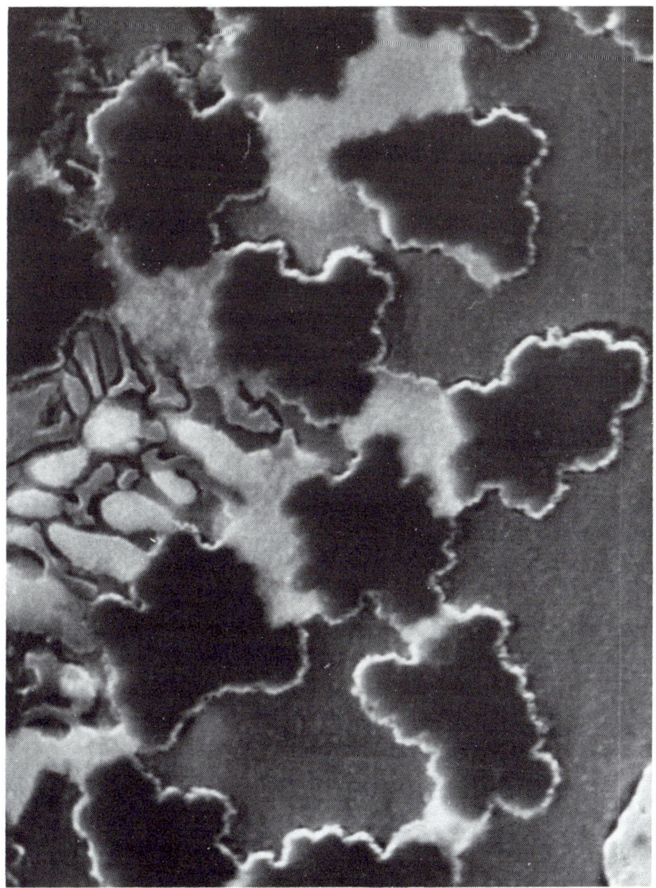

Fig. 35. $CuAl_2$ dendrite growth engulfing graphite fibers in Al–9Si–4Cu ternary-matrix composite.

was the $CuAl_2$ intermetallic within the Al–Si–Cu matrix, as shown in Fig. 35. Engulfment of the fibers by the dendritic matrix structure is evident. As a hard, brittle phase, the intermetallic phase was ineffectual in transferring loads to the fiber, and resulted in composites with very low strength levels.

The work previously described was all performed on batch-processed material, and the encouraging results justified a scale-up of the process. Growing demands for quantities of material as well as incentives for furthering the engineering development of the material motivated efforts toward development of a continuous process for fabrication of the composite. An apparatus was designed and constructed (Fig. 36) that is capable of con-

Fig. 36. Continuous wire-drawing facility for making graphite–aluminum composite strands.

tinuously producing infiltrated composite strand at a rate of 150 m/day. Considerable process development was required for conversion from the batch to continuous operation. Times, temperatures, speeds, pressures, and atmospheres had to be established for production of a composite with the same strength qualities previously obtained. Initially, a strand of Thornel-50

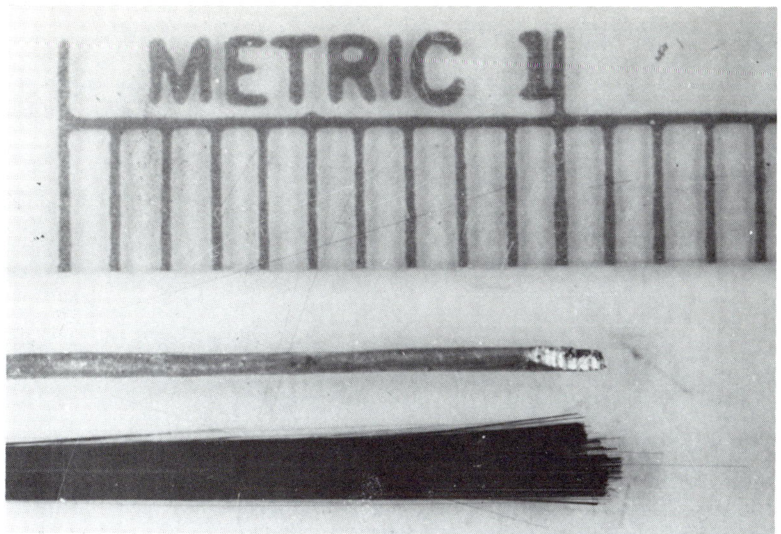

Fig. 37. Thornel-50 fiber and infiltrated strand.

(1440 filaments) was used to stabilize the process. A strand of the Thornel-50 yarn and the finished infiltrated strand are shown in Fig. 37. Since the initial work, bundles of eight strands of the yarn (11,520 filaments) are now processed through the cycle to yield a composite wire of about 0.15-cm diam. A transverse microstructure of the Thornel-50–356-aluminum alloy composite is shown in Fig. 38, where the irregular contour of the wire can be noted. In order to demonstrate the capability of the process with a PAN-based fiber, Hercules HM was drawn through to yield a composite with 356 aluminum (Fig. 39). The process is a complex one that involves many reactions at low and high temperatures. Occasionally, the kinetics of these reactions is altered by geometrical effects (surfaces), and infiltration of the fiber is blocked as shown in Fig. 40. When this occurs, this section of the strand is removed and discarded. Output of strand from the continuous-process apparatus is currently being coiled on a 65-cm-diam drum in order to store a specific lot of material more compactly. Sufficient lengths are then cut for hot-compacting into test specimens.

The corrosion behavior of the aluminum–graphite composite is being studied in distilled water and 3.5% NaCl solution at different temperatures for a time period of 1000 hr. Preliminary data in Table XI show that this solution further increases the corrosion rate as compared with distilled water. Increases in temperature are also shown to increase the corrosion rate. Comparison of the aluminum–graphite corrosion rate data with that of 356 aluminum alone, which is used as a standard and is the matrix material

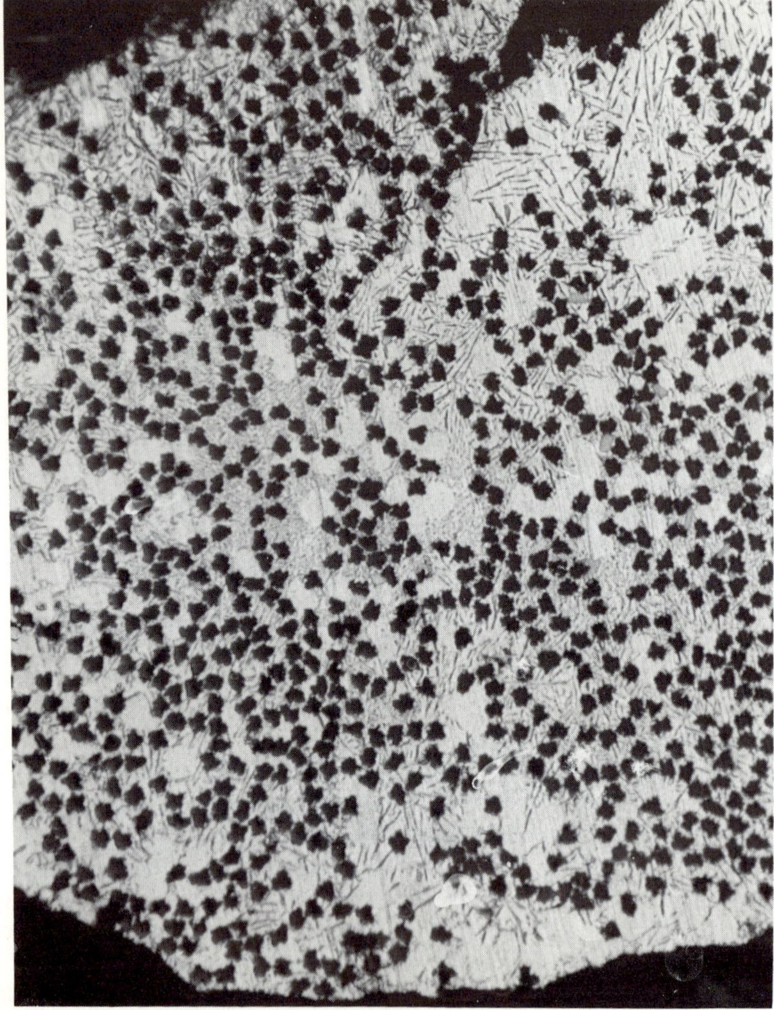

Fig. 38. Transverse microstructure of Thornel-50–356-aluminum-alloy composite.

in the composite, shows that the former corrodes one to four times faster than the latter, depending on environmental conditions. This increase is attributed to a galvanic coupling effect that occurs between the aluminum matrix and the graphite.

A series of studies by Goddard *et al.* (1972) was made to evaluate the potential of several processes for joining aluminum–graphite composites. Joints

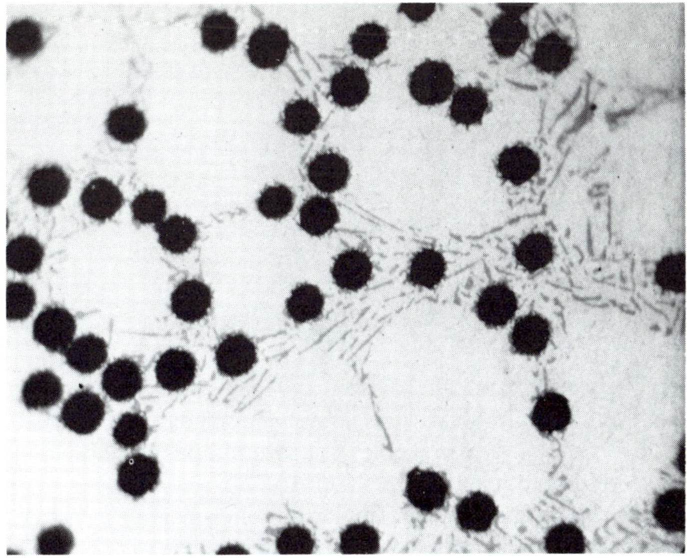

Fig. 39. Microstructure of aluminum–graphite composite continuously processed from Hercules HM fiber.

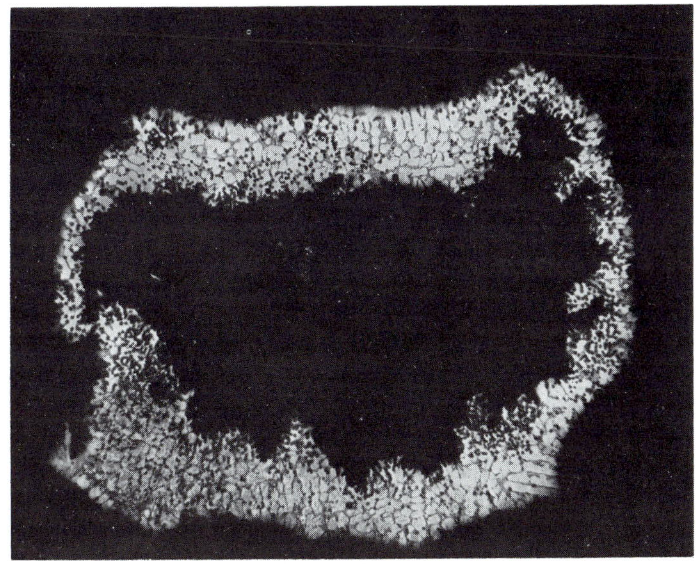

Fig. 40. Partially infiltrated graphite yarn.

TABLE XI

Corrosion Behavior of Aluminum–Graphite Composite for 1000 hr[a]

	356 aluminum		356 aluminum–25 v/o Thornel-50	
Environment	(23°C)	(50°C)	(23°C)	(50°C)
Distilled water	Nil	Nil	1.2	1.2
3.5% NaCl solution	1.1	4.9	4.7	9.8

[a]Mils per year (1mil = 25 μm).

were made between composite pieces, and also between composites and standard aluminum alloys. Various joining methods were evaluated, and the results are summarized in Table XII. Aluminum–graphite composites, with matrices of both pure aluminum and Al–7Zn, were successfully furnace-brazed to 6061 alloy sheet. A photomicrograph typical of a brazed joint is shown in Fig. 41. Standard 718 (Al–12Si) filler alloy was preplaced in the joint, and the parts were fluxless-brazed in a flowing argon atmosphere. Composites having pure aluminum matrices were also joined to each other. A modified filler alloy composed of a combination of 6061 and 718 alloys was used. Composites having Al–7Zn matrices were joined to each other by the same technique (Fig. 42), although the composition of the filler metal was not optimized. It appears that an aluminum alloy containing magnesium and silicon is necessary. Liquid-phase pressure-welding, or furnace-welding, promotes intimate contact between the fibers of the composites being joined and thus promotes a uniform fiber distribution in the joint area. However, difficulty

TABLE XII

Potential Joining Processes for Aluminum–Graphite Composites

PROCESS	RESULTS TO DATE	COMMENTS
FURNACE BRAZING	GOOD	• NO FLUX • INERT ATMOSPHERE • Al-Mg-Si ALLOY REQUIRED
RESISTANCE SPOT WELDING	FAIR	• CONTROL CRUSHING AND EXPULSION • WELDING TO ALLOYS DIFFICULT
GAS TUNGSTEN-ARC WELDING	FAIR	• CONTROL ALUMINUM CARBIDE FORMATION • ARC BREAKS UP SURFACE FILM

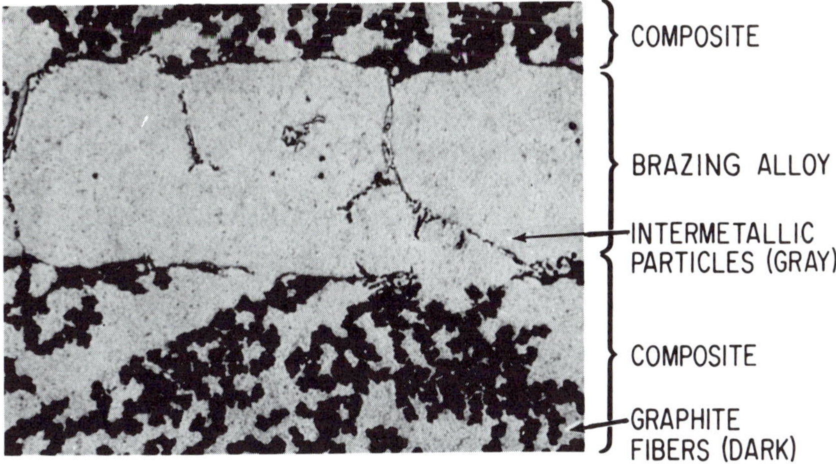

Fig. 41. Micrograph of Thornel-75–Al-7Zn composite brazed to 6061 aluminum alloy.

has been encountered in ensuring the complete mutual wetting of the two composite matrices. Resistance spot-welding of composites to 2219 (Al–6Cu) alloy sheet appears feasible, but the tendency for heat to localize in the composite causes problems. Gas tungsten-arc welding of composites also appears possible. In this case, the major problem is that severe overheating can cause aluminum carbide to form, which is detrimental to the corrosion properties of the composites.

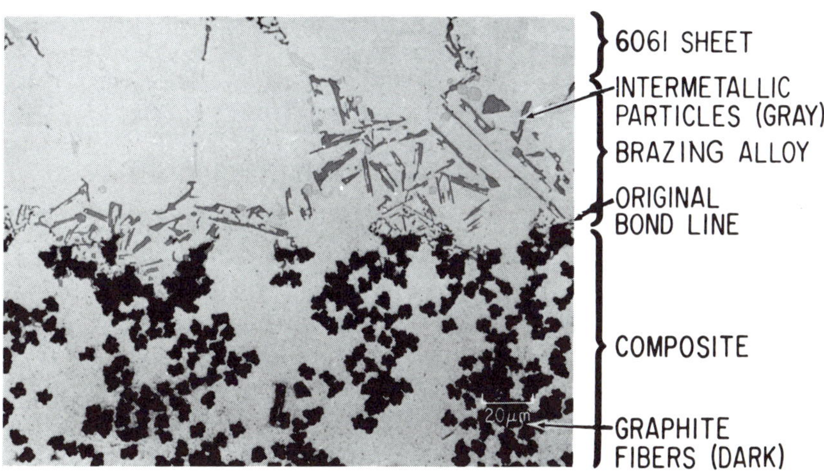

Fig. 42. Micrograph of Thornel-75–Al-7Zn composite brazed to itself.

C. *Nickel–Graphite*

Nickel- and cobalt-based alloys are the most widely used materials for high-temperature applications. The superalloys such as Rene 41, Inconel 100, Udimet 700, Mar-M-200, and TD Nickel maintain the best strength properties in the range of 1600°F for long periods of time in oxidizing atmospheres. The specific strengths of several of these alloys are plotted as a function of temperature in Fig. 43. Efforts to improve their specific strength through conventional alloying techniques have reached a plateau in the region of $0.5–0.8 \times 10^6$ cm ($0.2–0.3 \times 10^6$ in.). The temperature range above 1000°C is of particular interest because specific strengths and moduli higher than those currently available could be used effectively in gas turbines, hypersonic airframe structures, and rocket motors. Incorporation of high-strength graphite fibers into a nickel or nickel-alloy matrix appears to have great potential toward achieving this goal.

Calculated values from the rule of mixtures for nickel–graphite composites containing 40, 50, and 60 v/o Thornel-50 yarn are plotted in Fig. 43. Specific-strength values at 900°C extend from $1.5–3 \times 10^6$ cm ($0.6–1.1 \times 10^6$ in.), or a maximum of fivefold improvement. The effect of a higher-strength alloy matrix is shown in Fig. 44, where specific strength for a 50-v/o Thornel-50–80Ni-20Cr composite extends to 2.5×10^6 cm at 900°C. At 1300°C, there is a slight solid solubility of carbon in nickel that could provide a mechanism

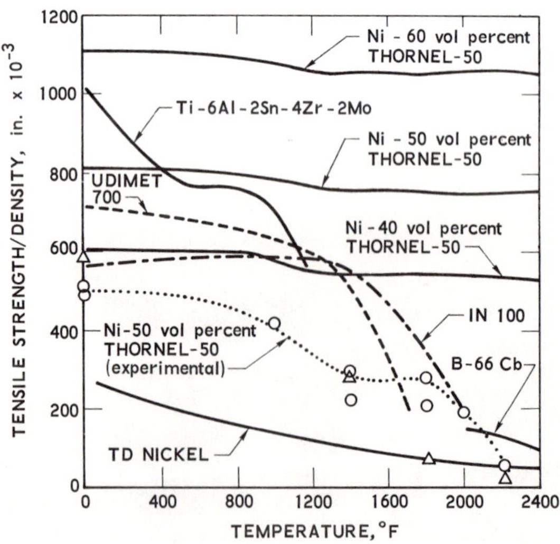

Fig. 43. Specific strength as a function of temperature for graphite–nickel composites and other high-temperature alloys.

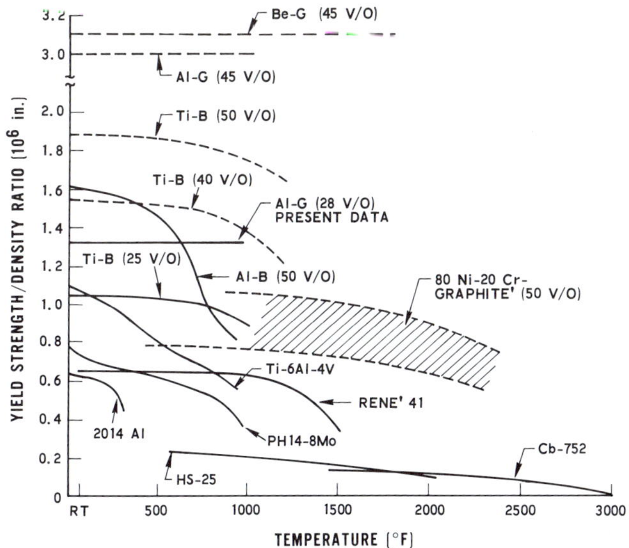

Fig. 44. Comparative properties of current materials (solid curves) and future materials predicted from the rule of mixtures (dashed curves).

for achieving good fiber–matrix bonding. However, carbon diffusion could cause microstructural instability.

Work in the United States on nickel–graphite systems has been carried out primarily in two laboratories: Sara and coworkers at Union Carbide, and Niesz and coworkers at Battelle. Sara and coworkers (Sara, 1968, 1969, 1971; Sara and Winter, 1971; Sara *et al.*, 1966) have investigated the fabrication and properties of nickel–graphite composites. Thornel-50 and Thornel-75 graphite fibers were electroplated with nickel on a continuous basis. A yarn was produced that was ribbonlike, continuous, flexible, free of twist, and each filament had a reasonably uniform deposition of metal. Detailed studies of the fabrication variables (pressure, temperature, and time) on the physical properties of the composites were conducted. Hot-pressing conditions were evaluated over the pressure range of 11–34 MN/m^2 (1500–5000 psi) at temperatures ranging from 750 to 1250°C for time periods of 5 min to 2 hr. Density, Young's modulus, and shear modulus approached their highest values at the upper range of these conditions. However, the best tensile-strength values were obtained when low forming pressures (14 MN/m^2) were used. The combined effect of low pressure and well-collimated fibers resulted in a minimum of fiber breakage and, thus, improved tensile-strength values ranging up to 800 MN/m^2 (115,000 psi) for composites containing 50 v/o of Thornel-50 yarn. The Young's modulus for the same composites was 240

GN/m^2 (35 × 10^6 psi). Other specimens fabricated with Thornel-75 yarn had tensile strength and modulus values of 830 MN/m^2 (120,000 psi) and 310 GN/m^2 (45 × 10^6 psi), respectively. The corresponding specific strength and modulus are 1.5 × 10^6 and 600 × 10^6 cm, respectively. The tensile specimens made from yarn having a modulus of 520 GN/m^2 failed in shear. The highest tensile-strength values were obtained when the tensile axis was parallel to the fiber axis. The tensile strength decreased with test angle as the failure mode changed from tensile failure of the fibers to shear failure at the matrix–fiber interface. The average transverse strength was 34 MN/m^2 (5000 psi), which is substantially below the value predicted on the basis of matrix tensile failure. The Young's modulus decreased as the test angle relative to the fiber direction increased; for the transverse direction, the modulus was 38 GN/m^2 (5.5 × 10^6 psi). The Young's modulus in unidirectional composites was lower than the values predicted by Whitney's micromechanics model for orthotropic filaments. This descrepancy apparently is not related to a decrease in fiber modulus as a result of processing or errors in the sonic measurements. The tensile strengths of composites in the direction of the fiber axis were in reasonable agreement with predicted values if consideration is given to the fact that hot-pressing reduced the average tensile strength of graphite filaments. For example, after pressing at 1050°C, the average tensile strength of Thornel-50 yarn decreased from 1850 to 1450 MN/m^2 (267,000 to 207,000 psi). This property was also reduced in Thornel-75 yarn by the fabrication process. Stress–strain measurements parallel to the fiber axis on unidirectional composites containing Thornel-75 graphite fibers showed that a 50 v/o fiber composite typically fails at 690 MN/m^2 (100,000 psi) and 0.25% strain. Plastic yielding of the matrix is evident at 0.05% strain. Failure in compression occurs at 690 MN/m^2 (100,000 psi). Compressive and tensile moduli are comparable.

Orthogonal laminates with three-, five-, and seven-ply configurations with orientations of 0, 45, and 90° were tested in tension to obtain the static modulus, stress–strain behavior, and ultimate tensile strength. These properties correlated well with predictions based on uniaxial properties. This correlation indicates that, to a first approximation, microcracks caused by the thermal-expansion differences of the laminate layers do not decrease the composite strength and modulus properties. The Young's modulus of nickel metal and composites containing Thornel-50 and Thornel-75 graphite fibers was measured to 1000°C. Short-time tensile-strength tests were made up to 1050°C on composites that contained Thornel-75 filaments. A tensile strength of 520 MN/m^2 (75,000 psi) was measured at 500°C, but this property decreased at the higher temperatures because of extensive fiber pullout. The trend for flexural strength was characterized by a maximum at 250°C; some specimens failed at 1250 MN/m^2 (180,000 psi) at this temperature. The stress-rupture characteristics were measured at 500°C. Failure occurred in less than 100 hr when stress exceeded 280 MN/m^2 (40,000 psi). The linear thermal expansion was measured between room temperature and 1000°C and found to be $0.5 \times 10^6/°C$ and $20 \times 10^6/°C$ in the longitudinal and transverse directions, respectively. The composite is highly anisotropic, and the thermal expansion is controlled by the thermal behavior of the graphite fiber. The effect of thermal cycling on the composite properties was determined for up to 1000 cycles and a maximum ΔT of 875°C. Considerable deformation of the composite occurred perpendicular to the fiber direction. Deformation is insignificant parallel to the fiber axis. The change in cross-sectional areas is accompanied by an increase in porosity and a decrease in all mechanical properties. This dimensional change is related to a "thermal ratchet" mechanism that is caused by thermal-expansion differences between the constituents, which causes plastic deformation in compression of the matrix at the higher temperature and a partial recovery of the deformation at lower temperatures. Improved interfacial bonding by means of TiC-coated graphite fibers eliminated this ratcheting. Furthermore, dimensional instability was not observed in composites containing nickel with 20 w/o of either chromium or iron.

Niesz and coworkers (Niesz, 1968; Niesz *et al.*, 1967, 1968; Kistler and Niesz, 1969) have studied both nickel- and cobalt-matrix–graphite-reinforced composites. The objective of their study was to evaluate the material for application to turbines in jet engines. Initial developments were conducted with cobalt in order to avoid the low-temperature eutectic and embrittling Ni_3P phase present in electroless nickel. Electroless plating was chosen rather than chemical-vapor deposition and electrodeposition because of ease in operation, high throwing power, and the ability to coat uniformly the individual fine fibers of Thornel with a 7-μm-diam cross section. Techniques were developed for pretreatment of the Thornel yarn in order to enhance adhesion of the metal. Excellent uniform coatings were obtained and dense bodies were fabricated through hot isostatic pressing.

Subsequently, nickel was substituted as a matrix under similar test conditions. Improvements included a specialized plating bath to yield phosphorus-free nickel and modification to the fabrication cycle to avoid filament damage. Again, excellent uniform coatings were obtained, and finished bodies were on the order of 98% of theoretical density. Test data for these composites, which were 50 v/o graphite, are shown as a function of temperature in Fig. 43, together with a comparison of other high-temperature alloys. The rule-of-mixtures strengths were not obtained, and the gap widened as temperature increased. Filament damage and poor metal–filament bonding were attributing factors. Other concerns of the investigators included nickel–carbon reaction at 1800°F (projected use temperature) and oxidation of the graphite to carbon monoxide by bulk diffusion of oxygen through the nickel as well as along the length of the filaments from exposed ends. After sufficient exposure at temperature, the filaments simply disappeared, leaving open pores in the metal.

In a series of investigations sponsored by Rolls Royce, Ltd., Jackson and Marjoram (1968, 1970) have studied electrodeposited nickel–graphite composites. Exposure of coated fibers at and above 1000°C caused severe degradation in strength, as shown in Fig. 45. This was attributed to the reaction of nickel to promote recrystallization of the graphite, which was characterized by a change in crystallite size and appearance of additional lines in an X-ray diffraction pattern. Similar effects were also observed, at much lower temperatures, however, with cobalt (700°C), nickel–chromium (500°C), platinum (700°C), and copper (700°C). Additional studies (Braddick *et al.*, 1971) on nickel–graphite were directed toward mechanical properties and fracture modes by means of optical microscopy and scanning electron microscopy. Dense bodies were fabricated by first electrodepositing nickel on 10,000-filament tow Courtaulds fiber followed by hot-pressing in flat dies. Very low strengths were attained and were again attributed to poor bonding, brittle nickel matrix, and broken fibers. As in previous investigations, oxidation of the fibers occurred rapidly at 600°C, with complete disappearance in 5 hr. This rate was faster than with graphite fibers alone, suggesting that nickel causes oxygen to diffuse through it in a highly activated monatomic form.

Baker *et al.* (1971) developed electrodeposition techniques for the production of composites of graphite with nickel, aluminum, lead, and copper. In the case of nickel, with a sulfamate bath, successful penetration of 1000–2500-filament tow graphite was accomplished. Difficulty was experienced with 5000-filament fiber. Fiber volume fractions up to 50 v/o were attained. Composites were fabricated by liquid-phase diffusion bonding using a copper–

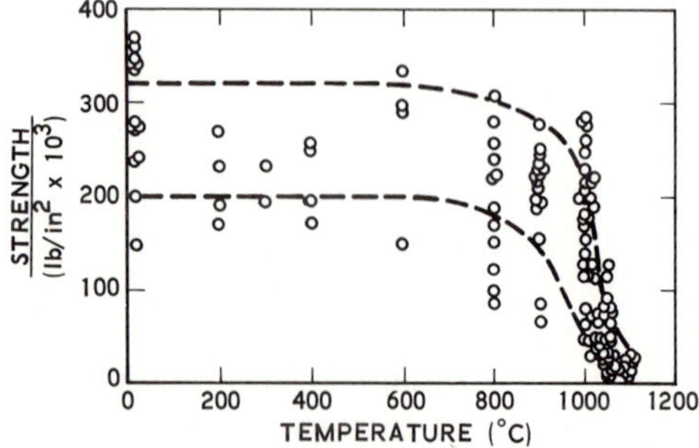

Fig. 45. Strength of nickel-coated graphite fibers after exposure at elevated temperatures for 24 hr.

silver-solder material. Properties were not reported. Tensile properties of electroplated Ni–10Co on Grafil HT carbon fibers were studied as a function of the coating thickness at room temperature by Perry *et al.* (1970). They found that the Young's modulus is defined by the rule of mixtures, and the fracture of the coated fibers is governed by the fibers themselves and not by the metal coating.

In summary, it has been shown that electrodeposition is an excellent technique for penetrating and coating high-filament graphite tow with nickel and nickel alloys. However, because of porosity, fiber damage, and poor interfacial bonding, problems have been encountered in the consolidation of these coated fibers into dense composites with rule-of-mixtures mechanical properties. High-temperature limitations are related to the degradation of fibers from nickel-induced recrystallization and oxidation by diffusion of monatomic oxygen through the nickel. The use of nickel-alloy matrices, graphite-fiber surface treatment, and liquid-phase hot-compacting should be considered in future research to alleviate some of these concerns.

D. Copper–Graphite

Interest in copper–graphite alloys and their application to the electrical industry dates back to 1923 (Wulff, 1942). These alloys have usually been prepared by powder-metallurgy methods with a variety of copper, graphite, and carbon powders. Higher strength, improved dimensional stability, resistance to arc erosion, better sliding friction, and lower-cost materials are some of the objectives that have motivated their development.

High-strength carbon fibers were not used to form copper–graphite alloys until 1961. Powders of cupronickel were mixed with chopped fibers in a ball mill and then sintered to form the composite. The as-sintered compacts showed low strengths and large amounts of porosity. Compacts that were hot-rolled or hot-forged after sintering had much lower amounts of porosity and increased strength but were not as strong as the matrix material alone.

In the preparation of cupronickel–graphite composites, Sara (1971) first electroplated high-strength fibers with a nickel coating and then electocladded copper on top of the nickel. The plated graphite yarns were then hot-compacted at 900°C. The low modulus of 180 GN/m^2 (25.9×10^6 psi) and tensile strength of 380 MN/m^2 (55,800 psi) obtained were attributed to delaminations, poor metal-matrix distribution, and porosity.

Electroless plating of 5000-filament carbon tows with 1-μm copper from a proprietary solution has been reported by Judd (1970). Separation of the filaments while being drawn through the bath caused incomplete coating at the center. No composites were prepared, and mechanical properties were not determined.

Howlett *et al.* (1971) also used electroplating for the preparation of copper–graphite composites. Type II carbon-fiber tows (10^4 filaments) were first plated with copper from a cyanide bath, resulting in a porous and uneven surface coating on the fibers at fiber volume fractions from 0.25 to 0.50. The composites, which were hot-pressed in an inert atmosphere, showed little porosity but some oxide particles in the matrix. In Fig. 46, the strengths of tensile samples machined from the compacts are compared with the rule-of-mixtures prediction. The samples from the first series of pressings show no strengthening. The second series yielded samples at low volume fraction that show some fiber reinforcement, but the strengthening degrades at higher fiber volume fractions, probably because of fiber interactions during pressing.

In studies of liquid-infiltration methods for composite fabrication, the effect of alloying additions to copper on the wetting of graphite has been investigated by Mortimer and Nicholas (1970, 1971), using sessile drop experiments. Two types of graphite samples, bulk HX 30 with a crystal size of 1000 Å and vitreous graphite with a crystal size of 15–25 Å, were selected for the study of this variable on the wetting properties. Twenty different alloys of 1 at. % copper each were made *in situ* and melted over the graphite in vacuum at 1145°C (mp Cu—1085°C). Only vanadium and chromium additions caused the copper to wet the vitreous graphite surface, whereas only chromium caused the alloy to wet the bulk graphite surface. Metallographic examination showed that wetting occurred only when the alloy additions formed coherent and adherent reaction products with the graphite. This type of carbide product was found most often with alloy additions from the fourth period of the periodic table. It was concluded that, for alloy

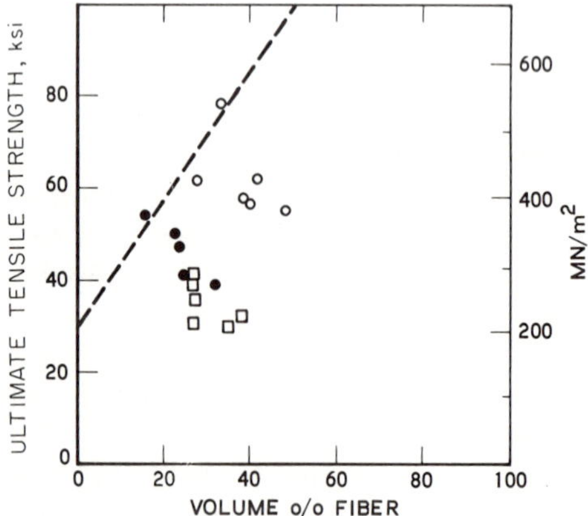

Fig. 46. Tensile strengths of reinforced electroplated and hot-pressed copper: □ series 1, ○ series 2, ● series 3.

ingredient X to promote wetting of metal M on the surface of graphite C, the interfacial free energy of M on XC must be less than the surface free energy of C.

Baker *et al.* (1971) prepared copper–graphite composites by first electroplating from a sulfate bath and then hot-compacting at 600°C. Fiber volume fractions of 0.30 to 0.50 were obtained with erratic tensile-strength results. Strengths of 490 MN/m^2 (71 ksi) were easily obtained, but some values were as high as 910 MN/m^2 (132 ksi) at room temperature and 560 MN/m^2 (81 ksi) at 400°C.

From these results, it can be speculated that high-strength copper materials can be fabricated with no accompanying major loss in electrical conductivity even at slightly elevated temperatures.

E. Magnesium–Graphite

Magnesium-matrix–graphite-fiber composites have been fabricated by a hot-pressing technique as reported by Levitt *et al.* (1972). Their wetting studies revealed that uncoated graphite fibers would not be wetted by molten magnesium directly. Therefore, they attempted to precoat the fibers. Graphite fibers precoated either with titanium by plasma spraying or physical-vapor deposition or with electroless nickel demonstrated good wetting by the molten magnesium. The physical-vapor-deposited titanium coating was chosen for sample fabrication studies because of its lower density and higher melting point, and because no brittle intermetallic compounds are formed between magnesium and titanium. The titanium coating was applied to Modmor I fibers mounted on a small frame in the chamber of an electron-beam welder. Titanium was vaporized and allowed to condense on the fiber surface to a coating thickness of approximately 2 μm. This created an increase in the fiber cross-sectional area greater than that of the fiber itself. With magnesium powder and titanium-coated fibers, composites were fabricated by hot-pressing at 670°C and 280 KN/m^2 (40 psi) pressure in an argon atmosphere. An ultimate strength of 230 MN/m^2 (33,600 psi) and modulus of 80.0 GN/m^2 (11.7×10^6 psi) were obtained for a fiber volume fraction of 11 v/o. These values are 75 and 88% of the respective rule-of-mixtures predictions. The low strengths, a sample of two, were attributed to porosity of the specimens and fiber misalignment. In addition, the composite contained 20 v/o titanium derived from the coating. The low fiber volume fraction was attributed to the thick nature of the titanium coating. It was concluded that a much thinner coating would be required in order to produce a viable composite system.

Magnesium–graphite samples have also been prepared by the batch-process liquid-metal-infiltration technique previously described. One-hundred-fifty

TABLE XIII

Summary of Mechanical Properties of Magnesium–Graphite Composites

Composite	Strength (psi)	Strength (MN/m^2)	Modulus of elasticity (10^{-6}psi)	Modulus of elasticity (GN/m^2)	Density (gm/cm^3)	Strength/density (10^{-6}cm)	Modulus/density (10^{-6}cm)
Mg–42 v/o Thornel-75	65,000	450	26.6	184	1.77	2.5	1000
Mg–ZK60A	50,000	345	6.5	45	1.80	1.9	250

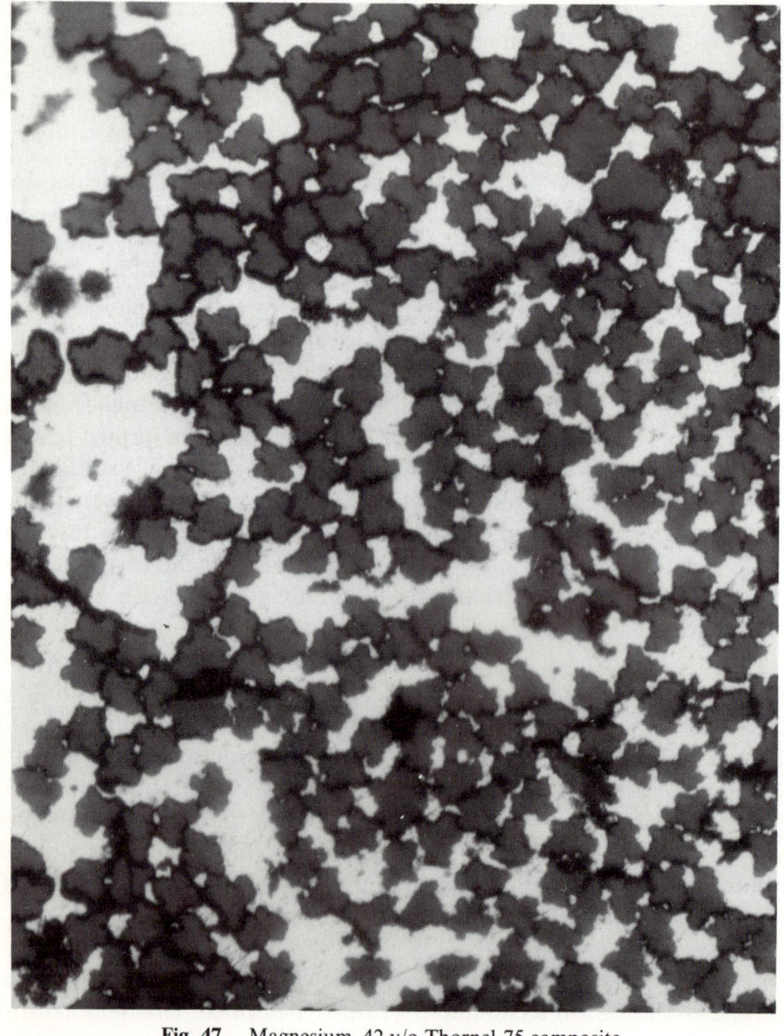

Fig. 47. Magnesium–42 v/o Thornel-75 composite.

strands of Thornel-75 graphite fiber were infiltrated with commercially pure magnesium. An ultimate tensile strength of 450 MN/m^2 (65,000 psi) and modulus of 185 GN/m^2 (26.6×10^6 psi) were obtained for an average fiber volume fraction of 42 v/o. These values are approximately 53 and 85 % of the respective rule-of-mixtures predictions. In Table XIII, the mechanical properties are summarized and compared with the strongest commercial magnesium-based alloy. The microstructure of this composite is shown in Fig. 47.

The data in Table XIII show that the specific modulus of the magnesium–graphite composite is four times that of the ZK60A magnesium-based alloy. The magnesium–graphite composite is in a preliminary stage of development, and further work is being directed toward the use of a high-strength magnesium alloy as a matrix. It is predicted that magnesium–graphite composites with strengths in excess of 700 MN/m^2 can be developed. On the basis of experience with aluminum–graphite composites, it is anticipated that the magnesium–graphite composite will have excellent mechanical properties at temperatures up to 600°C. The combination of the very high fiber strength and stiffness and low density of both the fibers and the magnesium matrix offers the possibility of a material with the highest specific strength of any known metallic structural material.

F. Lead–Graphite

Lead and lead alloys are of great importance in the chemical, battery, building, and bearing industries because of their high resistance to corrosion in many chemical environments, accoustical damping, and excellent antifrictional properties. However, the low strength of lead and its alloys imposes severe limitations on their use. A solution to this problem appears to be through graphite-fiber strengthening. Potential applications for fiber-reinforced lead include structural members in chemical processing plants, plates in lead–acid storage batteries, wall-panel sound insulation, and high-load-carrying-capacity, self-lubricated bearings. A major cause of failure in lead–acid batteries is buckling of the lead plates. Reinforcement of the plates with graphite fibers would prevent buckling. Such materials could lead to the design of improved lead–acid batteries of reduced size and weight for a given capacity. Lead and bulk graphite are excellent antifriction materials, but their use in many bearing applications is restricted by their lack of strength. Bearings made of lead–graphite composites may have superior properties to currently available bearing materials, such as phosphor bronze, and may ultimately be less expensive.

Lead–graphite composites have been successfully prepared by both liquid-infiltration and electroplating techniques. Hot-compacting has yielded

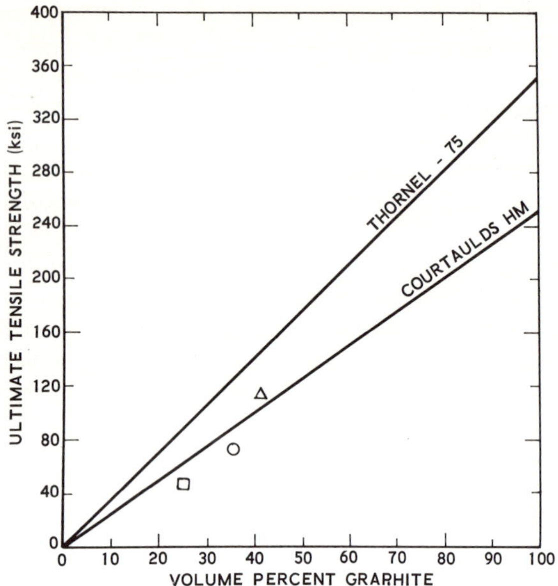

Fig. 48. Rule-of-mixtures relationship for lead–graphite composites: △ 41 v/o Thornel-75, ○ 35 v/o Courtaulds HM, □ 25 v/o Courtaulds HM.

dense bodies in both cases. The rule-of-mixtures relationship is shown in Fig. 48 for both a 350-ksi strength fiber (Thornel-75) and a 250-ksi strength fiber (Courtaulds HM) in a pure lead matrix.

Lead–graphite composites have been fabricated by first infiltrating bundles of fibers with molten lead and then hot-compacting into larger shapes. A transverse section of a lead-matrix, 35 v/o Courtaulds HM fiber composite is shown in Fig. 49. The uniaxial strength of this composite was found to be 497 MN/m^2 (72,000 psi), 82% of the rule of mixtures, and the tensile modulus 120 GN/m^2 (17.4×10^6 psi). Another data point was obtained with a 25 v/o Courtaulds HM fiber composite having a tensile strength of 290 MN/m^2 (42,000 psi). Tensile fractures of the composite at 150× and 1500× magnifications are shown in Fig. 50. Examination of the fracture surfaces and the absence of fiber pullout show that a strong bond exists between the lead matrix and the graphite fibers.

Similarly, with Thornel-75 graphite yarn, a 41 v/o composite yielded an ultimate tensile strength of 717 MN/m^2 (104,000 psi), 73% of the rule of mixtures, and a corresponding modulus of 200 GN/m^2 (29×10^6 psi). A transverse microstructure is shown in Fig. 51.

A comparison of the properties of the two composites with pure lead and a high-strength lead alloy is summarized in Table XIV. Marked improvement over pure lead is readily apparent. Of special note are the specific-strength

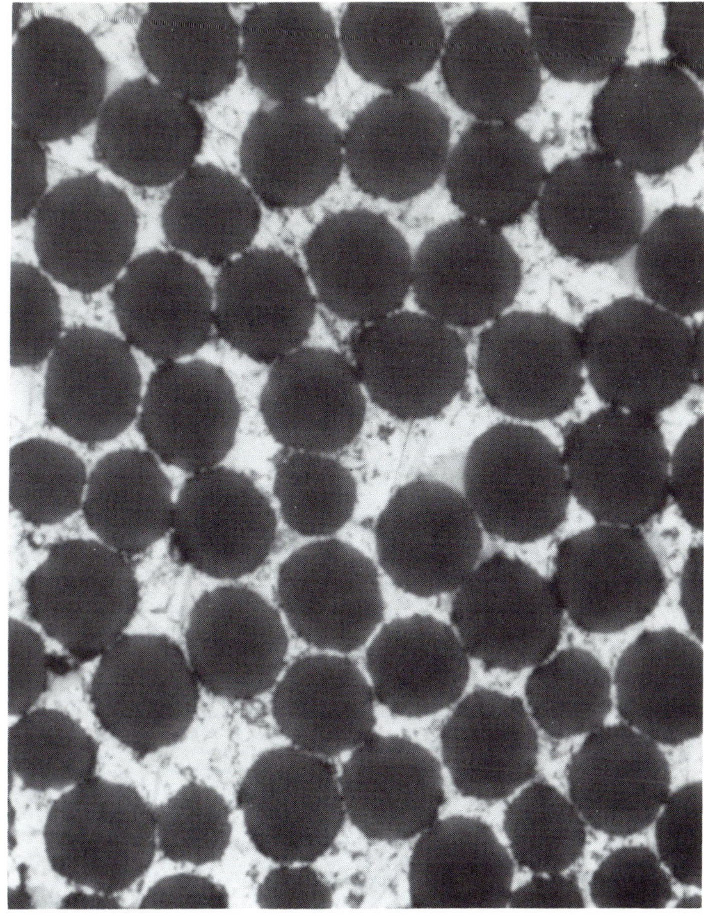

Fig. 49. Transverse section of lead-matrix–35 v/o Courtaulds HM fiber.

values, which put the lead composite in the category of the medium-strength steels.

The formation of lead–graphite composites by electroplating lead from a fluoroborate bath onto a 1000–2500-filament graphite tow has been described by Baker, *et al.* (1971). Hot-compacting was performed at 300°C in vacuum. For a 250-ksi fiber, composites with an ultimate tensile strength of 330 MN/m^2 (47.8 ksi), which is 62% of the rule of mixtures, were obtained.

From this work, it can be concluded that extremely high-strength, high-modulus, lead-based materials can be produced from composite technology, and it can be expected that they will ultimately gain wide usage throughout industry.

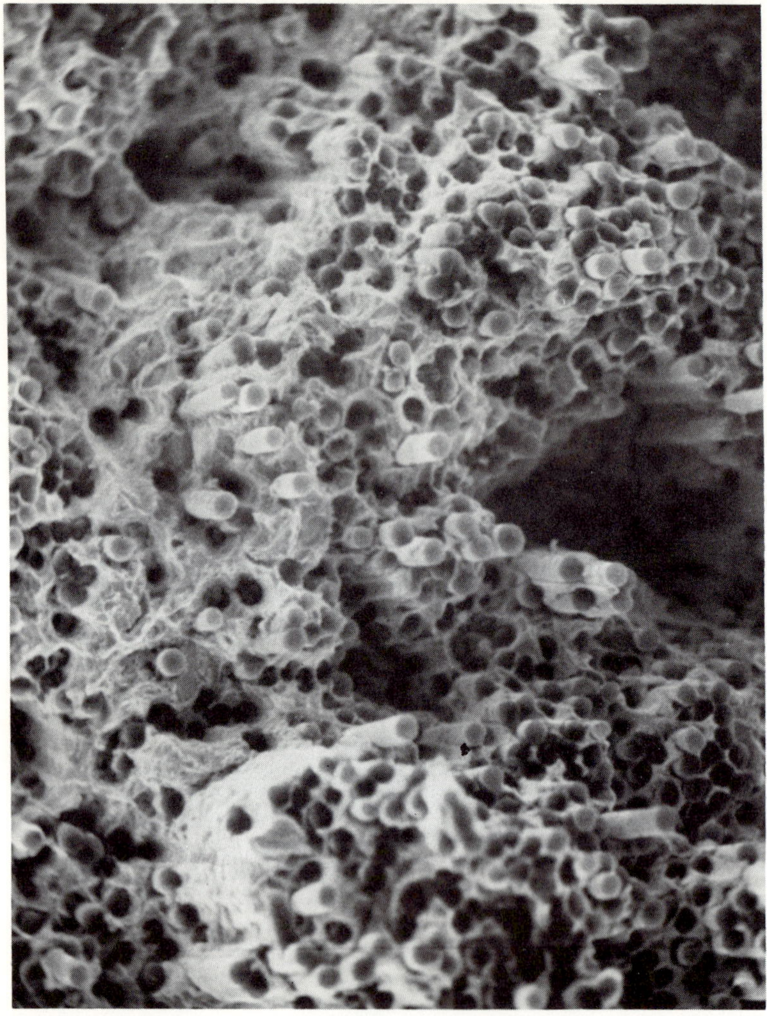

Fig. 50. Tensile fracture of lead-matrix–35 v/o Courtaulds HM fiber.

G. Zinc–Graphite

The single largest industrial use of zinc metal is in the galvanizing of iron and steel for corrosion protection. However, the major structural use of zinc is in zinc-alloy die castings in automotive, home-appliance, and machinery products. The most severe limitation in the use of zinc alloys is the loss of

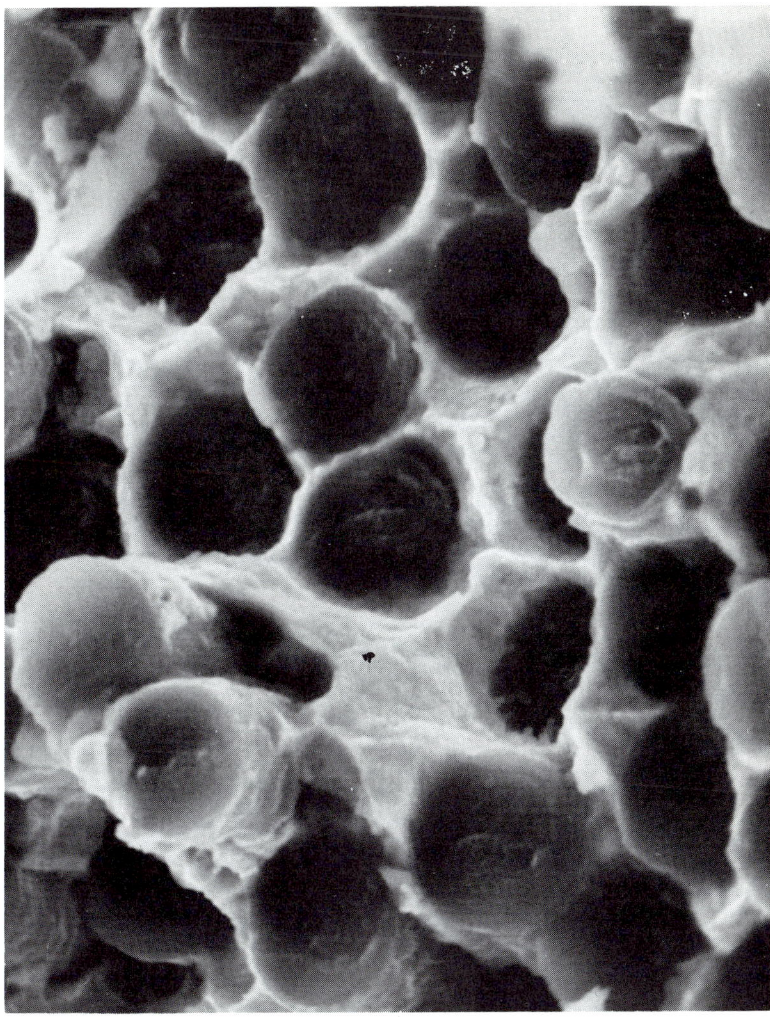

Fig. 50. (cont.)

strength at slightly elevated temperatures (200–300° F) and its poor creep resistance. Significantly, it is for such applications that zinc-graphite composites appear to be so attractive. A rule-of-mixtures relationship for a pure zinc matrix and Thornel-75 graphite fiber is shown in Fig. 52. It is apparent that in order to equal the strength level of commercial zinc die-casting alloys (276–324 GN/m^2 or 41–47 ksi), a fiber volume fraction of only 0.12 would be required.

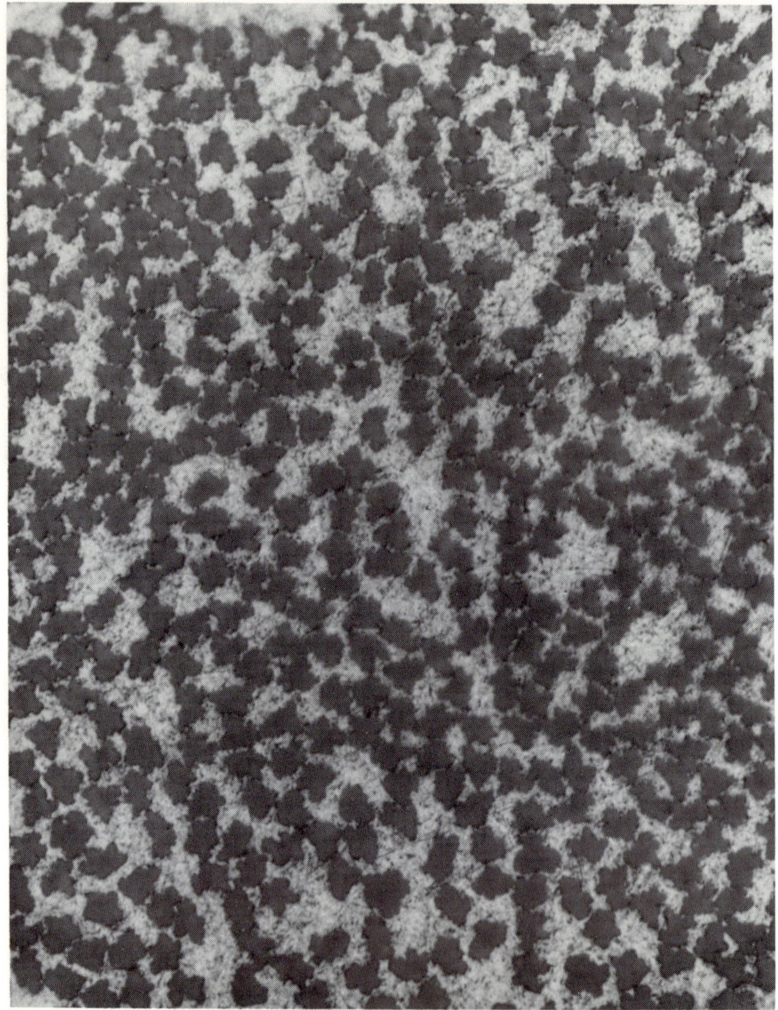

Fig. 51. Microstructure of lead–41 v/o Thornel-75 graphite composite.

From a liquid-metal infiltration technique coupled with hot-compacting, a fully dense zinc–graphite composite containing 35 v/o Thornel-75 yarn was obtained. From an average of four samples a tensile strength of 759 MN/m^2 (110 ksi), 88% of rule-of-mixtures predictions, and a tensile modulus of 117 GN/m^2 (16.9×10^6 psi) were obtained. A transverse microstruc-

TABLE XIV

Room-Temperature Properties of Lead–Graphite Composites

COMPOSITE	STRENGTH, lb/in.^{2}	MODULUS OF ELASTICITY, 10^{-6} lb/in.^{2}	DENSITY lb/in.^{3}	STRENGTH/ DENSITY, 10^{-6} in.	MODULUS/ DENSITY, 10^{-6} in.
PURE LEAD	2,000	2.0	0.41	0.005	4.9
LEAD-BASE BEARING ALLOY (75 Pb-15 Sb-10 Sn)	10,500	4.2	0.35	0.030	12.0
LEAD-GRAPHITE COMPOSITE 41 VOL % THORNEL-75 FIBERS	104,000	29.0	0.270	0.385	107.0
LEAD-GRAPHITE 35 VOL% COURTAULDS HM	72,000	17.4	0.28	0.26	62.3

ture of this composite is shown in Fig. 53. Excellent spacing of the fiber is indicated. A comparison of the properties of the zinc–graphite composite with the highest-strength zinc diecasting alloy (AG 40A) can be found in Table XV. A threefold improvement in strength and 70% increase in modulus are indicated. More striking are the specific values due to the much lower density of the composite. Again, as in the case of lead, these composite materials are the equivalent of steels.

With such a marked increase in tensile strength and a well-defined modulus (zinc alloys do not have one), these zinc-composite materials will undoubtedly find their way into many applications.

H. Beryllium–Graphite

Beryllium has great potential as a structural material in aerospace vehicles, but, unfortunately, it suffers from inadequacies in its mechanical properties. Beryllium–graphite composites may resolve some of these limitations, which include low elevated-temperature strength, low elevated-temperature stress-rupture strength, and very poor resistance to crack propagation. The specific strengths of various alloys and composites are plotted in Fig. 44 as a function of temperature. Also shown are predicted values for Ti–50 v/o B, Al–45 v/o G, Be–45 v/o G, and NiCr–50 v/o G. For short-time exposures to 1800°F, this beryllium–graphite composite is shown to have a specific strength four to

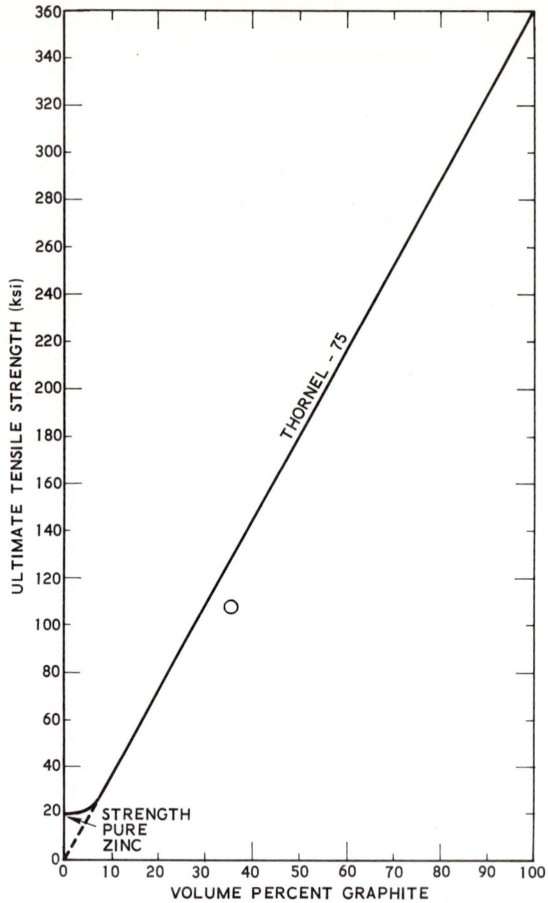

Fig. 52. Rule-of-mixtures relation for pure zinc matrix–Thornel-75 graphite fiber.

five times better than the predicted NiCr–G composite and 30 times better than most of the commercially available nickel superalloys such as MAR-M-200, Rene 41, and TD nickel–chrome. In addition, the predicted specific modulus for the composite is 15×10^8 cm (6×10^8 in.), as compared to only 1.5×10^8 cm (ten times greater) for the superalloys. On the basis of this potential, the development of beryllium–graphite composites is receiving attention.

In a preliminary study, fabrication of beryllium–graphite composites by diffusion bonding was attempted by Harrigan (1972). The rule-of-mixtures

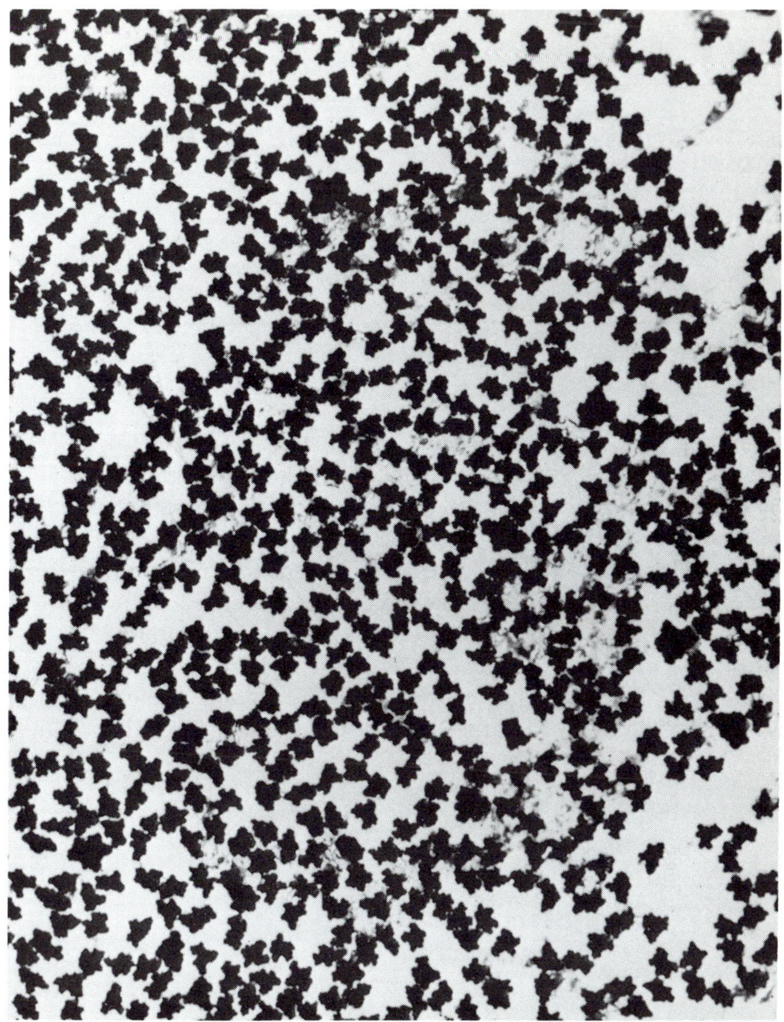

Fig. 53. Transverse microstructure of zinc–graphite composite.

relationship for strengthening beryllium sheet by using a 2.0-GN/m^2 (300-ksi) graphite filament (Hough) is shown in Fig. 54. It can be hypothesized that, with fabrication techniques similar to those for boron–aluminum, an 1100 MN/m^2 (160 ksi) composite could be produced for a 45 v/o fiber loading. Beryllium foils were coupled with the Hough monofilament and hot-

TABLE XV

Summary of Mechanical Properties of Zinc and a Zinc–Graphite Composite

System	Strength (psi)	Modulus of elasticity (10^{-6} psi)	Density (lb/in.^{3})	Strength/density (10^{-6} in.)	Modulus/density (10^{-6} in.)
Z–35 v/o Thornel/75	110,900	16.9	0.191	0.58	88.5
Alloy AG 40A	41,000	10.0	0.240	0.17	41.7

pressed with cycles similar to those used for boron–aluminum composites. Parameters studied were time, temperature, and pressure. Aluminum foil between the beryllium was evaluated as a diffusion enhancer. Bonds of the type shown in Fig. 55 were obtained. Severe delamination occurred frequently and was attributed to residual stresses. Fiber spacing appeared to be critical, and complete crushing of the fiber was often experienced. Although no satisfactory composite material was produced, directions for future research were delineated.

Goddard (1973) has reported on liquid infiltration as a method for incorporating beryllium into Thornel-50 yarn and Hercules HM fiber. Initial tests

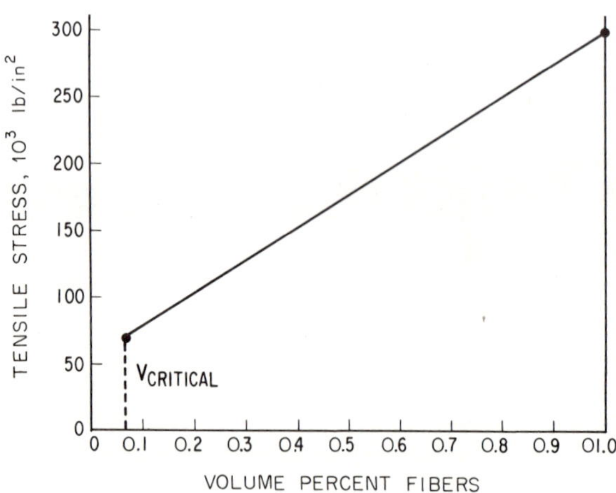

Fig. 54. Rule-of-mixtures relationship for strengthening PS-20 beryllium sheet with Hough filament.

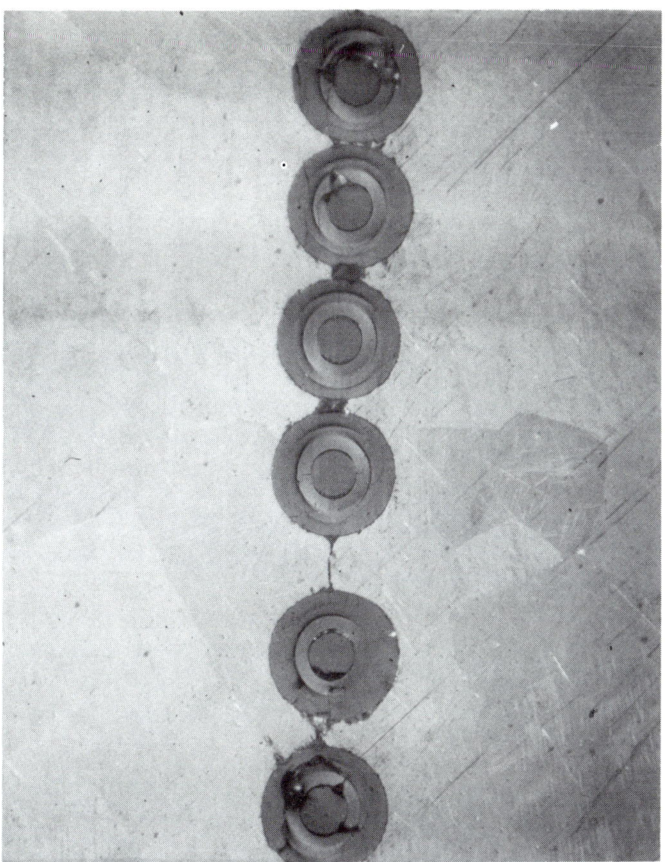

Fig. 55. Diffusion-bonded beryllium–Hough fiber composite.

were performed with aluminum-infiltrated Thornel-50. After very short exposures to molten beryllium (15–30 sec), the specimens were sectioned and studied metallographically. It was ascertained that the Thornel-50 yarn was not significantly attacked, either by the aluminum or the beryllium, to cause a significant decrease in cross section of the fiber (Fig. 56). Microprobe analysis and mechanical properties have not yet been reported.

On the other hand, Hercules untreated fiber was immersed in the molten beryllium for only 5 sec, and vigorous attack was observed (Fig. 57).

It can be concluded from this preliminary work that high-modulus graphite fibers must be treated or coated with a diffusion-barrier element in order to prevent attack by molten beryllium. Candidate elements with higher melting points than beryllium include iron, nickel, cobalt, and titanium.

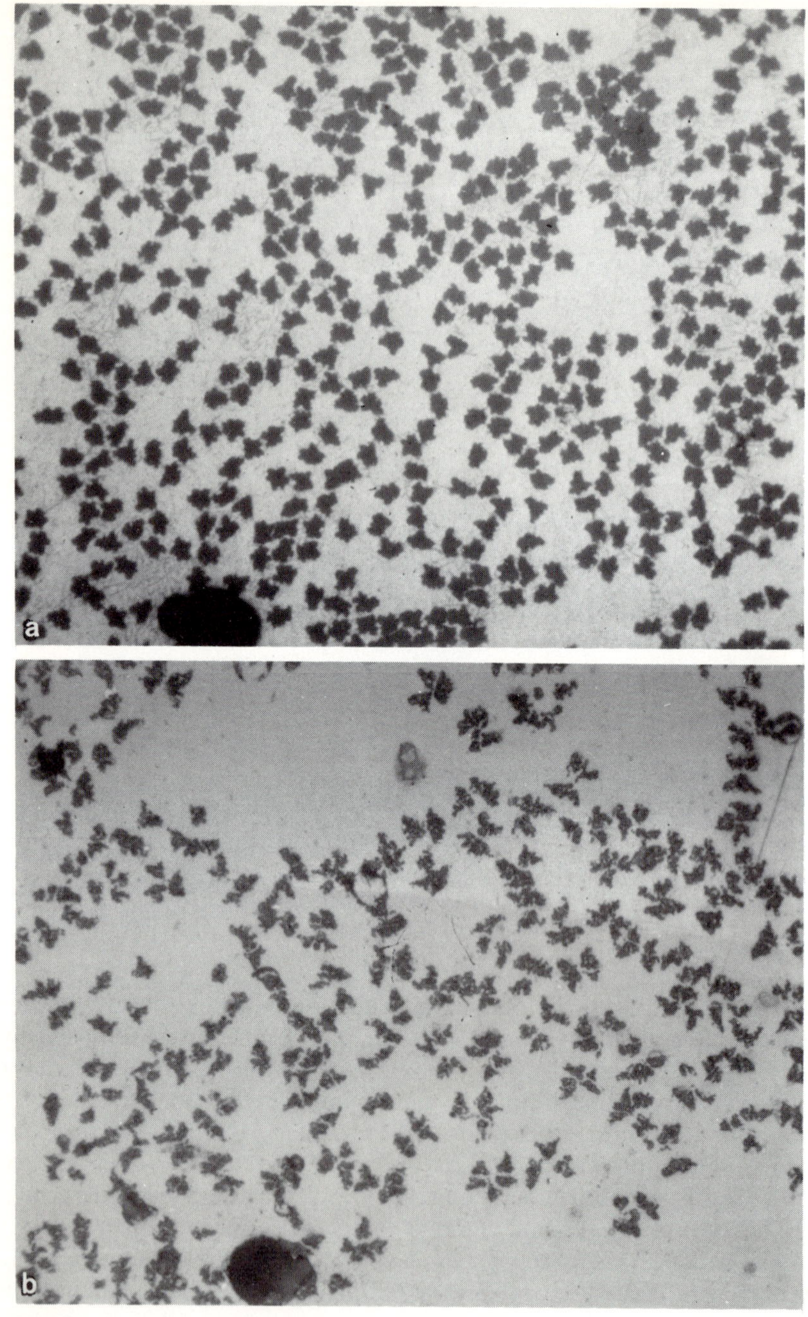

Fig. 56. Transverse microstructure of Thornel-50–aluminum composite (a) before and (b) after exposure to molten beryllium.

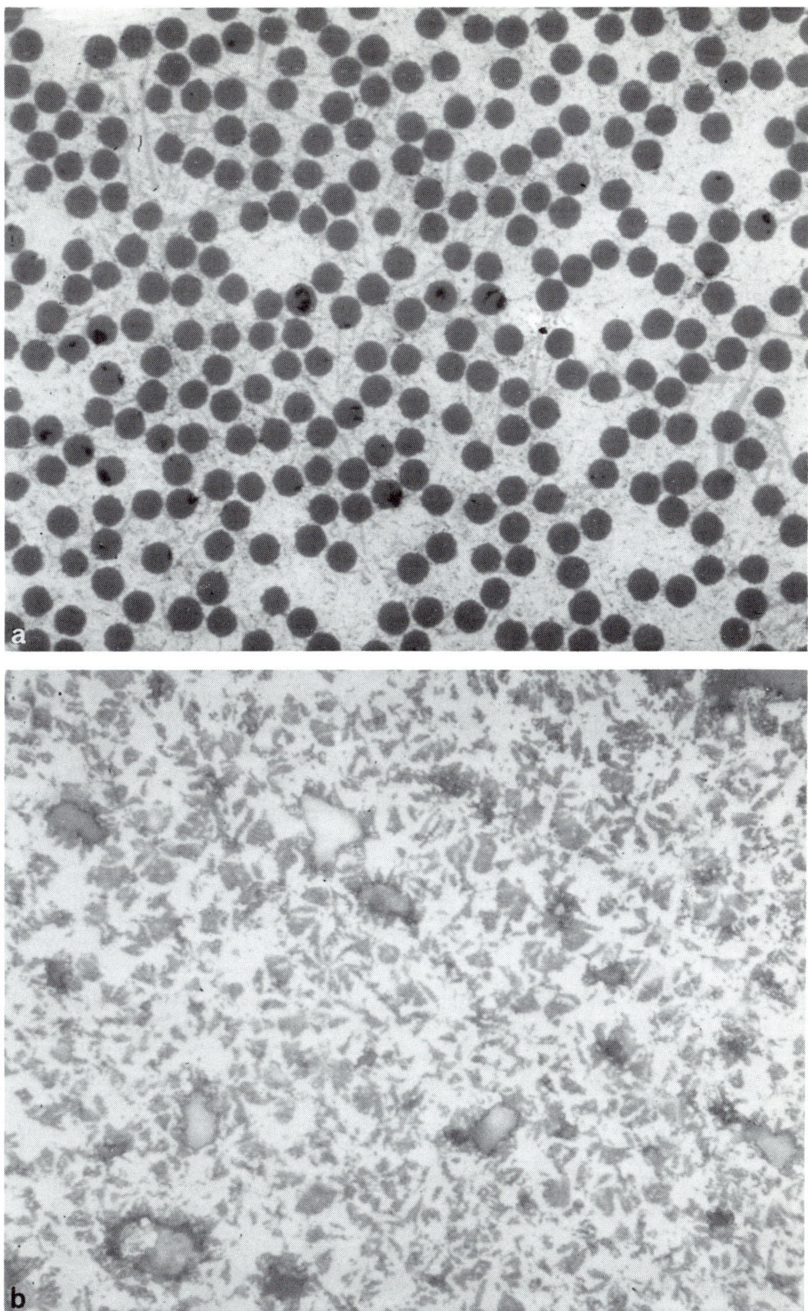

Fig. 57. Transverse microstructure of Hercules graphite fiber (a) before and (b) after immersion in molten beryllium.

Acknowledgments

The author wishes to acknowledge invaluable contributions from the following members of his staff: Dr. Roger T. Pepper, principal investigator in the development of aluminum–graphite composites, presently at Fiber Materials Inc., Biddeford, Maine; Dr. John Upp, mechanical testing of composites, presently at U.S. Polymeric Inc., Santa Ana, California; Mr. L. Roy Davis, fiber testing and microscopy, presently President, NETCO, Long Beach, California; Dr. W. Harrigan, secondary fabrication and elevated-temperature testing; Dr. D. Goddard, composite process development; Miss Pamela Sullivan, fiber testing and microscopy; and Mr. D. Rosenthal, chemical analysis.

References

An Annotated Bibliography on Subjects Related to Graphite-Fiber-Reinforced Metal-Matrix Composites (1969). AFML-TR-54, Lockheed Missiles and Space Co.

Bacon, R. (1960). *J. Appl. Phys.* **31**, 283.

Bacon, R., and Schalamon, W. A. (1968). *Amer. Chem. Soc. Polym. Reprints* **9**, 1338.

Bacon, R., and Silvaggi, A. (1968). *Carbon* **6**, 199.

Bacon, R., and Silvaggi, A. F. (1971). *Carbon* **9**, 321–325.

Bacon, R. (1966). High-Strength High-Modulus Carbon Fibers. AFML-TR-66-334, Air Force Mater. Lab., Wright-Patterson AFB, Ohio. Part I, AE-811531L (1966). Part II, AD-831999L (Dec., 1967). Part III, AD-845354 (May 1968). Part IV, AD-865296 (April, 1969). Part V (Dec., 1970). Part VI (July, 1971).

Badami, D., Joiner, J., and Jones, G. (1967). *Nature* (*London*) **215**, 386.

Baker, A., Martin, A., and Bache, R. (1971). *Composites* **2** (3).

Baker, A., Shipman, C., and Jackson, P. (1972a). *Fiber Sci. Technol.* **5** (3).

Baker, A., Cripwell P., Jackson, P., and Shipman, C., (1972b). *Fiber Sci. Technol.* **5** (4).

Bibliography of Fibers and Composite Materials 1969–1972 (1972). MCIC-72-09 Battelle Memorial Institute, Columbus, Ohio.

Blankenburgs, C. (1969). *J. Aust. Inst. Met.* **14** (4).

Braddick, D. M., Jackson, P. W., and Walker, P. J. (1971). *J. Mater. Sci.* **6**, 419–426.

Brown, H. (1968). Exploratory Development Leading to High-Strength High-Modulus Continuous Graphite Yarns. AFML-TR-68-75, AD-836668, Air Force Mater. Lab., Wright-Patterson AFB, Ohio.

Buschow, A., Esola, C. *et al.* (1969). Electroforming of Aluminum Composite Structures by Codeposition of High-Strength High-Modulus Fibers or Whiskers. CR-66749, Nat. Aeronaut. Space Administration.

"Composite Materials" (1965). Inst. of Metallurgists. American Elsevier, New York.

Davis, L. W., and Bradstreet, S. (1970). "Metal and Ceramic Matrix Composites." Cahners, Boston, Massachusetts.

DeLamotte, E., Phillips, K., Perry, A., and Killias, H. (1972). *J. Mater. Sci.* **7**, 346–349.

Didchenko, D. (1968). Carbon and Graphite Surface Properties Relevant to Fiber-Reinforced Composites. AFML-TR-68-45, Union Carbide Corp., Carbon Products Div., Lawrenceburg, Tennessee.

Diefendorf, J. (1969). Refractory Composite Research at RPI. In DMIC S-25.

Dietz, A. G. H. (1964). *Intern. Sci. Techn.* **58**, 42–52.

Evans, J. and Braddick, D. (1971). *Corros. Sci.* **11**, 611.

Ezekiel, H. (1971). High-Strength High-Modulus Graphite Fibers. AFML-TR-70-100, Air Force Mater. Lab., Wright-Patterson AFB, Ohio.

Fiber Composite Materials (1965). ASM.

Fiber-Reinforced Metal-Matrix Composites: 1964–1966 (1967). DMIC 241, Defense Mater. Inform. Center, Battelle Memorial Inst., Columbus, Ohio.

Fiber-Reinforced Metal-Matrix Composites: 1967 (1968). DMIC S-21, Defense Mater. Inform. Center, Battelle Memorial Inst., Columbus, Ohio.

Fiber-Reinforced Metal-Matrix Composites: 1968 (1969). DMIC S-27, Defense Mater-Inform. Center, Battelle Memorial Inst., Columbus, Ohio.

Fiber-Reinforced Metal-Matrix Composites: 1969–1970 (1971). DMIC S-33, Defense Mater. Inform. Center, Battelle Memorial Inst., Columbus, Ohio.

Flom, D., Gorowitz, B., Krutchen, C. M., Roberts, B. W., and Sliva, D. E. (1972). Exploratory Development of High-Strength High-Modulus Graphite Filaments. AFML-TR-72-24, Air Force Mater. Lab., Wright-Patterson AFB, Ohio.

Fourdeux, A., Perrett, R., and Ruland, W. O. (1970). General Structural Features of Carbon Fibers. Rep. 61/70, Union Carbide Eur. Res. Ass., Brussels, Belgium.

Galasso, F. (1969). "High-Modulus Fibers and Composites." Gordon and Breach, New York.

Galasso, F. (1972). *Ceram. Age.*, May, 26–30.

Goddard, D. (1973). Infiltration of Graphite Yarn with Beryllium. ATR-73(8136)-1, The Aerospace Corp. (to be published).

Goddard, D. M., Pepper, R. T., Upp, J. W., and Kendall, E. G. (1972). *Welding Res. Suppl. 178-s* **51** (4).

Harrigan, W. C. (1972). Development of Beryllium-Graphite Composites. TOR-0172(2760)-1, The Aerospace Corp., El Segundo, California.

Hough, R. L. (1970). Development of Manufacturing Process for Large-Diameter Carbon-Base Monofilaments by Chemical Vapor Deposition. CR-72770, Nat. Aeronaut. and Space Administration.

Hough, R., and Richmond, R. (1971). Improvement of Chemical Vapor Deposition Process for Production of Large-Diameter Carbon-Base Monofilaments. CR-120902, Nat. Aeronaut. and Space Administration.

Howlett, B., Minty, D., and Old, C. (1971). *Plast. Inst. (London)* **14**.

Jackson, P. W. (1969). *Met. Eng. Quart.* **9**, 22–30.

Jackson, P. W., and Marjoram, J. (1968). *Nature (London)* **218** (5136), 83–84.

Jackson, P. W., and Marjoram, J. R. (1970). *J. Mater. Sci.* **5**, 9–23.

Jackson, P., Braddick, D., and Walker, P. (1972). *Fiber Sci. Technol.* **5** (3), 219–236.

Johnson, W., and Watt, W. (1967). *Nature (London)* **215**, 384.

Judd, N. C. W. (1970). *Composites* **1** (6), 325–390.

Kelly, A., and Davies, G. (1965). *Met. Rev.* **10**, 37.

Kendall, E. G. (1968). *Metalworking News* July 1, 9–10.

Kendall, E. G., and Pepper, R. T. (1970a). Aluminum-Graphite Composites. TR-0066(9250-03)-3, The Aerospace Corp., El Segundo, California.

Kendall, E. G., and Pepper, R. T. (1970b). High-Temperature Properties of Aluminum—Graphite Composites. TR-0059(6250-10)-9, The Aerospace Corp., El Segundo, California.

Kendall, E. G. *et al.* (1971). A Preliminary Investigation of Joining Methods for Aluminum–Graphite Composites. TR-0172(9250-03)-1, The Aerospace Corp., El Segundo, California.

Kistler, C. W., and Niesz, D. (1969). Development of Carbon-Filament-Reinforced Metals. AD-854149, Battelle Memorial Inst., Columbus, Ohio.

Koppenaal, T., and Parikh, N. (1961). Fiber-Reinforced Metals and Alloys. ARF 2221-3, Armour Res. Foundation., Chicago, Illinois.

LaIaconna, F. P. (1971). Graphite Aluminum Composites. NASA/MSFC, private communication.

Larsen, J., Smith, T., and Erickson, P. (1971). Carbon-Fiber Surface Treatments. TR-71-165, Naval Ord. Lab., White Oak, Maryland.

Levitt, A., DiCesare, E. and Wolf, S., Fabrication and Properties of Graphite-Fiber-Reinforced Magnesium. TR-71-44 (November 1971); also, *Met Trans.* **3**, 2455 (1972).
Lynch, C., and Kershaw, J. (1972). "Metal-Matrix Composites." CRC Press, Cleveland, Ohio.
Manning, C., and Gurganus, T. (1969). *J. Amer. Ceram. Soc.* **52** (3), 115–118.
McDanels, D. L., Jech, R. W., and Weeton, J. W. (1965). *Trans. AIME* **233**, 636.
Mehan, R., and Gebhardt, J. (1969, 1970). Private communication, General Electric Co., November, 1969, January, 1970.
Meiser, M., and Davidson, J. (1969). Graphite Filament Reinforcement of an Iron–Aluminum Alloy. Paper presented at *Westec Conf.*
Metal-Matrix Composites (1968). STP-438, Amer. Soc. for Testing and Mater., Philadelphia, Pennsylvania.
Metal-Matrix Composites (1969). DMIC 243, Defense Mater. Inform. Center, Battelle Memorial, Inst., Columbus, Ohio.
Moreton, R., Watt, W., and Johnson, W. (1967). *Nature (London)* **213**, 690.
Morris, A. W. H. (1971). *Plastics Inst. (London)* **17**.
Morris, A., and Walles, K. (1971). U.K. Patent 1,250,830.
Morse, B. K. (1966). Graphite Fibers as Reinforcements in a Metal Matrix: A Definition and Discussion of the Problem. AFML-TR-66-346, Air Force Mater. Lab., Wright-Patterson AFB, Ohio.
Mortimer, D., and Nicholas, M. (1970). *J. Mater. Sci.* **5**, 149–155.
Munson, R. (1967). *Carbon* **5**, 471.
Naidich, Y. V., and Kolesnichenko, G. A. (1961). *Porosh. Met. SSR* **1** (6), 55–62.
Naidich, Y. V., and Kolesnichenko, G. A. (1963). *Porosh. Met.* **1**, 49–53; translated in *Sov. Powder Metal* **1**, 35–38.
Naidich, Y. V., and Kolesnichenko, G. A. (1964). *Porosh. Met.* **2**, 97–99.
Nicholas, M., and Mortimer, D. (1971). *Plast. Inst. (London)* **19**.
Niesz, D. E. (1968). Development of Carbon-Filament-Reinforced Metals. AD-836794, Battelle Memorial Inst., Columbus, Ohio.
Niesz, D. E., Fleck, J., and Kistler, C. (1967). Development of Filament-Reinforced Metals. AD-649509, Battelle Memorial Inst., Columbus, Ohio.
Niesz, D. E., Fleck, J., Kistler, C., and Machlin, I. (1968). Paper No. 68–338, *Proc. AIAA Struct. Mater. Conf., 9th.*
Pepper, R. T., Kendall, E. G., and Raymond, L. (1970a). The Infiltration of Graphite Yarn with Aluminum Alloys. TR-0066(9250-03)-2, The Aerospace Corp., El Segundo, California.
Pepper, R. T., Rossi, R. C., Upp, J. W., and Riley, W. C. (1970b). Development of an Aluminum–Graphite Composite. TR-0059(9250-03)-1, The Aerospace Corp., El Segundo, California.
Pepper, R. T., Rossi, R. C., Upp, J. W., and Riley, W. C. (1971a). *Amer. Ceram. Soc. Bull.* **50**, 484.
Pepper, R. T., Upp, J. W., Rossi, R. C., and Kendall, E. G. (1971b). *Met. Trans.* **2**, 117.
Perrett, R., and Ruland, W. (1969). *Carbon* **7**, 723.
Perry, A., DeLaMotte, E., and Phillips, K., (1970). *J. Mater. Sci.* **5**, 945–950.
Quackenbush, N. (1970). Large-Diameter Graphite–Carbon Composite Filament Development. CR-72769, Nat. Aeronaut. and Space Administration.
Rashid, M., and Wirkus, C. (1972). *Ceram. Bull.* **51** (11), 836.
Rauch, H. (1971). *Ceram. Age.*, May, 24–29.
Rhee, S. (1970). *J. Amer. Ceram. Soc.* **53** (7), 386–389.
Ross, J. (1971). *Mater. Res. Std.*, May, 11–15.
Sara. R. (1968). *SAMPE* **14**, II-4A-4.
Sara, R. V. (1969). U.S. Patent 3,473,900.
Sara, R. V. (1971). U.S. Patent 3,553,820.

Sara, R., and Winter, L. (1967). *Refractory Compos. Working Group Meeting, 13th.*

Sara, R., Volk, H., Nara, H., and Hanley, D. (1966). Integrated Research on Carbon Composite Materials. Part I, AFML-TR-66-310 (1966). Part II, AFML-TR-66-310 (1967). Part III, AFML-TR-66-310 (1969). Part IV, AFML-TR-66-310 (1969). Part V, AFML-TR-66-310 (1971).

Schmidt, D., and Hellman, R. (1969). U.S. Patent 3,462,288.

Watt, W., and Johnson, W. (1969). *Appl. Polym. Symp.* **9**, 215.

Watt, W. (1972). *Carbon* **10**, 121–143.

Widdows, D., and Nicholas, M. (1967). The Wetting of Graphite by Liquid Aluminum and Aluminum-Based Alloys. U.K.A.E.A., AERE-R5625, Harwell.

Wulff, J. (1942). "Powder Metallurgy." Amer. Soc. Metals, Cleveland, Ohio., pp. 21–22.

8

Boron-Reinforced Aluminum

KENNETH G. KREIDER

Institute of Applied Technology
National Bureau of Standards
Gaithersburg, Maryland

K. M. PREWO

United Aircraft Research Laboratories
East Hartford, Connecticut

I. Introduction

This chapter is intended to present a review of the current state of the art in the boron–aluminum family of composite materials. The chapter presents background material on the objectives leading to the development of the system, a discussion of the selection of materials and the most important composite fabrication techniques, a review of the physical and mechanical properties of the system, and a discussion of engineering considerations for use of boron–aluminum. The authors have intended to present the work from the viewpoint of the materials engineer rather than to emphasize composite mechanics, design, or applications.

A. Materials Engineering Needs

New requirements are being placed on materials by advances in engineering technology. An important requirement in the aerospace industry is the need for large, lightweight structures. Similar requirements are present throughout the transportation, communication, and manufacturing industries. Examples of these structures include the large, efficient air transports such as the Boeing 747, long-range communication antennae, large rockets for space payloads and the space vehicles themselves, and large, high-speed manufacturing processing equipment. One of the problems in designing these structures relates to the square–cube size relationship; that is, the strength and stiffness of a structure increases as the square of its dimension, whereas the weight increases with the cube of its linear dimension. Therefore, if mobility and efficiency are to be maintained, more sophisticated design and better materials are required to make larger structures.

The engineering materials that are used today, such as steel, aluminum, and glass, have similar elastic modulus-to-density ratios, and organic materials generally have lower stiffness ratios. In addition, the strength-to-density ratio of our high-strength metal alloys is also a serious limitation on structural design. An important objective of composite materials is to circumvent some of these limitations by using extremely strong, covalently bonded fibers such as boron combined with a matrix that allows structured fabrication and application.

B. Objectives in Composite Development

The objectives in developing the aluminum–boron composite system have included:

1. Developing the strongest and stiffest structures at lowest weight. The weight savings are critical in any system that moves, since power (or fuel) is required to move the mass. This weight problem is particularly acute in aerospace structures.

2. Developing fabrication techniques for large structures. This system will be most useful in large structures that have to move with respect to their surroundings. This objective relates to the weight problem also.

3. Developing dependable and durable materials and components. This quality is not only necessary for reliability but also significant with respect to cost.

4. The material system should take the maximum advantage of current art and technology. This objective in the development of the aluminum–boron system is important in order to achieve all the other objectives.

5. The techniques developed for fabrication and application of aluminum–boron composites should lead toward the lowest possible costs.

C. Advantages of Boron—Aluminum Composites

The aluminum–boron composite system combines the outstanding strength, stiffness, and low density of the boron fiber with the fabricability and engineering reliability of an aluminum-alloy matrix. Proposed applications of aluminum–boron composite materials include turbojet-engine fan blades, aircraft wing skins, stiffeners and spars for aerospace structures, and complete aerospace structural components. Section VI relates to materials engineering for these composite applications.

A comparison of the specific properties of boron–aluminum composites compared with boron–epoxy, high-tensile-strength-graphite–epoxy, and the high-strength titanium alloy Ti–6Al–4V is given in Table I. Both the unidirectionally reinforced and pseudoisotropic $0^\circ \pm 60^\circ$ reinforcement arrays are given. Most engineering applications require a reinforcement array that has properties intermediate to the two orientation arrays described above. The improvement in modulus-to-density ratio due to the reinforcing fiber is quite apparent in Table I. The modulus-to-density ratio of the boron fiber is approximately six times that of any of the standard engineering metals, including steel, aluminum, molybdenum, copper, and magnesium, and relates to the fact that the covalent bonding of boron is intrinsically stiffer than metallic bonding. The stiffness of the metallic bond is, however, much greater than that of the organic resin bond. Whereas a metallically bonded material has a modulus-to-density ratio of approximately 2.5×10^8 cm, that of organic resin bonding is more typically 2.5×10^7 cm. Therefore, the low resin stiffness leads to a low transverse modulus and low shear modulus

TABLE I

Properties of Multidirectionally Reinforced Composites

	Tensile strength 0° (10^3 psi)[a]	Tensile modulus 0° (10^6 psi)[a]	Tensile strength 90° (10^3 psi)[a]	Tensile modulus 90° (10^6 psi)[a]	Compressive strength 0° (10^3 psi)[a]
Boron–epoxy (N5505)					
0° ±45°	103	17	16.0	4.6	287
0° ±60°	68	11	42.0	11.0	226
Boron–aluminum					
0° ±60°	75	27	75	27	—
Boron–aluminum					
0°	200	33	20	20	>300
Graphite–epoxy (HMG-50/BP907)					
±30°	38	7.7	6.0	1.5	33
0° ±45°	68	12.0	9.4	1.7	—
7075-T6	75	10.0	75.0	10.0	—
Ti–6Al–4V	140–170	16.5	140–170	16.5	—

[a]Multiply psi × 7.03×10^{-4} for kg/mm^2

in resin composites. This advantage of boron–aluminum is carried over in multidirectionally reinforced composites, since plies not oriented in the principal stress direction still contribute substantially to the longitudinal modulus.

A second advantage of the relatively high-modulus matrix such as aluminum is observed under compressive loading. The high modulus in the matrix is important in preventing microbuckling of the fibers in the matrix, which has been reported by Corten (1967). The microbuckling problem in compression of fiber is also aggravated by small fiber diameters and is generally given as the primary reason for the low compressive strength of composites reinforced with very fine (7-μm) graphite fiber.

In addition to the more nearly isotropic elastic modulus of boron–aluminum composites as compared to resin composites, the off-axis strength of the boron–aluminum is also higher. The transverse tensile strength and shear strength of the composite can be equal to the matrix aluminum alloy strength, which is considerably higher than that available in resin matrix materials. Typical strength values for these materials are also given in Table I.

Other important physical and mechanical properties of boron–aluminum composites include the high electrical and thermal conductivity; ductility and toughness; abrasion resistance; ability to be coated, joined, formed, and heat treated; and noninflammability. The high-temperature capability and resistance to moisture are often important with respect to the durability of an engineering structure. Perhaps one most important asset of the aluminum–boron composite system relates to the present technology in the use of the aluminum matrix alloys in structural hardware. The experience in fabrication and application of aluminum alloy forms an excellent background for the utilization of boron–aluminum.

II. Materials Selection

The boron–aluminum composite system has received more attention with respect to research and development and has been proposed for far more structural applications than any other high-modulus, fiber-reinforced metal-matrix composite system. This section contains reasons for the choice of these two constituent materials and some background in the selection of the particular type of boron fiber and aluminum alloy. A discussion of the synergistic effect of the constituents, some shortcomings, and third-constituent additions is also presented.

The objectives in the selection of the constituents of the composite were primarily high modulus-to-density and strength-to-density ratios for structural applications. In addition, durability under structural engineering environments and fabricability are extremely important.

A. Fiber

The prime requisites of the reinforcing filament are high modulus-to-density ratio, high strength-to-density ratio, good reproducibility of properties, low cost, and ease of fabrication into the composite. The significant fiber properties with respect to these requisites are given in Table II. Aluminum-matrix composites have been made with each of these fibers. Each fiber has disadvantages when compared to the boron fiber as described in Table II. The glass fiber has an attractive price and strength factor, but its low modulus and its reactivity with aluminum are serious drawbacks and have been discussed by Baker and Cratchley (1966). The availability in yarn form aggravates this problem, since composite fabrication with yarn is best performed by infiltration of the bundle by a liquid, which is much more reactive than a solid aluminum alloy. The alumina fibers that have been discussed in detail by Mehan and Moore in Chapter 4 have inferior specific modulus and specific strength, and are very expensive. Silicon-carbide fiber is less reactive with aluminum than boron and is used as a coating on boron

TABLE II

Typical Properties of Commercially Available Reinforcement Filaments

	Diameter (μm)	Manufacturing technique	Average strength (GN/m^2)	Average strength (ksi)	Density (gm/cm^3)	Modulus (GN/m^2)
Boron	100–150	Chemical vapor deposition	34	500	2.6	400
SiC-coated B	100–150	Chemical vapor deposition	31	450	2.7	400
SiC	100	Chemical vapor deposition	27	400	3.5	400
Monofil-C	70	Chemical vapor deposition	20	300	1.9	150
B_4C	70–100	Chemical vapor deposition	24	350	2.7	400
B on C	100		24	350	2.2	
Graphite HS	7[a]	Pyrolysis	27	400	1.75	250
Graphite HM	7[a]	Pyrolysis	20	300	1.95	400
Al_2O_3	250	Melt withdrawal	24	350	4.0	250
S-glass	7[a]	Melt extrusion	41	600	2.5	80
Be	100–250	Wire drawing	13	200	1.8	250
Tungsten	150–250	Wire drawing	27	400	19.2	400
"Rocket wire"	50–100	Wire drawing	41	600	7.9	180

[a] 10^4 Fibers per tow.

(BORSIC†); but with a density 30% higher than boron and lower strength properties, it would not appear as attractive. The high-modulus graphite fibers—HS with high strength but moderate modulus and HM with the high modulus but moderate strength—appear attractive and have been treated in Chapter 7. The availability in the form of a yarn, however, is a serious drawback, since solid-state fabrication processes are normally inadequate to penetrate the 10,000-filament bundle, and the fiber is very reactive with molten aluminum alloys. Hard-drawn wires of the alph–beta titanium alloy Ti–6Al–4V and precipitation-hardened steel rocket wire NS-355 do not have competitive strength-to-density or modulus-to-density ratios because of their high density. Hard-drawn beryllium wire has attractive properties but is very expensive to produce.

High-strength, high-modulus boron fiber was first reported by Tally (1959), and the fiber produced today is basically similar. Prior boron production

† BORSIC is a registered trade mark of United Aircraft Corp.

from boron trichloride was intended primarily for the purpose of research studies on pure chemical boron. This work was summarized by Kohn *et al.* (1960) in a book entitled "Boron—Synthesis, Structure, and Properties." Development of the fiber to obtain high strength, uniform quality, and low cost was pursued by the United States Air Force and several American corporations with considerable interest. An excellent review of boron-fiber technology was made by Wawner (1967), although considerable advances have been made since then.

Boron fiber is synthesized by chemical-vapor deposition from the reduction of boron trichloride on a tungsten substrate in the presence of hydrogen. The tungsten wire is resistively heated to the temperature range 1100–1300° C and continuously pulled through the reactors to obtain the desired boron coating thickness. Galasso (1970) has reported that during the boron deposition, the tungsten substrate is converted to several borides. The boron structure appears amorphous in X-ray studies, but the surface appears granular as shown in Fig. 1. The present fiber is sold in quantity in two diameters (100 and 140 μm) and can be purchased with a 2-μm-thick coating of silicon carbide to improve the oxidation resistance. Figure 2 gives the distribution of strength

Fig. 1. Surface of boron fiber imbedded in aluminum. (Note "corn cob" grains.)

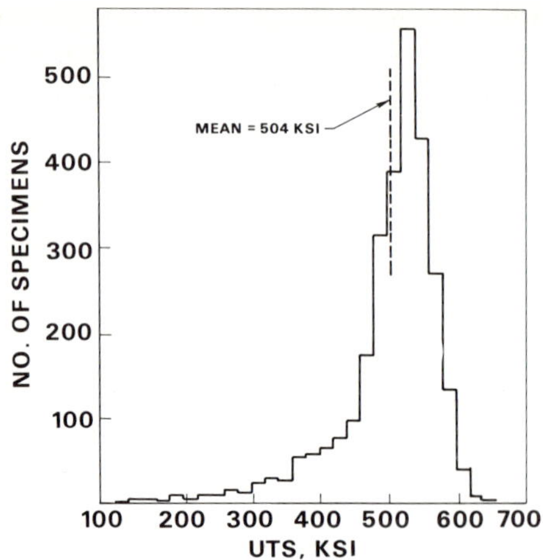

Fig. 2. Frequency distribution of individual ultimate tensile strength (UTS) values. $N = 2800$, standard deviation = 73 ksi, coefficient of variation = 15%.

properties realized in monofilament tensile tests of commercial boron. The boron fiber has an elastic modulus of 57–58 × 10^6 psi, and the stress–strain curve is elastic to failure. The room-temperature density is 2.55 gm/cm^3 for the 140-μm fiber. An interesting property of boron fiber is its excellent handleability and insensitivity to surface abrasion or corrosion, which relates to the high residual compressive stresses on the fiber surface.

Elevated-temperature properties have been studied, including tensile strength, creep and stress rupture, oxidation, and strength retention after high-temperature exposure. The high-temperature strength of the fiber is excellent; Herring (1966) found up to 75% strength retention at 1200°C. The creep properties as reported by Ellison and Boone (1967) were superior to those of tungsten at 1200 and 1500°F. However, exposure to oxygen has a

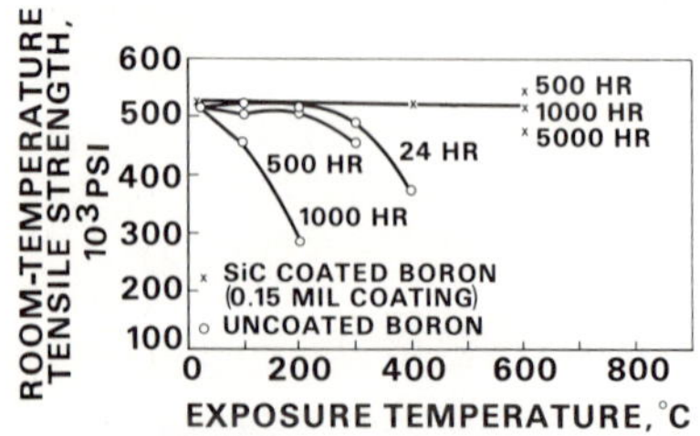

Fig. 3. Average room-temperature ultimate tensile strength of filaments after being heated in air. × SiC-coated boron, ○ uncoated boron.

severe degrading effect on the strength of the fiber after 24 hr at 300°C or shorter times at temperatures above 500°C. This degradation has been reported by Basche *et al.* (1968) and is depicted in Fig. 3. The degradation led to the development of a silicon-carbide coating, which prevents the oxidation degradation (Fig. 3).

B. Matrix

Aluminum matrix alloys for boron composites have been chosen for their combination of properties. Included in the desirable properties of the matrix are good bonding characteristics; high fracture toughness, which serves to arrest a crack at a broken or split fiber; the creep-forming capability to surround and bond to the fiber; high strength, and corrosion resistance. In composites to be used at elevated temperature, matrix creep resistance and oxidation resistance are important. In addition, the matrix should be weldable or brazeable, and for some applications the matrix should permit the use of composite creep-forming techniques.

Commercial aluminum alloys have been developed as wrought products, cast products, brazing products, and some such as sintered aluminum powder (SAP) for powder-metallurgy fabrication. The specific requirements of aluminum alloy development for the above processes do not coincide exactly with boron metal-matrix requirements; however, several alloys have been used successfully. Composite fabrication processes involving melt fabrication are generally performed better with casting alloys such as KO-1 (Al–5 Cu–1Ag–0.3Mg–0.2Ti) as described by Davies (1971). The best high-temperature properties above 250°C have been obtained with SAP-type alloys, and brazing operations are normally performed with brazing alloys. The largest quantity of aluminum-matrix composites, however, are bonded by solid-state hot-pressing, and the best matrix alloys for this purpose have been found to be the wrought alloys.

Several classes of wrought aluminum alloys have been used for composite fabrication using foils and prealloyed powders in plasma spraying. A list of aluminum-alloy properties is given in Table III. The 1000 and 3000 series, which are readily available, have excellent ductility and brazeability, but the low tensile and creep strengths are a disadvantage for off-axis composite properties. The 7000 (zinc) and 4000 (silicon) series alloys have been investigated but generally have lower than average fracture toughness. The 5000 series alloys such as 5052 and 5056 have good fracture toughness and have been used to fabricate strong boron-fiber composites.

The heat treatable alloys of 6061 and 2024 have received the most attention. The 2024 alloy Al–4.5 Cu–1 Mg has the advantages of its availability as

TABLE III

Properties of Aluminum-Matrix Materials

Alloy	Elastic modulus (GN/m^2)	Yield stress (0.2% offset) (10^7 N/m^2)[a]	Ultimate tensile strength (10^7 N/m^2)	Strain to failure (%)
1100	6.3	4.3	8.6	20
		3.5[b]	9.0[b]	35[b]
2024	7.1	12.8	24.0	13
		7.5[b]	18.7[b]	20[b]
5052	6.8	13.5	26.5	13
		9.0[b]	19.5[b]	25[b]
6061	7.0	7.7	13.6	16
		5.5[b]	12.5[b]	25[b]
Al–7Si	7.2	6.5	12.0	23

[a] 10^7 N/m^2 = 1400 psi.
[b] Wrought alloy properties.

foil or powder, the high strength available by age-hardening, good high-temperature creep strength, and the extensive technological experience in the use of 2024 in structural applications.

The 6061 alloy is one of the most commonly used aluminum structural alloys. It is heat treatable, forming a fine precipitate of magnesium silicide, and its tensile strength in this condition is in excess of 40,000 psi at room temperature. The low alloy content results in a high melting point, and high ductility results in low notch sensitivity. This alloy has been the most fully characterized in metal-matrix boron composite systems. Other important assets of the 6061 alloy include good corrosion resistance and very low stress-corrosion sensitivity. The high melting point is an advantage in brazing, and the brazing of 6061 is a well-established technology. This alloy is also tough at cryogenic temperatures and more easily formed than higher-strength alloys such as 2024 and 7075.

III. Composite Fabrication

Composite fabrication consists of assembling and bonding the components of the composite material into the form that is used in the manufacture of the composite part. Two general techniques are used: one consists of assembling and bonding the fiber and matrix simultaneously with the forming of the shaped part; the second technique consists of first manufacturing a composite preform such as broadgoods or a tape-form composite, followed by lamination and bonding of the preforms into the final shaped part. The former technique is analogous to the manufacture of cast hardware, the latter to casting an ingot followed by a forging operation to achieve the shape of the part.

Both the single-step and the two-step techniques have many of the same problems. These include mechanical problems such as accurate filament alignment, control of filament spacing and volume fraction, and minimization of porosity and flaws. Also important are thermodynamic and chemical kinetic problems such as oxidation, impurity pickup during bonding, and bonding and reactivity of the components (boron and the aluminum alloy).

A. Thermodynamics

The thermodynamics and chemical kinetics of the systems can be treated independently of the exact manufacturing technique. Basically, the thermodynamics of the system are simple and indicate two reactions that are destructive to the boron fiber surface. These reactions can be expressed as:

$$Al + B \longrightarrow AlB_2 \quad (1)$$

$$B + O \longrightarrow B_2O_3 \quad (2)$$
$$(mp = 577^\circ C)$$

The aluminum–boron reaction has been studied by Klein *et al.* (1972) and can be a primary degrading reaction on the fiber. This system is complicated by the compound AlB_{12}, but the formation of either compound might be expected to have a similar effect. A third degradation reaction can relate to leaching of the boron surface. The solubility of boron in molten aluminum has been reported to be 0.09 w/o at 730°C, whereas the solubility in solid aluminum is very small. A fourth reaction that is important during fabrication concerns the oxidation of aluminum.

The boron oxidation reaction has been studied by Basche *et al.* (1968). Their results, which are presented in Fig. 3, indicated a gradual degradation of axial fiber strength at temperatures up to 500°C but drastic degradation above 500°C. This may relate to melting of the oxide. Exposures in air at 300°C, however, caused significant degradation after 24 hr. Although the exact nature of the degradation has not been established, it is clear that the boron fiber is susceptible to damage by oxidation. The important empirical information relates to strength degradation and pertains to reaction rates. The most useful test for reactions has been found to be a fiber tensile test after reaction exposure.

Oxidation of the boron fiber can be circumvented during fabrication by reducing the partial pressure of oxygen. Bonding and consolidation of the composite in an inert atmosphere is also important in order that the oxidation of aluminum be minimized, since this reaction inhibits the manufacture of a high-quality, well-bonded, ductile matrix.

Fiber tensile tests have also indicated that the boron will not survive undamaged more than a 1-sec exposure to molten aluminum or a 1-hr exposure at 500°C. These reaction rates are, of course, affected by alloying ingredients in the aluminum and by partial pressure or activity of oxygen in contact with the fiber. However, the prime alloying ingredients in aluminum, such as copper, manganese, silicon, magnesium, nickel, and iron, are less reactive with boron than the aluminum is. The effect of alloying ingredients is far more profound in that they lower the melting point and increase the fluidity of the melt. Abrams and Davies (1971) have noted the severe effect of silicon, which increases the fluidity of the melt. The first reaction in molten aluminum is probably related to preferential leaching of the surface of boron rather than compound formation, which has been observed in the specimens that are more severely degraded. This leaching effect causes microcracks and is a function of the rate at which boron can be removed from the surface of the boron through the melt. This boron can be removed by diffusion in the liquid and convection currents. Each of these parameters is a function of the homologous temperature, and convection is related to the fluidity of the melt. Photomicrographs of degraded fibers are presented in Fig. 4. The degradation in Fig. 4a is probably related to dissolution of boron and can be compared to the unreacted filament. In Fig. 4b the replica of the interface between boron and aluminum is depicted by the electron microscope and illustrates the leaching effect on the smooth surface.

Empirical results have indicated that boron fiber can be bonded in the composite with the aluminum-alloy matrix (such as 6061) at 500°C and 10^{-3} Torr for periods less than 1 hr without significant fiber degradation. In order to close all porosity in the aluminum, approximately 5000–10,000 psi normal pressure will be required. This high-pressure requirement, coupled with the sensitivity of the fiber to degradation and the problem of brazing or heat-treating the part with the above-mentioned temperature limitations, has led to the development of a coating for the fiber to protect it from these three reactions: compound formation, oxidation, and melt leaching or liquid-metal corrosion. The development of coated boron fiber was also spurred on by the desire to use the fiber in other metal-matrix systems.

The coating was designed to be superior to the bare fiber in resistance to the three reactions discussed above, inexpensive to put on the fiber, and should not diminish the excellent strength properties of the fiber. Although many techniques were investigated, the two coatings that have been used

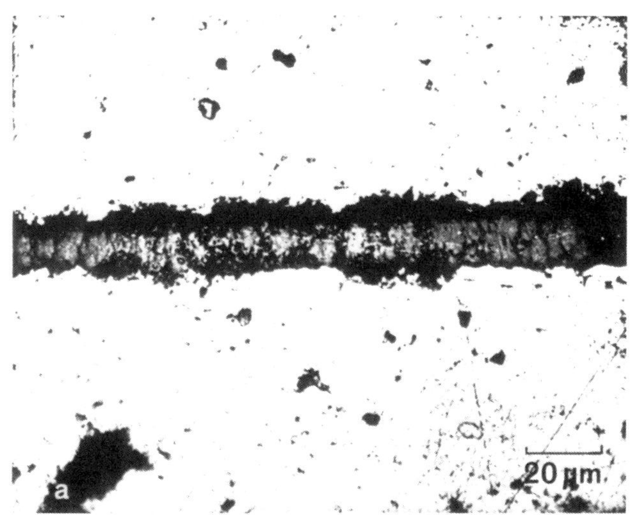

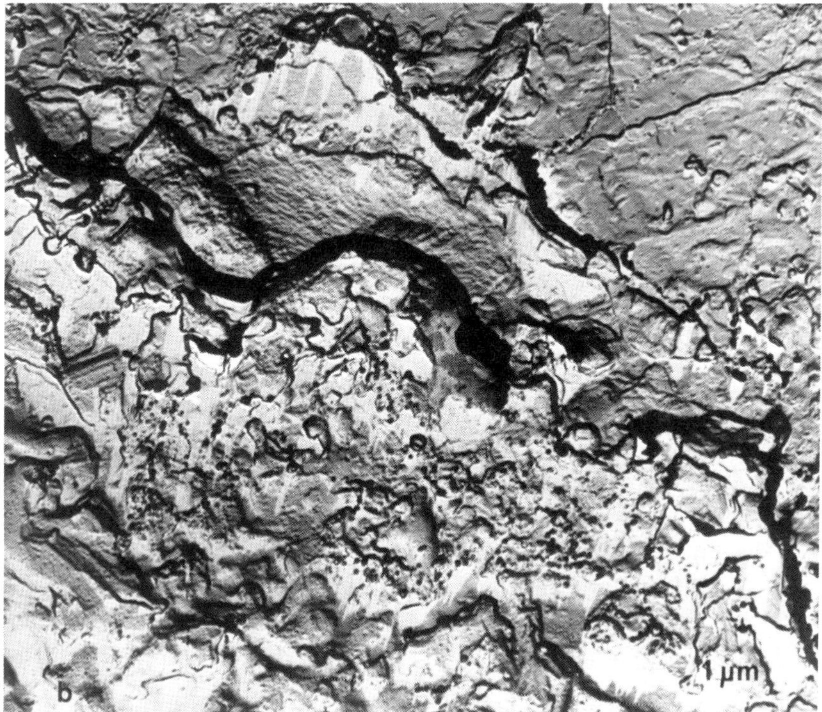

Fig. 4. Boron fiber degraded in aluminum: (a) heat-treated 1000 hr at 400°C in argon; (b) leaching effect on smooth surface.

successfully are a chemical-reaction coating of nitrided boron and a chemical-vapor-deposition coating of silicon carbide. The nitrided boron is inexpensive to put on and has been used to make high-strength boron–aluminum composites by melt infiltration.

The silicon-carbide coating on boron causes the fiber to have outstanding resistance to oxidation (Fig. 3). In addition, boride formation is completely inhibited, since the boron does not contact the aluminum. Aluminum does not form compounds with silicon, and the aluminum-carbide reaction is quite unfavorable thermodynamically in the presence of silicon carbide. The diffusion constants for the migration of boron or aluminum through silicon carbide are extremely low at 800°K, and a layer of 0.1 mil is an effective barrier, The third reaction of liquid-metal corrosion or leaching is more important and can take place. Although the carbon solubility is low, the mobility of the carbon in the melt should be very high, and therefore the diffusion (or migration) of silicon away from the silicon carbide should be the rate-limiting step:

$$\mathrm{Al} + \mathrm{SiC} \longrightarrow \mathrm{Si(Al)} + \mathrm{C(Al)} \tag{3}$$

Equation (3) expresses the dissolution of the compound silicon carbide. Thus, the rate of leaching or attack can depend on the fluidity of the aluminum-alloy melt which affects convection migration. The most obvious parameter to influence the fluidity is the volume fraction of solids in the melt at temperatures between the solidus and liquidus. This type of attack is not observed at temperatures below the solidus, and the attack is very slow until the melt has a very low fraction of solids in the alloy system 6061 and 2024. Other significant parameters, which are also important in brazing and casting operations include wetting and viscosity of the melt. Silicon, which is so useful in brazing and casting, has been reported by Abrams and Davies (1971) to be particularly active in accelerating the leaching attack.

The silicon-carbide coating does, however, permit boron–aluminum fabrication at considerably higher temperatures and correspondingly lower pressures, as reported by Basche *et al.* (1968). These higher temperatures not only permit lower (or uneven) pressures for fabrication, but also permit standard aluminum brazing cycles and heat treatments without fiber degradation. The hot pressing of complex shapes is much simpler if the die pressures do not have to be carefully controlled at the various angles of forces normal to the laminar surface.

B. Fabrication Processes

The fabrication process can be separated into three phases: collimation of the fiber, assembly and bonding of the components of the composite, and lamination of the part. Most of the boron–aluminum composite fabrication has been performed using a preform or intermediate composite. This preform may be a broadgoods product, a wide tape, or a small coated bundle.

Two of the most useful techniques employ very similar steps in fabrication. These are the hot-press diffusion-bonding of a "green" precursor of either a resin-bonded or plasma-sprayed tape. Either precursor can be collimated in a similar way, but the assembly and bonding of fiber and matrix is accomplished by a removable or "fugitive" resin binder in one case and by plasma spraying a portion of the matrix in the second case. After lamination and diffusion bonding, the composite is fully dense and has the properties desired for the component. The composite can be then bonded or brazed to other components for the hardware application. An alternative to the above technique is to diffusion bond a monolayer.

1. Composite Fabrication Techniques Using a Preform

A. FUGITIVE-BINDER TECHNIQUE. The fugitive-binder technique is a direct approach that requires little capital equipment or special skills and has been used by over a dozen fabricators. Although each fabricator may have individual techniques as to surface preparation of the aluminum foil or resin application, the steps in manufacture are basically similar.

The materials used in fabricating the preform include spools of boron fiber, aluminum-alloy foil, volatile resin that leaves no residue after evaporation (such as polystyrene), and a solvent as a vehicle for the resin. The thickness of the foil (usually 0.002 or 0.003 in.) is selected so that, combined with the proper fiber spacing, a given volume fraction of fiber is achieved. Since only certain aluminum-alloy foils are available, the process is limited to composites containing these alloys.

Two types of filament collimation are used: single-filament drum winding and multifilament collimation from a creel of fiber spools to a continuous tape array. The former technique has been used more frequently because of its simplicity. Drum winding (Fig. 5) produces a broadgoods that is sized as the drum width and circumference. Since a simple screw thread will provide the proper translation of the fiber spool in coordination with the drum rotation, very accurate spacing and control of tension can be achieved. Typically, the matrix aluminum-alloy foil is wrapped around the drum prior to winding. The binder can be applied by spraying the resin plus volatile solvent (which can be xylene for a polystyrene binder) on the bare foil. While the solvent is evaporating, the fiber is wound on the drum. A second coating of resin can be applied to the completed windings to strengthen the bonding of the fibers and the temporary resin matrix. Alternatively, the resin could be introduced by pulling the fiber through a resin bath, thus coating the fiber immediately prior to the winding placement. Important controls in the process include resin fluidity and room-temperature "setting," fiber winding

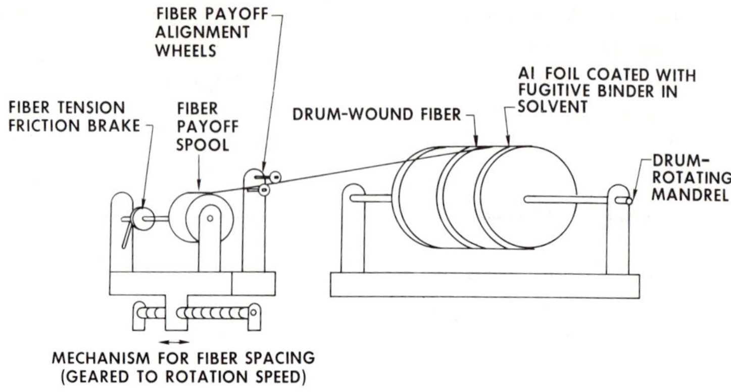

Fig. 5. Fiber drum winding.

placement accuracy, fiber tension uniformity, prevention of foil wrinkling, and strength and durability of the resin matrix.

The second fiber-collimation technique of continuous multifilament tape manufacture requires more sophisticated equipment. This type of equipment has been developed by boron–epoxy tape suppliers after extensive work on problems connected with feeding as many as 600 brittle fibers in a closely controlled array at 0.001-in. spacing. By this technique, coated foil is fed continuously onto the coated fiber array, and the styrene or acrylic is partially cured prior to coiling, as in the case with boron–epoxy tapes. The most popular tape width is 3 in., which is easily handled in coiled form and was selected as compromise between the number of seams (which increases with a narrower tape) and the waste factor (which tends to increase with a wider tape). The waste factor is important because of the high cost of the raw material and is the most important advantage of a tape compared with broadgoods, which require less labor in lamination. The waste factor is related to the preform geometry: end splices of tapes are not used because of the multiplicity of fiber ends and the accompanying stress concentration, whereas splices parallel to the fibers are much less harmful. This splice-type preference leads to the preference of a long and narrow preform shape to minimize waste. Broadgoods will probably be used where the waste can be designed to a minimum.

B. PLASMA-SPRAY TECHNIQUE. Plasma-sprayed boron–aluminum tapes are fabricated in the same way, except that instead of spraying a fugitive binder, a single layer of the matrix aluminum alloy is sprayed on the fiber–foil array, bonding the tape together. Since the plasma-sprayed alloy is chosen for the matrix as is the foil, both materials become part of the matrix. The aluminum-alloy powder is injected into the hot plasma gas stream and melted in the exothermic region. The molten particles are impacted on the fiber foil array and rapidly quenched well below the temperature at which fiber reactions take place. Oxidation can be prevented by surrounding the fibers with an inert (argon) atmosphere supplied from the plasma-torch effluent.

Some of the advantages of the process include the good fiber-spacing control, which is locked in by the matrix on the wound drum; the durability and strength of the as-sprayed tape; and the fact that the tape is ready for composite bonding as sprayed. The matrix is very porous and ridged over the fibers. This morphology requires further treatment such as hot pressing to achieve optimum properties; but, because of the bulk, the aluminum diffusion bonding is promoted as the plastic deformation breaks the oxide film when the tape porosity is collapsed. Typically a tape with 5.6-mil (0.140-mm) boron will be 0.28 mm thick "green" and 0.17 mm thick when bonded.

C. LAMINATION AND BONDING OF PREFORMS. The fugitive-binder and plasma-spray techniques are used to produce a "green" preform. This tape (or broadgoods) must be laminated and bonded to form the composite. With the fugitive-binder approach, the binder must be exhausted prior to hot-press bonding. This step is performed immediately prior to the high-pressure bonding, since the fiber and foil array must be held in place mechanically after the resin binder is removed. Typically, 1 hr at 400°C in vacuum is sufficient to remove all detectable residue of the polystyrene. Other resins such as acrylics have been chosen for higher strength, but polystyrene has a higher boil-off rate.

The lamination process consists of cutting the preform to shape and laying the pattern on the hot-press dies or platens. Typically, a shearing process is used for cutting the single plies because of its simplicity. The shear can resemble a sheet-metal shear, scissors, or "cookie cutters." The lay-up is usually "canned" if flat hot-press platens are used, since a controlled atmosphere is important to prevent oxidation of aluminum or boron at temperature. Photomicrographs of typical canned composites that are pressed between platens are given in Fig. 6. Figure 6a depicts a cross section of a composite with 50 v/o boron in a matrix of 6061 alloy prepared from plasma-sprayed tapes. Figure 6b depicts a cross section of a composite with 50 v/o fraction of BORSIC in a SAP-alloy matrix prepared by the fugitive-binder technique. If a complexly shaped part is to be made, shaped dies are used to press the lay-up, which may have many complexly shaped plies. With closed dies, a vacuum hot press has been used to form turbine-engine fan blades by Mangiapane *et al.* (1968). The cutting of the individual plies is shown in Fig. 7. The laminae are cut by a rolling shear analogous to a cookie cutter. These pieces will then be laminated and pressed in the blade-shaped dies. Either of

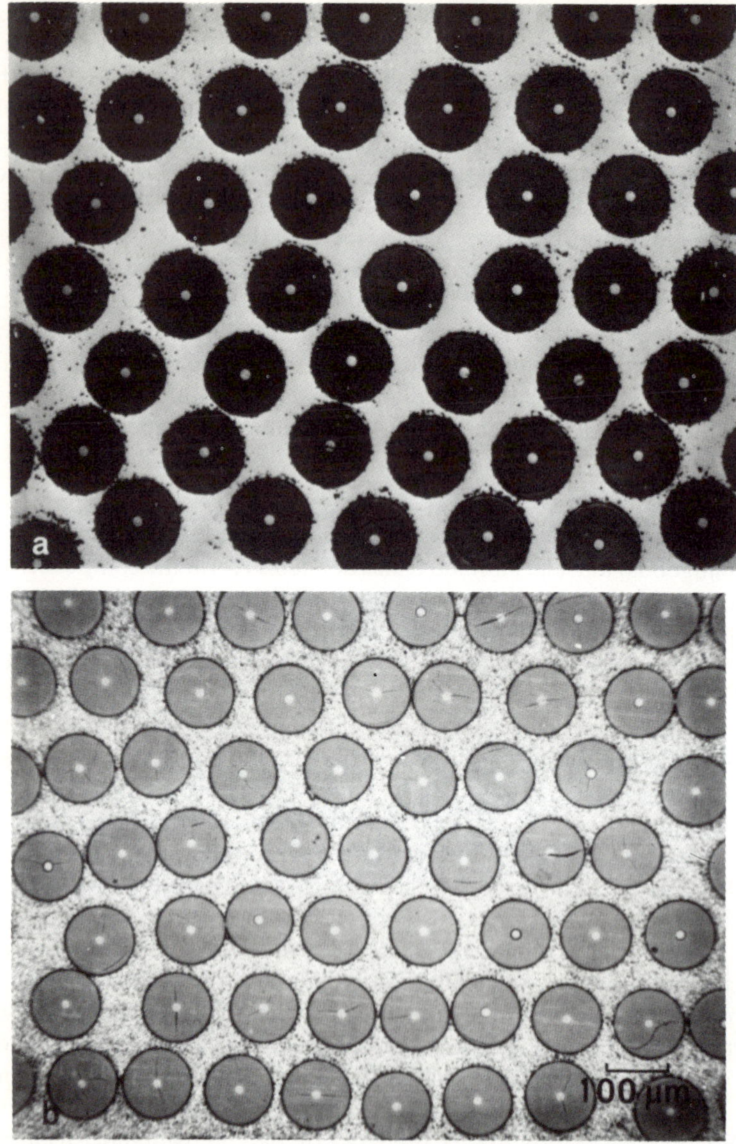

Fig. 6. Diffusion-bonded boron–aluminum composites hot pressed at 550°C: (a) bonded from plasma-sprayed 6061 tapes and boron fibers; (b) bonded from SAP foils and BORSIC fiber.

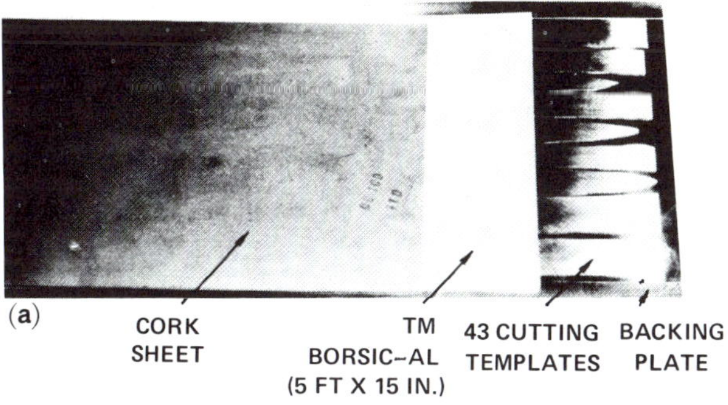

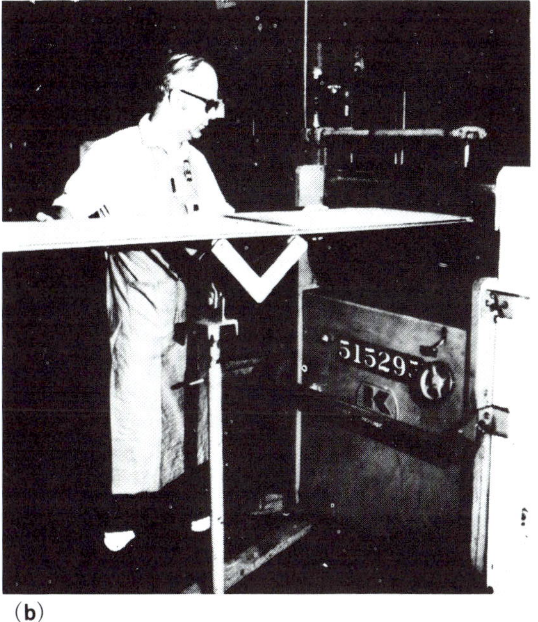

Fig. 7. Roll-cutting of composite fan-blade plies: (a) roll-cutting sandwich for composite fan blade; (b) roll-cutting the composite ply shapes.

these processes can also be used to hot-press long parts by leaving the ends of the dies or platens open and step-pressing the boron–aluminum composite. Step-pressing of the tapes has been used to manufacture long panels by means of platens, and to make long stiffeners such as L sections and hat sections by means of shaped dies.

The parameters of hot-pressing—temperature, pressure, time, and atmosphere—are very important to control, since degradation of the strength of the composite results from severe conditions and the diffusion-bonding of the aluminum is not achieved. The time–temperature window of useful conditions when using uncoated boron fibers is small. Studies of aluminum bonding by Mangiapane *et al.* (1968) have indicated minimum conditions of 4 hr at 485°C with 70.0 MN/m^2 (10,000 psi) pressure to ensure complete bonding in a vacuum of 10^{-5} torr. However, by promoting extensive plastic deformation, as is the case with the porous plasma-sprayed compaction or in extrusion, these times can be reduced considerably. Unfortunately, the exact determination of small unbonded regions is difficult to achieve with a nondestructive test (such as ultrasonic transmission). As was discussed above, the degradation of boron fiber is measurable at 500°C after 1 hr, so that most fabricators choose the temperature range 450–500°C if times of greater than $\frac{1}{2}$ hr are used. If considerably shorter times are used, higher temperatures are preferred. The pressure required is generally in excess of 70 MN/m^2 (10,000 psi), and a low partial pressure of oxygen (10^{-5}torr) is maintained for the longer times to prevent boron oxidation. High pressures, greater than 140 MN/m^2 (20,000 psi), have been observed to lead to fiber crushing, particularly when cross plies were used or with complexly shaped parts. Shapes such as turbine-engine fan blades present pressing surfaces whose normal is at least 40–50° from the pressing direction because of their twist. Also, during pressing in a closed die, some matrix-material migration and fiber twist and bending are required to exactly fill the cavity. These factors considerably reduce the maximum useful pressure for bonding the blade because the pressure gradients can break fibers or leave voids and unbonded regions.

There are several significant limitations in boron–aluminum bonding:

1. The time–temperature parameter is limited by the fiber-degradation problem.
2. The time–temperature–pressure parameter must be above a threshold value to ensure aluminum bonding and elimination of porosity, which is a combination creep- and diffusion-limited process.
3. High pressures lead to increased fiber breakage.
4. Complexly shaped closed-die-fabricated parts have considerable pressure gradients during the hot-pressing operation.
5. Careful atmosphere control is required to prevent boron oxidation.

If the boron fiber is coated with silicon carbide (BORSIC), most of these limitations are bypassed, since much higher temperatures can be used during fabrication without fiber degradation. This logic has been behind the choice of BORSIC for several large Air Force-sponsored turbine fan-blade development programs. Typical BORSIC–aluminum composite fabrication parameters include 500–600°C at 14.0–70.0 MN/m^2 (2000–10,000 psi) for 3–180 min.

Low-pressure processes for bonding have been developed that employ the presence of a liquid during the bonding comparable to pressurized 0.4–1.4 MN/m^2 (50–200 psi) liquid-phase sintering. These bonding processes take place between the liquidus and solidus temperatures in aluminum eutectic-alloy systems such as Al–Si, Al–Mg, and Al–Cu. The matrix does not generally have properties as good as those of the wrought alloy, and fiber degradation can be a problem even with the coated fibers.

Autoclave bonding techniques have also been employed in place of the single-action hot-pressing procedure. Both high-pressure 70 MN/m^2 (10,000 psi) and low-pressure 1.4 MN/m^2 (200 psi) bonding have been reported by Miller and Schaeffer (1971). These processes have the advantage of simpler die design and can be used to make more complex shapes than a single-action die set, such as I beams, T sections, and closed tubes. Excellent properties have been attained on over 1000 composite parts of various shapes using

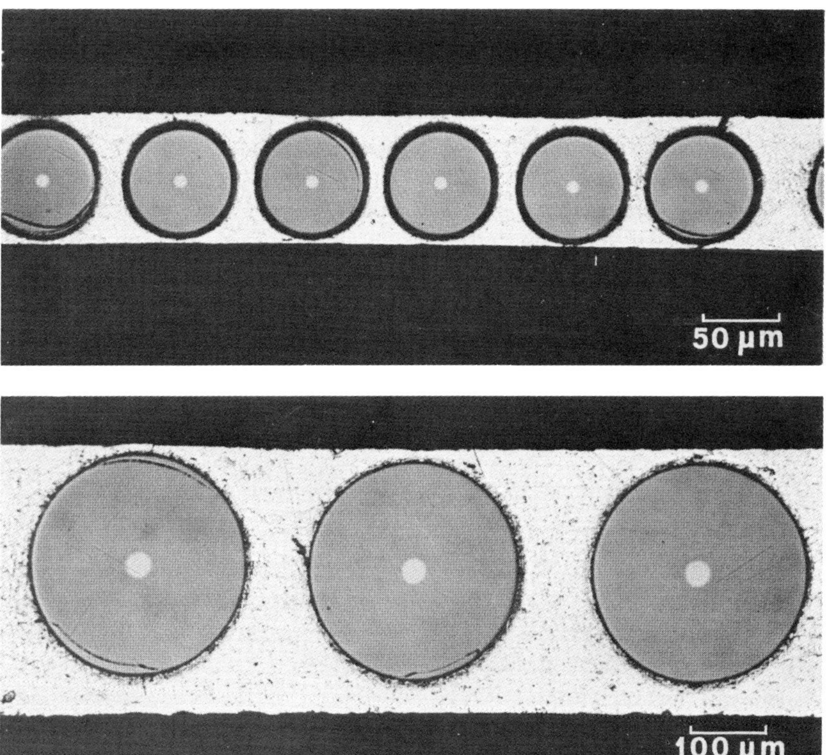

Fig. 8. 5.6-mil-boron–6061-aluminum preconsolidated monolayer tape.

plasma-sprayed tapes or consolidated tapes. However, Miller and Schaeffer concluded that severe restrictions are imposed on the high-pressure manufacturing cycle that tend to make the process very expensive. These include a complex fabrication cycle and problems in scaling up to large pieces. They favor the low-pressure process for large pieces.

D. BONDING OF CONSOLIDATED MONOLAYERS. A variation of the preform plus lamination and bonding technique described above includes the use of a fully consolidated preform composite. In this case, the preform, made by the fugitive-binder or the plasma-spray process, is hot-press bonded as a monolayer composite preform. The green tape or broadgoods is usually step-pressed into a monolayer composite that has identical properties to the final composite; a cross-sectional photomicrograph is given in Fig. 8. This fully dense preform can be diffusion-bonded (with or without a eutectic film melt present during bonding) or adhesively bonded to form the composite multilayer structure.

E. THIRD-COMPONENT ADDITIONS. During composite fabrication, it is possible to add third components to the boron and aluminum alloy in order to improve such properties as high-temperature transverse strength, erosion resistance, and toughness. The two most significant additions to date have been titanium foil (Ti–6Al–4V or Beta III) and high-strength rocket wire such as N5-355. Since the bonding conditions of the aluminum matrix to these tertiary constituents are the same as the self-bonding parameters, it is relatively easy to add the titanium foil or rocket wire to the preform and bond it in the composite. Photomicrographs of composite cross sections are given in Fig. 9. The preform is available with titanium foil substituted for the aluminum foil or rocket wire substituted for the fiber. Typical properties of the wire include 3.8 GN/m^2 (550,000 psi) tensile strength at 20°C and 2.8 GN/m^2 (400,000 psi) at 500°C, and the wire is not significantly annealed during hot-press cycles at 500 or 550°C.

2. *Other Boron–Aluminum Fabrication Processes*

Other processes have included electroforming, powder-metal compaction, casting, and filament winding combined with plasma spraying and sintering. The electroforming process can be performed by filament winding a mandrel and electroplating aluminum from an ether bath containing lithium-aluminum hydride and aluminum chloride (Whithers and Abrams, 1968). It was found that the aluminum does not deposit on the boron filament but rather grows preferentially in the space between the windings. One solution to this problem includes nickel plating the fiber prior to matrix deposition. The plating of multilayer windings with accurate filament spacings has been

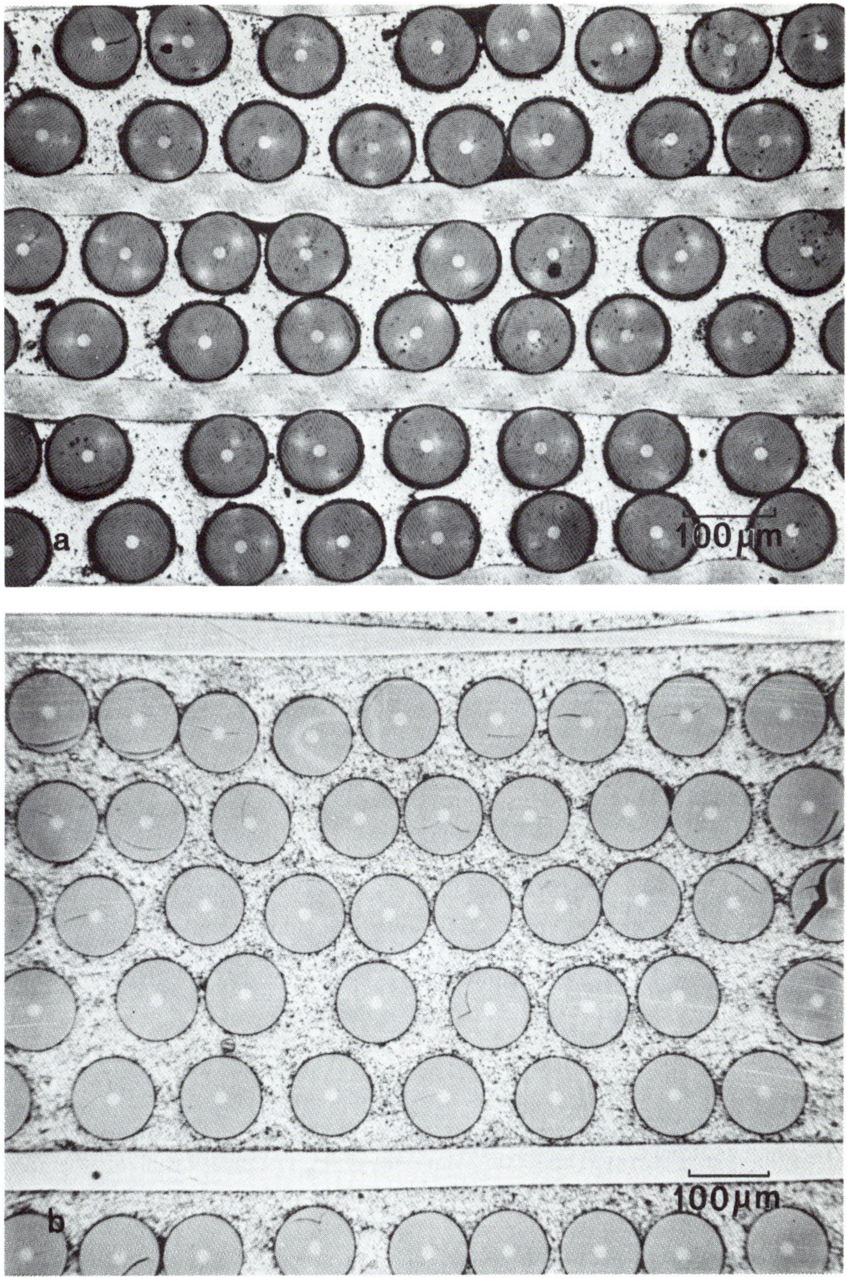

Fig. 9. Three-component composites: (a) BORSIC, aluminum, and titanium-foil composite (bonded from plasma-sprayed tapes); (b) BORSIC, aluminum, and AFC-77 steel wires (bonded from plasma-sprayed tapes).

found to be very difficult, but good monolayer tapes have been produced. Because of the relatively high cost of this technique, it has not been pursued as vigorously as others.

The powder-metal fabrication technique with boron fibers included the feeding of die-length fibers into a hot-press die set with layers of powdered aluminum surrounding the fiber. Some of the problems involved in the use of this technique are maintaining accurate fiber spacing, controlling fiber crushing at high filament volume fractions, and achieving good matrix-controlled properties under the pressing conditions required by the presence of the fibers. These problems have been discussed by Hirschorn and Sheehan (1969) and have limited the usefulness of powder metallurgy in this area.

The casting of aluminum–boron composites is complicated by the fiber-degradation reactions discussed above and by the general difficulty of making sound thin-wall aluminum castings. Because of the reaction between the boron fiber (coated and uncoated) with molten aluminum, the process has been performed by Camahort (1968) in times of less than 1 sec. The process has been limited to the infiltration of small bundles, where high strengths have been achieved. This limitation is probably related to the difficulties of manipulating large numbers of brittle fibers through an aluminum bath and maintaining accurate collimation. Vidoz *et al.* (1969) have used the infiltrated bundle for component fabrication by remelt bonding of long rod-shaped parts.

A technique of filament winding plus plasma spraying of the matrix has been used by Kreider and Leverant (1966a) to fabricate plates and large-diameter rings. The process yields a porous matrix that is fairly weak but can be improved by subsequent sintering or hot-pressing. Some early results on the testing of composites made by this procedure indicated the excellent high-temperature and fatigue strength of boron–aluminum.

IV. Secondary Processing

The term secondary processing is meant to include those manufacturing steps that are performed on basic composite shapes such as flat plates, beams, and tubes. Descriptions of the processes of forming, joining, machining, and heat-treating are presented. These processes would normally be performed by the component hardware manufacturer. Since most of the present applications are in the aerospace industry, most of this work has been performed by aerospace hardware manufacturers.

A. *Forming*

The forming of boron–aluminum composites relates to the forming of the constituents—a strong, nearly brittle fiber and a ductile, soft aluminum. The fiber has a truly elastic stress–strain behavior in a tensile test at room temperature and is very creep resistant at elevated temperatures. Since the maximum solid-state forming temperatures of the composite system are below 600° C, little plastic extension of the fibers is permissible without fiber fracture and very high fiber residual (elastic) stresses. On the other hand, at 400° C extensive creep deformation is possible in the aluminum matrix at very low stresses. The severe constraint placed on forming operations by the fiber has led to the practice of fabricating the part from the flexible preform sheets to the final shape during hot-pressing of the composite in most cases. This practice of fabrication to shape is widespread among manufacturers using lamination techniques with resin composites also. The constraint imposed on the composite by the fibers leads to maximum elongations to fracture in the axial direction of less than 1 %, and even lower elongation to fracture in the

Fig. 10. Creep-formed parts.

transverse direction in creep. High shear ductility and high creep elongation do permit creep-forming operations that require only matrix shear.

Weisinger (1970), Happe and Yeast (1969), Hersch and Duffy (1969), and Miller and Schaeffer (1971) have studied the formability of boron–aluminum composites in detail. They have found that by the minimizing of tensile stresses during creep forming, 4T and 5T bends can be formed in the composites. Examples of parts that were creep formed are shown in Fig. 10. The earlier creep-forming techniques actually used composites that had very low fiber concentration at a sharply curved region, but more recently the process has been used for homogeneous fiber concentrations. With the latter process, I-beams of 1.5-in. depth were creep formed to a radius of 60 in. in the lower flange. The creep-forming operation was performed gradually at 510°C for 8 hr.

B. Joining

The joining of the boron–aluminum composite to the attachments of the load-carrying structure is one of the most significant engineering areas in the application of the composite. Although the subject of composite joining implies a technology in many areas, this section is confined to the metallurgical bonding of the boron composites and will not treat composite joint design, the stress analysis of the joint, or aspects of aluminum joining that are well treated in the noncomposite literature.

The joining of aluminum–boron composites is an art based on the joining of aluminum, since no boron-to-boron bond is intended. Studies in joining aluminum–boron reported in the literature by Fleck *et al.* (1969), Weisinger (1970), Breinan and Kreider (1969), and Hersch and Duffy (1969). All have the intention of fabricating high-shear-strength joints of the matrix without degradation of composite mechanical properties. Since the joining of aluminum is a well-developed technology and the thermodynamic conditions for fiber degradation are known, the approach to the problem is well defined.

The use of large bonded areas in joints is required because of the extreme anisotropy of the mechanical properties of the composite. Since the axial strength of the unidirectionally reinforced composite may be 1.4–2.1 GN/m^2 and the shear strength of the joint may be typically 70–105 MN/m^2 (10–15,000 psi), a ratio of 20 : 1 in joint length to component thickness is implied. In addition, because of stress concentrations at the ends of the joint and peel stress, a considerable safety factor is appropriate in aluminum–boron joints. These factors preclude the use of butt joints for load carrying. Also, many other joint designs commonly used with isotropic materials are inappropriate. Scarf joints have been investigated by Breinan and Kreider (1969) and are useful if the aspect ratio is very small (approximately 3 in.). However, the most useful designs require large shear-stress transfer areas and increased section thickness to minimize stress-concentration effects. The boron–aluminum metal-bonded joint is actually very similar to the interlaminar shear zones of the composite itself, and similar fabrication techniques are required. The large joint areas make the use of fluxes highly questionable, since it becomes difficult to ensure complete removal.

The procedures for bonding include solid-state diffusion-bonding, brazing with braze filler alloy added, aluminum-welding techniques that require no added metal, and soldering. The solid-state diffusion-bonding of two aluminum–boron composites or an aluminum sheet to an aluminum–boron composite or aluminum–boron to titanium requires basically the same technology and equipment as the composite fabrication discussed earlier. High-strength joints with the shear strength of the matrix alloy can be achieved. Since high pressures are required for the bonding of aluminum to aluminum or to titanium alloys, this procedure is inconvenient and more expensive generally than brazing on a complexly shaped part, but it has been used for bonding boron–aluminum to titanium root blocks of turbine-engine fan blades. A boron-aluminum–titanium diffusion-bonded joint is presented in Fig. 11.

The soldering of the aluminum–boron compsite with high-zinc-bearing alloys such as the zinc–aluminum eutectic (MP = 380°C) has been used by Breinan and Kreider (1969) to obtain joint shear strengths in excess of 50 MN/m^2 (7000 psi). This joint does not have the inherent strength or ductility of the matrix aluminum alloy and may lack toughness for some applications. The authors used torch soldering and noted that in fluxless soldering, mechanical agitation of the joint was necessary in order to break up the oxide skin on the aluminum. This technique is probably less useful for typical applications than automated furnace brazing.

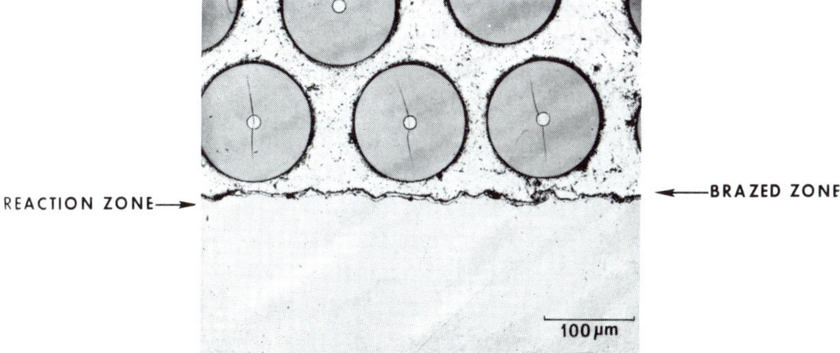

Fig. 11. Typical transverse brazed-joint microstructure (Borsic–6061 bonded to Ti–6Al–4V with brazing alloy 718).

Fluxless furnace brazing of boron–aluminum composites can be performed using standard aluminum brazing techniques if the cycle does not degrade the fiber. The standard technique is to insert a brazing foil between the parts to be joined and furnace braze with a contact pressure. With BORSIC fiber, alloy foils of 713 (Al–7 Si) or 718 (Al–12 Si) can be used, since the 590–610°C brazing cycle does not degrade the BORSIC fiber using a 6061 or 1100 matrix alloy. Boron fiber is severely degraded within a few minutes at those temperatures and presents more problems. Other brazing alloys with lower melting points such as 719 (Al–2.5Cu–9.5Si) may be more useful, particularly when available in foils. The use of high-speed, short brazing cycles would also improve the feasibility of using uncoated boron in composites that are to be brazed.

The dip-brazing process for aluminum–boron composite sheet has been reported by Miller and Schaeffer (1971). They have employed 718 brazing wire and dip brazed for 1 min at 590°C after a 540°C preheat. Excellent wetting and fillets are possible using the dip-brazing approach.

In summary, the braze-bonding of aluminum–BORSIC composites to aluminum or titanium is an attractive way of bonding the load-carrying composite into the structural end fittings. Both furnace-brazing foils and dip-brazing using wire filler have been successful. Shear strengths of the aluminum brazing alloys are near 20,000 psi at room temperature, and the joints can be fabricated by present technology and equipment.

The welding of aluminum alloys is used to achieved high-strength, corrosion-resistant, tough joints. In the welding of boron–aluminum, the most important added consideration is the degradation of fibers. This must be minimized by confining the heat-affected zone to as small a region as possible. Standard techniques to achieve this are rapid heat-up and cool-down, and using melt-pool geometries that have large depth-to-width ratios. An important secondary consideration in the welding of boron–aluminum is the residual stress state between the constituents. These joints can be annealed, but distortion due to the stresses should also be considered. Happe and Yeast (1969) evaluated tungsten–inert-gas, electron-beam, and spot welding. Hersch and Duffy (1969) have evaluated spot welding. They observed poor cross-tension strength in the spot welds and difficulties in distinguishing a sound nugget from a weak point using ultrasonic nondestructive testing techniques. Problems in controlling the location of the nugget arise from the variation of conductivity of the composite with the base aluminum alloy (2024). The spot-welding technique also has the disadvantage of having a low shear-stress transfer area. The geometrical disadvantage of welding, compared to brazing, which can incorporate large surfaces with very little penetration, is significant in the joining of boron–aluminum composites.

Alternative approaches such as mechanical fastening and adhesive bonding are also significant joining techniques for the composites. Adhesive bonding employs similar techniques to that used with the resin composites. Since the adhesives have considerably lower strengths than the brazing fillers, much lower shear stresses are permitted in the joints. Similarly, mechanical fastening can be used, with two primary reservations: the bearing allowables and pull-out stresses must be determined for the anisotropic composite material; and care must be taken to ensure that no significant damage is generated in the machining of the composite to accept the fastener.

C. Machining

Boron–aluminum composites are difficult to machine because of the high hardness (9 on the MOHS scale) of the boron fiber. However, in the monolayer form, sheet-metal shearing techniques are adequate. As the thickness of the composite is increased, the damage to the cut region increases rapidly. The problem of machining tensile coupons has generally been solved by either abrasive cutoff wheels or electron-discharge machining. The standard diamond-impregnated brass cutoff wheel yields a fine surface on the cut, and apparently the aluminum does not smear or damage an abrasive wheel of proper grit size because of the cleaning action of the boron itself.

One of the most extensive studies on the machining of aluminum–boron has been reported by Forrest and Christian (1970). They reported results on eight types of machining operations. Electron-discharge machining was favored because of the modest amount of damage, but the slow feed rates on this technique as well as on electrolytic grinding are a drawback. They found shearing and punching useful only on very thin sheets. Abrasive cutoff and grinding were considered acceptable processes, and diamond routing and diamond drilling were considered acceptable but did lead to excessive tool wear. Studies by Happe and Yeast (1969) indicated excellent results using ultrasonic machining. Later, Miller and Schaeffer (1971) concluded that the rotary ultrasonic machine using a diamond-impregnated core drill was the most effective drilling method available for producing holes in boron–aluminum composites. With each process, the quality of the cut surface, particularly with respect to crushed or split fibers, must be balanced against such cost factors as rate of cutting and tool life.

D. Thermomechanical Treatments

In the aluminum–boron system, several matrix alloys have been chosen that are strengthened by age hardening. This includes the aluminum–copper–magnesium alloy 2024, the aluminum–magnesium–silicon alloy 6061, and aluminum–zinc alloys such as 7178. The effect of the age hardening of the matrix alloy on the properties of the composite is complex because of the strong interaction with the fiber, which affects the residual stress state. However, standard heat treatments on these alloys have been found to be useful by Summer (1967), Hancock (1971), and Prewo and Kreider (1972a, b).

The standard T4 and T6 heat treatments for the three above alloys include a high-temperature, long-time solution heat treatment. The solution heat treatment for 2024 or 7178 does not normally cause degradation of boron fiber; however, the higher temperature used for 6061 is excessive and is not recommended for the uncoated-boron composite. Modification of the solution temperature has been used by Forrest and Christian (1970) to prevent serious fiber degradation. BORSIC fibers are not degraded during the standard solution heat treatment. A second modification to the heat treatment, which includes a liquid-nitrogen quench, has been used and later studied by Hancock and Swanson (1971). This quench changes the room-temperature residual-stress state between the fibers and matrix. These heat treatments are greatly affected by mechanical or thermal stress cycling and have not received the wide application in the composites that have been used with the base alloy.

Cold-rolling of the composites has been used to increase the axial and transverse strength. Getten and Ebert (1969) investigated rolling of aluminum–boron with the fibers perpendicular to the rolling direction. Modest amounts of increased axial strength of unidirectionally reinforced composites were noted, but the effect of the rolling on the transverse properties was not reported.

V. Mechanical Properties

A. *Elastic Constants*

Almost all current engineering machines and structures are designed to operate under loads within the elastic range of the materials involved. The elastic constants, which determine structural dimensions under load, are of considerable importance in these applications, since clearances and tolerances must be maintained during load application. In the aerospace industry, where metal-matrix composites are finding early application, the elastically controlled deflections of structural parts are of particular significance in the control of air flow. In addition, the elimination of destructive structural resonances in the dynamic performance of aircraft components, particularly rotating engine parts, is a vital design consideration. The selective use of boron–aluminum to strengthen or stiffen metal structures also depends on accurate knowledge of material elastic constants, since it is the ratios of component-material moduli that determine the distribution of load within the structure.

Unidirectionally reinforced boron–aluminum can be considered as an orthotropic material exhibiting transverse isotropy with five independent elastic constants. However, boron–aluminum is frequently used as a thin lamina, which is also the unit of construction for complex lay-ups. It is then treated as a thin orthotropic lamina in a state of plane stress, which requires only four independent elastic constants. These constants can be given as the axial elastic modulus E_{11}, the transverse elastic modulus E_{22}, the major Poisson's ratio ν_{12}, and the in-plane shear modulus G_{12}. A complete explanation of the constitutive relationship of composite materials is given by Ashton *et al.* (1969), where it is shown that the calculation of the elastic constants for composite materials can be performed with considerable accuracy using expressions frequently referred to as the Halpin–Tasi equations, using values of the elastic properties of the constituent materials. These general equations provide good agreement with the more specific results of other analytical studies. The elastic constants of the constituent boron fiber and aluminum matrices are given in Table IV.

The longitudinal elastic modulus E_{11} of boron–aluminum composites is predicted accurately by the rule-of-mixtures formulation. This equation is given as

$$E_{11} = E_F V_F + E_M V_M \tag{4}$$

where the subscripts F and M refer to fiber and matrix, and V designates volume fraction. The accuracy of this expression is typified by the data presented in Fig. 12 for BORSIC–aluminum. The transverse elastic modulus

TABLE IV

Elastic Constants

	Boron Fiber	Aluminum and its alloys[a]
Young's modulus E		
(10^6 psi)	55–60	10
(10^{10} N/m^2)	39–43	7.15
Poisson's ratio ν	0.21[b]	0.33
Shear modulus G		
(10^6 psi)	16–20	4
(10^{10} N/m^2)	11.5–14.3	2.9

[a]The elastic constants of aluminum and its alloys are substantially independent of alloy composition. The precise moduli are given by K. Van Horn (1967, Vol. II) for a number of alloys.

[b]Galasso (1969).

[c]Herring and Krishna (1966).

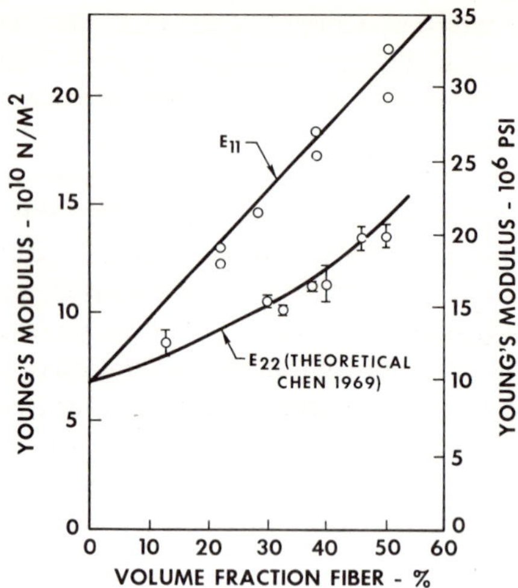

Fig. 12. Composite axial and transverse elastic moduli for 6061 aluminum reinforced with 100-μm BORSIC.

cannot be expressed as simply; however, good agreement between theory and experiment has been obtained by Adams and Doner (1967), Chen and Lin (1969), and Prewo and Kreider (1972a). Data are presented in Fig. 12. The anisotropy of elastic moduli is not very large (not as large as the anisotropy in strength) and, for the frequently used composite composition of 50 v/o fiber, the ratio of longitudinal to transverse modulus is approximately 3:2, as compared to much larger ratios for organic-matrix composites. It is notable that the specific axial and specific transverse elastic moduli of a 50 v/o boron–aluminum composite are approximately 300 and 200% greater, respectively, than the specific moduli of most current engineering metal alloys.

Both the axial and transverse moduli decrease with increasing temperature (Tsariff *et al.*, 1969; Lenoe, 1968; Kreider *et al.*, 1971). The major decrease in composite modulus is due to the decrease in matrix modulus, although a small decrease in fiber modulus also occurs (Ellison and Boone, 1967; Metcalfe and Schmitz, 1969).

The variation of four independent elastic constants of uniaxially reinforced BORSIC–aluminum composites due to rotation of the composite specimen axis with respect to fiber axis has been determined by Kreider *et al.* (1971). It was shown that the variation of these constants with angle is in good agreement with basic transformation equations described by Tsai (1966) and that measurements using both rotating and rigid grips provide similar results. The amount of shear coupling, which would affect the accuracy of the measurement, is small in BORSIC–aluminum.

Fewer data are available in the literature on the values of composite Poisson's ratio and shear modulus. A value of 0.23–0.28 has been shown to be appropriate for ν_{12}, while G_{12} is equal to approximately 6×10^{10} N/m^2 for composites containing 50 v/o fiber (Schaeffer and Christian, 1969; Kreider *et al.*, 1968, 1971).

Boron–aluminum composites will frequently be used in structures as multiaxially reinforced panels. For this reason, the elastic moduli of these arrays are also of considerable importance. Data obtained by Schaeffer and Christian (1969) for composites in the cross-ply (0–90°) and + 30° lay-up configurations indicate that both of these lay-ups provide Young's moduli substantially independent of specimen test orientation, i.e., quasi-isotropic in the plane of the fibers.

B. Composite Strength and Stress–Strain Behavior

The strength and overall stress–strain behavior of a nonhomogeneous orthotropic material such as boron–aluminum is, of necessity, complex. As in monolithic engineering materials, final structural composite performance is a result of original materials properties and composition as well as processing and fabrication history. The discrete nature of the distribution of fiber and matrix and the observable thermodynamics of the interaction between these phases, however, provide a metallurgical and structural system more amenable to quantitative analytical description than most previous engineering materials.

As has been discussed in a previous section, the boron–aluminum system comprises of a large family of possible fiber–matrix combinations. The list of possible aluminum matrices is limited only by the number of available engineering alloys. Each alloy, because of its physical and chemical properties, is capable of altering final composite characteristics. A list of alloys and some of their properties is presented in Table III. Each of the alloys enumerated has been used in the past as a matrix for boron fibers. The metallurgy of these alloys is well documented, so that the characteristics of each, when situated in the composite, can be predicted on the basis of metallurgical history.

Boron fiber, as has been noted, also can be obtained in various forms. It is currently available in several diameters and with or without a coating. Both of these factors affect the characteristics that the fiber brings to the composite. As examples, the larger 150-μm-diam boron fiber exhibits a higher transverse tensile strength than the 100-μm-diam fiber, while a SiC-coated fiber is more resistant to chemical attack than uncoated boron. In general, currently available boron fiber is characterized by an average axial strength of 35×10^9 N/m^2 (500×10^3 psi). Because it is a "brittle" material sensitive to flaws, it is also typified by a statistical spread in strength, which can be represented by a coefficient of variation of 5–25%. Both the average and distribution of strengths are reflected in final composite performance.

1. Axial Tension

The axial tensile behavior of boron–aluminum composites is dominated by the characteristics of the reinforcing fibers and the composite fiber content. Axial composite strength and failure strain are controlled by fiber properties, with less significant modifications in stress–strain behavior being related to matrix characteristics and residual stress state.

A. STRESS–STRAIN BEHAVIOR. The axial stress–strain behavior of unidirectionally reinforced boron–aluminum can be described by considering the individual and combined performance of the elastic–plastic matrix and the elastic fiber.

Axial stress–strain curves typical of the behavior of boron–aluminum are presented in Fig. 13. The behavior of both as-fabricated (F condition) and heat-treated (T-6 heat treatment after primary fabrication) are given in the figure. The T-6 heat treatment is based on the heat-treating procedures used

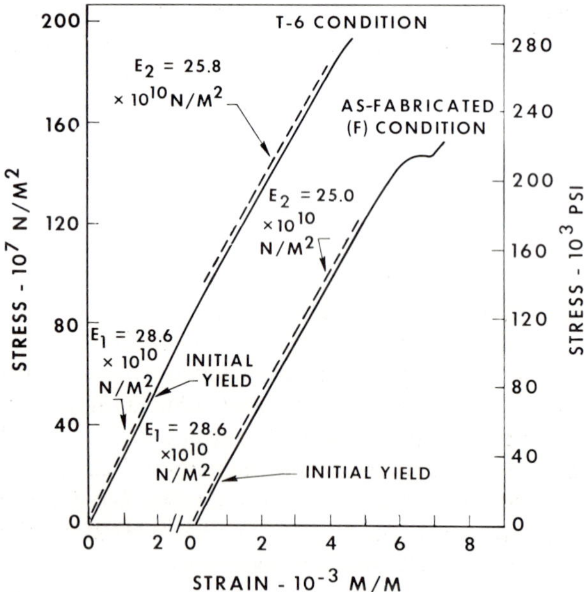

Fig. 13. Stress–strain curves for 2024 aluminum reinforced with 64 v/o 150-μm boron, tested in axial tension.

to strengthen unreinforced aluminum alloys and consists of an elevated-temperature solution heat treatment followed by a water quench and precipitation heat treatment. All temperatures and times at temperature must be chosen to avoid fiber-strength degradation by fiber–matrix reaction. Both stress–strain curves in Fig. 13 are characterized by an initial linear region (stage I), followed by a nonlinear region of transition prior to another linear region (stage II), and finally a nonlinear region prior to failure. This type of composite stress–strain behavior has been noted by many investigators for not only boron–aluminum but also other metal-matrix composite systems such as copper–tungsten by McDaniels *et al.* (1965), Stuhrke (1968), and Kreider *et al.* (1968, 1971). Stage I corresponds to the region of truly elastic behavior of both matrix and fiber and is terminated by the onset of plastic yielding of the matrix. The slope of stage I is equal to the primary elastic modulus of the composite and, as discussed in the previous section, can be related to the elastic moduli of the matrix and fiber using the rule of mixtures. The composite tensile stress at which primary linear behavior ends is dependent upon the proportional limit of the aluminum matrix material and the residual-stress state in the composite prior to tensile-stress application. In the nonlinear region that precedes stage II, the rate of work hardening of the matrix and the propagation of plastic deformation throughout the composite matrix determine the stress–strain curve shape. Stage II, although nearly linear, is not a region of truly elastic composite behavior, since it is representative of elastic-fiber and plastic-matrix deformation, the latter occurring at a nearly constant rate of work hardening. The deformation taking place in this region is not totally reversible. The slope of this stage can be calculated using Eq. (4), where the matrix contribution is determined by the rate of matrix work hardening. Because this is negligible compared to the fiber elastic modulus, the slope of stage-II behavior, sometimes referred to as the secondary modulus, can be accurately calculated by Eq. (5) for most fiber volume fractions:

$$E = E_F V_F + (d\sigma_M/d\varepsilon_M) V_M \approx E_F V_F \tag{5}$$

The final region of the composite stress–strain curve represents the failure process of the boron–aluminum composite. The shape and extent of this region relate to boron-fiber failure within the composite, which is accompanied by heightened specimen acoustic emission. The shape of the stress–strain curve in this final stage can vary considerably from specimen to specimen because of the existence of many possible combinations of events that can ultimately lead to final specimen rupture.

The role of residual stresses in boron–aluminum composite behavior is important in both uniaxial and multiaxially reinforced material. The composite residual-stress state is determined by the thermal and mechanical history of the composite. Residual stresses arise when boron–aluminum is heated or cooled because of the difference in matrix and fiber coefficients of thermal expansion (approximately 24×10^{-6} and 4.9×10^{-6} per °C, respectively). Thus, since boron–aluminum composites are fabricated at elevated temperatures, residual stresses are present after primary composite synthesis. On cooling from elevated temperature, the matrix is left in tension and the fibers in compression. The magnitudes of these stresses cannot be calculated purely on the basis of elastic behavior because of significant plastic relaxation of the matrix. This subject, together with thermal expansion of boron–aluminum, was treated in detail by Kreider and Patarini (1970). Composite behavior and the role of residual stresses are schematically represented in Fig. 14. The stress–strain behavior of a composite containing 50 v/o boron is presented with the stress–strain behavior of both fiber and matrix com-

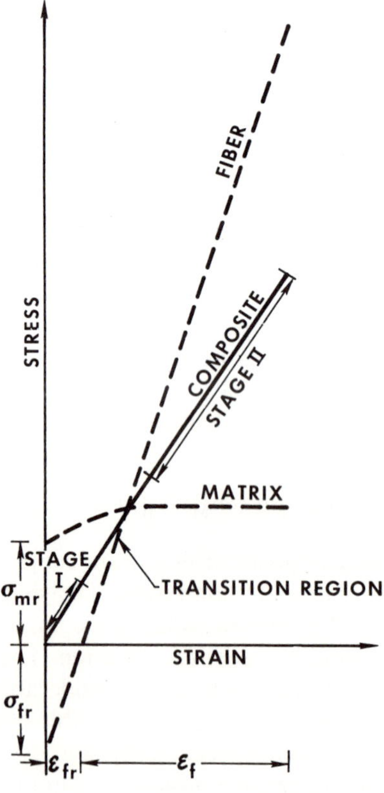

Fig. 14. Schematic of matrix, fiber, and composite stress–strain curves for aluminum composite reinforced with 50 v/o boron, tested in axial tension.

ponents. On cooling, the matrix has contracted plastically and elastically while the fibers have contracted only elastically. The matrix is left with a net average tensile residual stress σ_{Mr} equal in magnitude (for a 50 v/o-fiber composite) to the net compressive residual stress σ_{Fr} on the fiber. On tensile stressing of the composite, stage I, the transition stage, and stage II occur in sequence and correspond to matrix–fiber behavior of elastic–elastic, plastic-yield–elastic, and plastic–elastic, respectively. The net compressive residual stress on the boron fiber serves to increase, by ε_{Fr}, the total effective strain the fiber can undergo prior to composite failure. The effective fiber-failure strain obtained by tensile testing an array of unconstrained boron fibers is given in the figure by ε_F. The ultimate composite failure strain is thus also increased, since axial composite failure is controlled by boron-fiber failure. Increasing the net compressive residual strain on the fiber by heat-treatment procedures can therefore increase composite failure strain and ultimate strength over those values typical of the as-fabricated composite. Cheskis and Heckel (1968, 1970) have verified the above rationale of composite behavior using X-ray techniques to measure both fiber and matrix stress (for tungsten–aluminum) and matrix stress (for boron–aluminum) prior to and during composite tensile testing. Similarly, it has been shown that composite axial tensile strength can be significantly increased by thermal (Menke and Toth, 1970) as well as by thermomechanical treatments (Shimizu and Dolowy, 1969; Dolowy and Taylor, 1969; Ebert *et al.*, 1969). It should also be noted that the T-6 heat treatment also increases the matrix yield strength and ultimate strength, which can also affect composite axial tensile strength by changing the matrix-strength contribution as well as the critical load-transfer length and the ability of the matrix to distribute changes in stress state due to localized fiber breakage. These complex effects, combined with the ability to degrade the strength of boron fiber by reaction with the matrix at elevated temperature, can also result in no change or even a decrease in composite axial strength with heat treatment, as documented by Toth (1969) and Christian and Forest (1969).

B. AXIAL TENSILE STRENGTH. The high axial tensile strength of boron–aluminum composites is due to the high tensile strength of the reinforcing fibers. Other factors, such as matrix composition and residual stress, are of lesser, although still significant, importance. Figure 15 illustrates the dependence of composite strength on fiber content. Also indicated in the figure are the type of fiber and the date of presentation of the data. In general, for a given type of fiber, composite strength has increased with increasing fiber content. The highest values of strength are the more recently reported. This latter point is the result of improvement in the quality of boron and BORSIC fiber resulting from added production experience and research as well as the

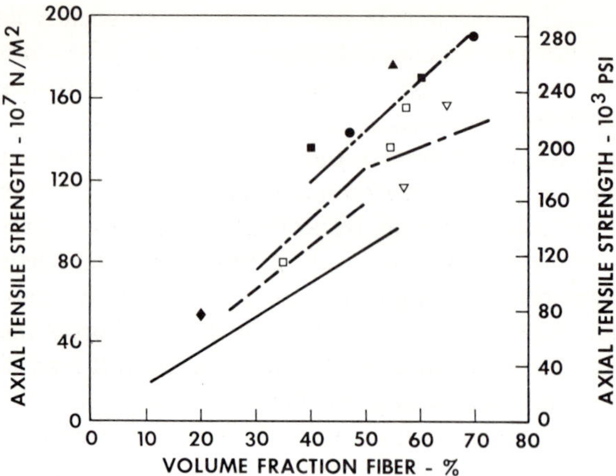

Fig. 15. Composite axial ultimate tensile strength as a function of fiber content.

- ● Kreider *et al.* (1971), 145-μm boron
- ▲ Prewo and Kreider (1971), 145-μm boron
- □ Kreider *et al.* (1972), 145-μm BORSIC
- ■ Menke and Toth (1971), 100μm boron
- ◆ Anthony and Chang (1968), 100-μm boron
- ▽ Ryder *et al* (1970), 100-μm nitrided boron
- — Kreider and Leverant (1966), 100-μm boron
- ---- Kreider *et al.* (1968), 100-μm BORSIC
- -- Menke and Toth (1970), 100-μm boron
- ---- Prewo and Kreider (1971), 100-μm boron

continued improvement of composite fabrication techniques. The effective strength characteristics of the reinforcing fiber in the fabricated structural part are the most significant fiber properties. This *in situ* strength can be much less than the as-received fiber strength because of degradation of the fiber by chemical reaction or fiber breakage.

The variation of composite axial tensile strength with fiber content has been found to be essentially linear for fibers having reproducible strengths. Large deviations from linearity have usually been due to variations in fiber strength from specimen to specimen. The linear dependence of axial composite strength on fiber volume fraction has suggested the use of a rule-of-mixtures approach to show the dependence of composite strength σ_C on fiber strength σ_F and matrix strength at the filament fracture strain σ_M:

$$\sigma_C = \sigma_F V_F + \sigma_M V_M \qquad (6)$$

The use of this equation to describe the failure criterion of the composite system can be very helpful; however, it can also be deceptive. The quantities σ_F and σ_M are complex functions of the fiber and matrix properties tested separately as well as the performance of these components *in situ* in the bonded composite. The matrix yield behavior is affected by the constraint of the surrounding fibers and, at the least, the fiber performance can be affected by the ability of the matrix to transfer loads over small fiber lengths. The fiber-fracture stress in the composite is frequently analyzed on the basis of the strength of a parallel bundle of fibers, referred to as a bundle strength (Corten, 1967; Tsai *et al.*, 1966). The bundle strength depends on both the average fiber strength and the distribution of strength as well as the fiber gauge length and has been used for the value of σ_F in calculations based on Eq. (6) (Shaefer and Christian, 1969) with fair success. Good agreement has also been achieved, however, by use of just the average fiber strength, which, in general, is greater than the bundle strength (Anthony and Chang, 1968). Another approach has been to relate composite strength to component properties and composite structure by including the effects of stress concentrations. Fiber breakage, occurring randomly throughout the composite, causes local stress concentrations, which then control composite failure (Rosen, 1964, 1970; Zweben and Rosen, 1970). This type of model has also been successful in describing the strength of boron–aluminum and agrees well

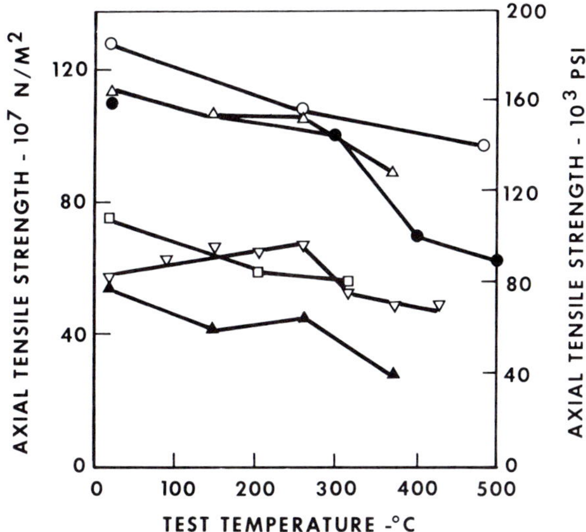

Fig. 16. Composite axial ultimate tensile strength as a function of test temperature.
○ Kreider *et al.* (1971), 48 v/o 145-μm boron–6061
▲ Schaefer and Christian 1969), 25 v/o 100-μm boron-6061
▽ Anthony and Chang (1968), 22 v/o 100-μm boron–1100/2024
□ Toth (1969), 25 v/o 145-μm boron–6061
△ Schaefer and Christian (1969), 50 v/o 100-μm boron–6061
● Kreider (1969), 50 v/o 100-μm boron–6061

with observations of the occurrence of enhanced fiber breakage in the presence of stress concentrations due to notches and broken fibers (Kreider *et al.*, 1971). In this latter approach the matrix achieves increased importance because of its role in transferring stress around local fiber breaks (see p. 22).

Another mechanistic representation of boron–aluminum composite failure has been suggested by Mullin *et al.* (1968). These authors suggested that the sudden release of elastic energy during initial fiber fracture can cause a shock wave capable of fracturing neighboring fibers. Herring (1972) has demonstrated the validity of this argument for a series of boron–aluminum specimens in which a chain reaction of fiber failure is initiated at a threshold stress that is dependent upon statistical fiber-strength distribution.

The dependence of boron–aluminum composite axial tensile strength on test temperature is illustrated in Fig. 16. This dependence is related primarily to the decrease of fiber strength with increasing test temperature (Veltri and Galasso, 1971) and also the occurrence of fiber–matrix reaction, which can degrade fiber strength. This latter point is more important at temperatures above 430°C and times longer than those used to perform the tensile tests represented by the data in the figure. The fact that elevated-temperature axial strength depends mainly on fiber strength points the way for improved composite high-temperature strength by improving fiber properties. Herring (1966) has shown that the elevated-temperature strength of boron can vary considerably because of variations in fabrication procedure and that boron fiber strengths of 2×10^9 N/m^2 at 1100°C have been obtained.

Variations in matrix strength and residual-stress relief can also affect the dependence of composite strength on test temperature. This is reflected in the change in fracture morphology of BORSIC–aluminum axial tensile specimens as shown in Fig. 17. At room temperature fracture surfaces are characterized by very little fiber pull-out in well-bonded composites. The fibers that do protrude from the average plane of fracture are frequently still covered with aluminum because of the strong fiber–matrix interface developed during bonding. At elevated temperature the composite surface roughness is increased and longer lengths of fiber protrude because of the larger shear-transfer length of the matrix at the higher temperatures. Another feature of the fracture surfaces of axial tensile specimens is the fragmented appearance of many of the fibers (Breinan and Kreider, 1970; Jones, 1969). The fragments are retained on the fracture surface by their bond to the aluminum matrix.

2. *Transverse Tension*

The transverse tensile properties of boron–aluminum composites are an important consideration in structural applications. In many cases the transverse composite tensile strength of a 50-v/o fiber-reinforced composite

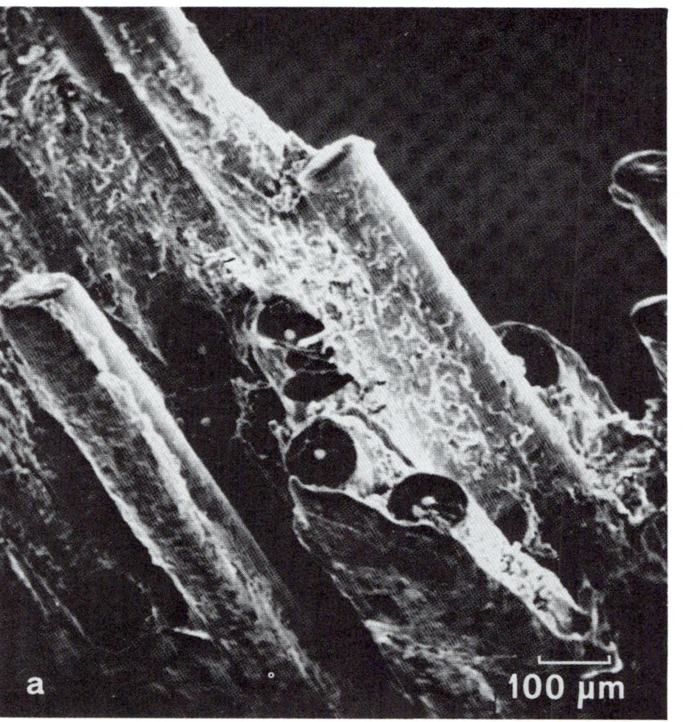

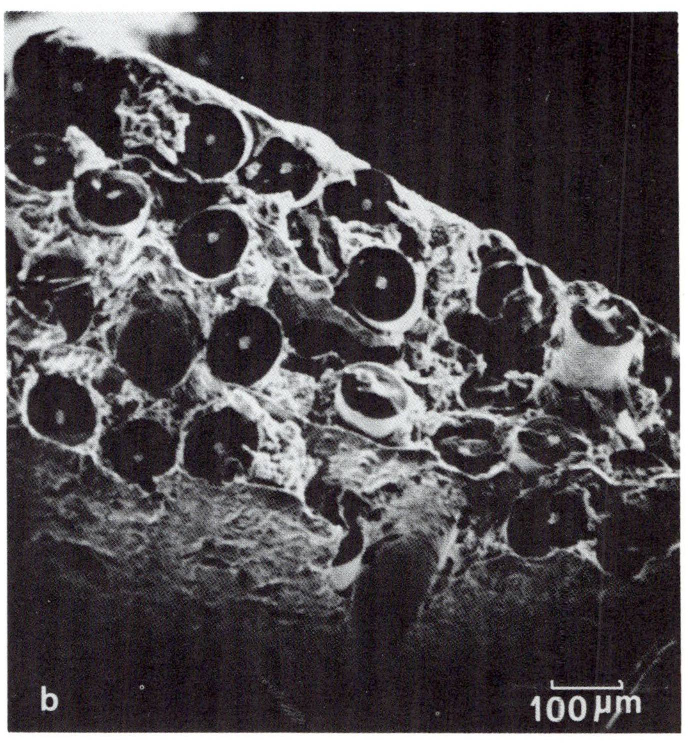

Fig. 17. Axial-tension fracture surfaces of BORSIC-reinforced 6061 aluminum: (a) 500°C, (b) 25°C.

is only 10–15% of the axial tensile strength, while the transverse elastic modulus is approximately 60% of the axial modulus. This large anisotropy in strength is a major reason for the use of multiaxial composite reinforcement. Transverse composite properties are a sensitive function of the properties of both the matrix and the fiber.

A. STRESS–STRAIN BEHAVIOR. The transverse tensile performance of all boron–aluminum composites can be divided into three general categories that are related to the morphology of the composite fracture surface. Type-I composite failure is characterized by a fracture surface consisting completely of matrix rupture, while type-II composite fracture surfaces consist of both matrix rupture and longitudinally split fibers. In type II, the fracture path is easiest through the fibers. Type-III fracture is typified by matrix rupture and fiber–matrix interfacial failure. This last class of failure will not be considered further here, since it has not, in the past, been typical of the performance of well-bonded boron–aluminum and has instead been an indication of an inadequately controlled composite fabrication technique. Interfacial failure does occur in the transverse tensile failure of other aluminum-matrix composites (Chen and Lin, 1969; Lin *et al.*, 1971) because of the inability to form a strong fiber–matrix bond. In the case of boron–aluminum, however, a bond is easily formed that can be stronger than the currently used aluminum-alloy matrices. An indication of the type of bond that can be formed has been demonstrated by Prewo and McCarthy (1972) and is shown in Fig. 18, where the interfacial region between a silicon-carbide-coated boron fiber and a 6061 aluminum matrix is shown in transmission electron microscopy. The interface is completely free of voids, and the aluminum alloy replicates every detail of the fiber surface.

Whether type-I or -II composite failure occurs is dependent upon the relative strengths of matrix and fiber when the composite is subject to transverse tensile loading. Several investigators (Adams and Doner, 1967; Adams, 1970; Greszczuk, 1971) have shown that a complex stress state is generated in the composite during transverse normal loading and that both fiber and matrix are subjected to localized principal stresses significantly higher than that applied to the composite. It is component performance under this stress state that determines fracture morphology. It was first suggested that composite transverse fracture would occur in the aluminum matrix (type I), because of the much higher tensile strength of the boron fibers. This assumption did not prove to be the general case (Kreider, 1969; Jones, 1969; Chen and Lin, 1969; Prewo and Kreider, 1972a) because boron and BORSIC fibers can exhibit much lower transverse strengths than axial tensile strengths. Transverse fiber strengths less than 10% of the axial fiber strength have been measured (Kreider and Prewo, 1972) and are related to the presence of radial

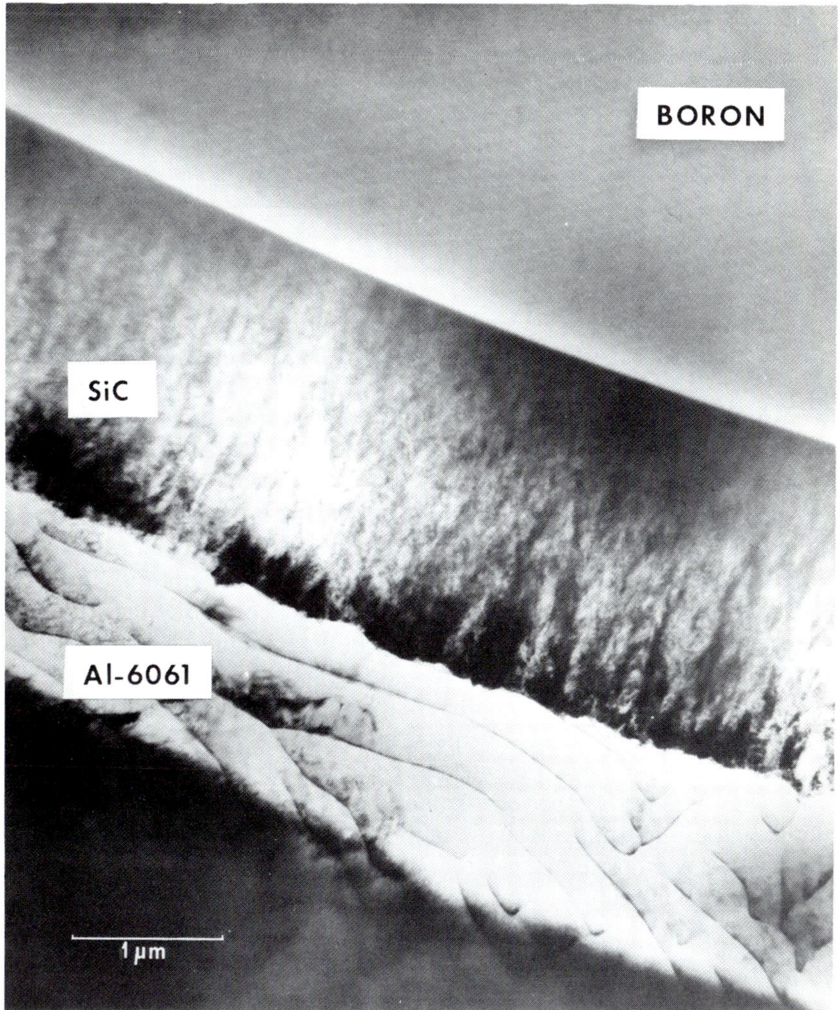

Fig. 18. Transmission electron microscopy photo showing interface of SiC-coated boron fiber in 6061 aluminum matrix.

flaws in 100-μm-diam boron and BORSIC. The larger 150-μm boron and BORSIC fibers tested in the same investigation exhibited significantly higher transverse strengths.

Composite stress–strain curves typical of type-I and -II composite performance are presented in Fig. 19. The composites reinforced with 150-μm BORSIC fiber failed primarily by matrix failure and are representative of

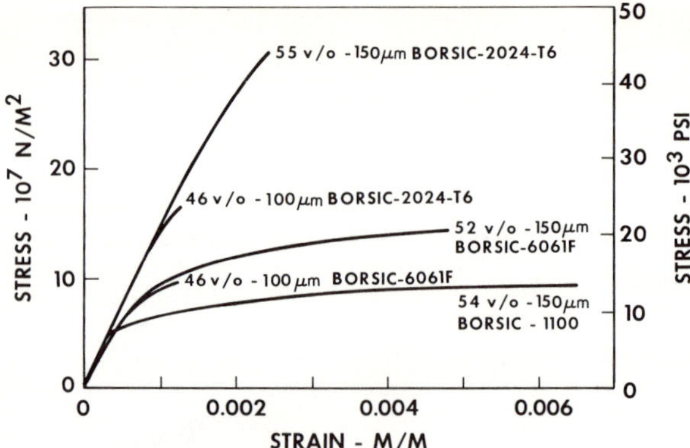

Fig. 19. Stress–strain curves for BORSIC-reinforced aluminum tested in transverse tension at 25°C.

type-I behavior. A typical fracture surface for these composites is illustrated in Fig. 20a. Some longitudinal fiber splitting is visible at the composite edges; however, the fracture is predominantly one of matrix rupture. The edge splitting is due to fiber damage introduced during composite machining to produce the tensile specimen. The 100-μm BORSIC-reinforced composite is representative of type-II performance; a typical fracture surface is given in Fig. 20b. Note that every fiber in the composite cross section is split along its axis. In general, composite elastic performance is the same for both types of composite behavior. Only after yield has occurred do marked differences appear. The ultimate fracture stress and failure strains of those composites exhibiting fiber splitting are less than those typified by matrix failure. In the case of type-I composites the major portions of the stress–strain curves, in the as-fabricated condition, occur after composite yield, and both the composite yield strength and tensile strength increase with increasing matrix strength. Composite failure strain also increases with increasing matrix failure strain.

B. TRANSVERSE TENSILE STRENGTH. The dependence of composite transverse tensile strength on unreinforced matrix strength and fiber type is presented in Fig. 21. The strength of the composites containing 150-μm BORSIC increases continuously with increasing matrix strength. The indicated near equality between composite and matrix strength agrees well with the predictions of Chen and Lin (1969) for composites containing fibers in either square or hexagonal arrays. It should be noted that composite heat treatment, to increase matrix strength, is very effective in increasing composite transverse strength. All of the 150-μm-BORSIC-reinforced composites failed predominantly by matrix rupture (type I). This type of behavior is also typical of 150-

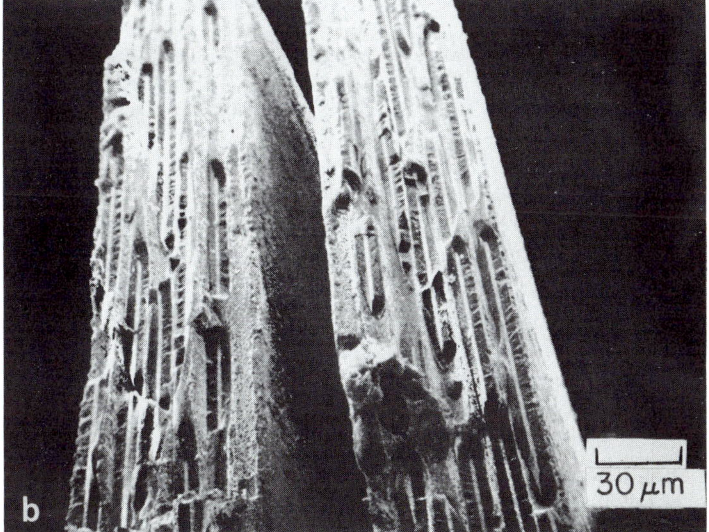

Fig. 20. BORSIC–aluminum transverse-tension fracture surfaces: (a) type I with predominance of matrix rupture; (b) type II with predominance of longitudinal fiber splitting.

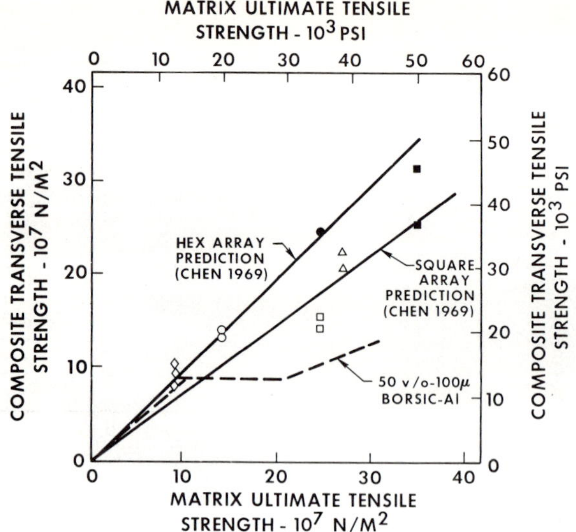

Fig. 21. Composite transverse tensile strength as a function of matrix strength and matrix composition.

○ 60 v/o 145-μm BORSIC–6061F
● 60 v/o 145-μm BORSIC–6061-T6
△ 57 v/o 145-μm BORSIC–5052/56
□ 56 v/o 145-μm BORSIC–2024F
■ 56 v/o 145-μm BORSIC–2024-T6
◇ 54 v/o 145-μm BORSIC–1100

μm-boron-reinforced composites over the same entire matrix strength range. In the case of 100-μm BORSIC, this equality between composite and matrix strength, and type-I fracture, extends up to a matrix strength of approximately 9.3×10^7 N/m^2. Matrices of strength greater than this cause fracture to occur predominantly through the BORSIC fibers, with an accompanying departure from the Chen and Lin prediction (Prewo and Kreider, 1972a). The dependence of composite strength on matrix strength also extends to the variations in the strength of a given matrix due to test temperature, as shown in Fig. 22.

The dependence of composite transverse tensile strength on the volume fraction of fiber is given in Fig. 23 for 150-μm BORSIC, 100-μm BORSIC, and 100-μm boron. Composite transverse strength is substantially independent of fiber content for those matrix–fiber combinations that are conducive to matrix failure, i.e., the transverse fiber strength is sufficiently high to prevent fiber splitting from occurring. The 150-μm-BORSIC–6061 composites exemplify this type of behavior, while the performance of the 100-μm-BORSIC–2024 composites is characterized by large amounts of fiber splitting and a decrease in composite strength with increasing fiber content. Data for the 100-μm-boron-fiber-reinforced composite are divided between both types

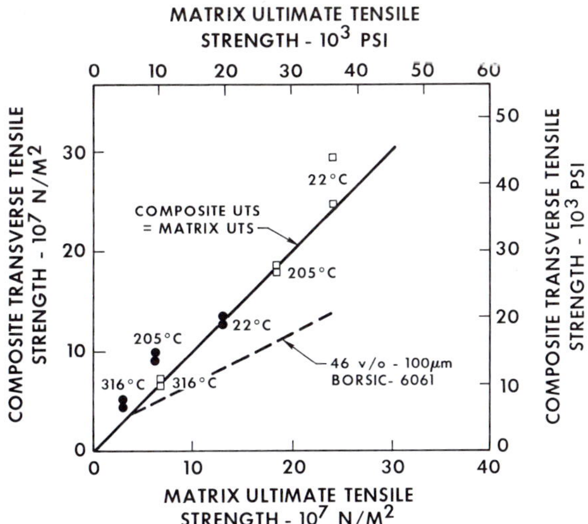

Fig. 22. Composite transverse tensile strength as a function of matrix strength and test temperature for 6061 aluminum reinforced with 60 v/o 150-μm BORSIC. ● as fabricated; □ T-6 condition.

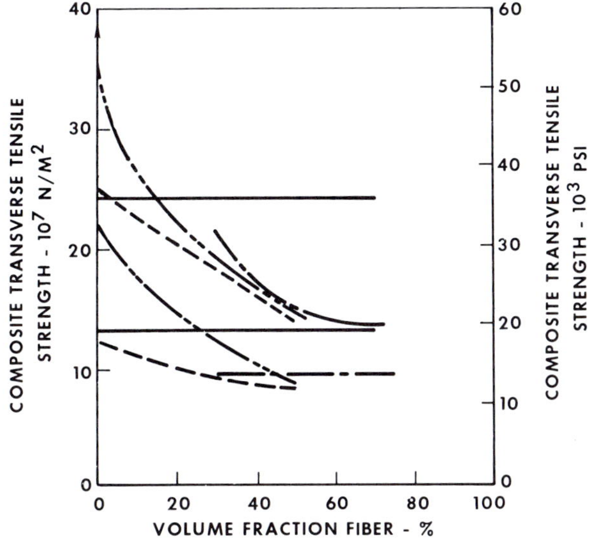

Fig. 23. Composite transverse tensile strength as a function of fiber content.

- – – – 100-μm boron–6061 (F and T-6), Lin *et al.* (1971)
- – · – 100-μm boron–6061 (F and T-6), Menke and Toth (1970)
- – · · – 100-μm boron–2024 (F and T-6), Prewo and Kreider (1972a)
- —— 145-μm BORSIC–6061 (F and T-6), Prewo and Kreider (1972a)

of behavior and illustrate that considerable differences in composite performance can occur despite a similarity in fiber type. This may relate to the observation that *in situ* transverse fiber strength depends both on inherent fiber strength and on the presence of flaws introduced by machining and other operations prior to test (Prewo and Kreider, 1972a).

3. Tensile Performance as a Function of Angle

The major portion of composite evaluation, in the past, has been performed by tensile testing uniaxially reinforced specimens parallel to or normal to the fiber axis. The variation of composite strength over the entire intervening range is also of importance and can be discussed in terms of the major theories of strength presently in use. These theories are given below and have been reviewed in detail by Azzi and Tsai (1965), Tsai (1966), and Ashton *et al.* (1969):

1. Maximum stress

$$\begin{aligned}\sigma_\theta &\le X/\cos^2\theta \\ &\le Y/\sin^2\theta \\ &\le S/\sin\theta\cos\theta\end{aligned}$$

2. Maximum strain

$$\begin{aligned}\sigma_\theta &\le X/(\cos^2\theta - \nu_{,12}\sin^2\theta) \\ &\le Y/(\sin^2\theta - \nu_{,21}\cos^2\theta) \\ &\le S/\sin\theta\cos\theta\end{aligned}$$

3. Maximum work

$$\frac{1}{\sigma_\theta{}^2} = \frac{\cos^4\theta}{X^2} + \frac{1}{S^2} - \frac{1}{X^2}\cos^2\theta\sin^2\theta + \frac{\sin^4\theta}{Y^2}$$

where σ_θ is the composite strength at angle θ to fibers, X the composite strength parallel to the fibers, Y the composite strength perpendicular to the fibers, and S the shear strength. The maximum-stress and maximum-strain theories assume that three independently operating failure modes occur, which depend on orientation: (1) fiber tensile failure is expected at low angles, (2) matrix shear failure is expected at intermediate angles, and (3) matrix plane-strain tensile failure is expected at angles close to 90°. The maximum-work theory considers interactions and assumes that failure will occur when the distortional energy exceeds a critical level.

A. STRESS–STRAIN BEHAVIOR. The tensile stress–strain curves for 100-μm-BORSIC–6061 composites are presented in Fig. 24. These curves indicate a continuous decrease in strength with increasing loading angle from the fiber direction. Composite strength has decreased by fully 25% after a rotation

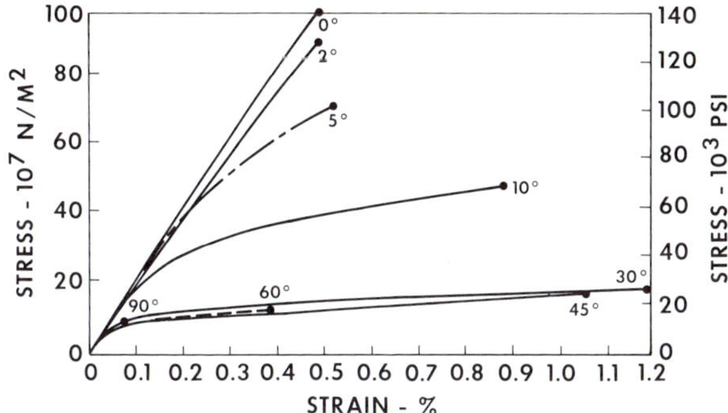

Fig. 24. Stress–strain curves for 6061 aluminum reinforced with 100-μm BORSIC, tensile tested at indicated angles to principal fiber axis.

of only 5° from the fiber axis. The general decrease in strength is accompanied by increasing composite failure strain, which reaches a maximum at the intermediate angle of 30° and then decreases again. This type of behavior has also been reported by Kreider and Marciano (1969) for 100-μm-BORSIC–6061, Toth (1969) for 100-μm-boron–6061 composites, and Tsariff *et al.* (1969) for 100-μm-boron–7178-T6. Toth (1969) has also shown that the general form of this behavior is typical of composite behavior at elevated temperatures of up to 600°F. In one investigation (Adsit and Forrest, 1969) some evidence was found to indicate that the strength of 100-μm-boron-6061 composites was slightly lower at 60° than 90° to the fiber axis; however, this was found to be true only of thin (0.020-in.) specimens, not thicker (0.080-in.) specimens tested at the same time, and is not common to the other investigations reported above.

The fracture path of the specimens enumerated in Fig. 24 changed from being nearly normal to the loading axis, at angles of up to 5°, to being usually parallel to the filament axis at the larger angles. Both fiber splitting and matrix shear occurred at the larger angles, with the latter feature dominant at the highest angles. Similar tests, performed on 5.7-mil BORSIC composites (Kreider *et al.*, 1971) exhibit the same trends in stress–strain behavior and fracture morphology, except that fiber splitting did not occur to any large extent at any test angle. This is in agreement with the previously discussed dependence of transverse fracture mode on fiber type.

B. TENSILE STRENGTH. The tensile strength for 150-μm-BORSIC–6061-T6 composites is presented as a function of test angle in Fig. 25. From this figure it can be seen that the maximum distortional energy criterion is capable of fitting the data over the entire range of angles tested. The other criteria of maximum strain or stress are not nearly as satisfactory, particularly at the

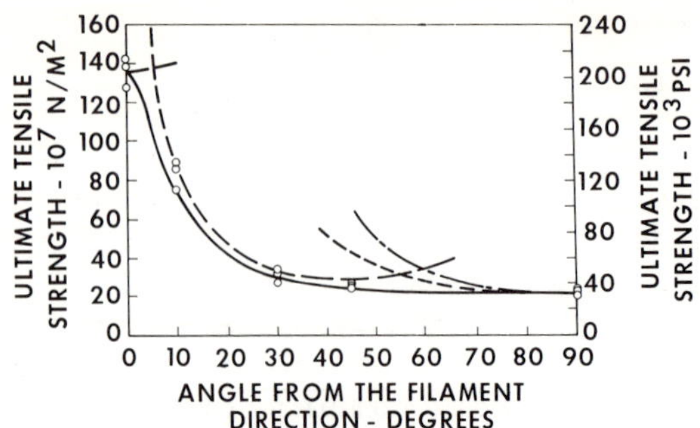

Fig. 25. Composite tensile strength as a function of angle to principal fiber axis for 6061 aluminum reinforced with 150-μm BORSIC, heat-treated to the T-6 condition.

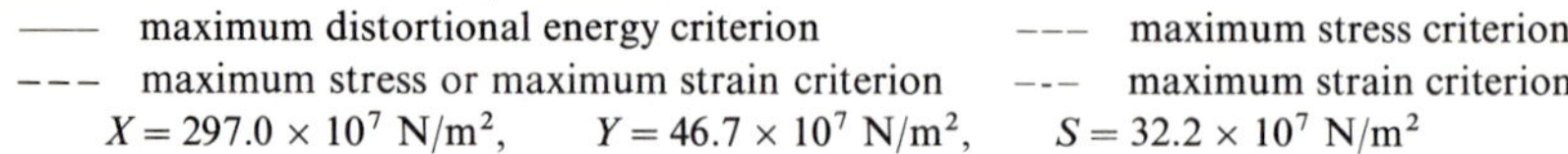

—— maximum distortional energy criterion   --- maximum stress criterion

--- maximum stress or maximum strain criterion   --- maximum strain criterion

$X = 297.0 \times 10^7$ N/m^2, $Y = 46.7 \times 10^7$ N/m^2, $S = 32.2 \times 10^7$ N/m^2

low angles, where the strength is observed to decrease with increasing angle. the agreement between the distortional-energy expression and experiment is equally good for composites whose failure mode includes longitudinal fiber splitting as well as other fiber–matrix combinations (Toth, 1969; Kreider and Marciano, 1969; Tsariff *et al.*, 1969). This universal applicability, without regard for failure mode, has made the distortional-energy failure formalism very useful in the description of boron–aluminum composite behavior.

4. Compression

In general, it has been found that the compression strength of boron–aluminum composites is equal to or superior to composite tensile strength. However, the compression strengths reported depend strongly on the type of test used to determine these strengths and the failure criterion used. Both structural (buckling) and material (brooming, shear, crushing) instability criteria have been used. The axial compressive strength of continuously cast BORSIC–aluminum tubes has been reported to reach 1.75×10^9 N/m^2 (Abrams and Davies, 1971), while a value of 2.14×10^9 N/m^2 was achieved for diffusion-bonded BORSIC–6061-aluminum (Kreider and Marciano, 1969), with failure occurring by brooming of the specimen ends. This latter value of compression strength was approximately equal to twice the axial ultimate strength of a tensile specimen of the same material. Similarly it has been shown that specimens compressed at 30, 60, and 90° to the principal fiber axis all exhibit compressive strengths in excess of the tensile strength.

C. Creep and Stress Rupture

Significant aerospace applications of boron–aluminum require the prolonged exposure of this material to stress at elevated temperature. Suitability for these applications, such as jet-engine fan blades and vanes, is strongly dependent upon creep and stress-rupture properties. The performance of boron–aluminum during prolonged exposure at high temperature is more complicated than that of most monolithic materials because of changes in residual stress patterns and reaction between fiber and matrix, as well as metallurgical processes in each separate component. As in the previously described case of tensile data, the properties of composite specimens tested colinear with the fiber axis are the highest. Off-axis composite properties decrease rapidly with deviation from this principal direction because of the increased role of unconstrained matrix shear.

The axial creep and stress rupture of unidirectionally reinforced boron–aluminum composites are superior to all currently available engineering alloys at temperatures of up to 500°C. This is due to the exceptional creep resistance of the boron fiber, as was illustrated by Ellison and Boone (1967).

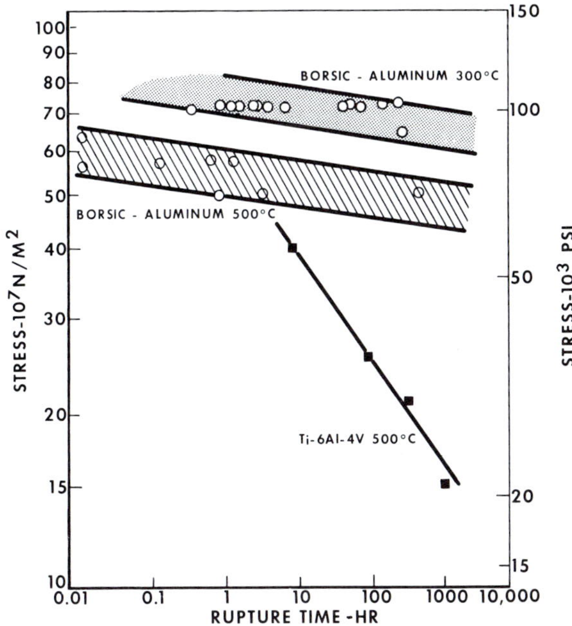

Fig. 26. Applied tensile stress as a function of time to rupture for 6061 aluminum reinforced with 100-μm BORSIC and for unreinforced Ti–6Al–4V.

These authors showed that no measurable fiber creep occurred at temperatures of up to 650°C, and that at 815°C boron fibers exhibited significantly lower creep rates than drawn tungsten wire. Creep and stress-rupture data have been reported for boron–aluminum and BORSIC–aluminum by Kreider and Leverant (1966a), Kreider *et al.* (1968), Toth (1969), Kreider and Breinan (1970), Anthony and Chang (1968), and Menke and Toth (1970, 1971). The excellent axial properties obtained by these authors are typified by the data of Fig. 26, which illustrates the stress for rupture of 100-μm-BORSIC–6061-aluminum as a function of rupture time at 300 and 500°C. The data are compared with 500°C data for the engineering alloy Ti–6Al–4V. The creep tests performed at these temperatures failed to reveal any significant composite plastic deformation. The largest permanent elongation noted was was 0.2%. It should be noted that these tests were performed with the entire specimen and grips heated to the test temperature. Tests performed on continuous-fiber-reinforced composites between cold grips can prejudice the resultant data by eliminating the need for transfer of load by the matrix at elevated temperature.

The importance of fiber–matrix compatibility in the stress-rupture performance of boron–aluminum has been demonstrated by Breinan and Kreider (1970). At a 400°C exposure temperature, it was demonstrated that

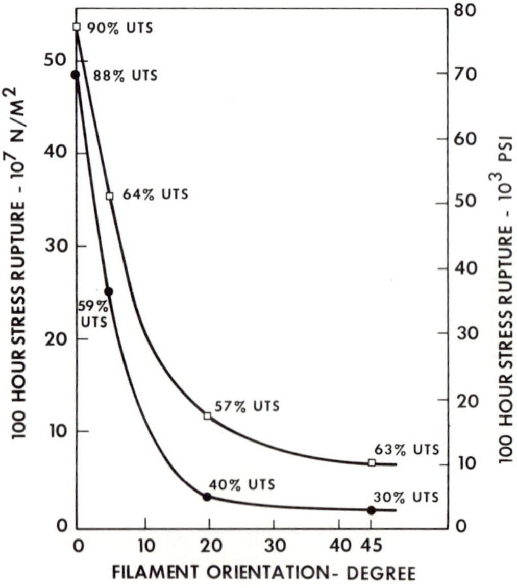

Fig. 27. Applied tensile stress to cause rupture at 100 hr as a function of angle to principal fiber axis for aluminum reinforced with 100-μm boron. Test temperature: ● 316°C, □ 205°C.

specimens reinforced with silicon-carbide-coated boron fiber exhibited twice the stress of uncoated-boron-reinforced aluminum for rupture at 100 hr. At shorter times, the effect was less significant than would be expected for degradation of fiber strength by fiber–matrix reaction.

The off-axis composite creep and stress-rupture performance decreases rapidly with increasing angle from the filament axis, as shown by Toth (1969) (Fig. 27). This is due to the low creep and stress-rupture properties of aluminum-alloy matrices. As in the case of composite tensile properties, the off-axis creep performance can be increased by cross-plying laminates, adding off-axis reinforcements, or altering matrix composition. The success of these last two approaches has been demonstrated by Breinan and Kreider (1973) for transverse creep and stress rupture (Fig. 28). The best stress-rupture performance of the three aluminum-alloy-matrix composites represented in the figure was associated with the heat-treated 2024 matrix. This matrix is also superior to the others at 300°C in the unreinforced condition. Further improvements in transverse creep and stress-rupture performance were obtained by the incorporation of titanium foils, steel wires, and foils of sintered aluminum powder (SAP).

D. Notched Tensile Strength and Fracture Toughness

Composite materials containing relatively high fractions of boron or other brittle high-modulus reinforcements exhibit nearly elastic behavior and a limited strain capability when loaded in the direction of fiber alignment. This occurs because, under isostrain conditions, the relatively high-modulus filaments carry the bulk of the load and largely determine the longitudinal modulus. In view of anticipated engineering applications of these materials in man-rated structures and the associated concern with material performance

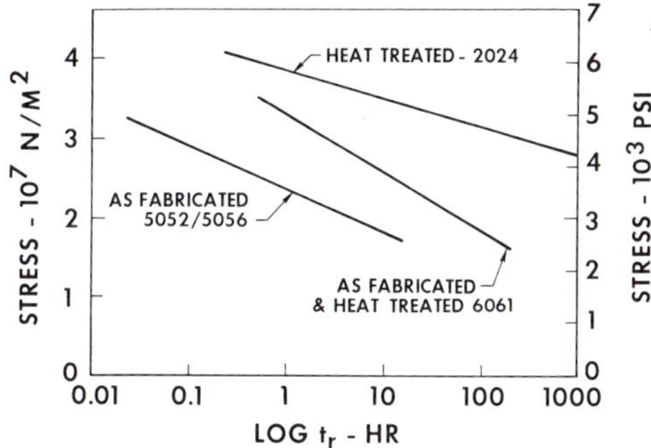

Fig. 28. Applied transverse tensile stress at 300°C as a function of time to rupture for aluminum reinforced with 100-μm BORSIC. (Matrix designations are given in the figure.)

in the presence of flaws, it is logical to inquire whether the fracture of these materials can be successfully treated within the framework of existing linear elastic fracture mechanics.

Linear elastic fracture theory is a development of the elastic stress-intensity factor concept and the Griffith–Irwin relationship (Tetelman and McEvily, 1967):

$$\sigma_f = (EG/\pi C)^{1/2} \quad \text{(infinite plate)}$$

where σ_f is the gross fracture stress, C the half crack length at the onset of unstable crack extension, E the elastic modulus, and G the crack extension force or energy per unit area of crack extension. This expression arises from equating the elastic energy release associated with incremental crack extension to the energy required to create new surface and perform local plastic work. The critical-stress system criterion is derived by describing the stress field ahead of the crack with elastic-stress-field equations that contain a $C^{-1/2}$ type singularity. This stress criterion is characterized by the stress intensity factor K:

$$K = \sigma(\pi C)^{1/2} \quad \text{(uniformly stressed infinite plate)}$$

where σ is the fracture stress, and C the half crack length. In the fully elastic case,

$$K = (EG)^{1/2} \quad \text{(plane stress)}$$

Fracture processes in fiber-reinforced composite materials are far more complex than in monolithic materials as a result of the complicating effects of anisotropy and inhomogeneity. Anisotropy has been treated by many writers, and it has been shown that in the general anisotropic case, the crack-displacement modes cannot be directly associated with the three fundamental modes of isotropic fracture mechanics. Under symmetric loading, for example, forward sliding will occur as well as opening. However, if the flaw is oriented along one of the principal directions of elastic symmetry, and propagation is in a direction collinear with the original crack, the situation becomes identical to the fracture criterion expressed by the Griffith–Irwin relationship (Wu, 1963, 1965, 1967, 1968a, b; Wu and Reuter, 1965; Sih, 1965; Paris and Sih, 1965).

Considerations of inhomogeneity are less well developed, since fracture mechanics are based on a continuum mechanics viewpoint. Ductile-matrix-composites present particular difficulty because, although they are nearly elastic and demonstrate limited strain capability, they are not truly brittle. The system 50 v/o BORSIC plus 6061-F aluminum, for example, undergoes transition from stage-I (fiber elastic, matrix elastic) to stage-II (fiber elastic, matrix plastic) behavior under uniaxial tension at approximately 1500

± 500 μin./in. (depending on the thermal and mechanical history). Thus, one-half of the volume yielded at a composite stress of about 50 ksi. If this system were notched, extensive matrix plasticity (at the notch tip) would surely occur. In fact, if $\pm 45°$ laminates are tensile tested, more than 10% is measured because the fiber does not impose the severe constraint of a 0° or 90° orientation. Under these conditions, it is not clear whether the stress singularity will be of the form $C^{-1/2}$. Expressions involving log C have appeared in some plasticity analyses by Weiss and Yakawa (1965) and Liebowitz (1968).

Two general types of composites have been encountered and reported in the literature that are reinforced with high-strength brittle fibers. One type has a ductile matrix and is well bonded to the fiber; the second has either a brittle matrix or a brittle, weak interfacial zone between the fiber and the matrix. Either of these nonhomogeneous materials could generate deformation and fracture modes that undermine key assumptions made in linear elastic fracture theory. In the case of the ductile matrix, a plastic shear deformation at relatively low stresses can uncouple the constraining influence of the notch. In the composite with brittle, weak interfaces or matrix, a shear crack collinear with the fibers and the loading direction can accomplish the same decoupling effect. The blunting of cracks by plastic shear deformation is to be greatly preferred to cracking, since the cracking seriously reduces many other properties in addition to axial tensile strength. Among the properties degraded by cracking are flexural and transverse tension modulus, resonance frequencies, in-plane shear strength, transverse tensile strength, fatigue resistance, torsional rigidity and strength, resistance to stress corrosion and corrosion, and numerous other related properties. These properties are much less affected by plastic shear deformation in the matrix.

The fracture toughness of boron– and BORSIC–aluminum has been investigated by Adsit and Forest (1969), Hancock and Swanson (1971), Olster and Jones (1971), Kreider and Dardi (1972), and Hoover and Allred (1972), and in each case it was found that the notch insensitivity and fracture toughness of the composites was outstanding. Both mechanisms of notch blunting (plastic deformation and interfacial splitting) were observed to occur in these investigations. It was shown that when BORSIC–aluminum specimens are poorly bonded, uniaxially reinforced axial tensile specimens are completely insensitive to notches oriented perpendicular to the fiber axis. This was due to the low transverse tensile strength of these composites (3100 psi) as compared to the axial strength (80,000 psi) and the occurrence of fiber–matrix interfacial failure. This is similar to observations on carbon–epoxy, S-glass–epoxy, and BORSIC–titanium, where in each case interfacial splitting also caused nearly complete notch insensitivity (Beaumont and Phillips 1972; Prewo and Kreider, 1972c). With increasing transverse composite strength the mechanism of notch blunting in boron–aluminum changed over to that of matrix plasticity collinear with the reinforcing fibers and an increased notch sensitivity. Even in this case, however, specimen notch sensitivity was less than that of an engineering alloy such as Ti–6Al–4V. A comparison of data is given in Fig. 29.

The values of fracture-toughness parameter *K* required to propagate cracks transverse to the fiber axis were found in every case to equal or exceed those of the unreinforced aluminum matrix alloys used. In addition, composite toughness increased with increasing fiber content. Although these values are encouraging, it has not yet been clearly established that these data can be treated in the same manner as those for current monolithic engineering alloys. This is best illustrated by the fact that the gross fracture strength of BORSIC–aluminum, as shown in Fig. 29, does not vary linearly with the inverse of the square root of flaw size. A power of -0.27, rather than -0.5, is more appropriate and relates to the outstanding ability of the composite to blunt the preexistent crack. It is also of interest to note that attempts at attaining a "sharpened" notch by fatiguing will be useless. This technique, used in the fracture-toughness testing of monolithic engineering alloys (Brown and Srawley, 1967), further isolates and blunts the preexistent flaw.

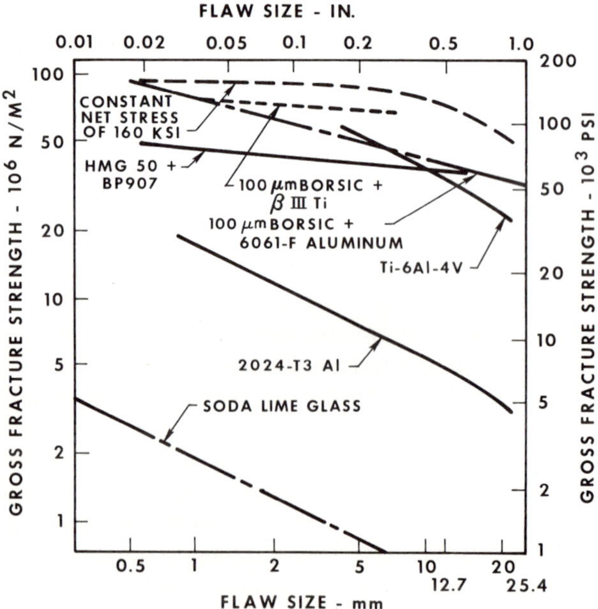

Fig. 29. Notched-specimen gross fracture strength as a function of machined flaw size. Fiber-reinforced and unreinforced materials are both represented.

E. Impact

The performance of boron–aluminum under impact loading is of considerable importance for many engineering applications. As in the case of monolithic materials, one of the major considerations in evaluation of impact resistance is the choice of the type of laboratory test to be used. This question is of particular importance in the evaluation of newly developed composite materials where a backlog of experience relating laboratory-specimen performance to service life is not available. The continued development of fracture mechanics is a prominent factor in the resolution of this difficulty. At present, however, a wide range of diverse test techniques is being used to classify composite impact performance for particular applications. It is beyond the scope of this chapter to reconcile the data obtained and to rank boron–aluminum on a generalized material impact scale. Some general conclusions, however, based on impact-test results, can be drawn with regard to fracture energy and fracture morphology.

The impact energy of BORSIC–aluminum composites has been studied by use of the Charpy impact test. This type of test, although developed for use on mild steel and in itself not sufficient to define fundamental material impact resistance, is capable of demonstrating composite performance in a widely recognized frame of reference. The high loading rates applied, substantial material section thickness, and fairly sharp notch acuity provide a severe test of material impact resistance. The strong dependence of impact energy on specimen orientation is demonstrated by the data of Fig. 30 (Kreider *et al.*, 1971). A factor-of-3 variation in energy absorbed is dependent upon the relative orientation of crack plane and fiber axis. The 1 orientation absorbs the greatest amount of energy because of the requirement of the crack to fracture each fiber in a tensile mode parallel to the fiber axis. This

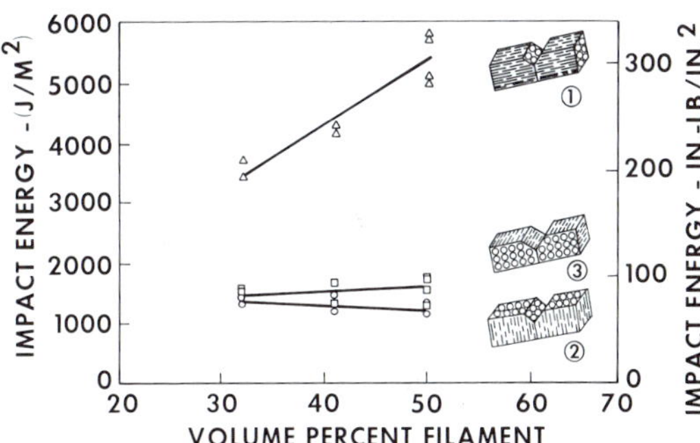

Fig. 30. Notched Charpy impact energy as a function of specimen orientation and volume fraction of fiber. Material is 6061 aluminum reinforced with 100-μm BORSIC.

mode of crack propagation requies a large amount of elastic energy to be supplied and generates regions of intense matrix plasticity around each fiber. Examination of a typical specimen fracture surface (Fig. 31) indicates the relevance of a plastic-flow mechanism, as does the increase in impact energy with increasing volume fraction of brittle (BORSIC) phase. Each of the fibers protruding from the fracture surface in Fig. 31 is coated with a layer of aluminum. The fiber–matrix interface was not the principal region of failure; instead, failure occurred by plastic deformation and rupture of a sheath of aluminum around each fiber. Each of the other specimen orientations, 2 and 3, requires much less energy for fracture and indicates little dependence on fiber content over the range of fiber volume fraction studied. Specimen performance in all orientations will, of course, differ significantly from the above at very low fiber contents due to the very high impact energy of the 6061 aluminum-alloy matrix phase. In both the 2 and 3 specimen orientations, the composite-specimen fracture surfaces consist primarily of longitudinally split fibers. This fracture morphology is similar to that described previously for the fracture of transverse tensile specimens. By using a fiber with a higher transverse tensile strength, 150-μm boron or 150-μm BORSIC, this fracture topography can be altered to one primarily characterized by matrix failure; however, no appreciable increase in specimen fracture energy accompanies this transition.

The 1 specimen configuration has been tested for a wide range of matrix alloys, fiber types, and volume fractions of fiber reinforcement. It was found that all of the data obtained followed a common dependence on composite properties of volume fraction of fiber (V_F), fiber diameter (d_F), fiber tensile strength (σ_{uF}), and matrix shear yield strength (τ_{My}). This dependence follows the formalism developed by Kelly (1966) to represent fiber pull-out and is shown in Fig. 32. The dependence on the expression ($V_F d_F \sigma_{uF}^2/\tau_{My}$) is apparent and was proposed by the authors to be due to matrix shear on planes parallel to the fiber axes and unrecovered elastic energy. Another result of this work was the determination of the dependence of composite impact energy on specimen geometry. The impact energy per unit area of type-1 specimens decreased with increasing ratio of notch depth to specimen thickness (dimension collinear with direction of crack propagation), while no dependence was found for a fourfold decrease in specimen width (dimension collinear with notch root) from the Charpy standard. This latter point illustrated that material constraint, due to section size, is effective at a specimen width of only 25 mm and is due to the effectiveness of the fibers in constraining matrix plasticity. The dependence of impact energy per unit area on notch depth is important for both engineering purposes and in the

Fig. 31. Fracture surface of type-1 notched Charpy impact specimen.

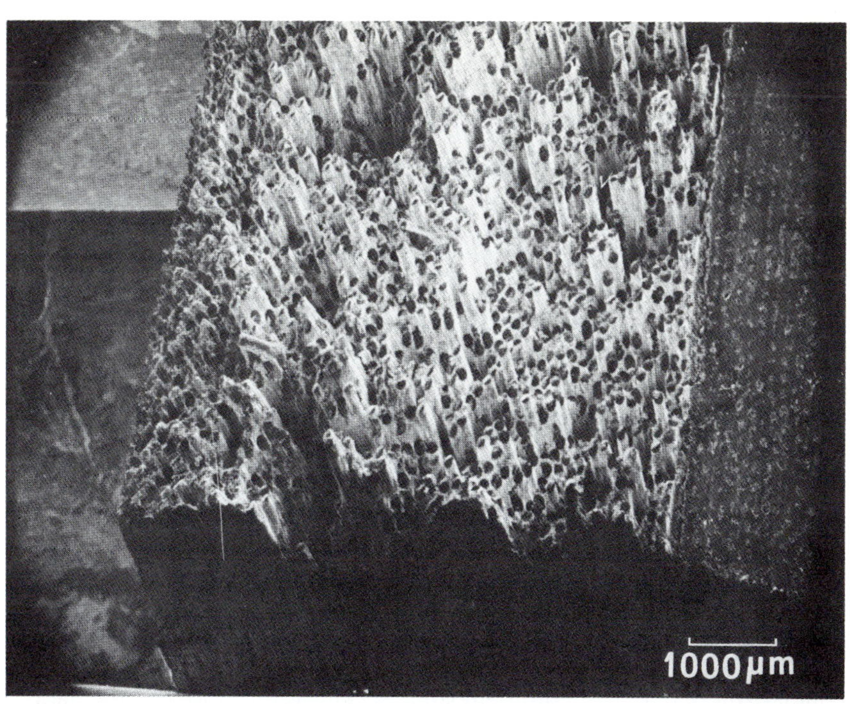
1000 μm

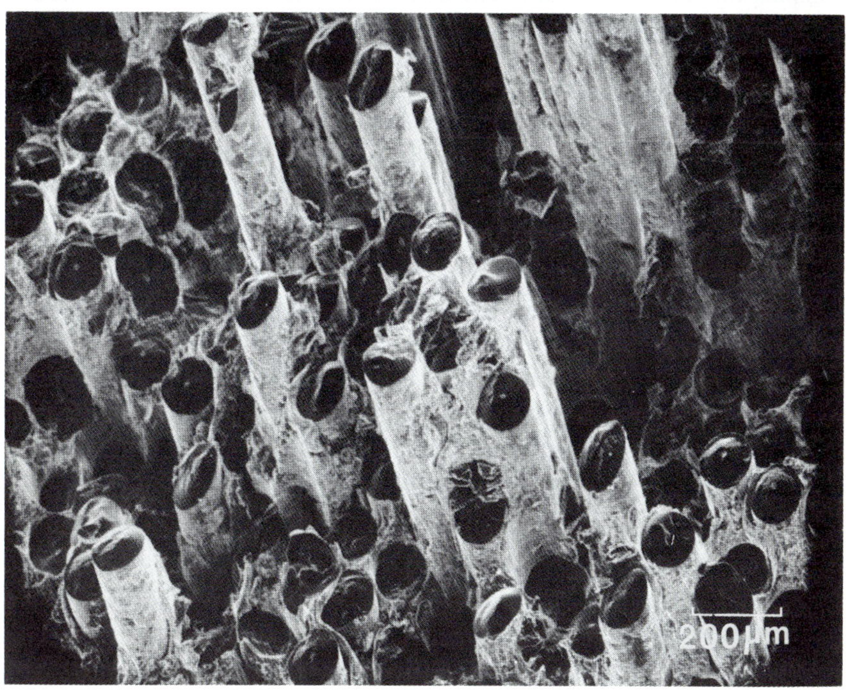
200 μm

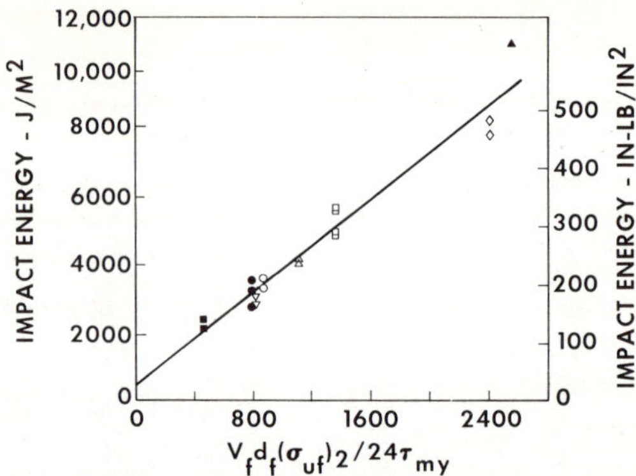

Fig. 32. Notched Charpy impact energy of type-1 specimens as a function of volume fraction of fiber V_F, fiber diameter d_F, fiber strength $\sigma_{\nu F}$, and matrix yield shear strength τ_{My}.

○ 32 v/o 100-μm BORSIC–6061-F
△ 41 v/o 100-μm BORSIC–6061-F
□ 50 v/o 100-μm BORSIC–6061-F
◇ 63 v/o 145-μm BORSIC–6061-F
▲ 63 v/o 200-μm BORSIC–6061-F
● 50 v/o 100-μm BORSIC–6061-T6
▽ 50 v/o 100-μm BORSIC–2024-F
■ 50 v/o 100-μm BORSIC–5052/5056-F

comparison of data from different sources. This is illustrated by the data of Jackson *et al.* (1971), who have also determined the impact energy of boron–aluminum. The impact energies per unit area obtained by these authors for type-1 specimens are considerably lower than those reported by Kreider *et al.* (1971); however, the ratio of notch depth to specimen thickness of the specimens used was much larger than that of a standard Charpy specimen. Comparison of impact energy per unit area between the above two investigations on the basis of similar axial composite strengths and similar specimen geometries shows good agreement for specimens in the fully annealed condition. Some disagreement still exists, however, in that Jackson noted no significant change in impact energy by specimen heat treatment while Kreider *et al.*, in accordance with Fig. 32, showed a large decrease in impact energy due to T-6 heat treatment.

Data reported by Prewo (1972) have shown that by control of matrix strength and using high-strength 5.6-mil-diam boron fibers, the impact energy of boron–aluminum can exceed that typical of some wrought engineering alloys. Values of greater than 30 J have been obtained for pure-aluminum-matrix composites and greater than 19 J for 6061-matrix composites. The results of this investigation also indicate that specimen impact energy can be varied by altering specimen fabrication procedures. This is similar to the results of Baker and Cratchley (1966), which indicated that a threefold increase in impact energy could be obtained in silica-fiber-reinforced aluminum by decreasing the hot-pressing temperature to avoid fiber-strength degradation and promote interfacial failure.

F. Fatigue

The performance of boron–aluminum in fatigue loading is excellent and is one of the major assets of this system to recommend it for structural applications. This performance is primarily due to the high fatigue resistance of the boron fibers. The aluminum matrix is also important, since it can effectively blunt local stress concentrations during fatigue cycling by plastic deformation. This is in marked contrast to the behavior of many monolithic engineering materials, whose sensitivity to flaws is severe during fatigue.

Boron–aluminum composite specimens have been tested in fatigue using cantilever bending (Kreider and Leverant, 1966a; Menke and Toth 1971) and axial loading (Kreider *et al.*, 1968, 1971; Toth, 1969; Shimizu and Dolowy, 1969; Schaefer and Christian, 1969; Menke and Toth, 1971). The former type of test configuration has the advantage of being capable of introducing compressive as well as tensile stresses into the composite without causing buckling. However, the disadvantages of a nonuniform and ambigous stress state as well as difficulty in defining a failure criterion make the results of this test difficult to interpret. The data to be discussed below were obtained by axial fatigue testing, which does not suffer from these disadvantages.

The data reported by Kreider *et al.* (1968) and Menke and Toth (1971) for axial fatigue of boron–aluminum are presented in Fig. 33 in the form of a Goodman diagram. The data indicate the excellent fatigue resistance of the material over a wide range of "A ratios" (ratio of alternating stress to mean

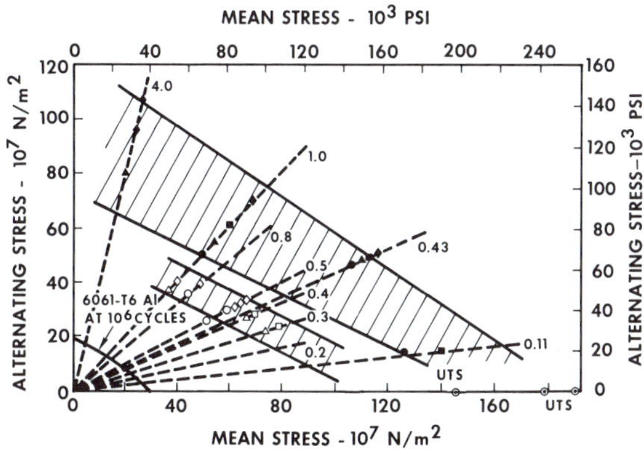

Fig. 33. Goodman diagram for boron-reinforced aluminum. Ratios of alternating to mean stress are given for dashed lines in figure.

Kreider		Menke
○	10^6 cycles	●
△	10^5–10^6	▲
□	10^4–10^5	■
◇	10^2–10^4	◆

stress). The major difference between the two sets of data is due to the difference in fiber strength and volume fraction, and hence axial composite strength, used in the investigation. This is related to the fact that the axial failure stress of boron–aluminum composite specimens tested in fatigue can be most effectively increased by increasing the composite axial tensile strength using increased fiber content and increased fiber strength. The high levels of stress sustainable under fatigue are due, primarily, to the high fatigue resistance of boron fiber. This resistance of boron to fatigue damage has been reported by Salkind and Patarini (1967). These tests, however, were performed in a bending mode and do not necessarily define fiber performance in axial fatigue, because fiber behavior is sensitive to the presence of flaws that may be present near the fiber core (neutral axis in bending) and thus fiber strength can be quite different in bending and in axial tension.

The mechanism of fatigue in axially reinforced specimens has been associated with both matrix deformation and progressive boron-fiber failure. Matrix deformation has been detected by X-ray measurement of matrix elastic stresses and by acoustic emission. Although composite specimens were tested in tension–tension fatigue, the aluminum matrix can be undergoing tension–compression fatigue. This is explained by the matrix plastically yielding in tension on the first tension stress cycles and then, to accommodate the total imposed change in applied stress and strain, going into compression on unloading of the specimen. The resultant residual matrix compressive stress is balanced by tensile stresses in the fibers. Fiber fracture during fatigue has also been detected by X-radiographic techniques, and it has been shown that progressive fiber fracture can occur at stress raisers at

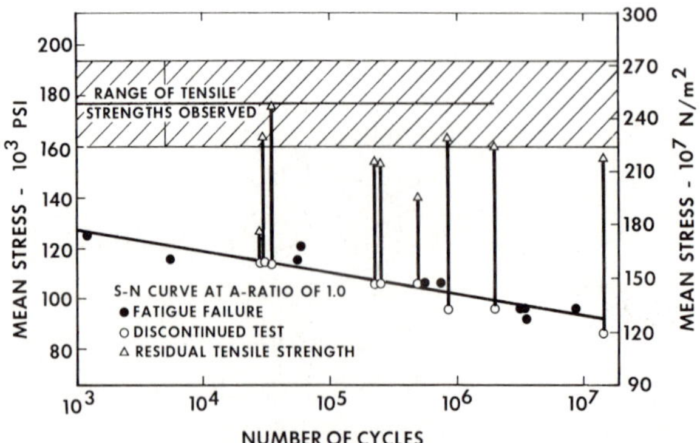

Fig. 34. Residual composite strength as a function of prior fatigue for 6061 aluminum reinforced with 100-μm boron. S–N curve at A-ratio of 1.0, ● fatigue failure, ○ discontinued test, △ residual tensile strength.

the specimen surface, such as machined fillets. It has also been shown that the extent of damage introduced by fatigue does not significantly decrease the residual composite axial tensile strength. Composite residual strength can equal the as-fabricated composite strength. This is illustrated by the findings of Menke and Toth (1971) as shown in Fig. 34 and indicates that final structural weakening of the composite occurs rapidly at the end of the specimen fatigue life.

The transverse tensile fatigue of uniaxially reinforced boron–aluminum has also been determined. Data for the low-cycle fatigue of BORSIC–aluminum are presented in Fig. 35. As in the case of the tensile behavior described previously, the 100-μm-BORSIC-reinforced specimens exhibited large amounts of longitudinal fiber splitting on transverse fatigue fracture surfaces, while the 150-μm-BORSIC specimens did not. It is interesting to note that the improvement in transverse tensile strength by use of the larger-diameter fiber is not fully maintained in fatigue. Specimens fatigued to 10^4 cycles, as indicated in Fig. 35, retained their full axial tensile strength when tensile tested.

Of particular interest is the fatigue performance of boron–aluminum in the presence of a notch. It has been shown by Schaefer and Christian (1969) and Shimizu and Dolowy (1969) that the axial fatigue of boron–aluminum specimens with reduced sections is associated with longitudinal shear of the specimens at the machined shoulders. Such reduced-section specimens thus effectively become parallel-sided specimens during fatigue prior to composite failure. This effect has been discussed by Kreider (1969) and is an indication

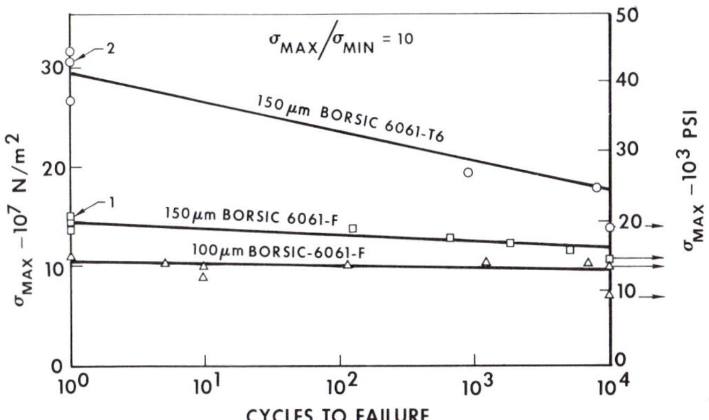

Fig. 35. Transverse tension–tension fatigue of BORSIC-reinforced 6061 composites. Specimens 1 and 2 were fatigued at the indicated stress and number of cycles and then tensile tested to determine the ultimate strengths indicated at 10° cycles. $\sigma_{max}/\sigma_{min} = 10$.

Fig. 36. Transmission X-radiograph of notch-tip regions in 6061 aluminum reinforced with 150-μm BORSIC; specimen fatigued in tension. White lines correspond to tungsten-boride fiber cores.

of the significant ability of the matrix to diminish the effects of specimen stress concentrators. Further evidence of crack blunting has been reported by Menke and Toth (1971) and Kreider *et al.* (1971) and is illustrated by Figs. 36 and 37. Figure 36 consists of two radiographs taken of a uniaxially reinforced BORSIC–6061-aluminum specimen. The radiographs were taken at the tips of a notch that was put in the center of the specimen. The white lines in the figure are images of the 0.005-in.-diam tungsten-boride cores of the BORSIC fibers. Radiographs of these same notch-tip regions prior to any mechanical testing did not reveal the presence of any local fiber breakage; however, after 1347 tension–tension fatigue cycles, the fiber breakage illustrated in Fig. 36 was detected. Additional fatigue cycling did not cause any further detectable fiber damage. Instead, longitudinal matrix shear occurred at the edge of the damaged regions. Tensile testing of the notched specimen after a total of 3800 cycles resulted in the fracture shown in Fig. 37. The notch had been effectively blunted during fatigue, and the net section stress of the composite at failure was equal to that of an unnotched specimen. This ability of boron–aluminum to prevent the propagation of flaws perpendicular to the fiber axis may prove to be one of the material's most important characteristics in structural application. It is similar, on a microscale, to the current use of stringers to inhibit crack propagation in plate structures.

G. Tricomponent Composites

In the previous sections it has been shown that the structural performance of boron–aluminum composites decreases rapidly with increasing angular deviation from the composite principal fiber axis. Compensation for this

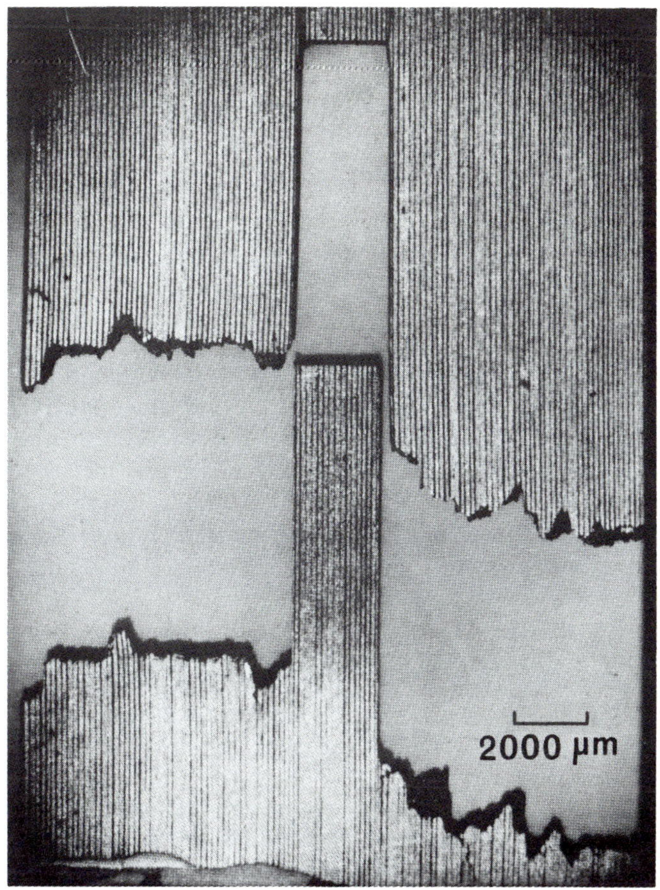

Fig. 37. Fractured center-notched specimen of 6061 aluminum reinforced with 150-μm BORSIC. Specimen is the same as that shown in Fig. 36. Note the fracture path deviating from the original notch plane.

decrease has been accomplished through the use of multiaxially reinforced configurations. Another approach is to use additional reinforcement components such as steel wire or titanium.

Aluminum-matrix composites reinforced with boron and steel wire have been shown to provide improved composite performance by several investigators. In each case the steel wires have been placed at 90° to the boron-fiber axis. Christian (1970, 1971) and Kreider *et al.* (1971) have shown that transverse composite strength can be increased significantly by adding only a few percent of steel fiber. In addition, it was demonstrated that composite formability and handleability were improved when the steel fibers were placed in the exterior lamina of a composite lay-up. This exterior steel-fiber–

aluminum sheath also increased the strength of joints between composite panels formed by spot welding. The tensile strength of 100-μm-BORSIC–stainless-steel–aluminum as a function of test temperature is presented in Fig. 38. The addition of 6 v/o stainless-steel fiber at 90° to the BORSIC-fiber axis has increased transverse composite strength by more than a factor of 2 over the entire temperature range investigated. The composite transverse tensile failure strain was also increased to a level of 1.1 %, as compared to approximately 0.2 % for standard 100-μm-BORSIC–aluminum. This increase in transverse properties has been accomplished without loss in axial (parallel to the boron fibers) strength.

The addition of stainless-steel fibers also increases the impact energy of composite specimens. The Charpy impact energy of a composite specimen containing 5.5 v/o stainless-steel fiber, 42 v/o 100-μm BORSIC, and the balance 6061 aluminum was found to be 6.1 J. The steel fiber was oriented perpendicular to the plane of the machined notch, and the BORSIC was oriented in the plane of the notch and parallel to the notch root. The above value represents an approximately fourfold increase in impact energy over that typical of specimens of the same orientation not containing stainless steel.

Titanium foil, used in place of aluminum foil in plasma-sprayed tapes, has also proven a useful addition to BORSIC–aluminum composites (Miller

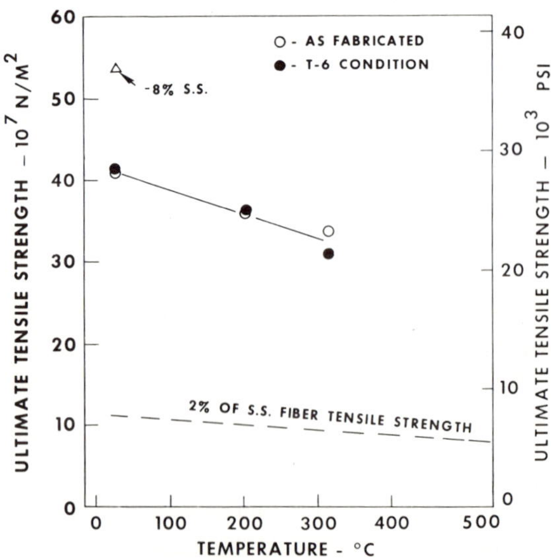

Fig. 38. Tensile strength of 6061 aluminum reinforced with 50-μm stainless-steel wire and 100-μm BORSIC. The tensile axis is parallel to the steel wires, while the BORSIC fibers are oriented at 90° to tensile axis. ○ As fabricated, ● T-6 condition.

TABLE V

Comparison of Metal-Matrix Composites

Material (v/o)	Material (component)	Relative Density (%)	Axial UTS (10^7 N/m^2)	Axial UTS (10^3 psi)	Transverse UTS (10^7 N/m^2)	Transverse UTS (10^3 psi)
50	100-μm BORSIC					
50	6061 Al	100	118	165	14	20
40	100-μm BORSIC					
30	Ti–6Al–4V	120	107	150	46	65
30	6061 Al					
39	100-μm BORSIC					
17	Ti–6Al–4V	110	100	140	22	31
44	2219 Al				14[a]	19[a]
45	100-μm BORSIC					
17	Ti–6Al–4V	110	118	165	24	33
38	6061 Al					
44	100-μm BORSIC					
21	Beta III Ti	115	116	163	38	53
35	6061 Al					

[a]At 400°C.

and Schaeffer 1971; Kreider *et al.*, 1971). Table V presents the data illustrating the large increases in transverse strength attainable. Also given are the modest increases in material density resulting from this titanium-foil addition. Even at 400°C, a composite transverse strength of 1.36×10^7 N/m^2 was obtained for a specimen containing 17 v/o titanium. The titanium foils, as in the case of steel wire, increase the handleability of BORSIC–aluminum as well as impact energy. It has been shown that the titanium foils also increase composite shear strength by a factor of 2, to 3.2×10^7 N/m^2 over that of BORSIC–aluminum or BORSIC–steel–aluminum.

VI. Engineering of Boron–Aluminum for Applications

Composite materials such as boron–aluminum require special considerations and material engineering to be suitable for specific applications. Structural applications of boron–aluminum to date have been developed primarily in three general areas, which require somewhat different approaches. These applications include: (1) complexly shaped parts, which are diffusion bonded between shaped dies; (2) large panels, which are either low-pressure bonded or diffusion bonded in steps; and (3) long beams and stiffeners.

The first type—the shaped part—can be typified by turbine-engine fan blades, propeller spars, and hollow compression sheets. Typically, large, flat panels are used as aircraft or spacecraft skins. Beams have been used to stiffen panels in aerospace structures or as ribs, spars, and structural supports.

In order to illustrate some of the considerations involved in using boron–aluminum for these applications, two particular examples have been chosen: turbine-engine fan blades, which have been developed by Shultz *et al.* (1971); and a large airframe skin panel, which has been proposed for spacecraft by Forrest and Christian (1970) and Miller and Schaeffer (1971). These have been chosen not only because they represent major classes of applications but also because these applications have been studied in the greatest depth.

Boron–aluminum turbine-engine fan-blade development has been pursued by the three large American turbine-engine manufacturers and by suppliers of fan blades, and has been strongly encouraged and supported by the Air Force Materials Laboratory. This interest has been generated primarily because of the following significant design considerations for the blades:

1. Increased stiffness of boron–aluminum fan blades of the same airfoil shape compared to the presently used titanium alloy permits the elimination of the necessary part-span shrouds on the blades. These shrouds or "bumpers" are used on titanium blades because the low intrinsic elastic modulus of the titanium would lead to excessive vibratory amplitude without them. The composite blades have been designed and tested without the shrouds, thus yielding higher aerodynamic efficiency.
2. Lower density of the composite material compared to the titanium leads to significant weight savings not only in the blade but also in the supporting structure. Many military engines are basically rated on a thrust-to-weight ratio for performance purposes; therefore, the weight savings are important.
3. The increased strength-to-density ratio of the composite can permit higher tip speeds of the blades and thus higher efficiencies in the compressor. Higher tip speeds can lead to fewer stages and therefore lower weights.
4. The material must be sufficiently strong and durable in environments of supersonic fans, which include temperatures of up to 300°C.

Several factors are involved in the materials selection for this application. The system must be well characterized and have good strength, toughness, and high-temperature strength. In addition, the material should be resistant to corrosive attack in the presence of sea water. These factors have led to the choice of 6061 alloy as the matrix material for most of the fan-blade applications. Also, the composite system should permit consistent high-quality part fabrication even with complex curvatures, complex fiber arrays, fabrication pressure gradients in the dies, and complexly shaped ply dropoffs. These fabrication considerations have led to the choice of diffusion-bonding the part at very high temperatures in order to minimize shear creep stresses in the matrix and subsequently to the choice of BORSIC fiber rather than boron. This fiber choice permits the higher temperatures during bonding without concomitant fiber degradation.

Materials characterization necessary for composite application as fan blades includes the standard strength and modulus tests plus evaluation of fatigue damage, axial and torsional creep resistance, impact properties, resistance to damage on thermal cycling, erosion and corrosion resistance, and fracture toughness. Combinations of the above test environments, such as the relationship between impact (from foreign-object damage) and fatigue damage, are also very important. The severity of the turbine-engine fan-blade environment requires extensive materials characterization, and these specialized testing programs have been described by Shultz *et al.* (1971). Although engine testing provides the ultimate testing of the material, no effective short-cut in materials characterization has been found effective for the incorporation of a new material in the engine.

The fabrication of turbine-engine fan blades has been discussed by Shultz *et al.* (1971), Kreider and Breinan (1970), and others. The process has consisted of the lamination and diffusion-bonding of precut plasma-sprayed tapes. The general fabrication process has been described earlier; however, there have been some unique developments in blade fabrication. In order that the complex shape of each ply in the laminate be cut efficiently and exactly, a new technique called roll cutting was developed and is depicted in Fig. 7. This process consists of laminating the boron aluminum sheet with soft sandwich covers of a soft material like cork and a precut form. As the sandwich is passed through the rollers, the appropriate pieces are cut in this cookie-cutter fashion.

Another specialized technique was developed for joining the root attachment and platform to the airfoil. The airfoil is diffusion-bonded to wedge-shaped inserts and the "fir-tree"-shaped compressive root fixture, which also includes the gas-sealing platform. This assembly is all diffusion-bonded prior to root machining in order that exact dimensions and balancing can be achieved.

Turbine-engine fan blades are also subjected to a series of specialized tests prior to engine testing. The vibration resonance frequency in bending and torsion (including harmonics and the fatigue properties) are determined on "shaker tables." Other nondestructive tests include ultrasonic inspection for lamination flaws, debonds, and voids; X-ray inspection for fiber alignment, crossovers, voids, and breaks; and laser holography to determine uniformity of vibratory response. The holographic inspection displays perturbation in contrast to normal vibratory conformity that may be due to fabrication or design defects.

Spin tests are used as proof testing of the blades in rotating force fields. The spin test then may be followed by a rig test that simulates that portion of the engine that can be rated for performance, such as the fan–compressor rig.

The second composite structural application chosen as an example is the large airframe skin panel. Although the hardware appears simple in concept, the size precludes some standard fabrication techniques, and the load fixturing in the actual application require considerable design, materials characterization, fabrication, and testing considerations. The panel is generally of hollow design with stiffeners or honeycomb separating the two skins.

Panel design criteria normally include material properties such as modulus-to-weight and strength-to-weight ratios. Special emphasis may be placed on compressive, bearing, and shear properties. Also, the effects of cutouts and joints on the material properties all relate to the optimization of the design. The design goal is generally the lowest possible weight and cost consistent with reliability and durability. In boron–aluminum composite systems, the prevailing choice has been 6061 plus boron fiber; however, 1100 alloys have been chosen for certain fabrication techniques, and BORSIC may be required if subsequent brazing operations are desired.

Materials characterization has emphasized studies with flexure of honeycomb beams—thus investigating tension, compression, and the shear bonding of the face sheets simultaneously—and the analysis of the material with the presence of rivet holes, cutouts, and concentrated load points.

The most significant fabrication problem pertains to the size of the panels. Diffusion-bonding pressures of 3000–5000 psi would lead to prohibitive forces for panels 4 × 8 ft. Therefore, two alternative approaches have been developed: low-pressure bonding and step pressing. Low-pressure bonding is generally performed in an autoclave in an analogous manner to that used for resin composite fabrication; step pressing is performed between platens. These processes were described earlier, in Section III.

Complex rig tests have been developed for testing large stiffened panels. These tests include mechanisms for simulating monotomic stresses and bending moments in addition to aerodynamic loadings both in static and cycling modes.

References

Abrams, E. F., and Davies, L. G. (1971). *In Proc. 1971 SAMPE Con., Space Shuttle Mater., Huntsville.*

Adams, D. F. (1970). *J. Compos. Mater.* **4**, 310.

Adams, D. F., and Doner, D. R. (1967). *J. Compos. Mater.* **1**, 152.

Adsit, N. R., and Forest, J. D. (1969). *In* Composite Materials: Testing and Design, ASTM-STP-460, p. 108. Amer. Soc. Test. Mater., Philadelphia, Pennsylvania.

Anthony, K. C., and Chang, W. H. (1968). *Trans. ASM* **61**, 550.

Ashton, J. E., Halpin, J. C., and Petit, P. H. (1969). *In* "Primer on Composite Materials: Analysis." Technomic, Stamford, Connecticut.

Azzi, V. D., and Tsai, S. W. (1965). *Exp. Mech.* **Sept.**, 283.

Baker, A. A., and Cratchley, D. (1966). *Appl. Mater. Res.* **April**, 92.

Baker, A. A., and Jackson, P. W. (1968). *Glass Tech.* **9** (2), 36.

Basche, M., Fanti, R., and Galasso, F. (1968). *Fib. Sci. Tech.* **1**, 1.

Beaumont, P. W. R., and Phillips, D. C. (1972). *J. Compos. Mater.* **6**, 32.

Breinan, E. M., and Kreider, K. G. (1969). *Met. Eng. Quart.* **Nov.**

Breinan, E. M., and Kreider, K. G. (1970). *Met. Trans.* **1**, 93.

Breinan, E. M., and Kreider, K. G. (1973). *Met. Trans.* (to be published).

Brown, W. F., and Srawley, J. E. (1967). Plane Strain Fracture Toughness Testing of High-Strength Metallic Materials, ASTM-STP-410.

Camahort, J. (1968). *J. Compos. Mater.* **1**, 104.

Chen, P. E., and Lin, J. M. (1969). *Mat. Res. Std.* **August**, 29.

Cheskis, H. P., and Heckel, R. W. (1968). *In* Metal Matrix Composites, ASTM-STP 438, 76–91.

Cheskis, H. P., and Heckel, R. W. (1970). *Met. Trans.* **1**, 1931.

Christian, J. (1970). ASME Publ. 70-GT-46.

Christian, J. (1971). *In Proc. 1971 SAMPE Conf., Space Shuttle Mater., Huntsville.*

Christian, J., and Forest, J. (1969). ASM Tech. Rep. No. P-9-5512.

Corten, H. B. (1967). *In* "Modern Composite Materials" (Broutman, L. J. and Krock, R. H., eds.). Addison-Wesley, Reading, Massachusetts.

Davies, L. G. (1971). Space-Shuttle Materials. *SAMPE.* North Hollywood, California.

Dolowy, J. F., and Taylor, R. J. (1969). *Proc. SAMPE Meeting 1969* 369.

Ebert, L. J., Hamilton, H. L., and Hecker, S. S. (1969). *Trans. ASM* **62**, 740.

Ellison, E. G., and Boone, D. H. (1967). *J. Less Common Metals* **13**, 103.

Fleck, J. N., Smith, E. G., and Laber, D. (1969). *J. Compos. Mater.* **3**, 699.

Forrest, J. D., and Christian, J. L. (1970). *Met. Eng. Quart.* **June**.

Galasso, F. (1970). "High Modulus Fibers and Composites." Gordon and Breach, New York.

Getten, J. R., and Ebert, L. J. (1969). *Trans. ASM* **62**, 869.

Greszczuk, L. (1971). *AIAA J.* **9**, 1274.

Hancock, J. R., and Swanson, G. D. (1971). *In* Composite Materials: Testing and Design, ASTM-STP-497, p. 299.

Happe, R. A., and Yeast, A. J. (1969). Materials and Processes for the 70s, V15 SAMPE, North Hollywood, California.

Herring, H. (1966). Selected Mechanical and Physical Properties of Boron Filaments. NASA Tech. Note TN-D-3202.

Herring, H. W. (1972). Fundamental Mechanisms of Tensile Fracture in Aluminum Sheet Unidirectionally Reinforced with Boron Filament. NASA Tech. Rep. NASA-TR-R-383.

Herring, H. W., and Krishna, V. (1965). Shear Moduli of Boron Filaments. UARL Rep. 66–4876.

Hersch, M. S., and Duffy, J. R. (1969). *J. Am. Welding Soc.*
Hirschhorn, J. S., and Sheehan, J. E. (1969). *Trans. ASM* **62**, 804.
Hoover, W. R., and Allred, R. E. (1972). *In* "Failure Modes in Composites" (I. Toth, ed.). AIME.
Jackson, P. W., Baker, A. A., and Braddick, D. M. (1971). *J. Mater. Sci.* **6**, 427.
Jones, R. C. (1969). *In* Composite Materials: Testing and Design, ASTM-STP-460. p. 512.
Kelly, A. (1966). "Strong Solids." Oxford Univ. Press (Clarendon), London and New York.
Kelly, A., and Davies, G. J. (1965). *Met. Rev.* **10**, 1, 37.
Klein, M. J., Metcalfe, A. C., and Gulden, M. E. (1972). AFML-TR-72-226.
Kohn, J. A., Nye, W. F., and Gaule, K. (1960). "Boron Synthesis, Structure and Properties." Plenum Press, New York.
Kreider, K. G. (1969). *In* Composite Materials: Testing and Design, ASTM-STP-460, p. 203.
Kreider, K. G., and Breinan, E. M. (1970). *Metal Prog.* 104
Kreider, K. G., and Dardi, L. (1972). *In* "Failure Modes in Composites" (I. Toth, ed.). AIME.
Kreider, K. G., Dardi, L., and Prewo, K. (1971). Metal-Matrix Composite Technology. AFML-TR-71-204.
Kreider, K. G., and Leverant, G. R. (1966a). Boron-Fiber Metal-Matrix Composites by Plasma Spraying, AFML-TR-66-219.
Kreider, K. G., and Leverant, G. R. (1966b). *SAMPE.* Vol. 10, North Hollywood, California.
Kreider, K. G., and Marciano, M. (1969). *Trans. Met. Soc.* **245**, 1279.
Kreider, K. G., and Patarini, V. M. (1970). *Met. Trans.* **1**, 3431.
Kreider, K. G., and Prewo, K. M. (1972). *In* Composite Materials: Testing and Design, ASTM-STP-497, p. 539.
Kreider, K. G., Schile, R., Breinan, E., and Marciano, M. (1968). Plasma-Sprayed Metal-Matrix Fiber-Reinforced Composites, AFML-TR-68-119.
Lenoe, E. M. (1968). *In* Metal-Matrix Composites, ASTM-STP-438.
Liebowitz, H. (1968). *In* "Fracture," Vol. 2, p. 287. Academic Press, New York.
Lin, J. M., Chen, P. E., and DiBenedetto, A. T. (1971). *Polym. Eng. Sci.* **11**, 344.
Mangiapane, J. A., Gray, D. F., Sattar, S. A., and Timoshenko, J. A. (1968). AIAA Paper 68-1037.
McDaniels, D. L., Jech, R. W., and Weeton, J. W. (1965). *Trans. Met. Soc.* **233**, 636.
Menke, G. D., and Toth, I. J. (1970). Time Dependent Mechanical Behavior of Composite Materials, AFML-TR-70-174.
Menke, G. D., and Toth, I. J. (1971). Time Dependent Mechanical Behavior of Composite Materials, AFML-TR-71-102.
Metcalfe, A., and Schmitz, G. (1969). *In Proc. Gas Turbine Conf., Cleveland.*
Miller, M., and Schaeffer, W. (1971). *In Proc. 1971 SAMPE Conf. Space Shuttle Mater., Huntsville.*
Mullin, J., Berry, J. M., and Gatti, A. (1968). *J. Compos. Mater.* **2**, 82–103.
Olster, E. F., and Jones, R. C. (1971). *In* "Composite Materials: Testing and Design," ASTM-STP-497, p. 189.
Paris, P. C. and Sih, G. C. (1965). Stress Analysis of Cracks, ASTM-STP-281, pp. 30–83.
Prewo, K. M. (1972). *J. Compos. Mater.* **4**,
Prewo, K. M., and Kreider, K. G. (1972a). *Met. Trans.* **3**, 2201.
Prewo, K. M., and Kreider, K. G. (1972b). *J. Compos. Mater.* **6**, 338.
Prewo, K. M., and Kreider, K. G. (1972c). *Proc. Int. Conf. Titanium, 2nd*, AIME.
Prewo, K. M., and McCarthy, G. (1972). *J. Mater. Sci.* **7**, 919.
Rosen, B. W. (1964). *J. AIAA* **2**, 1985.
Rosen, B. W. (1970). *Proc. Roy. Soc.* **319A**, 79.
Ryder, C. G., Vidoz, A. E., Crossman, F. W., and Camahort, J. L. (1970). *J. Compos. Mater.* **4**, 264.

Salkind, M., and Patarini, V. (1967). *Trans. AIME* **239**, 1268.

Schaefer, W. H., and Christian, J. L. (1969). Evaluation of the Structural Behavior of Filament-Reinforced Metal-Matrix Composites, Vol. III. AFML-TR-69-36.

Schultz, W. J., Mangiapane, J. A., and Stargardter, H. (1971). ASME Paper 71-GT-90.

Shimizu, H. and Dolowy, J. F. (1969). *In* Composite Materials: Testing and Design, ASTM-STP-460 p. 192.

Sih, G. C. (1965). *Int. J. Fracture Mech.* **3**, 189–203.

Stuhrke, W. F. (1968). *In* Metal Matrix Composites, ASTM-STP-438, p. 108.

Summer, W. (1967). *Advan. Struct. Compos. SAMPE.* **12**, AC-19.

Talley, C. P. (1959). *J. Appl. Phys.* **39**, 1114.

Tetelman, A. S., and McEvily, A. J. (1967). "Fracture of Structural Materials." Wiley, New York.

Toth, I. J. (1969). An Exploratory Investigation of the Time Dependent Mechanical Behavior of Composite Materials, AFML-TR-69-9.

Tsai, S. W. (1966). Mechanics of Composite Materials, Parts I and II. Tech. Rep. AFML-TR-66-149.

Tsai, S., Adams, D., and Doner, D. (1966). Effect of Constituent Material Properties on the Strength of Fiber-Reinforced Composite Materials. Defense Documentation Centre Rep. AD 638 922.

Tsariff, T. C., Herman, M., and Sippel, G. R. (1969). *Symp. Metal Matrix Compos.* Battelle Memorial Inst., Columbus, Ohio, p. 38.

Van Horn, K. R. (1966). "Aluminum," Vols. 1–3. Amer. Soc. Metals, Metals Park, Ohio.

Veltri, R. D., and Galasso. F. S. (1971). *J. Amer. Ceram. Soc.* **54**, 319.

Vidoz, A. E., Camahort, J. L., and Crossman, F. W. (1969). *J. Compos. Mater.* **3** (2), 254.

Wawner, F. E. (1967). *In* "Modern Composite Materials" (Broutman, L. J., and Krock, R. H., eds.). Addison-Wesley, Reading, Massachusetts.

Weisinger, M. D. (1970). ASM Paper W70-52.

Weiss, V., and Yakawa, S. (1965). Critical Appraisal of Fracture Mechanics., ASTM-STP-381, pp. 1–29.

Whithers, J. W., and Abrams, E. F. (1968). *Plating J.* 605.

Wu, E. M. (1963). Application of Fracture Mechanics of Orthotropic Plates. Univ. of Illinois, TAM Rep. 248, Naval Res. Lab., Contract No. Nonr. 2947(02)X.

Wu, E. M. (1965). A Fracture Criterion for Orthotropic Plates under the Influence of Compression and Shear. Univ. of Illinois, TAM Rep. 283, Bur. Naval Weapons, Contract No. NOW-0204-d.

Wu, E. M. (1967). *J. Appl. Mech.* **34**, 967-974.

Wu, E. M. (1968a). Discontinuous Mode of Crack Extension in Unidirectional Composites. Univ. of Illinois, TAM Rep. 309.

Wu, E. M. (1968b). "Fracture Mechanics of Anisotropic Plates" (S. W. Tsai *et al.*, eds.), pp. 20–43. Compos. Mater. Workshop, Technomic Publ.

Wu, E. M., and Reuter, R. C. (1965). Crack Extension in Fiberglass Reinforced Plastics. Univ. of Illinois, TAM Rep. 275, Bur. Naval Weapons, NOW-64-0178-d.

Zweben, C., and Rosen, B. W. (1970). *J. Mech. Phys. Solids* **18**, 189.

Author Index

Numbers in italics refer to the pages on which the complete references are listed

C

D

E

F

G

H

I

J

K

T

U

V

W

Subject Index

C

D

E

L

M

N

O

P

Q

R

S

T

A 4
B 5
C 6
D 7
E 8
F 9
G 0
H 1
I 2
J 3